NUMERICAL MATHEMATICS AND SCIENTIFIC COMPUTATION

Series Editors

A. M. STUART   E. SÜLI

NUMERICAL MATHEMATICS AND SCIENTIFIC COMPUTATION

Books in the series

Monographs marked with an asterisk (*) appeared in the series 'Monographs in Numerical Analysis' which is continued by the current series.

For a full list of titles please visit
http://www.oup.co.uk/academic/science/maths/series/nmsc

* J. H. Wilkinson: *The Algebraic Eigenvalue Problem*
* I. Duff, A. Erisman, and J. Reid: *Direct Methods for Sparse Matrices*
* M. J. Baines: *Moving Finite Elements*
* J. D. Pryce: *Numerical Solution of Sturm–Liouville Problems*

C. Schwab: *p- and hp- Finite Element Methods: Theory and Applications in Solid and Fluid Mechanics*
J. W. Jerome: *Modelling and Computation for Applications in Mathematics, Science, and Engineering*
A. Quarteroni and A. Valli: *Domain Decomposition Methods for Partial Differential Equations*
G. Em Karniadakis and S. J. Sherwin: *Spectral/hp Element Methods for CFD*
I. Babuška and T. Strouboulis: *The Finite Element Method and its Reliability*
B. Mohammadi and O. Pironneau: *Applied Shape Optimization for Fluids*
S. Succi: *The Lattice Boltzmann Equation for Fluid Dynamics and Beyond*
P. Monk: *Finite Element Methods for Maxwell's Equations*
A. Bellen and M. Zennaro: *Numerical Methods for Delay Differential Equations*
J. Modersitzki: *Numerical Methods for Image Registration*
M. Feistauer, J. Felcman, and I. Straškraba: *Mathematical and Computational Methods for Compressible Flow*
W. Gautschi: *Orthogonal Polynomials: Computation and Approximation*
M. K. Ng: *Iterative Methods for Toeplitz Systems*
M. Metcalf, J. Reid, and M. Cohen: *Fortran 95/2003 Explained*
G. Em Karniadakis and S. Sherwin: *Spectral/hp Element Methods for Computational Fluid Dynamics, Second Edition*
D. A. Bini, G. Latouche, and B. Meini: *Numerical Methods for Structured Markov Chains*
H. Elman, D. Silvester, and A. Wathen: *Finite Elements and Fast Iterative Solvers: with Applications in Incompressible Fluid Dynamics*
M. Chu and G. Golub: *Inverse Eigenvalue Problems: Theory, Algorithms, and Applications*
J.-F. Gerbeau, C. Le Bris, and T. Lelièvre: *Mathematical Methods for the Magnetohydrodynamics of Liquid Metals*
G. Allaire and A. Craig: *Numerical Analysis and Optimization: An Introduction to Mathematical Modelling and Numerical Simulation*
K. Urban: *Wavelet Methods for Elliptic Partial Differential Equations*
B. Mohammadi and O. Pironneau: *Applied Shape Optimization for Fluids, Second Edition*
K. Boehmer: *Numerical Methods for Nonlinear Elliptic Differential Equations: A Synopsis*
M. Metcalf, J. Reid, and M. Cohen: *Modern Fortran Explained*
J. Liesen and Z. Strakoš: *Krylov Subspace Methods: Principles and Analysis*

Krylov Subspace Methods
Principles and Analysis

Jörg Liesen
Technical University of Berlin

Zdeněk Strakoš
Charles University in Prague

UNIVERSITY PRESS

Great Clarendon Street, Oxford, OX2 6DP,
United Kingdom

Oxford University Press is a department of the University of Oxford.
It furthers the University's objective of excellence in research, scholarship,
and education by publishing worldwide. Oxford is a registered trade mark of
Oxford University Press in the UK and in certain other countries

© Jörg Liesen and Zdeněk Strakoš 2013

The moral rights of the authors have been asserted

First Edition published in 2013

Impression: 1

All rights reserved. No part of this publication may be reproduced, stored in
a retrieval system, or transmitted, in any form or by any means, without the
prior permission in writing of Oxford University Press, or as expressly permitted
by law, by licence or under terms agreed with the appropriate reprographics
rights organization. Enquiries concerning reproduction outside the scope of the
above should be sent to the Rights Department, Oxford University Press, at the
address above

You must not circulate this work in any other form
and you must impose this same condition on any acquirer

British Library Cataloguing in Publication Data

Data available

Library of Congress Cataloging in Publication Data

Library of Congress Control Number: 2012945094

ISBN 978-0-19-965541-0

Printed and bound by
CPI Group (UK) Ltd, Croydon, CR0 4YY

Links to third party websites are provided by Oxford in good faith and
for information only. Oxford disclaims any responsibility for the materials
contained in any third party website referenced in this work.

To Anja and to Iva

This seems to be a very elementary problem without
deeper meaning. However, one meets this task again
and again in the electric industry and in all kinds
of oscillation problems.
A short while ago, I found a rather elegant solution.
The reason why I am strongly drawn to such
approximation mathematics problems is not
the practical applicability of the solution,
but rather the fact that a very "economical"
solution is possible only when it is very "adequate".
To obtain a solution in very few steps
means nearly always that one has found a way
that does justice to the inner nature of the problem.
        CORNELIUS LANCZOS in a letter to Albert Einstein on March 9, 1947.

Your remark on the importance of
adapted approximation methods makes very
good sense to me, and I am convinced
that this is a fruitful mathematical aspect,
and not just a utilitarian one.
        EINSTEIN'S reply to Lanczos on March 18, 1947[1].

---

1. We thank the Albert Einstein Archives, The Hebrew University of Jerusalem, Israel, which holds the copyrights on Einstein's letters, for giving us access to the correspondence between Lanczos and Einstein. The two corresponded in German, and the above are our translations of the following original quotes:
"Das scheint ja ein reichlich elementares Problem ohne tiefere Bedeutung zu sein. Doch stösst man in der elektrischen Industrie und bei allen möglichen Schwingungsproblemen immer wieder auf diese Aufgabe. Vor kurzem ist mir eine recht elegante Lösung geglückt. Der Grund weshalb mich solche Probleme der Approximationsmathik [spelling error by Lanczos; most likely he meant to write "Approximationsmathematik"] stark anziehen, ist nicht die praktische Verwendbarkeit der Lösung, sondern vielmehr der Umstand, dass eine stark "ökonomische" Lösung nur möglich ist, wenn sie auch stark "adequat" ist. Mit wenig Schritten zum Resultat zu kommen bedeutet fast immer, dass man einen Weg gefunden hat, der dem inneren Wesen des Problems gerecht wird." (Lanczos to Einstein, Albert Einstein Archives document number 15–313.)
"Ihre Bemerkung über die Bedeutung angepasster Approximations-Methoden leuchtet mir sehr ein, und ich bin überzeugt, dass dies ein fruchtbarer mathematischer Gesichtspunkt ist, nicht nur ein utilitaristischer." (Einstein to Lanczos, Albert Einstein Archives document number 15–315.)

PREFACE

Quite frequently in life, the most elegant solution to a problem also turns out to be the most efficient one for practical purposes. This heuristic observation certainly applies to Krylov subspace methods. The algorithms devised by Magnus Hestenes, Eduard Stiefel, Cornelius Lanczos, and others in the early 1950s for iteratively solving large and sparse linear algebraic systems and eigenvalue problems can hardly be more elegant and aesthetically pleasing. Yet, these algorithms and their numerous later variants and extensions are nowadays used widely and successfully throughout science and engineering. Because of their overwhelming success in applications, Krylov subspace methods are counted among the 'Top 10 Algorithms' of the 20th century [116, 140].

Not surprisingly, several first-rate books describing Krylov subspace methods are available. They have been written by excellent communicators and leading researchers in the field, including Bernd Fischer [183] (republished by SIAM as a 'Classics in Applied Mathematics' title in 2011), Anne Greenbaum [272], Gérard Meurant [452, 453], Yousef Saad [543], and Henk van der Vorst [636]. Last but not least we mention the closely related book by Gene Golub and Gérard Meurant [246]. These books, as well as the other books and most survey papers on the subject we are aware of, reflect the current state-of-the-art, which is the outcome of explosive algorithmic developments over the last few decades. Such developments were necessary because of tremendous challenges raised by an ever growing variety of application problems. For many years, investigations on *how* to solve problems computationally and the derivation of new methods and their algorithmic realisations have dominated the analysis of existing approaches.

Our aim with this book is to complement the existing literature by focusing on *mathematical fundamentals* of Krylov subspace methods rather than their algorithmic details, and on *addressing the why* more than the how. In the quote given above, Lanczos announces to Einstein that he has found a new elegant algorithm (he refers to his nowadays classical method for computing eigenvalue approximations) and, more importantly, he explains his main motivation for working on iterative methods. He is attracted to them because they can only be made to work efficiently when they uncover a deeper truth, namely the 'inner nature of the problem'. The utilitarian viewpoint of practical applicability is secondary to him, and apparently to Einstein as well, who in his answer to Lanczos points out the 'fruitful mathematical aspect' involved in Lanczos' thinking. In our book we try to explore precisely this aspect, which in the past has been overshadowed by the algorithmic view.

In the process of writing we went back to the early papers by Krylov (1931), Gantmacher (1934), Lanczos (1950 and 1952), Hestenes and Stiefel (1952), and others from that period. These authors presented many close relationships of their methods with mathematical concepts beyond the realm of what is known today as 'matrix computations'. Examples include quadrature methods, orthogonal polynomials, continued fractions, moments, projections, and invariant subspaces. Reading these original works was a fascinating experience.

Our feeling is nicely expressed by the following quote from the German mathematician Eduard Study (1862–1930): 'Mathematics is neither the art of calculation nor the art of avoiding calculations. Mathematics, however, entails the art of avoiding superfluous calculations and conducting the necessary ones skilfully. In this respect one could have learned from the older authors.' [601, p. 4] (our translation). The fact that reading older authors is not just a matter of studying history has also been stressed by the English writer Joseph Rudyard Kipling (1865–1936), author of *The Jungle Book* and Nobel Prize winner for literature in 1907, who wrote in his essay 'The Uses of Reading' (1928) that 'it is only when one reads what men wrote long ago that one realises how absolutely modern the best of the old things are.' Lanczos formulated a similar point of view even more strongly in his essay 'Why Mathematics?' from 1967: 'But to hail our times as the originator of an entirely new science, which need not bother with the past and has the right to construct everything from scratch, betrays a dangerous short-sightedness which can lead to a dissolution of mathematical research into an empty play with words.'

Returning back to the original sources led us to examine a wide range of areas and their interconnections, and to study many results developed in the 19th and the early 20th centuries. Examples are such classics as continued fractions with their relationship to orthogonal polynomials, quadrature and minimal partial realisation in the works of Gauss (1814), Jacobi (1826), Christoffel (1857, 1858), Chebyshev (1855, 1859), Markov (1884), Stieltjes (1884, 1894), and many others. Further classical results we looked at and include in this book are Cauchy's interlacing theorem (1829), Jacobi's reduction to tridiagonal form (1848), Jordan's canonical form (1870), Stieltjes' moment problem (1894), and the Riemann–Stieltjes integral representation of operators by von Neumann (1927, 1932) going back to Hilbert (1906, 1912), who praised the work of Stieltjes.

In order to make the book as self-contained as possible, we have included complete proofs of many stated results. In addition, we have tried to the best of our abilities to give references to the original sources. Throughout the book we have used framed boxes to highlight points in the development that we consider particularly important. In these boxes we usually skip some technical details in order to focus on the main message of the corresponding mathematical results or questions.

To keep the project manageable for ourselves, we have focused on methods for solving linear algebraic systems, and we left aside the equally interesting area of

Krylov subspace based eigenvalue solvers. However, many results presented in this book are relevant for such solvers as well, since they are usually based on the same principles as the methods for linear algebraic systems (with the Lanczos and Arnoldi algorithms as the basic building blocks).

As indicated above, we strongly believe that a mathematical theory can be better understood when it is viewed in its historical context. We therefore discuss the original developments in extensive historical notes. In our opinion, the knowledge of early developments can also help in understanding very recent computational developments. The outcome of the historically motivated approach therefore is practical and readily applicable. It shows what can and what cannot be expected from Krylov subspace methods *today*. Moreover, it challenges some common 'modern' views that have been articulated in the justification of 'practical', though mathematically questionable, approaches.

When solving real-world and large-scale problems, Krylov subspace methods must always be combined with acceleration techniques. The goal is to improve the behaviour of Krylov subspace methods, and the techniques are (somewhat imprecisely) called 'preconditioning'. Construction of preconditioners is usually based on some specific properties of the real-world problem or on empirical (which does not mean simple!) observations and heuristics. Successful application of preconditioning often requires an extensive and deep theoretical knowledge from many areas combined with skilful implementation of graph-theoretical ideas. Therefore it is sometimes viewed as 'a combination of art and science'. In this book we do not explicitly consider preconditioning techniques. (They are studied, for example, in [272, Part II], [452, Chapters 8–10], and [543, Chapters 9–14]; see also the survey [54].) Nevertheless, we believe that our book also contributes to the area of preconditioning. Since most of the presented analysis is applicable to preconditioned systems, it applies, assuming exact arithmetic, to preconditioned Krylov subspace methods. Moreover, many results can be modified in order to describe finite precision computations with preconditioning. Most important of all, we believe that a better understanding of the fundamentals of Krylov subspace methods is a prerequisite for establishing an analytic base on which a theory of preconditioning can be developed.

OVERVIEW OF THE BOOK

The book contains five chapters. The first chapter, *Introduction*, introduces the general setting of the book and the context of solving a real-world problem via the stages of modelling, discretisation, and computation. We also recall the richness of ideas related to the above mentioned original works of Hestenes, Stiefel, and Lanczos. Many of the mathematical topics addressed in the related works are closely examined in Chapters 2–4, which form the theoretical core of our book. In these chapters we consider Krylov subspace methods from different points of view, and we make links between these viewpoints.

Chapter 2, *Krylov Subspace Methods*, focuses on the idea of *projections*. The so-called 'finite termination property' then naturally leads to the introduction of the

Krylov subspaces. We characterise the major methods CG, MINRES, GMRES, and SYMMLQ in terms of their projection properties. Using these properties and the standard approaches for generating Krylov subspace bases (namely the Lanczos and Arnoldi algorithms), we derive algorithmic descriptions of some Krylov subspace methods.

In Chapter 3, *Matching Moments and Model Reduction View*, we consider the ideas of *moments* and *model reduction*, starting from (a simplified version of) Stieltjes' classical moment problem. We discuss important related concepts, ranging from the Gauss–Christoffel quadrature and orthogonal polynomials to continued fractions. Through Jacobi matrices we find the matrix computations analogies in Krylov subspace methods, and we characterise the methods in terms of their moment matching properties.

The central concept of Chapter 4, *Short Recurrences for Generating Orthogonal Krylov Subspace Bases*, is the *invariant subspace*. We discuss how the length of a Krylov sequence is related to the Jordan canonical form of the given matrix. The main goal of the chapter is to explain when a Krylov sequence can be orthogonalised with a short recurrence. The main result in this context motivates the general distinction in the area of Krylov subspace methods between methods for Hermitian and non-Hermitian matrices.

Some of the general ideas and relationships presented in the first four chapters are illustrated in Chapter 5, *Cost of Computations Using Krylov Subspace Methods*. The chapter starts with a general discussion of the concept of computational cost and related issues, including the difference between direct and iterative methods, particular computations and complexity, and the concept of convergence in general. We then focus on the major methods CG and GMRES, and analyse their exact and finite precision behaviour. We summarise and present many results published previously (scattered throughout many papers), while making no claim for completeness of coverage. Some results and views presented here were not previously published, or were just briefly mentioned without an extensive treatment. Moreover, we feel that there is a need to pose and to investigate new questions in relation to application areas. For example, the questions of measuring the error and evaluating the cost when solving practical problems with Krylov subspace methods cannot be resolved, in our opinion, within the field of matrix computations alone. The fact that they need a much wider context is one of the challenges that is formulated in Chapter 5. The chapter ends with a discussion of some open questions, omitted topics, and an outlook.

Summing up, our goal is neither to give a classification of all existing Krylov subspace methods, nor to review or reference all existing approaches. Rather, we want to identify the major ideas and thoroughly analyse the resulting major methods, with algorithmic details presented only when they are relevant for the exposition. (For additional algorithmic descriptions we give appropriate references.) We thus attempt to be *analytic rather than algorithmic* and *focused rather than encyclopedic*. We hope that the readers of this book might find this approach stimulating for further analytic investigations of Krylov subspace methods as well as for using these methods more effectively in the future.

ACKNOWLEDGEMENTS

The book has been written out of our experience in research and teaching, and in discussing results with professionals in the field as well as in application areas. Many friends, students, and colleagues have helped us over the years with comments and suggestions for improvements of all kinds. It is our pleasure to thank, in particular, Michele Benzi, Hanka Bílková, Jurjen Duintjer-Tebbens, Martin Gander, André Gaul, Tomáš Gergelits, Anne Greenbaum, Lek Heng Lim, Robert Luce, Gerard Meurant, Chris Paige, Jan Papež, Miroslav Rozložník, Olivier Sète, Daniel Szyld, Petr Tichý and Miroslav Tůma. In both our personal and scientific lives, the late Gene Golub played an important role. His views have deeply influenced our understanding of the subject and thus the presentation in this book.

We gratefully acknowledge the Deutsche Forschungsgemeinschaft and the Grant Agencies of the Czech Republic and of the Academy of Science of the Czech Republic for support of the research that has led us to the publication of this book. Jörg Liesen was supported by the Emmy Noether and Heisenberg programmes of DFG, and Zdeněk Strakoš was supported by the projects IAA100300802, M100300901, and GAČR 201/09/0917. Furthermore, we thank the editors and staff at Oxford University Press for their interest in the topic and for their encouragement to get the book finished. Endre Süli, Clare Charles, Viki Mortimer, and Keith Mansfield were extremely helpful in this process.

In the writing we greatly benefited from the MATLAB computing language and interactive environment, which was used in all numerical experiments in this book, the LaTeX document preparation system, and the MathSciNet database of the American Mathematical Society.

Most of all, we would like to thank our families and our wives Anja and Ivanka for their lasting support. Without their love, patience, care, and encouragement this work would not have been done.

While working on this book we often felt, perhaps even more strongly than at other times, what is written in 1 Cor 4:7:

What do you possess that you have not received?

Jörg Liesen and Zdeněk Strakoš
August 2012

CONTENTS

1. Introduction 1
   1.1. Solving Real-world Problems 2
   1.2. A Short Recollection of Ideas 7
   1.3. Specification of the Subject and Basic Notation 10

2. Krylov Subspace Methods 12
   2.1. Projection Processes 12
   2.2. How Krylov Subspaces Come into Play 19
   2.3. Mathematical Characterisation of Some Krylov Subspace Methods 22
   2.4. The Arnoldi and Lanczos Algorithms 26
       2.4.1. Arnoldi and Hermitian Lanczos 26
       2.4.2. Non-Hermitian Lanczos 31
       2.4.3. Historical Note: The Gram–Schmidt Method 33
   2.5. Derivation of Some Krylov Subspace Methods 36
       2.5.1. Derivation of CG Using the Hermitian Lanczos Algorithm 36
       2.5.2. CG and the Galerkin Finite Element Method 42
       2.5.3. CG and the Minimisation of a Quadratic Functional 47
       2.5.4. Hermitian Indefinite Matrices and the SYMMLQ Method 54
       2.5.5. Minimising the Residual Norm: MINRES and GMRES 57
       2.5.6. Further Krylov Subspace Methods 61
       2.5.7. Historical Note: A. N. Krylov and the Early History of Krylov Subspace Methods 64
   2.6. Summary and Outlook: Linear Projections and Nonlinear Behaviour 69

3. Matching Moments and Model Reduction View 71
   3.1. Stieltjes' Moment Problem 73
   3.2. Model Reduction via Orthogonal Polynomials 76
       3.2.1. Derivation of the Gauss–Christoffel Quadrature 77
       3.2.2. Moment Matching and Convergence Properties of the Gauss–Christoffel Quadrature 84
       3.2.3. Historical Note: Gauss' Fundamental Idea, an Early Application, and Later Developments 87

3.3. Orthogonal Polynomials and Continued Fractions 89
    3.3.1. Three-term Recurrence and the Interlacing Property 89
    3.3.2. Continued Fractions 96
    3.3.3. The Gauss–Christoffel Quadrature for Analytic Functions 101
    3.3.4. Summary of the Previous Mathematical Development 103
    3.3.5. Historical Note: Chebyshev, Markov, Stieltjes, and the Moment Problem 104
    3.3.6. Historical Note: Orthogonal Polynomials and Three-term Recurrences 107
3.4. Jacobi Matrices 108
    3.4.1. Algebraic Properties of Jacobi Matrices 109
    3.4.2. The Persistence Theorem, Stabilisation of Nodes and Weights in the Gauss–Christoffel Quadrature 121
    3.4.3. Historical Note: The Origin and Early Applications of Jacobi Matrices 130
3.5. Model Reduction via Matrices: Hermitian Lanczos and CG 136
3.6. Factorisation of the Matrix of Moments 142
3.7. Vorobyev's Moment Problem 145
    3.7.1. Application to the Hermitian Lanczos Algorithm and the CG Method 146
    3.7.2. Application to the Non-Hermitian Lanczos Algorithm 148
    3.7.3. Application to the Arnoldi Algorithm 151
    3.7.4. Matching Moments and Generalisations of the Gauss–Christoffel Quadrature 153
3.8. Matching Moments and Projection Processes 156
3.9. Model Reduction of Large-scale Dynamical Systems 160
    3.9.1. Approximation of the Transfer Function 161
    3.9.2. Estimates in Quadratic Forms 165

4. Short Recurrences for Generating Orthogonal Krylov Subspace Bases 168
   4.1. The Existence of Conjugate Gradient Like Descent Methods 169
   4.2. Cyclic Subspaces and the Jordan Canonical Form 172
      4.2.1. Invariant Subspaces and the Cyclic Decomposition 173
      4.2.2. The Jordan Canonical Form and the Length of the Krylov Sequences 185
      4.2.3. Historical Note: Classical Results of Linear Algebra 188
   4.3. Optimal Short Recurrences 189
   4.4. Sufficient Conditions 193
   4.5. Necessary Conditions 196
   4.6. Matrix Formulation and Equivalent Characterisations 204
   4.7. Short Recurrences and the Number of Distinct Eigenvalues 209
   4.8. Other Types of Recurrences 211
      4.8.1. The Isometric Arnoldi Algorithm 211
      4.8.2. $(s+2,t)$-term Recurrences 215
      4.8.3. Generalised Krylov Subspaces 219
   4.9. Remarks on Functional Analytic Representations 222

Contents xv

5. Cost of Computations Using Krylov Subspace Methods 227
   5.1. Seeing the Context is Essential 228
   5.2. Direct and Iterative Algebraic Computations 238
       5.2.1. Historical Note: Are the Lanczos Method and the CG Method Direct or Iterative Methods? 241
   5.3. Computational Cost of Individual Iterations 245
   5.4. Particular Computations and Complexity 248
   5.5. Closer Look at the Concept of Convergence 250
       5.5.1. Linear Stationary Iterative Methods 250
       5.5.2. Richardson Iteration and Chebyshev Semi-iteration 252
       5.5.3. Historical Note: Semi-iterative and Krylov Subspace Methods 254
       5.5.4. Krylov Subspace Methods: Nonlinear Methods for Linear Problems 257
   5.6. CG in Exact Arithmetic 258
       5.6.1. Expressions for the CG Errors and Their Norms 260
       5.6.2. Eigenvalue-based Convergence Results for CG 265
       5.6.3. Illustrations of the Convergence Behaviour of CG 271
       5.6.4. Outlying Eigenvalues and Superlinear Convergence 275
       5.6.5. Clustered Eigenvalues and the Sensitivity of the Gauss–Christoffel Quadrature 280
   5.7. GMRES in Exact Arithmetic 285
       5.7.1. Ritz Values and Harmonic Ritz Values 286
       5.7.2. Convergence Descriptions Based on Spectral Information 289
       5.7.3. More General Approaches 294
       5.7.4. Any Nonincreasing Convergence Curve is Possible with Any Eigenvalues 297
       5.7.5. Convection–diffusion Example 303
       5.7.6. Asymptotic Estimates for the GMRES Convergence 306
   5.8. Rounding Errors and Backward Stability 308
       5.8.1. Rounding Errors in Direct Computations 311
       5.8.2. Historical Note: Wilkinson and the Backward Error Concept 315
       5.8.3. The Backward Error Concept in Iterative Computations 316
   5.9. Rounding Errors in the CG Method 320
       5.9.1. Delay of Convergence 320
       5.9.2. Delay of Convergence can Invalidate Composite Convergence Bounds 326
       5.9.3. Maximal Attainable Accuracy 328
       5.9.4. Back to the Poisson Model Problem 331
   5.10. Rounding Errors in the GMRES Method 338
       5.10.1. The Choice of the Basis Affects Numerical Stability 338
       5.10.2. Does the Orthogonalisation Algorithm Matter? 340
       5.10.3. MGS GMRES is Backward Stable 343
   5.11. Omitted Issues and an Outlook 345

References 349
Index of Historical Personalities 385
Index of Technical Terms 388

1
Introduction

This chapter can be considered a *prologue* to the book. It consists of three sections that establish the general setting and which give background information that is useful to know before starting to read the other chapters.

The first section presents an overview of a typical solution process of real-world problems in the computationally oriented sciences and engineering. The three stages in this process are modelling, discretisation, and computation. This book deals with iterative methods and their applications to solving problems in matrix computations. At first sight its topic therefore appears to be restricted to the last stage of the solution process. However, as we argue in the first section of this chapter and at many other places in this book, the computational stage and the iterative methods should always be considered in the context of the overall solution process. We consider this point of utmost importance for the future development of the field.

In the second section, we further motivate our approach to the subject through its historical roots and the original sources. We very much share the views expressed by Bultheel and van Barel in the Preface of their book *Linear Algebra, Rational Approximation, and Orthogonal Polynomials* [87]. Starting from the initial observation that the same principles and techniques are often developed and used independently in different fields of mathematics, they give a tour through numerous interconnected topics developed over many centuries and that are covered in their book (from the algorithm of Euclid through continued fractions, formal power series, Padé approximants, the Lanczos algorithm, Toeplitz and Hankel matrices to minimal partial realisation, and other applications in linear systems theory and signal processing). In the second section of this chapter we provide a similar overview of the evolution of topics studied in our book by briefly recalling some of the historical lines of research that are closely related to the modern field of Krylov subspace methods. This overview is also helpful for putting the historical notes later in this book into their proper historical context.

The last section specifies the main notation used in the book, and it gives some hints on prerequisites and general complementary literature.

1.1 SOLVING REAL-WORLD PROBLEMS

Numerical solving of real-world problems typically consists of several stages, each corresponding to some specific areas of science and mathematics. Typically, after describing a problem in a mathematical language and its proper reformulation and discretisation, the resulting finite-dimensional problem has to be solved. In nonlinear problems, numerical solution is based on some form of linearisation, and the 'computational kernels' solve linear algebraic problems that are usually formulated using matrices.

Motivated by applications in computationally oriented sciences and engineering, we now describe in more detail a fairly general setting that will occur frequently in this book. In this setting a part of reality is described (in mathematical abstraction) by a system of integral and/or differential equations. After choosing a proper formulation of the mathematical model, the existence and uniqueness of the analytic solution is investigated. Subsequently, the continuous problem is discretised, for example using the finite element method. A discrete approximation to the analytic solution is then constructed as a linear combination of a finite number of basis functions; the correspondence between the discrete solutions and the analytic solution has to be established. Finally, the coefficients determining the discrete approximation are computed by solving a linear algebraic problem.

The stages in the solution process we have just described are shown schematically in Figure 1.1 and Figure 1.2 (in areas where some stages do not fully apply, the figures could be modified accordingly). Any solution process starts and ends at the real-world problem stage. Going down the structure represents constructing an approximate solution. Going up represents an interpretation of the computed results, which should always include understanding of the errors. The whole process of solving real-world problems numerically, sometimes labelled as *scientific computing*,[1] combines tools from the areas of the given application, applied mathematics, numerical analysis, optimisation, numerical methods, matrix computations, numerical linear algebra, and computer science.

---

1. It is interesting and useful to search for the origin of the names of the closely connected disciplines called *numerical analysis*, *numerical* or *scientific computing*, *computational mathematics*, and *computational sciences*. These names reflect developments of the broad fields of science involving mathematics, computer science, and applications with the purpose of solving real-world, problems. Very thoughtful essays [46, 165], [622] (reprinted in [625, pp. 321–327]), and [624], as well as historical perspectives described in [451, 472], are recommended to any reader interested in this topic.

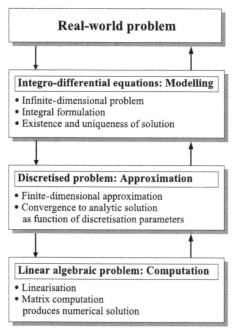

Figure 1.1 Typical stages in solving a real-world problem that is described by integro-differential equations. Here, *verification* deals with the last three stages, while *validation* deals with the adequacy of the mathematical model, i.e. with the relevance of the last three stages for the real-world problem.

All stages in the solution process are accompanied by errors. The main types of errors are approximation errors of the model, discretisation errors of the finite-dimensional formulation, and truncation and/or rounding errors of the numerical solution of the linear algebraic problem. The analysis of errors in the part of the process starting and ending with the mathematical model is called *verification* in the literature on partial differential equations (PDEs). Hence the purely mathematical problem of verification is concerned with estimating and controlling the numerical approximation error. The goal is to verify whether the equations constituting the mathematical model were solved correctly (modulo an acceptable inaccuracy). *Validation* of a mathematical model, on the other hand, asks whether the mathematical model and its numerical solution describe the real-world problem. This clearly involves questions beyond mathematics alone, as demonstrated, for example, in [30, 31, 485]. An enlightening discussion of different sides of verification and validation can also be found in the September/October 2004 issue of the journal *Computing in Science and Engineering*. In particular, the article by Roache [531] in that issue (see also [529]) deals in detail with the verification and validation concepts in solving nonlinear PDEs, and it addresses common misunderstandings related to them. Furthermore, it discusses handling uncertainty in

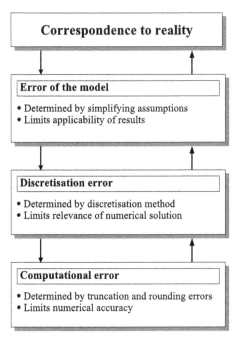

Figure 1.2 Structure of errors in solving a real-world problem. Note that one often also has to deal with *data uncertainty*, i.e. incomplete or inaccurate knowledge of data that determines the mathematical model (e.g. physical constants).

mathematical models and numerical errors, and distinguishes between verification of codes and verification of calculations.

The discussion in [531] follows the detailed description of many issues related to verification and validation given by Roache in the monograph [530]. Reading his book is truly inspirational as it summarises leading expertise with verification and validation of enormously complex codes and computations used for solving extremely demanding real-world problems. We will refer to some particular points in several places throughout the book. There is, however, one issue in which the focus of our book principally differs from the approach used in verification. In particular, as clearly formulated in [530, p. 36]: '*Most importantly, incomplete iterative convergence will corrupt Verification*'. This essentially means that the part of the error due to stopping the iterative algebraic solver is in the verification process *excluded from consideration*. This is supposed to be done (without a rigorous justification that such an approach will always work) by using sufficiently tight stopping criteria. This argumentation is documented, for example, by the following quotes:

- 'It is important that iterative convergence criteria be stringent so as not to confuse the incomplete iteration (residual) error with discretization error.' [530, p. 72]

- 'The basic idea, of course, is to make the iteration residual negligible with respect to the discretization error.' [511, p. 424]
- 'IICE [incomplete iterative convergence error] can be decreased to the level of roundoff error by using sufficiently tight stopping criteria.' and 'By using tight iterative tolerances and well-conditioned test problems, we can ensure that the numerical error is nearly the same as the discretization error.' [389, p. 15]

The situation is nicely summarized by the next quote from the monograph of Knupp and Salari [389, p. 31]: 'There may be incomplete iterative convergence (IICE) or round-off-error that is polluting the results. If the code uses an iterative solver, then one must be sure that the iterative stopping criteria is sufficiently tight so that the numerical and discrete solutions are close to one another. Usually in order-verification tests, one sets the iterative stopping criterion to just above the level of machine precision to circumvent this possibility.'

We by no means argue against the practice outlined above (although, as follows from our book, it may not always be easy to satisfy the stringent stopping criteria). We wish to point out, however, that verification and practical computations do look at some particular parts of the solution process differently, and some conditions which are applied in verification can not be applied in practical computations. This is in full agreement with [530] and the other literature referenced in this paragraph.

In order to use human and computer resources efficiently, the whole solution process must be well balanced. For example, an unrealistic model or inappropriate discretisation may lead to unsolvable computational problems. Difficulties on the computational stage, however, sometimes occur even if the best known modelling and discretisation techniques have been used. It is truly difficult to solve some very large systems of linear algebraic equations (having millions of unknowns), or to find good approximate solutions to ill-posed problems (such as in image restoration and reconstruction). In such cases, an inadequate approach to solving computational problems can waste an enormous amount of computer time (and money) or may lead to numerical disasters and meaningless numerical solutions.

Each stage of the solution process requires its own knowledge and expertise. A mistake at any of the stages can hardly be compensated for by excellence at the others. To pure mathematicians and to some numerical analysts, discretisation represents the key point; it transforms a continuous 'mathematical' problem to a finite-dimensional 'numerical' problem. Discretisation should certainly be given proper attention. Improper discretisation (discretisation error) can lead to *qualitative differences* between the investigated phenomena and the solution of the finite-dimensional discretised problem, as in numerical chaos [46, p. 19] or in solving convection–diffusion problems [84, 158, 159]. It should be noted, however, that improperly handled computational errors can have similar effects. Uncontrolled computational errors can also lead to qualitatively wrong solutions. Nevertheless, the process of error estimation is in the PDE literature typically only concerned with a-priori and a-posteriori analysis of discretisation errors (for an example of a more general discussion we refer to [675]). The paradoxical nature of such an exclusion of computational errors from the analysis has been pointed out (with respect to rounding errors) by Parlett in his essay devoted to the work of Wilkinson [505, pp. 19–20]. In this context it is interesting to recall the work of Babuška [29], which calls for embedding the study of numerical stability in linear algebra into a study of numerical stability for general numerical methods.

It is worth pointing out that the step from the real-world problem to the mathematical model represents a kind of *reduction* of the reality, so that it can be analysed and computationally handled. Discretisation represents another reduction from a (continuous) infinite-dimensional problem to a discrete and finite-dimensional system of equations. In our book we focus on the last stage of the solution process, i.e. on solving the linearised (finite-dimensional) problem by the methods of matrix computations. More specifically, our book is devoted to iterative methods in matrix computations, and mostly to Krylov subspace methods for solving linear algebraic systems. As presented in this book (based on the work of many other authors), computations using Krylov subspace methods represent nothing but a further (matching moments) model reduction. We recall this important point later.

In various sources it is estimated that between seventy and ninety percent of all computational time in scientific computing is spent on solving linear algebraic problems. Progress in development, application, and understanding of methods in this area therefore has a wide impact on many areas of research and on technology development. As described above, a proper approach to solving linear algebraic problems arising from real-world applications should always consider the connection to the original mathematical model and its discretisation. Moreover, analogously to the so-called 'Method of Manufactured Solutions' for code verification, described by Roache in [531, pp. 33–35], it is highly desirable to interpret the inaccuracies in the solution process (including the algebraic errors) in terms of a meaningful modification of the mathematical model. Proper understanding of the interactions between all stages in solving any real-world problem is fundamental for keeping the right perspective while working within any specific stage. Isolating the algebraic stage (compare, e.g. the simplified flow diagram for a typical PDE code given in [389, p. 36, Figure 4.1]) leads to misunderstandings and misconceptions. For example, sometimes the computational stage is considered a routine application of a solver from some software package, and, as mentioned above, the computational errors (both the truncation and the roundoff parts) are completely ignored, as though computers would give accurate solutions. In other cases, stopping criteria and convergence evaluation of iterative solvers are based on unrealistic assumptions, which can not be applied to practical computations.

These issues cannot be resolved by any single piece of work. We point out that they should, however, be taken into account whenever possible. We believe that they are of principal importance. They motivate many views about iterative solving of linear algebraic problems presented in this book.

Finally, we point out that linear algebraic problems arise in many areas where the setting described above does not necessarily apply. For example, such problems are inherent parts of numerical optimisation, computational statistics, information retrieval, and image processing. While the setting may be different, the general point about considering the linear algebraic problems to be a part of a broader context still remains of principal importance for the solution process.

1.2 A SHORT RECOLLECTION OF IDEAS

The best way to guide the mental development of the individual is to let him retrace the mental development of the race – retrace its great lines, of course, and not the thousand errors of detail.[2]

In our book we aim at (what we consider) the current state-of-the-art in the analysis of Krylov subspace methods. Nevertheless, as described in the Preface, we believe that modern mathematics should not be separated from its historical roots. Research and teaching *today* can benefit greatly from original sources. In the spirit of the above quote we present here, at the very start of the book, an overview of the main lines of mathematical research that eventually led or are related to the field of Krylov subspace methods as we know it today. This overview is provided in Figure 1.3, which shows the flow of major mathematical ideas, with corresponding names and dates ranging over approximately 300 years, roughly from 1650 to 1950. We have intentionally stopped with the appearance in the early 1950s of the papers by Lanczos and Hestenes and Stiefel, which mark the starting point of the field of Krylov subspace methods as it is known nowadays. Developments prior to 1650 are described, for example, in [76, 77, 87, 384]. Each column in Figure 1.3 shows the development of a certain major topic. The choice of topics as well as the items in each of them is, of course, strictly limited. Further details with many other references are presented throughout the book.

As in any academic field, the style and the language in mathematics is constantly changing. To make Figure 1.3 comprehensible for contemporary readers we have used modern terminology throughout. This simplification and the presentation of clearly separated lines of thought conceals that the actual development was much less straightforward. As Lanczos pointed out in [409, p. 409],

> ... the evolution of science does not occur in steady growth but in fruitful jumps, initiated by sudden flashes of ingenuity which are not different from the manner of artistic creation.

In this sense each box in Figure 1.3 represents one (or more) flashes of ingenuity. Referring to the history of physics, Lanczos continued in [409, p. 410] that the flash itself is not enough, but that

> ... [a]fter the big jump a period of consolidation must come, in which the new insight is systematized and brought into harmony with the rest of the edifice.

---

2. This quote is from the memorandum on the teaching of mathematics [3] signed in 1962 by 75 leaders of the field, including Lars Ahlfors, Richard Bellman, Richard Courant, Garrett Birkhoff, Kurt Otto Friedrichs, Hermann Goldstine, Peter Lax, Max Schiffer, Abraham Taub, and André Weil.

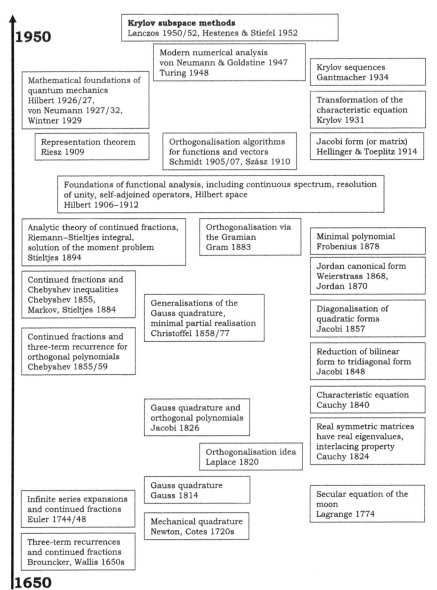

Figure 1.3 Lines of mathematical research closely related to the field of Krylov subspace methods.

This is certainly true also in mathematics, where an ingenious idea often appears as a raw diamond which is subsequently polished and refined. Figure 1.3 displays numerous examples. One of them is Gauss' quadrature method of 1814, which Jacobi related to the idea of orthogonal polynomials in 1826 and which was significantly extended (close to the modern form we use in this book) by Christoffel in 1858 and 1877; cf. the corresponding column in Figure 1.3. Another example is the development of the analytic theory of continued fractions, formulated by

Stieltjes in 1894 on the top of the work of many authors who fundamentally contributed to that subject.

Important flashes of ingenuity for the subject matter of this book occurred in the late 1940s and early 1950s. Though they cannot be attributed to a single person or event, they manifest themselves in the publication of four fundamental papers. Three of the four papers were written by Lanczos (unfortunately, one of them has been largely overlooked), and the fourth one jointly by Hestenes and Stiefel. The papers were written at the Institute for Numerical Analysis (INA) at the University of California, Los Angeles, in the late 1940s and early 1950s. They represent the culmination of the two INA projects on computational methods for solving linear algebraic systems and eigenvalue problems. Figure 1.4 shows these papers' titles and the mathematical areas that were *explicitly addressed* in at least one of them. We observe that the papers are related to a large part of what is contained in the 300-year development illustrated in Figure 1.3. In an exemplary interdisciplinary spirit, Lanczos, Hestenes, and Stiefel systematised and brought into harmony a fascinating variety of topics. But they did not only collect and reorder known results. They connected these results with the modern topic of *numerical analysis*, including issues like cost of computation, rounding errors, and rate of convergence, which have been important ever since. Thereby they produced a 'next big jump' in the scientific progress.

Confirming Lanczos' observation, it took a period of consolidation until Krylov subspace methods finally emerged as a field of their own in the early 1970s. This particular piece of history will be discussed in detail in Section 5.5.3 of the book. That section is one of the historical notes that can be found throughout the book, and which allow interested readers to retrace the development of specific topics. A complete list of these notes is given in Table 1-1.

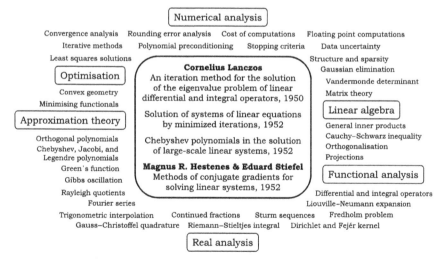

Figure 1.4 The four fundamental papers of Lanczos and Hestenes and Stiefel, and mathematical topics mentioned in these works.

Table 1-1. TITLES, SECTION NUMBERS AND STARTING PAGES OF THE HISTORICAL NOTES IN THIS BOOK

Historical note	Section (first page)
The Gram–Schmidt method	2.4.3 (p. 33)
A. N. Krylov and the early history of Krylov subspace methods	2.5.7 (p. 64)
Gauss' fundamental idea, an early application, and later developments	3.2.3 (p. 87)
Chebyshev, Markov, Stieltjes and the moment problem	3.3.5 (p. 104)
Orthogonality and three-term recurrences	3.3.6 (p. 107)
The origin and early applications of Jacobi matrices	3.4.3 (p. 130)
Classical results of linear algebra	4.2.3 (p. 188)
Are the Lanczos method and the CG method direct or iterative methods?	5.2.1 (p. 241)
Semi-iterative and Krylov subspace methods	5.5.3 (p. 254)
Wilkinson and the backward error concept	5.8.2 (p. 315)

1.3 SPECIFICATION OF THE SUBJECT AND BASIC NOTATION

Our main interest in this book is the analysis of Krylov subspace methods for solving linear algebraic systems. We write these systems in the form

$$Ax = b. \qquad (1.3.1)$$

The matrix A is an element of $\mathbb{F}^{N \times N}$, where either $\mathbb{F} = \mathbb{R}$ (the real numbers) or $\mathbb{F} = \mathbb{C}$ (the complex numbers). Unless otherwise noted, the square matrix A is considered without any further specific assumptions. Some methods or results require special assumptions, e.g. that A is nonsingular, symmetric, or Hermitian. If required, such assumptions will always be stated explicitly.

If $\mathbb{F} = \mathbb{R}$, i.e. A is a real matrix, we usually consider only real right-hand sides, $b \in \mathbb{R}^N$ and real initial approximations, $x_0 \in \mathbb{R}^N$, giving a real initial residual, $r_0 \equiv b - Ax_0 \in \mathbb{R}^N$. Statements like 'for any r_0' should then be read 'for any $r_0 \in \mathbb{R}^N$', the occurring polynomials have real coefficients, and the conjugate transpose, denoted by v^* for a vector v and M^* for a matrix M, coincides with the transpose.

For a given symmetric or Hermitian positive definite matrix, $B \in \mathbb{F}^{N \times N}$, we denote the *B-inner product* on $\mathbb{F}^N$ by

$$(v, w)_B \equiv w^* B v, \quad \text{for all } v, w \in \mathbb{F}^N.$$

The corresponding *B-norm for vectors* is given by

$$\|v\|_B \equiv (v, v)_B^{1/2}, \quad \text{for all } v \in \mathbb{F}^N,$$

and the associated *B-norm for matrices* is

$$\|M\|_B \equiv \max_{v \in \mathbb{F}^N \setminus \{0\}} \frac{\|Mv\|_B}{\|v\|_B}, \quad \text{for all } M \in \mathbb{F}^{N \times N}.$$

Two vectors $v, w \in \mathbb{F}^N$ are *B-orthogonal* if $(v, w)_B = 0$, which we alternatively write as $v \perp_B w$.

For $B = I_N$ (the identity matrix; we often write I when its size is clear from the context) we obtain the *Euclidean inner product* and the corresponding *Euclidean vector norm* with the associated matrix *spectral norm* (both these norms are sometimes referred to as *2-norm*). We then typically skip the dependence on B in the notation and simply write (v, w), $\|v\|$, $\|M\|$, and $v \perp w$.

Further notation will be introduced and explained at appropriate places throughout the book.

We have tried to make the book as self-contained as possible by including detailed motivations and derivations of the main concepts and proofs of many stated results. Nevertheless, the material does not start from completely elementary parts. It requires background knowledge of linear algebra as well as real and complex analysis. The results we refer to throughout the book can be found, in addition to the references given in the text, in other good textbooks. A knowledge of basic numerical methods, including matrix computations, methods of interpolation, numerical integration, and the numerical solution of ordinary and partial differential equations is also beneficial. An introduction to some of these subjects can be found in the first chapters of Saad's book on iterative methods [543]. That book can also be used as a reference text for implementations of methods which are not stated in detail in our book.

2

Krylov Subspace Methods

As described in the Preface, our book is more analytic than algorithmic, and more focused than encyclopedic. This is reflected by the approach in this chapter to the central subject of our investigation. Under natural conditions and assumptions we derive the main ideas of Krylov subspace methods from a general *projection framework*. Within this general mathematical framework we characterise some relevant special cases and give mathematical descriptions of the most important Krylov subspace methods.

The characterisation of Krylov subspace methods as projection methods evolved rather late in the historical development of the field. It was pioneered by Saad in a series of papers in the early 1980s [541, 542]; see his book [543] for a survey. By then, Krylov subspace methods had been around for more than 30 years. Saad's work provided a clear and simple, yet effective way to describe the basics of Krylov subspace methods, which has been adopted by many authors of survey articles and books on this topic; see, e.g. [79, 154, 155, 570, 636]. In later chapters of our book we will complement the projection view of Krylov subspace methods by other characterisations and approaches, in particular by the ideas of moments and model reduction.

2.1 PROJECTION PROCESSES

Consider a linear algebraic system $Ax = b$ with $A \in \mathbb{F}^{N \times N}$ and $b \in \mathbb{F}^N$. Suppose that in an iterative process the *nth approximation* x_n, $n = 1, 2, \ldots$, is of the form

$$x_n \in x_0 + \mathcal{S}_n, \qquad (2.1.1)$$

where $x_0 \in \mathbb{F}^N$ is a given *initial approximation* (x_0 may be the zero vector) and $\mathcal{S}_n \subseteq \mathbb{F}^N$ is some n-dimensional subspace, called the *search space*. Let x be a solution of the given linear algebraic system (which is uniquely defined if and only if A is nonsingular). Then the vector $x - x_n$ is the *error* corresponding to the nth approximation x_n, or briefly the *nth error*. Using (2.1.1) we see that

$$x - x_n \in (x - x_0) + \mathcal{S}_n,$$

and thus

$$x - x_0 = z_n + (x - x_n) \quad \text{for some } z_n \in \mathcal{S}_n. \tag{2.1.2}$$

In (2.1.2) the *initial error* $x - x_0$ is approximated by the vector z_n from the subspace $\mathcal{S}_n$, with the inaccuracy of this approximation given by the nth error $x - x_n$.

Since $\mathcal{S}_n$ has dimension n, we have n degrees of freedom to construct x_n (or, equivalently, the vector z_n). In order to determine x_n we therefore generally need n constraints. We will impose these on the *residual* corresponding to the nth approximation x_n, briefly the *nth residual*, which is defined by

$$\begin{aligned} r_n &\equiv b - Ax_n \\ &= b - A(x_0 + z_n) = (b - Ax_0) - Az_n \in r_0 + A\mathcal{S}_n. \end{aligned} \tag{2.1.3}$$

The vector $r_0 \equiv b - Ax_0$ is called the *initial residual*. We will require that r_n be orthogonal to a given n-dimensional subspace $\mathcal{C}_n \subseteq \mathbb{F}^N$, called the *constraints space*,

$$r_n \perp \mathcal{C}_n. \tag{2.1.4}$$

Here the orthogonality is meant with respect to the Euclidean inner product, i.e. $(r_n, z) \equiv z^* r_n = 0$ holds for all $z \in \mathcal{C}_n$. Instead of (2.1.4) we also write $r_n \in \mathcal{C}_n^\perp$.

Note that $r_n = A(x - x_n)$. Hence (2.1.4) and a multiplication of (2.1.2) from the left with A give the relations

$$r_0 = Az_n + r_n, \quad Az_n \in A\mathcal{S}_n, \quad r_n \in \mathcal{C}_n^\perp. \tag{2.1.5}$$

These relations say that the initial residual r_0 is approximated by the vector Az_n from the subspace $A\mathcal{S}_n$, with the inaccuracy of this approximation given by the nth residual r_n, which is in the orthogonal complement of the constraints space $\mathcal{C}_n$.

If $\mathbb{F}^N$ is the direct sum of $A\mathcal{S}_n$ and $\mathcal{C}_n^\perp$,

$$\mathbb{F}^N = A\mathcal{S}_n \oplus \mathcal{C}_n^\perp, \tag{2.1.6}$$

then for any $r_0 \in \mathbb{F}^N$ the corresponding vectors Az_n and r_n in (2.1.5) are uniquely determined, and r_0 can be written as

$$r_0 = Az_n + r_n = r_0 \big|_{A\mathcal{S}_n} + r_0 \big|_{\mathcal{C}_n^\perp}, \tag{2.1.7}$$

where $r_0|_{A\mathcal{S}_n}$ and $r_0|_{\mathcal{C}_n^\perp}$ denote the uniquely defined parts of r_0 in the subspaces $A\mathcal{S}_n$ and $\mathcal{C}_n^\perp$, respectively. If (2.1.6) holds, then

$$Az_n = r_0|_{A\mathcal{S}_n}$$

represents a *projection* of r_0 onto $A\mathcal{S}_n$ and orthogonal to $\mathcal{C}_n$, with the complement of this projection given by the (uniquely determined) nth residual r_n.

Remark 2.1.1

The projection process we have described above is an instance of the so-called *Petrov–Galerkin method*. An analogous method is used, for example, in the discretisation of variational problems of the form,

$$\text{Find } u \in \mathcal{S} \text{ such that } a(u,v) = f(v) \text{ for all } v \in \mathcal{C},$$

where $\mathcal{S}$ and $\mathcal{C}$ are infinite-dimensional (Hilbert) spaces, a is a bilinear form on $\mathcal{S} \times \mathcal{C}$, and f is a linear functional on $\mathcal{C}$. If $\mathcal{S}_h$ and $\mathcal{C}_h$ are finite-dimensional subspaces of $\mathcal{S}$ and $\mathcal{C}$, respectively, the discretised variational problem is:

$$\text{Find } u_h \in \mathcal{S}_h \text{ such that } a(u_h, v_h) = f(v_h) \text{ for all } v_h \in \mathcal{C}_h.$$

The finite-dimensional spaces $\mathcal{S}_h$ and $\mathcal{C}_h$ correspond to the search and constraints spaces in our development above. If $u \in \mathcal{S}$ solves the infinite-dimensional variational problem, and $u_h \in \mathcal{S}_h \subset \mathcal{S}$ solves the finite-dimensional one, then for every $v_h \in \mathcal{C}_h \subset \mathcal{C}$ the error $u - u_h \in \mathcal{S}$ satisfies

$$a(u - u_h, v_h) = a(u, v_h) - a(u_h, v_h) = f(v_h) - f(v_h) = 0.$$

Hence the error is 'orthogonal' to the constraints space $\mathcal{C}_h$ with respect to the bilinear form a. This corresponds to the orthogonality condition (2.1.4), which can be written as $a(x - x_n, z) = 0$ for all $z \in \mathcal{C}_n$, where $a(w,y) \equiv y^*Aw$ for all $w, y \in \mathbb{F}^N$. A specific example of such a method and its relation to iterative methods for solving linear algebraic systems will be given in Section 2.5.2. A nice description of the general framework of Petrov–Galerkin methods is given in [515, Chapter 5].

Let us now consider a matrix representation of the projection process (2.1.1)–(2.1.4). We denote by S_n and C_n two $N \times n$ matrices, with their respective columns forming (arbitrary) bases of the n-dimensional subspaces $\mathcal{S}_n$ and $\mathcal{C}_n$. Then (2.1.1) translates into

$$x_n = x_0 + S_n t_n, \qquad (2.1.8)$$

for some vector $t_n \in \mathbb{F}^n$. In order to determine t_n we write the orthogonality constraints (2.1.4) as

$$0 = C_n^* r_n = C_n^*(b - Ax_n) = C_n^* r_0 - C_n^* AS_n t_n,$$

or, equivalently,

$$C_n^* A S_n t_n = C_n^* r_0. \tag{2.1.9}$$

The system (2.1.9) is called the *projected system*, and $C_n^* A S_n$ is the *projected matrix*. The key idea of the projection approach for solving linear algebraic systems can be formulated as follows:

Instead of the (possibly) large system $Ax = b$ of order N, we solve in step n of the projection process the projected system $C_n^* A S_n t_n = C_n^* r_0$ of order n, and the goal is to obtain a good approximation $x_n = x_0 + S_n t_n$ for $n \ll N$.

The projected system can be viewed as a *model reduction* of the original system $Ax = b$. This aspect will be considered in detail in Chapter 3.

When the solution t_n of the projected system is uniquely determined, i.e. when $C_n^* A S_n$ is nonsingular, we say that the projection process (2.1.1)–(2.1.4) is *well defined at step n*. We next show that the projection process (2.1.1)–(2.1.4) is well defined at step n if and only if the condition (2.1.6) holds. For interpretations and some details concerning the more general situation when (2.1.6) does not hold we refer to [154, Remarks 2.5 and 3.7].

Theorem 2.1.2
Let $A \in \mathbb{F}^{N \times N}$ and let S_n and C_n be two n-dimensional subspaces of $\mathbb{F}^N$. Consider the projection process (2.1.1)–(2.1.4), represented in matrix form by (2.1.8)–(2.1.9). Then the matrix $C_n^* A S_n$ is nonsingular if and only if (2.1.6) holds.

Proof
If (2.1.6) holds, then the subspace AS_n has dimension n, and hence the matrix AS_n has (full) rank n. Now if the matrix $C_n^* A S_n$ is singular, then $C_n^* A S_n z = 0$ for some $z \neq 0$. In particular, $AS_n z \in C_n^\perp$. Since additionally $AS_n z \in AS_n$, (2.1.6) gives $AS_n z = 0$, which contradicts the fact that AS_n has rank n. Therefore $C_n^* A S_n$ is nonsingular.

On the other hand, let $C_n^* A S_n$ be nonsingular. Since $C_n^* A S_n z = 0$ implies $z = 0$, the columns of AS_n are linearly independent and $C_n^\perp \cap AS_n = \{0\}$. Since the subspaces AS_n and $C_n^\perp$ have dimensions n and $N - n$, respectively, (2.1.6) holds. □

Theorem 2.1.2 implies that the question of whether a projection process is well defined at step n depends only on the choice of the subspaces S_n and C_n, and not on their bases. In particular, if the subspaces S_n and C_n satisfy (2.1.6), then for *any* choice of the bases the corresponding matrix $C_n^* A S_n$ is nonsingular.

If $C_n^* A S_n$ is nonsingular, then (2.1.9) yields

$$t_n = (C_n^* A S_n)^{-1} C_n^* r_0,$$

so that

$$x_n = x_0 + S_n t_n = x_0 + S_n (C_n^* A S_n)^{-1} C_n^* r_0, \tag{2.1.10}$$

and
$$r_n = b - Ax_n = (I - AS_n(C_n^*AS_n)^{-1}C_n^*)r_0 = (I - P_n)r_0, \qquad (2.1.11)$$

where
$$P_n \equiv AS_n(C_n^*AS_n)^{-1}C_n^*.$$

The matrix P_n represents a *projector* because $P_n^2 = P_n$. For all vectors $v \in \mathbb{F}^N$,
$$P_n v \in AS_n \quad \text{and} \quad (I - P_n)v \in C_n^\perp. \qquad (2.1.12)$$

This means that P_n projects onto AS_n and orthogonally to C_n. Assuming (2.1.6), i.e. assuming that x_n is uniquely determined, the projector defined by (2.1.12) is unique. Since it depends only on the subspaces AS_n and C_n, and not on the choice of their bases, P_n has the form
$$P_n = W_n(C_n^*W_n)^{-1}C_n^*,$$

where the columns of the matrix W_n form an arbitrary basis of AS_n.

Note that (2.1.11) can be written as
$$r_0 = P_n r_0 + r_n,$$

which represents the decomposition of r_0 into $P_n r_0 \in AS_n$ and $r_n \in C_n^\perp$; cf. (2.1.7). If $AS_n = C_n$, then this decomposition is *orthogonal* because its two components are mutually orthogonal, i.e.
$$\text{if} \quad \mathbb{F}^N = AS_n \oplus C_n^\perp \quad \text{and} \quad AS_n = C_n, \quad \text{then} \quad r_n^* P_n r_0 = 0.$$

In this case the unique orthogonal projector onto $AS_n = C_n$ is given by
$$P_n = W_n(W_n^*W_n)^{-1}W_n^*.$$

Moreover the orthogonal decomposition of r_0 implies
$$\|r_0\|^2 = \|P_n r_0 + r_n\|^2 = \|P_n r_0\|^2 + \|r_n\|^2 \geq \|r_n\|^2,$$

where $\|v\| \equiv (v,v)^{1/2}$ is the Euclidean norm.

If, on the other hand, $AS_n \neq C_n$, then the decomposition of r_0 is *oblique*, and the corresponding matrix
$$P_n = W_n(C_n^*W_n)^{-1}C_n^*$$

represents the unique oblique projector onto AS_n and orthogonal to C_n.

Definition 2.1.3: Assume that (2.1.6) holds, i.e. that the projection process (2.1.1)–(2.1.4) is well defined at step n. When $AS_n = C_n$, we call the projection process (2.1.1)–(2.1.4) orthogonal. When $AS_n \neq C_n$, we call the projection process (2.1.1)–(2.1.4) oblique.

Projection Processes

We illustrate an orthogonal and an oblique projection process in the following example.

Example 2.1.4
Consider the linear system $Ax = b$ with

$$A = \begin{bmatrix} 1 & 2 \\ 0 & -1 \end{bmatrix}, \quad b = \begin{bmatrix} 2 \\ 1 \end{bmatrix}.$$

We choose $x_0 = [0, 0]^T$, so that $r_0 = b$. Furthermore, we choose $\mathcal{S}_1 = \operatorname{span}\{r_0\}$, giving $A\mathcal{S}_1 = \operatorname{span}\{Ar_0\} = \operatorname{span}\{[4, -1]^T\}$.

For the orthogonal projection process with $\mathcal{C}_1 = A\mathcal{S}_1$ we have

$$P_1 = \frac{1}{17}\begin{bmatrix} 16 & -4 \\ -4 & 1 \end{bmatrix} = P_1^*, \quad r_1 = (I - P_1)r_0 = \frac{1}{17}\begin{bmatrix} 6 \\ 24 \end{bmatrix}.$$

Figure 2.1 shows the resulting orthogonal decomposition of $r_0 = P_1 r_0 + r_1$, where $P_1 r_0$ is the orthogonal projection of r_0 onto $A\mathcal{S}_1$. Moreover, we can observe that $\|r_1\| \leq \|r_0\|$ (even $\|r_1\| < \|r_0\|$ in this example).

Figure 2.2 shows the oblique decomposition of r_0 resulting from the oblique projection process with $\mathcal{C}_1 = \operatorname{span}\{[1, 1]^T\} \neq A\mathcal{S}_1$. In this case

$$P_1 = \frac{1}{3}\begin{bmatrix} 4 & 4 \\ -1 & -1 \end{bmatrix} \neq P_1^*, \quad r_1 = (I - P_1)r_0 = \begin{bmatrix} -2 \\ 2 \end{bmatrix},$$

and $P_1 r_0$ is the oblique projection of r_0 onto $A\mathcal{S}_1$ and orthogonal to $\mathcal{C}_1$. Moreover, in this example we can observe that the property $\|r_1\| \leq \|r_0\|$ does not hold.

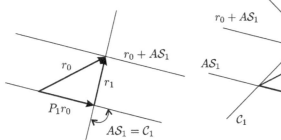

Figure 2.1 Orthogonal projection of r_0 onto $A\mathcal{S}_1 = \mathcal{C}_1$.

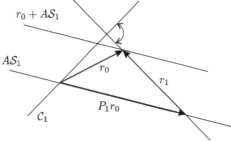

Figure 2.2 Oblique projection of r_0 onto $A\mathcal{S}_1$ and orthogonal to $\mathcal{C}_1 \neq A\mathcal{S}_1$.

Remark 2.1.5
In our exposition, projections are determined by the corresponding decompositions and vice versa. We feel that the terminology should reflect this relationship. Therefore we distinguish between the cases $A\mathcal{S}_n = \mathcal{C}_n$ (orthogonal projection and decomposition) and $A\mathcal{S}_n \neq \mathcal{C}_n$ (oblique projection and decomposition). This deviates from terminology that is widely used in the literature on Krylov subspace methods. In this literature, a distinction is often made between the cases $\mathcal{S}_n = \mathcal{C}_n$ and $\mathcal{S}_n \neq \mathcal{C}_n$. Moreover, in some publications projection processes (2.1.1)–(2.1.4) with $\mathcal{S}_n = \mathcal{C}_n$ are called orthogonal, and they are called oblique when $\mathcal{S}_n \neq \mathcal{C}_n$.

Our Definition 2.1.3 departs from these conventions. Such steps are always risky, and should be adopted only when they are well justified. We believe that in this case the proposed change, though troublesome, is strictly forced by mathematical consistency and clarity. Our view seems to be in agreement with the approach in [154, 155], where the authors distinguish between the general case of possibly different $A\mathcal{S}_n$ and $\mathcal{C}_n$ (called *OR approach*) and the special case $A\mathcal{S}_n = \mathcal{C}_n$ (called *MR approach*). In [543, Sections 5.2.1 and 5.2.2] Krylov subspace methods corresponding to the case $A\mathcal{S}_n = \mathcal{C}_n$ are called *residual projection methods*; methods for Hermitian positive definite (HPD) matrices A and with $\mathcal{S}_n = \mathcal{C}_n$ are called *error projection methods*.

The following theorem gives some practically important conditions under which the projection process (2.1.1)–(2.1.4) is well defined at step n, i.e. under which (2.1.6) holds (cf. Theorem 2.1.2).

Theorem 2.1.6
Let $A \in \mathbb{F}^{N \times N}$, and let $\mathcal{S}_n$ be an n-dimensional subspace of $\mathbb{F}^N$. The projection process (2.1.1)–(2.1.4) is well defined at step n, i.e. (2.1.6) is satisfied, if any of the following conditions holds:

(1) A is Hermitian positive definite, and $\mathcal{C}_n = \mathcal{S}_n$,
(2) A is Hermitian and nonsingular, and $\mathcal{C}_n = A^{-1}\mathcal{S}_n$,
(3) A is nonsingular, and $\mathcal{C}_n = A\mathcal{S}_n$.

Proof
First note that in each case the constraints space $\mathcal{C}_n$ is n-dimensional.
(1) Let $z \in A\mathcal{S}_n \cap \mathcal{C}_n^\perp = A\mathcal{S}_n \cap \mathcal{S}_n^\perp$. Since $z = Ay$ for some $y \in \mathcal{S}_n$, and $z \in \mathcal{S}_n^\perp$, we get $0 = y^*Ay$, which implies $y = 0$ because A is HPD. Therefore $z = 0$, so that $A\mathcal{S}_n \cap \mathcal{C}_n^\perp = \{0\}$. Since the subspaces $A\mathcal{S}_n$ and $\mathcal{C}_n^\perp$ of $\mathbb{F}^N$ have dimensions n and $N - n$, respectively, (2.1.6) must hold. The proof of (2) is analogous.
(3) Here, trivially, $A\mathcal{S}_n \oplus \mathcal{C}_n^\perp = A\mathcal{S}_n \oplus (A\mathcal{S}_n)^\perp = \mathbb{F}^N$. □

Of course, item (2) only yields a reasonable method for solving $Ax = b$ when the use of A^{-1} can be avoided. To achieve this, one can consider a search space of the form $\mathcal{S}_n = A\widehat{\mathcal{S}}_n$, where $\widehat{\mathcal{S}}_n$ is another n-dimensional subspace of $\mathbb{F}^N$ (see item (2) in Theorem 2.3.1 below for an example).

The nth approximation x_n generated by the projection process (2.1.1)–(2.1.4) solves the given linear algebraic system $Ax = b$ if and only if $r_n = b - Ax_n = 0$. If the process is well defined at step n, we see from (2.1.11) that $r_n = 0$ holds if and only if $r_0 = P_n r_0$, and since P_n projects onto $A\mathcal{S}_n$, this is equivalent to $r_0 \in A\mathcal{S}_n$. In the following lemma we state a useful sufficient condition for this property.

Lemma 2.1.7
Suppose that the projection process (2.1.1)–(2.1.4) is well defined at step n, i.e. that (2.1.6) holds. If $r_0 \in \mathcal{S}_n$ and $A\mathcal{S}_n = \mathcal{S}_n$, then $r_n = 0$.

Proof
If $r_0 \in \mathcal{S}_n$ and $A\mathcal{S}_n = \mathcal{S}_n$, we have $r_0 \in A\mathcal{S}_n$, and therefore $r_n = r_0 - P_n r_0 = 0$. □

The lemma gives a suffucent condition for the *finite termination property* of a projection process: a solution of the given linear algebraic system is obtained in a finite number of steps. To guarantee this property it seems like a good idea to start the projection process with the search space $S_1 = \text{span}\{r_0\}$, and to build up a *nested sequence of search spaces*,

$$S_1 \subset S_2 \subset S_3 \subset \cdots,$$

such that these spaces eventually satisfy $AS_n = S_n$ for some n. This idea naturally leads to Krylov subspaces, which we introduce in the following section.

2.2 HOW KRYLOV SUBSPACES COME INTO PLAY

In his 1931 paper [398], Krylov was not thinking in terms of projection processes, and he was not interested in solving a linear system. Motivated by an application in the analysis of oscillations of mechanical systems (see Section 2.5.7 for details), he constructed a method for computing the minimal polynomial of a matrix. Algebraically, his method is based on the following important fact.

Given $A \in \mathbb{F}^{N \times N}$ and a nonzero vector $v \in \mathbb{F}^N$, consider the *Krylov sequence generated by A and v*,

$$v, Av, A^2v, \ldots.$$

There then exists a uniquely defined integer $d = d(A, v)$, so that the vectors $v, \ldots, A^{d-1}v$ are linearly independent, and the vectors $v, \ldots, A^{d-1}v, A^d v$ are linearly dependent. We always have $d \geq 1$ since v is nonzero, and $d \leq N$ since the $N+1$ vectors $v, Av, \ldots, A^N v \in \mathbb{F}^N$ must be linearly dependent.

By construction, there exist scalars $\gamma_0, \ldots, \gamma_{d-1}$ with

$$A^d v = \sum_{j=0}^{d-1} \gamma_j A^j v. \tag{2.2.1}$$

Here $A^d v$ is either the zero vector (and hence $\gamma_0 = \cdots = \gamma_{d-1} = 0$), or $A^d v$ is a nontrivial linear combination of the linearly independent vectors $v, \ldots, A^{d-1}v$. We can write (2.2.1) as

$$p(A)v = 0, \quad \text{where } p(\lambda) \equiv \lambda^d - \sum_{j=0}^{d-1} \gamma_j \lambda^j.$$

The polynomial $p(\lambda)$ is called the *minimal polynomial of v with respect to A*. Its degree $d = d(A, v)$ is called the *grade of v with respect to A*. The terminology used here had already been formulated by Gantmacher in his 1934 paper [215] on the algebraic analysis of Krylov's method.[1] We will study the minimal polynomial of a

---

1. Gantmacher wrote this paper very early in his career. It was only his fourth of ultimately 46 publications (among them several books). The paper was presented at the Second All-Union Congress of mathematicians of the Soviet Union, held in Leningrad (St. Petersburg) from June 24–30, 1934.

given vector with respect to A in great detail in Chapter 4. Remark 4.2.12 contains a brief algebraic description of Krylov's original method.

Example 2.2.1
If

$$A = \begin{bmatrix} 0 & 0 & 0 \\ 1 & 0 & 0 \\ 0 & 1 & 0 \end{bmatrix}, \quad B = \begin{bmatrix} 0 & 0 & 1 \\ 1 & 0 & 0 \\ 0 & 1 & 0 \end{bmatrix}, \quad v = \begin{bmatrix} 1 \\ 0 \\ 0 \end{bmatrix},$$

then

$$Av = \begin{bmatrix} 0 \\ 1 \\ 0 \end{bmatrix}, \quad A^2v = \begin{bmatrix} 0 \\ 0 \\ 1 \end{bmatrix}, \quad A^3v = \begin{bmatrix} 0 \\ 0 \\ 0 \end{bmatrix},$$

and hence A^3v is a trivial linear combination of the previous linearly independent vectors v, Av, A^2v. The minimal polynomial of v with respect to A is given by $p(\lambda) = \lambda^3$, and $d(A, v) = 3$.

On the other hand,

$$Bv = \begin{bmatrix} 0 \\ 1 \\ 0 \end{bmatrix}, \quad B^2v = \begin{bmatrix} 0 \\ 0 \\ 1 \end{bmatrix}, \quad B^3v = \begin{bmatrix} 1 \\ 0 \\ 0 \end{bmatrix},$$

and thus B^3v is a nontrivial combination of the previous linearly independent vectors v, Bv, B^2v. The minimal polynomial of v with respect to B is given by $p(\lambda) = \lambda^3 - 1$, and $d(B, v) = 3$.

Krylov's observation can be rephrased in the following way. For each matrix A and vector v, the (nested) sequence of *Krylov subspaces* defined by

$$\mathcal{K}_n(A, v) \equiv \text{span}\{v, Av, \ldots, A^{n-1}v\}, \quad \text{for } n = 1, 2, \ldots, \tag{2.2.2}$$

will eventually stop to grow and become invariant under A. In particular, if d is the grade of v with respect to A, then $A^d v \in \mathcal{K}_d(A, v)$, cf. (2.2.1), and hence

$$A\mathcal{K}_d(A, v) \subseteq \mathcal{K}_d(A, v).$$

For technical reasons we define $\mathcal{K}_0(A, v) = \{0\}$. Basic facts about Krylov subspaces are given in the following lemma.

In December 1934 Gantmacher entered the Steklov Mathematical Institute in Moscow, which had just become an independent institute in April of the same year. He presented his doctoral thesis to the Institute in 1938. For many years Gantmacher collaborated with Krein, work that culminated in 1941 with the monograph [217]. Gantmacher also wrote a nowadays classical monograph on matrix theory, which appeared in 1953 [216]. Both monographs have been translated into several languages. Further biographical information and a complete list of Gantmacher's publications can be found in the obituary [4].

Lemma 2.2.2

Let A be a square matrix, and let v be a vector of grade $d \geq 1$ with respect to A. Then the following hold:

(1) $\dim \mathcal{K}_n(A, v) = n$ for $n = 1, \ldots, d$.
(2) $d - 1 \leq \dim A\mathcal{K}_d(A, v) \leq d$, and if A is nonsingular, then $A\mathcal{K}_d(A, v) = \mathcal{K}_d(A, v)$.
(3) If $A\mathcal{K}_d(A, v) = \mathcal{K}_d(A, v)$, then $v \in \text{Range}(A)$.

Proof
(1) By definition, the vectors $v, \ldots, A^{n-1}v$ are linearly independent for $n = 1, \ldots, d$.
(2) The $d - 1$ vectors $Av, \ldots, A^{d-1}v \in A\mathcal{K}_d(A, v) \subseteq \mathcal{K}_d(A, v)$ are linearly independent, so that $d - 1 \leq \dim A\mathcal{K}_d(A, v) \leq d$. Now let A be nonsingular, and suppose that

$$\sum_{j=1}^{d} \beta_j A^j v = 0,$$

for some scalars $\beta_1, \ldots, \beta_d$. Multiplying both sides of this equation by A^{-1} yields

$$\sum_{j=1}^{d} \beta_j A^{j-1} v = 0,$$

which implies that $\beta_1 = \cdots = \beta_d = 0$ because of the linear independence of the vectors $v, \ldots, A^{d-1}v$. Hence $\dim A\mathcal{K}_d(A, v) = d$, and $A\mathcal{K}_d(A, v) = \mathcal{K}_d(A, v)$.
(3) We have $v \in \mathcal{K}_d(A, v) = A\mathcal{K}_d(A, v) \subseteq \text{Range}(A)$. □

For a simple illustration of item (2) in Lemma 2.2.2 consider Example 2.2.1. The matrix A is singular, $d(A, v) = 3$, but since $A^3 v = 0$, $A\mathcal{K}_3(A, v)$ is a proper subspace of $\mathcal{K}_3(A, v)$.

We can now put together the pieces of the preceding development.

Theorem 2.2.3

Consider the projection process (2.1.1)–(2.1.4) and let the search spaces be given by the Krylov subspaces (2.2.2) with $v = r_0$, i.e.

$$\mathcal{S}_n = \mathcal{K}_n(A, r_0), \quad \text{for } n = 1, 2, \ldots. \tag{2.2.3}$$

If r_0 is of grade d with respect to A, then $r_0 \in \mathcal{S}_1 \subset \mathcal{S}_2 \subset \cdots \subset \mathcal{S}_d = \mathcal{S}_{d+j}$ for all $j \geq 0$. Moreover, if A is nonsingular and the projection process is well defined at step d, then $r_d = 0$.

Proof
We only have to show the last statement. If A is nonsingular, then item (2) in Lemma 2.2.2 guarantees that $A\mathcal{S}_d = \mathcal{S}_d$, and if in addition the projection process is well defined at step d, Lemma 2.1.7 implies that $r_d = 0$. □

In a nutshell, consider a (well-defined) projection process with search spaces given by Krylov subspaces generated by a nonsingular matrix A and an initial residual r_0. When the Krylov subspace has become invariant under A, the process terminates with a zero residual. The importance of invariant subspaces in the context of Krylov subspaces will be a recurring theme in this book.

2.3 MATHEMATICAL CHARACTERISATION OF SOME KRYLOV SUBSPACE METHODS

A projection process of the form (2.1.1)–(2.1.4), where the search spaces are Krylov subspaces generated by A and the initial residual r_0, is called a *Krylov subspace method*. Different choices of constraints spaces then yield different Krylov subspace methods, each having its own characteristic mathematical properties.

One may also consider search spaces that are related to $\mathcal{K}_n(A, r_0)$, for example $\mathcal{S}_n = A\mathcal{K}_n(A, r_0)$, or $\mathcal{S}_n = \mathcal{K}_n(A^*, r_0)$. In addition, one may consider methods that search their iterates in Krylov subspaces, but that are not characterised by a projection property. Should such methods still be called Krylov subspace methods? We consider this a matter of taste rather than scientific significance. According to our taste, the answer to this question is 'yes'. Consequently, we will consider methods that are based on search spaces $\mathcal{S}_n = \mathcal{K}_n(A, r_0)$ and a projection property, as well as related methods, as long as they build up nested sequences of search spaces using matrix–vector multiplications. Using this general framework, the possibilities for mathematical definitions of Krylov subspace methods are virtually unlimited. This fact contributes to the still ongoing and almost overwhelming construction of different methods. Attempts at systematic classifications have been made, for example, in [17] and [86].

It is important to realise that the road from a mathematical description, in particular the specification of search and constraints spaces, to a numerically sound implementation of a Krylov subspace method can be long and winding. There are often several different implementations of a single mathematical description, with greatly varying numerical properties. Moreover, some mathematical descriptions that look promising at first sight have no efficient and numerically stable implementation. We will discuss this in detail later in this book. In this section we will continue to focus on mathematical properties and exact arithmetic.

When looking at the Krylov sequence $r_0, Ar_0, A^2r_0, \ldots$, one immediately recalls the classical *power method*. Starting with the vector $w_0 = r_0$, this method is defined by the following steps (see, e.g. [508, p. 63]):

> For $n = 1, 2, \ldots$
> Compute $w_n = Aw_{n-1}$.
> Normalise $w_n = w_n / \|w_n\|$.
> End

Under certain conditions, the sequence of vectors w_n, $n = 0, 1, 2, \ldots$, converges towards an eigenvector corresponding to the dominant (i.e. largest in magnitude) eigenvalue of A.

The vector w_n is a scalar multiple of $A^n r_0$, and hence the power method in effect computes a Krylov subspace basis. Unlike the power method, however, where all previously computed information is discarded in every step, the Krylov subspace idea is to *save all information along the way*. The approximate solution x_n is then found from a multi-dimensional subspace rather than just the last computed direction. Considering a simple three-dimensional example, Parlett described the essential difference between the Krylov subspace idea and the power method with the words 'A plane is better than its axes' [508, Example 12.1.1]. Clearly, one can most often find better approximations in the plane spanned by r_0 and Ar_0, i.e. the Krylov subspace $\mathcal{K}_2(A, r_0)$, than by just considering either of its two axes.

Instantaneously we are faced with the question of the quality of the approximate solutions found in Krylov subspaces; in other words, the question of *convergence* of Krylov subspace methods. As mentioned above, the Krylov subspace $\mathcal{K}_n(A, r_0)$ is nothing but the space formed by all linear combinations of the vectors generated by n steps of the power method applied to A and r_0. Intuitively, the Krylov subspaces therefore tend to contain the *dominant information* of A with respect to r_0. While the precise relations are often complicated and depend on the specific method, the idea of closely approximating an A-invariant subspace as quickly as possible naturally appears in this context.

In general, the question of convergence is very difficult to answer. No universal approach to the convergence analysis of Krylov subspace methods is known that would give a satisfactory and complete quantitative insight. The convergence behaviour depends on various properties of the system matrix A, as well as the initial residual r_0, and is often studied in the context of each method (or class of methods) separately. However, a common property of all Krylov subspace methods is the *orthogonality* condition (2.1.4). This property can sometimes be related to a certain *optimality* property, which in turn is a starting point for the investigation of convergence properties of specific methods.

The following theorem gives the mathematical description of several important Krylov subspace methods in terms of the search and constraint spaces, and it shows the corresponding optimality properties.

Theorem 2.3.1
Consider the projection process (2.1.1)–(2.1.4) for solving a linear algebraic system $Ax = b$, with an initial approximation x_0, and let the initial residual $r_0 = b - Ax_0$ be of grade $d \geq 1$ with respect to A. Then the following hold:

(1) *If A is HPD and $\mathcal{S}_n = \mathcal{C}_n = \mathcal{K}_n(A, r_0)$, $n = 1, 2, \ldots$, then the projection process is well defined at every step n until it terminates at step d. It is characterised by the orthogonality property*

$$x - x_n \perp_A \mathcal{K}_n(A, r_0), \quad \text{or} \quad x - x_n \in \mathcal{K}_n(A, r_0)^{\perp_A},$$

where

$$\mathcal{K}_n(A, r_0)^{\perp_A} \equiv \left\{ w \in \mathbb{F}^N : (v, w)_A \equiv w^* A v = 0, \ \forall v \in \mathcal{K}_n(A, r_0) \right\}.$$

The equivalent optimality property is

$$\|x - x_n\|_A = \min_{z \in x_0 + \mathcal{K}_n(A, r_0)} \|x - z\|_A,$$

where $\|v\|_A \equiv (v, v)_A^{1/2} \equiv (v^*Av)^{1/2}$ is the A-norm of the vector $v \in \mathbb{F}^N$. *(Mathematical characterisation of the Conjugate Gradient (CG) method [313].)*

(2) *If A is Hermitian and nonsingular,* $\mathcal{S}_n = A\mathcal{K}_n(A, r_0)$ *and* $\mathcal{C}_n = A^{-1}\mathcal{S}_n = \mathcal{K}_n(A, r_0), n = 1, 2, \ldots$, *then the projection process is well defined at every step n until it terminates at step d. It is characterised by the orthogonality property*

$$x - x_n \perp A\mathcal{K}_n(A, r_0).$$

The equivalent optimality property is

$$\|x - x_n\| = \min_{z \in x_0 + A\mathcal{K}_n(A, r_0)} \|x - z\|.$$

(Mathematical characterisation of the SYMMLQ method [496].)

(3) *If A is nonsingular,* $\mathcal{S}_n = \mathcal{K}_n(A, r_0)$, *and* $\mathcal{C}_n = A\mathcal{S}_n$, $n = 1, 2, \ldots$, *then the projection process is well defined at every step n until it terminates at step d. It is characterised by the orthogonality property*

$$r_n \perp A\mathcal{K}_n(A, r_0), \quad \text{or} \quad x - x_n \in \mathcal{K}_n(A, r_0)^{\perp_{A^*A}}.$$

The equivalent optimality property is

$$\|r_n\| = \min_{z \in x_0 + \mathcal{K}_n(A, r_0)} \|b - Az\|, \quad \text{or}$$

$$\|x - x_n\|_{A^*A} = \min_{z \in x_0 + \mathcal{K}_n(A, r_0)} \|x - z\|_{A^*A}.$$

(Mathematical characterisation of the Minimal Residual (MINRES) method [496] and the Generalised Minimal Residual (GMRES) method [544].)

Proof
The fact that the projection processes in (1) and (3) are well defined at each step n until termination with $r_d = 0$, follows from items (1) and (3) in Theorem 2.1.6, respectively, and Theorem 2.2.3.

The projection process in (2) is well defined at each step $n = 1, 2, \ldots, d$ because of item (2) in Theorem 2.1.6. Since A is nonsingular, we have $\mathcal{S}_d = A\mathcal{K}_d(A, r_0) = \mathcal{K}_d(A, r_0)$ according to item (2) in Lemma 2.2.2. Now Lemma 2.1.7 yields $r_d = 0$.

It remains to show the three orthogonality and optimality characterisations.

(1) The orthogonality property of the projection process is $r_n \perp \mathcal{K}_n(A, r_0)$, which is equivalent to $x - x_n \perp_A \mathcal{K}_n(A, r_0)$. Let $x_n \in x_0 + \mathcal{K}_n(A, r_0)$ be the (uniquely determined) nth approximation satisfying this orthogonality condition. Then for any $z \in x_0 + \mathcal{K}_n(A, r_0)$,

$$\|x-z\|_A^2 = \| \underbrace{(x-x_n)}_{\in \mathcal{K}_n(A,r_0)^{\perp_A}} - \underbrace{(z-x_n)}_{\in \mathcal{K}_n(A,r_0)} \|_A^2 = \|x-x_n\|_A^2 + \|z-x_n\|_A^2$$
$$\geq \|x-x_n\|_A^2.$$

Hence the vector x_n minimises $\|x-z\|_A$ over the affine space $x_0 + \mathcal{K}_n(A, r_0)$. On the other hand, the above inequality is strict unless $z = x_n$, and thus the unique minimiser of $\|x-z\|_A$ is $z = x_n$.

(2) The orthogonality property of the projection process is $r_n \perp \mathcal{K}_n(A, r_0)$, which is equivalent to $x - x_n \perp A\mathcal{K}_n(A, r_0)$ (here we have used that A is Hermitian). If $x_n \in x_0 + A\mathcal{K}_n(A, r_0)$ is the (uniquely determined) nth approximation satisfying the orthogonality condition, and $z \in x_0 + A\mathcal{K}_n(A, r_0)$ is arbitrary, then

$$\|x-z\|^2 = \| \underbrace{(x-x_n)}_{\in A\mathcal{K}_n(A,r_0)^{\perp}} - \underbrace{(z-x_n)}_{\in A\mathcal{K}_n(A,r_0)} \|^2 = \|x-x_n\|^2 + \|z-x_n\|^2$$
$$\geq \|x-x_n\|^2,$$

and the argument can be finished as in case (1).

(3) The orthogonality property of the projection process is $r_n \perp A\mathcal{K}_n(A, r_0)$, which is equivalent to $x - x_n \perp_{A^*A} \mathcal{K}_n(A, r_0)$. If $x_n \in x_0 + \mathcal{K}_n(A, r_0)$ is the (uniquely determined) nth approximation satisfying the orthogonality condition, and $z \in x_0 + \mathcal{K}_n(A, r_0)$ is arbitrary, then

$$\|x-z\|_{A^*A}^2 = \| \underbrace{(x-x_n)}_{\in \mathcal{K}_n(A,r_0)^{\perp_{A^*A}}} - \underbrace{(z-x_n)}_{\in \mathcal{K}_n(A,r_0)} \|_{A^*A}^2$$
$$= \|x-x_n\|_{A^*A}^2 + \|z-x_n\|_{A^*A}^2$$
$$\geq \|x-x_n\|_{A^*A}^2,$$

and again the argument can be finished as in case (1). □

Algorithms that implement the methods characterised in this theorem will be described in Section 2.5.

Since the minimisation problems in items (1), (2), and (3) in this theorem are defined over (affine) spaces of increasing dimensions, the corresponding sequences of norms, $\|x - x_n\|_A$, $\|x - x_n\|$, and $\|r_n\|$, respectively, are *nonincreasing* for $n = 0, 1, 2, \ldots, d$. We will show in Section 2.5.3 (see Theorem 5.6.1) that the norms $\|x - x_n\|_A$ in (1) always decrease strictly monotonically. On the other hand, the norms $\|r_n\|$ in (3) can in some cases stagnate; see Section 5.7.4 for details.

The theorem shows the equivalence of orthogonality and optimality properties of certain Krylov subspace methods. The following result represents a generalisation of this equivalence to a general subspace of $\mathbb{F}^N$.

Theorem 2.3.2
Let $B \in \mathbb{F}^{N \times N}$ be an HPD matrix, and let $z \in \mathbb{F}^N$ be given. Furthermore, let $\mathcal{U}$ be an n-dimensional subspace of $\mathbb{F}^N$. Then the following two statements are equivalent:

(1) $z_n \in \mathcal{U}$ satisfies $z - z_n \perp_B \mathcal{U}$.
(2) $z_n \in \mathcal{U}$ satisfies $\|z - z_n\|_B = \min_{u \in \mathcal{U}} \|z - u\|_B$.

Proof
The proof is similar to the proof of the optimality properties in Theorem 2.3.1. If $z_n \in \mathcal{U}$ is (uniquely) determined by $z - z_n \perp_B \mathcal{U}$ and $u \in \mathcal{U}$ is arbitrary, then

$$\|z-u\|_B^2 = \|(z-z_n)-(u-z_n)\|_B^2 = \|(z-z_n)\|_B^2 + \|(u-z_n)\|_B^2 \geq \|(z-z_n)\|_B^2.$$

Hence the vector z_n minimises the B-norm $\|z - u\|_B$ over the subspace $\mathcal{U}$. On the other hand, the above inequality is strict unless $u = z_n$, which shows that the (unique) minimiser is $u = z_n$. □

With the appropriate choices of search and constraint spaces, the three cases of Theorem 2.3.1 correspond to $B = A$, $B = I$, and $B = A^*A$ in Theorem 2.3.2.

In the next section, we will discuss some practically relevant methods for generating bases for the Krylov subspaces.

2.4 THE ARNOLDI AND LANCZOS ALGORITHMS

From the mathematical point of view, each Krylov subspace method is completely determined by the choice of search and constraint spaces. The key role for the *numerical behaviour* of a method is played by the choice of bases for these spaces. As discussed in Section 2.3, let the Krylov sequence converge towards a dominant eigenvector of A. Hence the vectors in this sequence eventually become very close, even numerically identical. Inevitably this leads to loss of information. To prevent this, and to transfer as much information as possible from the original linear algebraic system into the projected system, practically sound methods must use *well conditioned bases* (at best close to orthonormal) for the Krylov subspaces.

We point out that there is a purely mathematical reason for using orthogonal Krylov subspace bases, which is not inspired by finite precision computations. As shown below, the computation of orthogonal Krylov subspace bases is related to orthogonal (or unitary) transformations of the matrix A. In many cases this matrix represents data that are contaminated by errors. When A is transformed unitarily, these errors are not amplified in any unitarily invariant norm $\| \cdot \|_*$. Moreover, if in a projection method the bases we work with are orthonormal, and thus the matrices S_n and C_n have orthonormal columns, then the projected matrix satisfies $\|C_n^* A S_n\|_* \leq \|A\|_*$.

In the following we will describe the most important algorithms for generating orthogonal Krylov subspace bases. *In our description we continue to assume exact arithmetic.*

2.4.1 Arnoldi and Hermitian Lanczos

Although historically things were developed differently, it is convenient to first consider what is now called the *Arnoldi algorithm* [16]. It can be viewed as a variant of the Gram–Schmidt orthogonalisation method applied to the Krylov sequence in order to generate an orthonormal basis of the Krylov subspace (see Section 2.4.3 for historical details on the Gram–Schmidt method).

Algorithm 2.4.1 is the *classical Gram–Schmidt implementation* of the Arnoldi algorithm. Starting with a vector $v_1 = v/\|v\|$, step n of the algorithm generates a vector v_{n+1} which results from the orthogonalisation of the vector Av_n (rather than $A^n v_1$) with respect to the previously generated orthonormal vectors $v_1, \ldots, v_n$. If the algorithm does not stop at the step n, we get

$$v_{n+1} = \varphi_n(A) v_1, \qquad (2.4.1)$$

where $\varphi_n(\lambda)$ is a polynomial of exact degree n, and

$$(v_{n+1}, v_j) = v_1^*(\varphi_{j-1}(A))^* \varphi_n(A) v_1 = 0, \quad j = 1, \ldots, n, \qquad (2.4.2)$$

where $(\cdot, \cdot)$ denotes the Euclidean inner product. The algorithm terminates when $h_{n+1,n} = 0$, which happens if and only if $Av_n \in \mathrm{span}\{v_1, \ldots, v_n\} = \mathcal{K}_n(A, v)$. By construction, this means $n = d$, where d is the grade of the initial vector v with respect to A. The vector $\widehat{v}_{d+1}$ vanishes, and therefore the final Krylov subspace $\mathcal{K}_d(A, v)$ is A-invariant.

Because of numerical instabilities, the classical Gram–Schmidt implementation of the Arnoldi algorithm is rarely used. The most common implementation used in practical computations is based on the *modified Gram–Schmidt orthogonalisation*, stated in Algorithm 2.4.2. Algorithms 2.4.1 and 2.4.2 are mathematically equivalent, which means that in the absence of rounding errors they yield identical vectors, $v_1, \ldots, v_d$.

In Algorithm 2.4.2 the ith intermediate result $\widehat{v}_{n+1,i}$ of the orthogonalisation at step n is the result of orthogonalising Av_n against the first i previous vectors, $v_1, \ldots, v_i$. Since the orthogonalisation is performed *recursively*, Algorithm 2.4.2 has much less potential for concurrent performance of arithmetic operations than Algorithm 2.4.1. This can play a role on modern parallel computer architectures, where concurrency of computations must be taken into account. Details about parallel numerical algorithms can be found in, e.g. [141, 264, 512].

Algorithm 2.4.1 Arnoldi algorithm, classical Gram–Schmidt implementation

---

Input: matrix $A \in \mathbb{F}^{N \times N}$, nonzero initial vector $v \in \mathbb{F}^N$. Let d be the grade of v with respect to A.
Output: orthonormal vectors $v_1, \ldots, v_d$ with $\mathrm{span}\{v_1, \ldots, v_n\} = \mathcal{K}_n(A, v)$ for $n = 1, \ldots, d$.

Initialise: $v_1 = v / \|v\|$.

For $n = 1, 2, \ldots$

$\widehat{v}_{n+1} = Av_n - \sum_{i=1}^{n} h_{i,n} v_i$, where $h_{i,n} = (Av_n, v_i), i = 1, \ldots, n$,
$h_{n+1,n} = \|\widehat{v}_{n+1}\|$, if $h_{n+1,n} = 0$ then stop,
$v_{n+1} = \widehat{v}_{n+1} / h_{n+1,n}$.

End

Algorithm 2.4.2 Arnoldi algorithm, modified Gram–Schmidt implementation

Input: matrix $A \in \mathbb{F}^{N \times N}$, nonzero initial vector $v \in \mathbb{F}^N$. Let d be the grade of v with respect to A.
Output: orthonormal vectors $v_1, \ldots, v_d$ with span $\{v_1, \ldots, v_n\} = \mathcal{K}_n(A, v)$ for $n = 1, \ldots, d$.

    Initialise: $v_1 = v / \|v\|$.

    For $n = 1, 2, \ldots$
        $\widehat{v}_{n+1,0} = Av_n$,
        For $i = 1, 2, \ldots, n$
            $\widehat{v}_{n+1,i} = \widehat{v}_{n+1,i-1} - h_{i,n} v_i$, where $h_{i,n} = (\widehat{v}_{n+1,i-1}, v_i)$,
        End
        $\widehat{v}_{n+1} = \widehat{v}_{n+1,n}$,
        $h_{n+1,n} = \|\widehat{v}_{n+1}\|$, if $h_{n+1,n} = 0$ then stop,
        $v_{n+1} = \widehat{v}_{n+1} / h_{n+1,n}$.

    End

We now write the Arnoldi algorithm (in either implementation) in terms of matrices. Let $V_n \equiv [v_1, \ldots, v_n]$ have the orthonormal basis vectors $v_1, \ldots, v_n$ as its columns, and let

$$H_{n,n} \equiv \begin{bmatrix} h_{1,1} & h_{1,2} & \cdots & h_{1,n} \\ h_{2,1} & h_{2,2} & \cdots & h_{2,n} \\ & \ddots & \ddots & \vdots \\ & & h_{n,n-1} & h_{n,n} \end{bmatrix}$$

be the $n \times n$ unreduced upper *Hessenberg matrix*[2] containing the recurrence coefficients $h_{i,j}$ as its entries ('unreduced' means that no subdiagonal entry $h_{i,i-1}$ vanishes). After n steps of the algorithm we have

$$AV_n = V_n H_{n,n} + h_{n+1,n} v_{n+1} e_n^T, \quad n = 1, 2, \ldots, \quad (2.4.3)$$

where $V_n^* V_n = I_n$, $V_n^* v_{n+1} = 0$, and e_n is the nth column of I_n. At the dth iteration we get $h_{d+1,d} = 0$, and (2.4.3) becomes

$$AV_d = V_d H_{d,d}. \quad (2.4.4)$$

---

2. Named after Karl Adolf Hessenberg (1904–1959), who was supposedly the first to use these matrices on p. 23 of the Technical Report [311] published in 1940.

Here $H_{d,d}$ can be interpreted as the matrix representation of the orthogonal restriction of the linear operator A to the A-invariant subspace $\mathcal{K}_d(A, v)$ in the basis $v_1, \ldots, v_d$. Indeed,

$$V_d^*(V_d V_d^* A) V_d = V_d^* A V_d = H_{d,d}.$$

The Arnoldi algorithm can be viewed as a tool for the *unitary reduction of A to upper Hessenberg form*: when the algorithm stops with $h_{d+1,d} = 0$ and $d < N$, the reduction can be continued (if needed) by starting another Arnoldi algorithm with A and an initial vector v_{d+1} which is orthogonal to $v_1, \ldots, v_d$. This ultimately yields a unitary matrix V and an upper Hessenberg matrix H (not necessarily unreduced), such that $AV = VH$.

Now suppose that A is Hermitian. Then the upper Hessenberg matrix $V_d^* A V_d = H_{d,d}$ must also be Hermitian, and thus it must be *tridiagonal*. The full recurrence for the vector $\widehat{v}_{n+1}$ in Algorithm 2.4.1 then reduces to the *three-term recurrence*

$$\widehat{v}_{n+1} = A v_n - h_{n,n} v_n - h_{n-1,n} v_{n-1}.$$

For a Hermitian matrix A it therefore suffices to orthogonalise the vector $A v_n$ against the two most recent orthonormal vectors v_n and v_{n-1}; the orthogonality against all other vectors $v_{n-2}, \ldots, v_1$ is automatically guaranteed. The resulting algorithm is called the *Hermitian Lanczos algorithm*. Though attributed to Lanczos [405, 406], this algorithm represents nothing but the matrix-algebra formulation of the classical Stieltjes algorithm for computing orthogonal polynomials [104, 583, 585, 584, 586, 588]. This relationship will be thoroughly described in Chapter 3.

Algorithm 2.4.3 is an implementation of the Hermitian Lanczos algorithm which is based on two coupled two-term recurrences. As a consequence of Paige's landmark analysis of different implementations [489, 490], this has become the

---

Algorithm 2.4.3 Hermitian Lanczos algorithm

Input: Hermitian matrix $A \in \mathbb{F}^{N \times N}$, nonzero initial vector $v \in \mathbb{F}^N$. Let d be the grade of v with respect to A.
Output: orthonormal vectors $v_1, \ldots, v_d$ with span $\{v_1, \ldots, v_n\} = \mathcal{K}_n(A, v)$ for $n = 1, \ldots, d$.

Initialise: $v_0 = 0, \delta_1 = 0, v_1 = v / \|v\|$.

For $n = 1, 2, \ldots$

$u_n = A v_n - \delta_n v_{n-1},$
$\widehat{v}_{n+1} = u_n - \gamma_n v_n,$ where $\gamma_n = (u_n, v_n),$
$\delta_{n+1} = \|\widehat{v}_{n+1}\|,$ if $\delta_{n+1} = 0$ then stop,
$v_{n+1} = \widehat{v}_{n+1} / \delta_{n+1}.$

End

standard version of the Hermitian Lanczos algorithm; cf., e.g. [508, Algorithm 13.1.1, Chapter 13], [272, Chapter 2.5], or [543, Chapter 6.6]. It uses recursive orthogonalisation analogous to the modified Gram–Schmidt orthogonalisation procedure.

The matrix form of the Hermitian Lanczos algorithm is given by

$$AV_n = V_n T_n + \delta_{n+1} v_{n+1} e_n^T, \quad n = 1, 2, \ldots, \qquad (2.4.5)$$

where $V_n^* V_n = I_n$ and $V_n^* v_{n+1} = 0$; see (2.4.3). The recurrence coefficients are the entries of the Hermitian tridiagonal matrix

$$T_n \equiv H_{n,n} = \begin{bmatrix} \gamma_1 & \delta_2 & & & \\ \delta_2 & \gamma_2 & \delta_3 & & \\ & \ddots & \ddots & \ddots & \\ & & \delta_{n-1} & \gamma_{n-1} & \delta_n \\ & & & \delta_n & \gamma_n \end{bmatrix}. \qquad (2.4.6)$$

We point out that $\delta_{j+1} = \|\widehat{v}_{j+1}\| > 0$, $j = 1, \ldots, n-1$. Thus, the matrix T_n in (2.4.6) is a real symmetric tridiagonal matrix that has positive elements on its first sub- and super-diagonal, regardless of A being real or not. Since such matrices will play a major role later in this book, we give the following formal definition.

Definition 2.4.1: *A Jacobi matrix is a real symmetric tridiagonal matrix of the form (2.4.6) with positive off-diagonal elements.*

Remark 2.4.2
There exist different definitions of the term *Jacobi matrix* in the literature. For example, Gantmacher [216, Vol. 2, p. 99] calls any real tridiagonal matrix a Jacobi matrix, while for Householder [346, p. 86] the term refers to a tridiagonal matrix with real diagonal elements and the products of corresponding off-diagonal elements being non-negative (cf. the final part of Section 3.4.1 in our book). A definition like ours has been used, for example, in [115, p. 72]. Properties of Jacobi matrices will be studied in detail in Section 3.4 below. In particular, the historical note in Section 3.4.3 will describe the origin and early applications of the Jacobi matrices.

Let us get back to the Hermitian Lanczos algorithm. At the dth iteration step we get $\delta_{d+1} = 0$, and (2.4.5) becomes

$$AV_d = V_d T_d, \qquad (2.4.7)$$

where T_d can be interpreted as the matrix representation of the orthogonal restriction of the linear operator A to the A-invariant subspace $\mathcal{K}_d(A, v)$ in the basis $v_1, \ldots, v_d$.

The Lanczos algorithm can be viewed as a *unitary reduction of a Hermitian matrix to tridiagonal form*. If the algorithm stops with $\delta_{d+1} = 0$ and $d < N$, then, analogously to the Arnoldi algorithm, the reduction can be continued by starting another Lanczos algorithm with A and an initial vector v_{d+1} that is orthogonal to $v_1, \ldots, v_d$.

2.4.2 Non-Hermitian Lanczos

For a Hermitian matrix A the full recurrence in the Arnoldi algorithm reduces to a three-term recurrence. Hence in the Hermitian case an orthogonal Krylov subspace basis can be generated by an algorithm that has constant work and storage requirements per iteration. A fundamental result in the theory of Krylov subspace methods, first obtained by Faber and Manteuffel [176], is that for a general matrix A there exists *no* Arnoldi-like algorithm that would generate an orthogonal Krylov subspace basis, with a recurrence of a fixed length per iteration. This result will be studied in Chapter 4.

In order to keep work and storage requirements constant throughout the iteration for a general non-Hermitian matrix A, one can relax the orthogonality requirement. It turns out that well conditioned non-orthogonal bases can often lead to efficient and practically useful methods. One approach is to generate two sets of mutually *biorthogonal* basis vectors (instead of one set of orthogonal basis vectors) using the *non-Hermitian Lanczos algorithm* stated in Algorithm 2.4.4. This algorithm, originally proposed by Lanczos in [405, 406], is also known as the *Lanczos biorthogonalisation algorithm* or the *two-sided Lanczos algorithm*. For Hermitian matrices A and with initial vectors $v = w$ it is mathematically equivalent to the Hermitian Lanczos algorithm.

Algorithm 2.4.4 Non-Hermitian Lanczos algorithm

Input: matrix $A \in \mathbb{F}^{N \times N}$, initial vectors $v, w \in \mathbb{F}^N$ with $w^* v \neq 0$.
Output: vectors $v_1, \ldots, v_n$ that span $\mathcal{K}_n(A, v)$ and vectors $w_1, \ldots, w_n$ that span $\mathcal{K}_n(A^*, w)$, so that $(v_i, w_j) = \delta_{i,j}$ for $i, j = 1, \ldots, n$.

Initialise: $v_0 = w_0 = 0, \beta_1 = \delta_1 = 0, v_1 = v / \|v\|, w_1 = w / \overline{(v_1, w)}$.

For $n = 1, 2, \ldots$

$$\widehat{v}_{n+1} = Av_n - \gamma_n v_n - \beta_n v_{n-1}, \quad \text{where } \gamma_n = (Av_n, w_n),$$
$$\widehat{w}_{n+1} = A^* w_n - \overline{\gamma}_n w_n - \delta_n w_{n-1},$$
$$\delta_{n+1} = \|\widehat{v}_{n+1}\|, \quad \text{if } \delta_{n+1} = 0 \text{ then stop},$$
$$v_{n+1} = \widehat{v}_{n+1} / \delta_{n+1},$$
$$\beta_{n+1} = (v_{n+1}, \widehat{w}_{n+1}), \quad \text{if } \beta_{n+1} = 0 \text{ then stop},$$
$$w_{n+1} = \widehat{w}_{n+1} / \overline{\beta}_{n+1}.$$

End

Remark 2.4.3
The scaling factors δ_{n+1} and β_{n+1} in Algorithm 2.4.4 are chosen so that $\|v_{n+1}\| = 1$ and $(v_{n+1}, w_{n+1}) = 1$. This is the same choice as in [272, Chapter 5.1]. Several other choices of scaling factors have been used in the literature; see, e.g. [543, Chapter 7.1].

Also note that it is possible to formulate the non-Hermitian Lanczos algorithm using A^T instead of A^* and by conjugating the recurrence for the vectors w_n; see, e.g. [203, Algorithm 3.1]. This formulation leads to vectors $v_1, \ldots, v_{n+1}$ and $w_1, \ldots, w_{n+1}$ that satisfy $w_i^T v_j = 0$ for $i \neq j$. Thus, even when the initial data A, v_1, w_1 are not real, the computed vectors will be biorthogonal with respect to the Euclidean inner product on $\mathbb{R}^N$, which represents a bilinear form on $\mathbb{C}^N$.

We now state the basic properties of the non-Hermitian Lanczos algorithm. Proofs of these properties by elementary inductions are given, e.g. in [272, Chapter 5.1] and [543, Chapter 7.1]. Suppose that the algorithm does not stop in steps 1 through n, i.e. that $\delta_2 \beta_2 \neq 0, \ldots, \delta_{n+1} \beta_{n+1} \neq 0$. Then the vectors $v_1, \ldots, v_{n+1}$ and $w_1, \ldots, w_{n+1}$ form a biorthogonal system, i.e. $(v_i, w_j) = \delta_{i,j}$ for $i, j = 1, \ldots, n+1$. Moreover,

$$\text{span}\{v_1, \ldots, v_{n+1}\} = \mathcal{K}_{n+1}(A, v), \quad \text{span}\{w_1, \ldots, w_{n+1}\} = \mathcal{K}_{n+1}(A^*, w),$$

and

$$v_j = \varphi_{j-1}(A) v_1, \quad w_j = \psi_{j-1}(A^*) w_1, \quad j = 1, \ldots, n+1,$$

where $\varphi_{j-1}(\lambda)$ and $\psi_{j-1}(\lambda)$ are polynomials of exact degree $j - 1$. Denoting by $T_{n,n}$ the $n \times n$ tridiagonal matrix containing the recurrence coefficients,

$$T_{n,n} = \begin{bmatrix} \gamma_1 & \beta_2 & & & \\ \delta_2 & \gamma_2 & \beta_3 & & \\ & \ddots & \ddots & \ddots & \\ & & \delta_{n-1} & \gamma_{n-1} & \beta_n \\ & & & \delta_n & \gamma_n \end{bmatrix},$$

the matrix form of the non-Hermitian Lanczos algorithm is given by

$$\begin{aligned} AV_n &= V_n T_{n,n} + \delta_{n+1} v_{n+1} e_n^T, \\ A^* W_n &= W_n T_{n,n}^* + \beta_{n+1}^* w_{n+1} e_n^T, \\ W_n^* A V_n &= T_{n,n}, \\ W_{n+1}^* V_{n+1} &= I_{n+1}, \quad \|v_1\| = \cdots = \|v_{n+1}\| = 1. \end{aligned} \quad (2.4.8)$$

The algorithm terminates at step n, when $\delta_{n+1} = 0$ or $\beta_{n+1} = 0$, which can happen in two different situations.

When $\widehat{v}_{n+1} = 0$ or $\widehat{w}_{n+1} = 0$, the algorithm has found an invariant subspace. If $\widehat{v}_{n+1} = 0$, then span$\{v_1, \ldots, v_n\}$ is A-invariant, while if $\widehat{w}_{n+1} = 0$, then span$\{w_1, \ldots, w_n\}$ is A^*-invariant. This situation is often called a *lucky breakdown* (or *benign breakdown* [506]) of the non-Hermitian Lanczos algorithm, because the computation of invariant subspace information often represents a highly desirable result (cf. our previous remarks on invariant subspaces in the context of Krylov subspace methods).

When $\beta_{n+1} = (v_{n+1}, \widehat{w}_{n+1}) = 0$, while the two vectors $\widehat{w}_{n+1}$ and v_{n+1} are both nonzero, no invariant subspace of A or A^* has been computed. This situation is often called a *serious breakdown*. An analysis was given in [538] (see also [347, p. 34]) and in Wilkinson's book [670, pp. 389–391]. Wilkinson used a nonsymmetric 3×3 matrix with integer entries, eigenvalues 1, 2, 3, and very well conditioned eigenvectors to conclude that the potential for serious breakdowns 'is not associated with any shortcoming in the matrix A. It can happen even when the eigenproblem of A is very well conditioned. We are forced to regard it as a specific weakness of the Lanczos method itself.' As pointed out by Taylor in his PhD thesis [610] (see also Parlett's beautiful expository paper [506]), the serious breakdown, although it does not give an invariant subspace of A, has some remarkable properties. In particular, at the occurrence of a serious breakdown with $\widehat{w}_{n+1}^* v_{n+1} = 0$, while $\widehat{w}_{n+1} \neq 0$ and $v_{n+1} \neq 0$, the eigenvalues of the tridiagonal matrix $T_{n,n}$ are eigenvalues of A.

Several attempts have been made to overcome the problem of breakdowns in the Lanczos method. Most notably, so-called *look-ahead versions* of the non-Hermitian Lanczos algorithm were proposed by Parlett, Taylor, and Liu in 1985 [509], and by Freund, Gutknecht, and Nachtigal in the early 1990s [204, 286, 287, 468]; see also [81, 82]. Although they neither cure all possible serious breakdowns, nor can guarantee that the computed biorthogonal bases are always well conditioned, look-ahead techniques can be very useful in practical computations. Some details on their use in the context of iterative methods for linear algebraic systems are given in the discussion of the QMR method in Section 2.5.6.

2.4.3 Historical Note: The Gram–Schmidt Method

As mentioned in Section 2.4.1, the Arnoldi and Lanczos algorithms are special variants of the *Gram–Schmidt method* for orthogonalising a given set of vectors with respect to a given inner product. In this section we will look briefly at the history of this method.

A method that is mathematically equivalent to computing the QR factorisation of a matrix via the modified Gram–Schmidt method was given by Laplace in 1820 [412]. Laplace did not formulate his results using matrix notation or the language of matrix factorisation, which were both invented much later. He was concerned with computing the mass of the planets Jupiter and Saturn from a system of normal equations that was obtained from astronomical measurements. In the same context, Laplace devised a method that can be interpreted as the Cholesky factorisation of a matrix. Parts from Laplace's work were translated and discussed from a contemporary point of view by Langou in 2009 [411].

The modern history of the Gram–Schmidt orthogonalisation method starts with Schmidt's dissertation of 1905 [553] and the almost identical paper of 1907 [554]. In these works Schmidt stated what is known today as the classical Gram–Schmidt method. He used this method to construct n orthonormal functions, $\psi_1(x), \ldots, \psi_n(x)$, from n given real, continuous, and linearly independent functions, $\varphi_1(x), \ldots, \varphi_n(x)$. Orthonormality was meant with respect to the inner product $(f, g) = \int_a^b f(x)g(x)dx$, defined by the Riemann integral. In [553, pp. 5–6] and [554, p. 442] Schmidt wrote the method as follows:

$$\psi_1(x) = \frac{\varphi_1(x)}{\sqrt{\int_a^b (\varphi_1(y))^2 dy}}$$

$$\psi_2(x) = \frac{\varphi_2(x) - \psi_1(x)\int_a^b \varphi_2(z)\psi_1(z)dz}{\sqrt{\int_a^b \left(\varphi_2(y) - \psi_1(y)\int_a^b \varphi_2(z)\psi_1(z)dz\right)^2 dy}}$$

$$\vdots$$

$$\psi_n(x) = \frac{\varphi_n(x) - \sum_{\varrho=1}^{\varrho=n-1} \psi_\varrho(x)\int_a^b \varphi_n(z)\psi_\varrho(z)dz}{\sqrt{\int_a^b \left(\varphi_n(y) - \sum_{\varrho=1}^{\varrho=n-1} \psi_\varrho(y)\int_a^b \varphi_n(z)\psi_\varrho(z)dz\right)^2 dy}}.$$

He called the functions $\psi_1(x), \ldots, \psi_n(x)$ a *normalised orthogonal system of functions* ('normiertes und orthogonales Funktionensystem').

In a footnote Schmidt pointed out that Gram had given 'essentially the same formulas' in a paper of 1883 [263]. In this paper, which is a (condensed) German version of his Danish dissertation of 1879 [262], Gram dealt with the expansion of real functions using the method of least squares. From a given set of linearly independent vectors (or functions) $X_1(x), \ldots, X_n(x)$, he derived a set of orthogonal vectors (or functions) $\Phi_1(x), \ldots, \Phi_n(x)$, by setting up what is known today as the *Gram matrix* and its determinant (often called the *Gramian*). In [263, equation (8)] he stated the formula

$$\Phi_n(x) = \begin{vmatrix} p_{11} & p_{12} & \cdots & p_{1,n-1} & X_1 \\ p_{21} & p_{22} & \cdots & p_{2,n-1} & X_2 \\ \cdot & \cdot & \cdot & \cdot & \cdot \\ p_{n1} & p_{n2} & \cdots & p_{n,n-1} & X_n \end{vmatrix}, \qquad (2.4.9)$$

where (in modern terminology) p_{ik} is the inner product of X_i and X_k. Without explicitly using terms like 'inner product' or 'orthogonalisation', Gram orthogonalised the given $X_1(x), \ldots, X_n(x)$ with respect to weighted sums in the discrete cases, and weighted Riemann integrals in the continuous case. Consistently with his application, he called the resulting orthogonal functions $\Phi_n(x)$ *expansion functions* ('Entwickelungsfunctionen'). Gram was aware of previous related results of Chebyshev, Heine, Liouville, Sturm, and others. He mentioned links of the expansion functions to continued fractions, and he particularly built upon Chebyshev's 1855 paper [104] (cf. the historical notes in Sections 3.3.5 and 3.3.6). An orthogonalisation method à la Gram in a modern and compact notation can be found in Gantmacher's book [216, Chapter IX, Section 6]. Gram spent his professional career in the Danish insurance business; he never held a university position.

Nevertheless he produced important results in pure and applied mathematics (for example in probability theory, numerical analysis, and number theory). His thesis ranks among the founding works of the Scandinavian statistical school.

While Gram's method came more than 20 years earlier, Schmidt's description in [553, 554] is an exemplary case of mathematical elegance and completeness. Schmidt was an aesthete, who strived for the optimal presentation. In his *Antrittsrede* at the Prussian Academy of Science, delivered in July 1919, Schmidt said that 'in recollection of the great difficulties I had when reading mathematical works, I always made an effort to simplify proofs' [555, p. 565] (our translation). It is tempting to picture Schmidt at his desk, working through Gram's paper and mining out his own version of the Gram–Schmidt method. Schmidt's paper [554] was the first of a sequence of three, published between 1907 and 1908 in the Mathematische Annalen. These and the related sequence of six papers entitled *Grundzüge einer allgemeinen Theorie der linearen Integralgleichungen* (1904–1910) by Schmidt's advisor Hilbert are considered to be milestones in the development of modern functional analysis (also see the historical note on Jacobi matrices in Section 3.4.3). The orthogonalisation method was only a technical tool stated at the very beginning of Schmidt's theory. Yet, it remains to this day Schmidt's and Gram's most cited result.

It is interesting to note that Lanczos in his paper of 1950 traced back the orthogonalisation idea (of vectors) not to Schmidt or Gram, but to a different source when he wrote [405, p. 266]:

> The idea of successive orthogonalization of a set of vectors was probably first employed by O. Szász, in connection with a determinant theorem of Hadamard.

The mentioned Otto Szász studied mathematics at the University of Budapest and the Institute of Technology of Budapest. During his undergraduate years he went to Göttingen (in 1908), where he attended lectures by Hilbert, Klein, Minkowski, Toeplitz, and others. His paper [603], which Lanczos referred to in [405], was probably written after his return to Budapest. It appeared in a Hungarian journal in 1910, and a German translation (containing an additional section) appeared three years later [605]. In modern terms, Szász gave an algorithm for successively orthogonalising the rows of a (general) matrix $A \in \mathbb{C}^{N \times N}$, and hence he effectively obtained the QR factorisation of A. He applied this factorisation to prove the Hadamard determinant inequality, i.e. $|\det(A)| \leq \prod_{j=1}^{N} \|a_j\|$, where $a_1, \ldots, a_n \in \mathbb{C}^{1 \times N}$ are the N rows of A.

Although Szász did not cite Schmidt's works on orthogonalisation [553, 554] in [603, 605], he used and cited them frequently in later papers; see his Collected Works, which contain 112 of his 134 mathematical papers [606]. He finished his doctoral studies at the University of Budapest under the supervision of Fejér in 1911. In September 1911 Lanczos started his studies at the same university. Szász' dissertation (published in Hungarian in 1912 [604]) and his early academic career were strongly influenced by contemporary mathematical developments in Göttingen. In a similar spirit to that of Toeplitz (see the historical note in Section 3.4.3) he took an algebraic approach to Hilbert's newly developed functional analysis. This also led him to study continued fractions; 11 papers in [606] are devoted to this subject. (We will discuss continued fractions in detail in Chapter 3.) Szász and Lanczos knew each other very well. Before emigrating to the USA, they both spent many years in the faculty of the University of Frankfurt in Germany (Szász from 1914 to 1933 and Lanczos from 1924 to 1931). A description of the life and work of Lanczos is given in [230].

2.5 DERIVATION OF SOME KRYLOV SUBSPACE METHODS

The main goal of this section is to show how the Krylov subspace methods that are characterised mathematically in Theorem 2.3.1 can be implemented.

We will start with the CG method. Since its introduction by Hestenes and Stiefel in 1952 [313], the method has been the subject of numerous discussions and interpretations. In Sections 2.5.1–2.5.3 we will discuss the CG method and its mathematical properties from three different points of view. The first presents the method's derivation from the projection process in case (1) in Theorem 2.3.1 and the Hermitian Lanczos algorithm. Although it reveals nicely the algebraic interconnections, this derivation should not be used for explaining CG to the newcomer. In our opinion, students should be taught CG as described in Section 2.5.2 (motivation of the method by the PDE context) and Section 2.5.3 (step-by-step motivated algebraic derivation based on minimisation). There exist a number of additional derivations and interpretations of the method that we will not discuss in this book. For example, Vorobyev showed how the method arises in the context of the method of moments for self-adjoint operators in Hilbert spaces [651, Chapter III, Section 4]. Vorobyev's work and its relation to Krylov subspace methods will be further studied in Chapter 3.

After dealing with the CG method we will discuss the closely related (and more general) SYMMLQ method (case (2) in Theorem 2.3.1) in Section 2.5.4 and the MINRES and GMRES methods (case (3) in Theorem 2.3.1) in Section 2.5.5. In Section 2.5.6 we will give a brief survey of Krylov subspace methods that are not characterised in Theorem 2.3.1.

2.5.1 Derivation of CG Using the Hermitian Lanczos Algorithm

Consider a linear algebraic system $Ax = b$ with an HPD matrix $A \in \mathbb{F}^{N \times N}$ and a right-hand side vector $b \in \mathbb{F}^N$. Let x_0 be an initial approximation to the solution x and let $r_0 = b - Ax_0$ be the corresponding initial residual. The CG method is an implementation of the projection process of case (1) in Theorem 2.3.1, i.e. of the relations

$$x_n \in x_0 + \mathcal{K}_n(A, r_0) \quad \text{and} \quad r_n \perp \mathcal{K}_n(A, r_0). \tag{2.5.1}$$

We assume in the following that the grade of the initial residual r_0 with respect to A is larger than n. Thus, the projection process is well defined and has not found the solution at step n.

In order to implement (2.5.1), we use the basis of the Krylov subspace $\mathcal{K}_n(A, r_0)$ generated by the Hermitian Lanczos algorithm (Algorithm 2.4.3) applied to A and the initial vector $v \equiv r_0$. The initial step of this algorithm is $v_1 = r_0 / \|r_0\|$, and after n steps we obtain the matrix relation

$$AV_n = V_n T_n + \delta_{n+1} v_{n+1} e_n^T; \qquad (2.5.2)$$

see (2.4.5). The columns of the matrix V_n form the orthonormal basis of the nth Krylov subspace $\mathcal{K}_n(A, r_0)$ and the projected matrix

$$T_n = V_n^* A V_n \in \mathbb{R}^{n \times n} \qquad (2.5.3)$$

is a symmetric tridiagonal matrix with positive off-diagonal elements; see (2.4.6). For any nonzero vector $z \in \mathbb{F}^n$,

$$z^* T_n z = z^* V_n^* A V_n z = \|V_n z\|_A > 0,$$

so that T_n is symmetric positive definite (SPD). Hence T_n can be factorised (in a numerically stable way) as

$$T_n = L_n D_n L_n^T, \quad \text{where } L_n \equiv \begin{bmatrix} 1 & & & \\ \mu_1 & 1 & & \\ & \ddots & \ddots & \\ & & \mu_{n-1} & 1 \end{bmatrix} \in \mathbb{R}^{n \times n}, \qquad (2.5.4)$$

with $\mu_j \neq 0$, $j = 1, \ldots, n-1$, and $D_n \equiv \mathrm{diag}(d_1, \ldots, d_n) \in \mathbb{R}^{n \times n}$ with $d_j > 0$, $j = 1, \ldots, n$; see, e.g. [250, Chapter 4].

The approximate solution x_n of the projection process (2.5.1) is of the form $x_n = x_0 + V_n t_n$ for some vector $t_n \in \mathbb{F}^n$. This vector is determined by the orthogonality condition in (2.5.1), i.e.

$$0 = V_n^* r_n = V_n^*(b - Ax_n) = V_n^* r_0 - V_n^* A V_n t_n = V_n^*(\|r_0\| v_1) - T_n t_n$$
$$= \|r_0\| e_1 - T_n t_n,$$

or, equivalently,

$$T_n t_n = \|r_0\| e_1;$$

cf. (2.1.9). The approximate solution can be written as

$$x_n = x_0 + V_n T_n^{-1} (\|r_0\| e_1) \qquad (2.5.5)$$
$$= x_0 + (V_n L_n^{-T})(\|r_0\| D_n^{-1} L_n^{-1} e_1). \qquad (2.5.6)$$

A straightforward calculation using (2.5.2) and (2.5.5) shows that the corresponding residual is given by

$$r_n = -(\delta_{n+1} \|r_0\| e_n^T T_n^{-1} e_1) v_{n+1} \in \mathrm{span}\{v_{n+1}\}. \qquad (2.5.7)$$

Consequently, the residual vectors $r_0, \ldots, r_n$ are scalar multiples of the corresponding Lanczos basis vectors $v_1, \ldots, v_{n+1}$, and they form an orthogonal basis of the Krylov subspace $\mathcal{K}_{n+1}(A, r_0)$.

We will now simplify (2.5.6) and derive a three-term recurrence for generating the approximation x_n. To this end we define the matrix

$$\widehat{P}_n = [\widehat{p}_0, \ldots, \widehat{p}_{n-1}] \equiv V_n L_n^{-T}.$$

Equivalently,

$$V_n = \widehat{P}_n L_n^T = [\widehat{p}_0, \ldots, \widehat{p}_{n-1}] \begin{bmatrix} 1 & \mu_1 & & \\ & \ddots & \ddots & \\ & & 1 & \mu_{n-1} \\ & & & 1 \end{bmatrix},$$

and hence the columns of the matrix $\widehat{P}_n$ are determined recursively as

$$\widehat{p}_j = v_{j+1} - \mu_j \widehat{p}_{j-1}, \quad j = 0, 1, \ldots, n-1, \tag{2.5.8}$$

where $\mu_0 \equiv 0$ and $\widehat{p}_{-1} \equiv 0$. Since $v_1 = r_0/\|r_0\|$, we see that

$$\operatorname{span}\{\widehat{p}_0, \ldots, \widehat{p}_{n-1}\} = \operatorname{span}\{v_1, \ldots, v_n\} = \mathcal{K}_n(A, r_0).$$

Moreover,

$$\widehat{P}_n^* A \widehat{P}_n = L_n^{-1} V_n^* A V_n L_n^{-T} = D_n, \tag{2.5.9}$$

which shows that the columns of $\widehat{P}_n$ form an *A-orthogonal* or *conjugate* basis of the Krylov subspace $\mathcal{K}_n(A, r_0)$.

We next consider the vector

$$\widehat{c}_n = \begin{bmatrix} c_n^{(1)} \\ \vdots \\ c_n^{(n)} \end{bmatrix} \equiv \|r_0\| D_n^{-1} L_n^{-1} e_1$$

in (2.5.6). This vector is the uniquely defined solution of the linear algebraic system $L_n D_n \widehat{c}_n = \|r_0\| e_1$. Observe that this system can be written as

$$\begin{bmatrix} L_{n-1} D_{n-1} & 0 \\ \mu_{n-1} d_{n-1} e_{n-1}^T & d_n \end{bmatrix} \begin{bmatrix} c_n^{(1)} \\ \vdots \\ c_n^{(n-1)} \\ c_n^{(n)} \end{bmatrix} = \|r_0\| e_1.$$

Therefore, for each $n \geq 2$ the first $n-1$ entries of the vector $\widehat{c}_n$ are given by the previous vector $\widehat{c}_{n-1}$ and

$$\widehat{c}_n = \begin{bmatrix} \widehat{c}_{n-1} \\ c_n^{(n)} \end{bmatrix}, \quad \text{where } c_n^{(n)} = -\frac{\mu_{n-1} d_{n-1} c_{n-1}^{(n-1)}}{d_n}.$$

Using the previous relations in (2.5.6) yields

$$\begin{aligned} x_n = x_0 + \widehat{P}_n \widehat{c}_n &= (x_0 + \widehat{P}_{n-1} \widehat{c}_{n-1}) + c_n^{(n)} \widehat{p}_{n-1} \\ &= x_{n-1} + c_n^{(n)} \widehat{p}_{n-1}. \end{aligned} \quad (2.5.10)$$

Here the vector $\widehat{p}_{n-1}$ represents a *search direction* that is added (after multiplication with $c_n^{(n)}$) to the previous approximation x_{n-1} in order to obtain the new approximation x_n. The vectors $\widehat{p}_0, \ldots, \widehat{p}_{n-1}$ are therefore called *direction vectors* of the CG method. The residual corresponding to the new approximation x_n is of the form

$$r_n = b - A x_n = r_{n-1} - c_n^{(n)} A \widehat{p}_{n-1}. \quad (2.5.11)$$

Up to this point we explicitly require the factorisation (2.5.4). This, however, can be avoided by considering a rescaling of the direction vectors, which is based on the fact that

$$\widehat{p}_j \in \text{span}\{v_{j+1}, \widehat{p}_{j-1}\} = \text{span}\{r_j, \widehat{p}_{j-1}\};$$

cf. (2.5.7) and (2.5.8). We set $p_0 \equiv r_0$ and instead of (2.5.8) we compute direction vectors of the form

$$p_j = r_j + \omega_j p_{j-1}, \quad j = 1, \ldots, n, \quad (2.5.12)$$

where the scalars $\omega_1, \ldots, \omega_n$ are chosen so that the vector p_j is A-orthogonal to the vector p_{j-1}. It is easy to prove that such p_j is a scalar multiple of $\widehat{p}_j, j = 1, \ldots, n$. (The statement holds trivially for $j = 0$.) Since $\widehat{p}_1$ is the linear combination of r_1 and p_0 that is A-orthogonal to p_0, and the same holds for p_1, the statement is true for $j = 1$. Assume that it holds for all indices up to $j - 1$. Then both $\widehat{p}_j$ and p_j are linear combinations of r_j and p_{j-1} that are A-orthogonal to p_{j-1}. This finishes the induction step and proves the statement for a general j.

Thus, $p_0, \ldots, p_n$ are rescaled versions of $\widehat{p}_0, \ldots, \widehat{p}_n$ and they are mutually A-orthogonal; see (2.5.9). In particular, the A-orthogonality of the vectors p_{n-1} and p_n, i.e. $p_{n-1}^* A p_n = 0$, holds if and only if

$$\omega_n = -\frac{p_{n-1}^* A r_n}{p_{n-1}^* A p_{n-1}}. \quad (2.5.13)$$

We next use the vectors $p_0, \ldots, p_n$ instead of $\widehat{p}_0, \ldots, \widehat{p}_n$ for generating the approximation x_n and the corresponding residual r_n. Instead of (2.5.10) and (2.5.11) we then get

$$x_n = x_{n-1} + \alpha_{n-1} p_{n-1}, \tag{2.5.14}$$

and

$$r_n = r_{n-1} - \alpha_{n-1} A p_{n-1}, \tag{2.5.15}$$

where the scalar α_{n-1} is determined uniquely by the condition that the vectors r_{n-1} and r_n are mutually orthogonal. The equation $r_{n-1}^* r_n = 0$ holds if and only if

$$\alpha_{n-1} = \frac{r_{n-1}^* r_{n-1}}{r_{n-1}^* A p_{n-1}} = \frac{\|r_{n-1}\|^2}{p_{n-1}^* A p_{n-1}}. \tag{2.5.16}$$

To obtain the second equality we have used that $r_{n-1} = p_{n-1} - \omega_{n-1} p_{n-2}$ (see (2.5.12) for $j = n-1$) and that the vectors p_{n-2} and p_{n-1} are A-orthogonal. Note that, unless the algorithm stops with $r_{n-1} = 0$, we have $\alpha_{n-1} > 0$.

Finally, we rewrite (2.5.15) in the form

$$A p_{n-1} = -\frac{1}{\alpha_{n-1}} (r_{n-1} - r_n).$$

Inserting this into (2.5.13), and using (2.5.16) together with the orthogonality of the residual vectors leads to

$$\omega_n = -\frac{(Ap_{n-1})^* r_n}{p_{n-1}^* A p_{n-1}} = -\frac{1}{\alpha_{n-1}} \frac{r_{n-1}^* r_n - r_n^* r_n}{p_{n-1}^* A p_{n-1}} = \frac{\|r_n\|^2}{\|r_{n-1}\|^2}. \tag{2.5.17}$$

Combining the previous recurrences (2.5.12) and (2.5.14)–(2.5.17) gives the implementation of the CG method stated in Algorithm 2.5.1. This implementation is identical (except for minor notational differences) to the one given in the classical paper by Hestenes and Stiefel; see [313, p. 411, equations (3:1a)–(3:1f)].

In the following theorem we state the orthogonality properties of the CG residual and direction vectors that we noted in the algorithm's derivation.

Theorem 2.5.1
Suppose that the CG method (Algorithm 2.5.1) does not stop with the exact solution in steps 1 through n. Then the residual vectors $r_0, \ldots, r_n$ form an orthogonal basis of the Krylov subspace $\mathcal{K}_{n+1}(A, r_0)$, and the direction vectors $p_0, \ldots, p_n$ form an A-orthogonal basis of the Krylov subspace $\mathcal{K}_{n+1}(A, r_0)$.

In Algorithm 2.5.1 the residual and direction vectors are generated by two coupled two-term recurrences. It is possible to decouple these recurrences and work

Algorithm 2.5.1 The CG method

Input: HPD matrix $A \in \mathbb{F}^{N \times N}$, right-hand side vector $b \in \mathbb{F}^N$, initial approximation $x_0 \in \mathbb{F}^N$, stopping criterion, maximal number of iterations $n_{\max}$.
Output: Approximation x_n after the algorithm has stopped.

Initialise: $r_0 = b - Ax_0$, $p_0 = r_0$.

For $n = 1, 2, \ldots, n_{\max}$

$$\alpha_{n-1} = \frac{\|r_{n-1}\|^2}{p_{n-1}^* A p_{n-1}},$$

$$x_n = x_{n-1} + \alpha_{n-1} p_{n-1},$$

$$r_n = r_{n-1} - \alpha_{n-1} A p_{n-1},$$

Stop the iteration when the stopping criterion is satisfied,

$$\omega_n = \frac{\|r_n\|^2}{\|r_{n-1}\|^2},$$

$$p_n = r_n + \omega_n p_{n-1}.$$

End

---

with three-term recurrences instead. In particular, the difference of two consecutive direction vectors can be written as

$$p_n - p_{n-1} = (r_n + \omega_n p_{n-1}) - (r_{n-1} + \omega_{n-1} p_{n-2})$$
$$= (r_n - r_{n-1}) + (\omega_n p_{n-1} - \omega_{n-1} p_{n-2}).$$

Using that

$$r_n - r_{n-1} = -\alpha_{n-1} A p_{n-1}$$

leads to the three-term recurrence

$$p_n = -\alpha_{n-1} A p_{n-1} + (1 + \omega_n) p_{n-1} - \omega_{n-1} p_{n-2}.$$

Similarly, one can show that

$$r_n = -\alpha_{n-1} A r_{n-1} + \left(1 + \frac{\alpha_{n-1} \omega_{n-1}}{\alpha_{n-2}}\right) r_{n-1} - \frac{\alpha_{n-1} \omega_{n-1}}{\alpha_{n-2}} r_{n-2}.$$

Three-term recurrence implementations of the CG method were described early on by Rutishauser [163, Chapter II]; see also the book of Hageman and Young [292]. They are of interest for parallel computations, but they have some numerical disadvantages. In particular, their maximal attainable accuracy is more vulnerable to rounding errors than the implementation using coupled two-term recurrences; see [289] for a discussion and [457, Section 5.4] for a brief summary.

2.5.2 CG and the Galerkin Finite Element Method

In this section we will show that the optimality property of the CG method arises naturally in the context of the Galerkin finite element method[3] for solving self-adjoint and positive definite elliptic partial differential equations. Our exposition is based in part on [160, Chapters 1 and 2].

Consider, for simplicity of the exposition, the following Poisson equation model problem for an open and bounded domain $\Omega \subset \mathbb{R}^2$ and a given function f defined on Ω (see [160, (1.13)–(1.14)]).

Find a function $u \equiv u(\xi_1, \xi_2) : \overline{\Omega} \to \mathbb{R}$ that satisfies the equation

$$-\Delta u = f \quad \text{in } \Omega, \qquad (2.5.18)$$

and the given boundary conditions

$$u = g_{\mathcal{D}} \text{ on } \partial\Omega_{\mathcal{D}} \quad \text{and} \quad \frac{\partial u}{\partial n} = g_{\mathcal{N}} \text{ on } \partial\Omega_{\mathcal{N}}, \qquad (2.5.19)$$

where $\partial\Omega_{\mathcal{D}} \cup \partial\Omega_{\mathcal{N}} = \partial\Omega$, $\partial\Omega_{\mathcal{D}} \cap \partial\Omega_{\mathcal{N}} = \emptyset$, and $\int_{\partial\Omega_{\mathcal{D}}} ds \neq 0$.

Here $\partial u/\partial n$ denotes the directional derivative in the direction normal to the boundary $\partial\Omega$ (conventionally pointing outwards). The last assumption on the boundary is made to avoid that (2.5.19) represents a pure Neumann condition.

Assuming sufficient smoothness (see [75, 236]), consider the Sobolev space

$$\mathcal{H}^1(\Omega) \equiv \left\{ u : \Omega \to \mathbb{R} : u, \frac{\partial u}{\partial \xi_1}, \frac{\partial u}{\partial \xi_2} \in \mathcal{L}^2(\Omega) \right\},$$

where $\mathcal{L}^2(\Omega) \equiv \{w : \Omega \to \mathbb{R} : \|w\|_{\mathcal{L}^2(\Omega)} < +\infty\}$ is the standard Lebesgue space, and the spaces

$$\mathcal{H}^1_{\mathcal{D}} \equiv \{u \in \mathcal{H}^1(\Omega) : u = g_{\mathcal{D}} \text{ on } \partial\Omega_{\mathcal{D}}\},$$
$$\mathcal{H}^1_0 \equiv \{v \in \mathcal{H}^1(\Omega) : v = 0 \text{ on } \partial\Omega_{\mathcal{D}}\}.$$

Using Green's identity, any function $u(\xi_1, \xi_2)$ that satisfies (2.5.18) and (2.5.19) also satisfies the following weak formulation (see [160, (1.17)]).

Find a function $u \in \mathcal{H}^1_{\mathcal{D}}$, such that

$$\int_\Omega \nabla u \cdot \nabla v = \int_\Omega v f + \int_{\partial\Omega_{\mathcal{N}}} v g_{\mathcal{N}} \quad \text{for all } v \in \mathcal{H}^1_0. \qquad (2.5.20)$$

---

3. Named after the Russian engineer Boris Grigoryevich Galerkin (1871–1945). The corresponding work of Galerkin refers to the earlier results of the Swiss physicist and mathematician Walther Ritz (1878–1909); see footnote 4 in Section 2.5.3. The whole story of the origins of the finite element method and of the relationship between the works of Ritz, Galerkin, Courant, and others can be found in the very nice review paper by Gander and Wanner [212].

Using the bilinear form $a : \mathcal{H}^1(\Omega) \times \mathcal{H}^1(\Omega) \to \mathbb{R}$,

$$a(u,v) \equiv (\nabla u, \nabla v) \equiv \int_\Omega \nabla u \cdot \nabla v,$$

and the linear functional $\ell : \mathcal{H}^1(\Omega) \to \mathbb{R}$,

$$\ell(v) \equiv (f,v) + (g_\mathcal{N}, v)_{\partial\Omega_\mathcal{N}} \equiv \int_\Omega vf + \int_{\partial\Omega_\mathcal{N}} v g_\mathcal{N},$$

the weak formulation (2.5.20) can be restated as follows.
Find a function $u \in \mathcal{H}^1_\mathcal{D}$, such that

$$a(u,v) = \ell(v) \quad \text{for all } v \in \mathcal{H}^1_0. \tag{2.5.21}$$

In the weak formulation the solution $u \in \mathcal{H}^1_\mathcal{D}(\Omega)$ is 'tested against' all functions $v \in \mathcal{H}^1_0(\Omega)$. Therefore $\mathcal{H}^1_\mathcal{D}(\Omega)$ and $\mathcal{H}^1_0(\Omega)$ are called the solution and the test space, respectively. In the Galerkin finite element discretisation of (2.5.21) the test space is replaced by a finite-dimensional subspace. The latter is typically constructed by partitioning Ω into a finite number of non-overlapping subdomains (defining the finite element mesh; for example a triangulation), and by considering basis functions ϕ_i that are continuous on Ω, equal to polynomials on each subdomain, and nonzero only on a few subdomains. In the following we consider any given N-dimensional subspace

$$\mathcal{S}^h_0 \equiv \operatorname{span}\{\phi_1, \ldots, \phi_N\} \subset \mathcal{H}^1_0.$$

In order to satisfy the Dirichlet boundary condition in (2.5.19) one typically uses an additional function $u_\mathcal{D}$ that interpolates the boundary data $g_\mathcal{D}$ on $\partial\Omega_\mathcal{D}$. We will assume that this interpolation is exact. Instead of the infinite-dimensional problem (2.5.21) one then considers the following finite-dimensional problem.
Find a function $u_h \in \mathcal{S}^h_\mathcal{D} \equiv u_\mathcal{D} + \mathcal{S}^h_0$, such that

$$a(u_h, v_h) = \ell(v_h) \quad \text{for all } v_h \in \mathcal{S}^h_0. \tag{2.5.22}$$

Any solution u_h of this problem can be written as

$$u_h = u_\mathcal{D} + \sum_{j=1}^N \zeta_j \phi_j. \tag{2.5.23}$$

In order to determine the (real) coefficients $\zeta_1, \ldots, \zeta_N$, the right-hand side of (2.5.23) is substituted for u_h in (2.5.22). The resulting equation

$$\sum_{j=1}^N a(\phi_j, v_h) \zeta_j = \ell(v_h) - a(u_\mathcal{D}, v_h)$$

must hold for all $v_h \in S_0^h$. Equivalently, it must hold for $v_h = \phi_i$, $i = 1, 2, \ldots, N$. This leads to the linear algebraic system

$$Ax = b, \quad \text{where}$$

$$A = [a_{ij}], \quad a_{ij} = a(\phi_j, \phi_i), \quad x = \begin{bmatrix} \zeta_1 \\ \vdots \\ \zeta_N \end{bmatrix}, \qquad (2.5.24)$$

$$b = \begin{bmatrix} b_1 \\ \vdots \\ b_N \end{bmatrix}, \quad b_i = \ell(\phi_i) - a(u_D, \phi_i).$$

This linear algebraic system is called the *Galerkin system*, the symmetric matrix A is called the *stiffness matrix*, the right-hand side b is called the *load vector*, and the function (2.5.23) obtained by solving (2.5.24) is called the *Galerkin solution*. The bilinear form $a(u, v)$ is symmetric, and therefore A is a real symmetric $N \times N$ matrix.

Using the basis functions ϕ_i, which by construction have local supports, the solution $u \in \mathcal{H}_D^1$ of the weak formulation (2.5.21) is approximated by the *locally composed discretised solution*, $u_h \in S_D^h$. In order to make A sparse, and thus to make the algebraic problem of possibly very large dimension solvable, we cannot refrain from the locality of the basis functions. The globality of the approximation is then restored through the algebraic system (2.5.24) for computing the coefficients of the linear combination of the locally supported basis functions. This observation is of great importance. The global role of (2.5.24) must be taken into account when comparing the discretisation error and the algebraic error. In particular, as we will show in Section 5.1, one should carefully distinguish between evaluating the global and the local errors, which play an essential role when considering adaptivity and mesh refinement.

The assumption $\int_{\partial \Omega_D} ds \neq 0$ implies that $a(u, v)$ represents an inner product on the space $\mathcal{H}_0^1$ (this can be shown using the Poincaré–Friedrichs inequality; see [160, p. 37]). The inner product induces the *energy norm* in $\mathcal{H}_0^1$, which can be written as

$$\|\nabla u\| \equiv a(u, u)^{1/2} = \left(\int_\Omega \nabla u \cdot \nabla u \right)^{1/2}. \qquad (2.5.25)$$

Here

$$\|w\| \equiv (w, w)^{1/2} \equiv \left(\int_\Omega w \cdot w \right)^{1/2}$$

denotes the $\mathcal{L}^2(\Omega)$-norm of a (vector-valued) function $w : \Omega \to \mathbb{R}^2$. Consequently, the (symmetric) matrix A is positive definite. In particular, A is invertible and thus the solution of the linear algebraic system (2.5.24) and the discretised formulation (2.5.22) are uniquely determined.

The energy norm is relevant in many applications. For example, [236, Section 2.2.1] describes a class of mechanical problems with homogeneous Dirichlet boundary conditions, for wich the solution u of the weak formulation (2.5.21) minimises the potential energy of the modelled system. Moreover, at the solution u the first variation of the potential energy is zero in every direction (therefore the term *variational form*).

Consider two arbitrary functions $v_h, w_h \in \mathcal{S}_0^h$. Using $\Phi = [\phi_1, \phi_2, \ldots, \phi_N]$ we can write these functions as $v_h = \Phi y$, $w_h = \Phi z$ for some coefficient vectors $y, z \in \mathbb{R}^N$. The inner product of the two functions is then given by

$$a(v_h, w_h) = \int_\Omega \nabla v_h \cdot \nabla w_h = z^* A y = (y, z)_A.$$

(For consistency of notation we write here the Hermitian transpose instead of the regular transpose, although the vectors are real.) In particular,

$$\|\nabla v_h\|^2 = a(v_h, v_h) = y^* A y = (y, y)_A = \|y\|_A^2. \qquad (2.5.26)$$

This means that the energy norm of the function v_h in the space $\mathcal{S}_0^h \subset \mathcal{H}_0^1$ is equal to the *algebraic energy norm* (or A-norm) of its coefficient vector $y \in \mathbb{R}^N$.

Since $\mathcal{S}_0^h \subset \mathcal{H}_0^1$, the solution $u \in \mathcal{H}_D^1$ of (2.5.21) satisfies

$$a(u, v_h) = \ell(v_h) \quad \text{for all } v_h \in \mathcal{S}_0^h.$$

On the other hand, since the Dirichlet boundary conditions are interpolated by u_D exactly, we have $\mathcal{S}_D^h \subset \mathcal{H}_D^1$, which means that the constructed approximation $u_h \in \mathcal{S}_D^h$ is *conforming*. (For examples of nonconforming finite elements, see, e.g. [75, Section 10.3].) This approximation satisfies

$$a(u_h, v_h) = \ell(v_h) \quad \text{for all } v_h \in \mathcal{S}_0^h.$$

Subtracting this equation from the analogous equation for u shows that the *discretisation error* $u - u_h \in \mathcal{H}_0^1$ has the *Galerkin orthogonality property*

$$a(u - u_h, v_h) = 0 \quad \text{for all } v_h \in \mathcal{S}_0^h. \qquad (2.5.27)$$

In other words, the discretisation error is orthogonal to the finite-dimensional Galerkin subspace $\mathcal{S}_0^h$ with respect to the energy inner product defined by $a(u, v)$.

Equivalently, the function $u_h \in \mathcal{S}_D^h$ is the best approximation to the solution $u \in \mathcal{H}_D^1$ from $\mathcal{S}_D^h$ with respect to the energy norm, i.e.

$$\|\nabla(u - u_h)\| = \min_{w_h \in \mathcal{S}_D^h} \|\nabla(u - w_h)\|. \qquad (2.5.28)$$

The equivalence of the orthogonality property (2.5.27) and the optimality property (2.5.28) is analogous to the equivalences of (1) and (2) in Theorem 2.3.2.

Now suppose that the solution x of the Galerkin system (2.5.24) is approximated by $x_n \in \mathbb{R}^N$. Then the corresponding function

$$u_h^{(n)} \equiv u_D + \Phi x_n \in S_D^h$$

is an approximation to the solution u of (2.5.21), and we have the equation

$$\underbrace{u - u_h^{(n)}}_{\text{total error}} = \underbrace{(u - u_h)}_{\text{discretisation error}} + \underbrace{(u_h - u_h^{(n)})}_{\text{algebraic error}}.$$

The following theorem shows the relationship between the energy norms of the total error, the discretisation error and the algebraic error.

Theorem 2.5.2
Using the previous notation,

$$\left\|\nabla(u - u_h^{(n)})\right\|^2 = \|\nabla(u - u_h)\|^2 + \|x - x_n\|_A^2. \qquad (2.5.29)$$

Proof
The squared energy norm of the total error is given by

$$\begin{aligned}\left\|\nabla(u - u_h^{(n)})\right\|^2 &= a\left(u - u_h^{(n)}, u - u_h^{(n)}\right) \\ &= a\left(u - u_h + u_h - u_h^{(n)}, u - u_h + u_h - u_h^{(n)}\right) \\ &= \left\|\nabla(u - u_h)\right\|^2 + \left\|\nabla(u_h - u_h^{(n)})\right\|^2 + 2a\left(u - u_h, u_h - u_h^{(n)}\right) \\ &= \left\|\nabla(u - u_h)\right\|^2 + \|x - x_n\|_A^2.\end{aligned}$$

In the last equality we have used that $a(u - u_h, u_h - u_h^{(n)}) = 0$ due to the Galerkin orthogonality condition (note that $u_h - u_h^{(n)} \in S_0^h$), and the fact that $\|\nabla(u_h - u_h^{(n)})\| = \|x - x_n\|_A$; see (2.5.26). $\square$

Using zero Dirichlet boundary conditions $g_D = 0$ on $\partial\Omega_D$, we set $u_D = 0$ and get $u_h \in S_0^h$. Consequently,

$$\begin{aligned}\|\nabla(u - u_h)\|^2 &= \|\nabla u\|^2 + \|\nabla u_h\|^2 - 2a(u, u_h) \\ &= \|\nabla u\|^2 + \|\nabla u_h\|^2 - 2a(u - u_h, u_h) - 2\|\nabla u_h\|^2.\end{aligned}$$

The Galerkin orthogonality implies $a(u - u_h, u_h) = 0$, and hence

$$\|\nabla(u - u_h)\|^2 = \|\nabla u\|^2 - \|\nabla u_h\|^2. \qquad (2.5.30)$$

In this case the squared energy norm of the discretisation error is nothing but the difference between the squared energy norms of the exact solution and the discretised solution, respectively, so that in fact $\|\nabla u_h\|^2$ approaches $\|\nabla u\|^2$ from below; see, e.g. [160, Section 1.5, relation (1.61) and Problem 1.11 on p. 63].

Equation (2.5.29) shows that the (squared) energy norm of the total error $u - u_h^{(n)}$ consists of two distinct components:

- the (squared) energy norm of the discretisation error $u - u_h$,
- the (squared) energy norm of the algebraic error $u_h - u_h^{(n)}$.

This illustrates our point made in Chapter 1, that in the solution of real-world problems that are modelled by partial differential equations, all stages of the solution process should be in balance. If the quality of the computed approximation is evaluated using the energy norm of the error, then (2.5.29) seems to suggest that it makes little sense to compute an approximation x_n whose algebraic error norm is *significantly below* the norm of the discretisation error (both measured in the corresponding energy norms). Moreover, as mentioned above, one has to take into account the local distribution of the error. Evaluating the balance between the discretisation error and the algebraic error (i.e. giving a sensible meaning to the term 'significantly below') can therefore be a challenging problem. We will return to this point in Section 5.1.

If the approximate solution x_n of the Galerkin system (2.5.24) is sought using a given *(affine) subspace* of $\mathbb{R}^N$, then the approximation $u_h^{(n)}$ is an element of the *corresponding (affine) subspace* of $\mathcal{S}_\mathcal{D}^h$. In order to get the best possible approximation $u_h^{(n)}$ in the sense of the minimal energy norm of the functional error $\|\nabla(u - u_h^{(n)})\|$, the corresponding x_n must have the minimal distance from the true solution x of (2.5.24) with respect to the induced algebraic energy norm over the given (affine) finite-dimensional subspace, i.e. $\|x - x_n\|_A$ must be minimal.

In Krylov subspace methods the given affine finite-dimensional subspace is $x_0 + \mathcal{K}_n(A, r_0)$. Hence the CG method, which minimises the algebraic energy norm over this subspace, naturally leads to the best possible approximation. Therefore the CG method is 'tailor-made' for the Galerkin finite element context and for solving the problem (2.5.21) and its discretised version (2.5.22).

2.5.3 CG and the Minimisation of a Quadratic Functional

We will now show that the discretised weak formulation (2.5.22) can be reformulated as a minimisation problem on $\mathcal{S}_\mathcal{D}^h$. Using the bilinear form $a(u, v)$ and the linear functional $\ell(v)$ we define the *quadratic functional* $\mathcal{F} : \mathcal{S}_\mathcal{D}^h \to \mathbb{R}$ by

$$\mathcal{F}(w) \equiv \frac{1}{2}a(w, w) - \ell(w). \qquad (2.5.31)$$

One can now consider the following minimisation problem:
Find a function $w_h \in \mathcal{S}_\mathcal{D}^h$, such that

$$\mathcal{F}(w_h) \leq \mathcal{F}(v_h) \quad \text{for all } v_h \in \mathcal{S}_\mathcal{D}^h. \qquad (2.5.32)$$

The next theorem shows the equivalence of the problems (2.5.22) and (2.5.32).

Theorem 2.5.3
The two problems (2.5.22) and (2.5.32) have the same uniquely determined solution.

Proof
Let $u_h \in S_D^h$ be the (uniquely determined) solution of the problem (2.5.22). Let $v \in S_0^h$ be an arbitrary given function, and define $v_h \equiv u_h + v \in S_D^h$. Then

$$\begin{aligned}
\mathcal{F}(v_h) = \mathcal{F}(u_h + v) &= \frac{1}{2}a(u_h + v, u_h + v) - \ell(u_h + v) \\
&= \frac{1}{2}\left(a(u_h, u_h) + a(v, u_h) + a(u_h, v) + a(v, v)\right) - \ell(u_h) - \ell(v) \\
&= \mathcal{F}(u_h) + (a(u_h, v) - \ell(v)) + \frac{1}{2}a(v, v) \\
&\geq \mathcal{F}(u_h),
\end{aligned}$$

where in the last inequality we have used that $a(u_h, v) - \ell(v) = 0$ (since u_h solves (2.5.22)) and that $a(v, v) \geq 0$ (since a is positive definite). Since the function v can be chosen arbitrarily, we have $\mathcal{F}(v_h) \geq \mathcal{F}(u_h)$ for all $v_h \in S_D^h$, and thus u_h solves the minimisation problem (2.5.32).

On the other hand, let w_h be any solution of the problem (2.5.32), and let an arbitrary function $v_h \in S_0^h$ be given. For $\epsilon \in \mathbb{R}$ we define the (continuous) function $f(\epsilon) \equiv \mathcal{F}(w_h + \epsilon v_h)$. Then

$$f(0) = \mathcal{F}(w_h) \leq \mathcal{F}(w_h + \epsilon v_h) = f(\epsilon) \quad \text{for all } \epsilon \in \mathbb{R},$$

i.e. $f(\epsilon)$ has a global minimum at $\epsilon = 0$. From the definition of $f(\epsilon)$ and the symmetry of $a(u, v)$ we get

$$f(\epsilon) = \frac{1}{2}a(w_h, w_h) + \epsilon a(w_h, v_h) + \frac{\epsilon^2}{2}a(v_h, v_h) - \ell(w_h) - \epsilon \ell(v_h),$$

and hence $f'(\epsilon) = a(w_h, v_h) + \epsilon a(v_h, v_h) - \ell(v_h)$. The right-hand side is equal to zero if and only if

$$\epsilon = \frac{\ell(v_h) - a(w_h, v_h)}{a(v_h, v_h)}.$$

But since a global minimum of $f(\epsilon)$ occurs at $\epsilon = 0$, we must have $a(w_h, v_h) = \ell(v_h)$. Since v_h can be chosen arbitrarily, w_h indeed solves the problem (2.5.22), and thus the solution of (2.5.22) is uniquely determined. □

Approximating the solution of the infinite-dimensional weak formulation (2.5.21) by solving the finite-dimensional minimisation problem (2.5.32) (instead of the discretised weak formulation (2.5.22)) is sometimes referred to as the

Ritz method or the *Rayleigh–Ritz method*;[4] see, e.g. [75, Section 2.5] or [519, Chapter 10].

One can easily generalise the previous theorem to a general symmetric and positive definite bilinear form $a(u, v)$ and a linear functional $\ell(v)$, both defined on a finite-dimensional Hilbert space. Moreover, one can generalise the result to show the equivalence of the infinite-dimensional weak formulation (2.5.21) and a corresponding infinite-dimensional version of the minimisation problem (2.5.32). In the infinite-dimensional case, however, additional assumptions on the bilinear form and the linear functional are required in order to guarantee the existence of a unique solution of (2.5.21). As shown by the famous Lax–Milgram Theorem, sufficient conditions are that the bilinear form be continuous and coercive, and the linear functional continuous. Details can be found in [75, Section 2.7] or [519, Section 5.5].

Consider any function of the form

$$w_h = u_\mathcal{D} + \Phi z \in \mathcal{S}_\mathcal{D}^h.$$

Then the quadratic functional $\mathcal{F}(w)$ defined in (2.5.31) satisfies

$$\mathcal{F}(w_h) = \frac{1}{2} a(w_h, w_h) - \ell(w_h) = \mathcal{F}(u_\mathcal{D}) + \frac{1}{2} z^* A z - z^* b;$$

see (2.5.24) and (2.5.26). The term $\mathcal{F}(u_\mathcal{D})$ is a constant that does not depend on the choice of z. Hence solving the minimisation problem (2.5.32), i.e. the Ritz method, is equivalent to minimising the right-hand side of the previous equation over $z \in \mathbb{R}^N$.

This is a particular example of a general class of optimisation problems which involve the minimisation of a quadratic functional of the form

$$\mathcal{F}(z) \equiv \frac{1}{2} z^* A z - z^* b, \qquad (2.5.33)$$

where $A \in \mathbb{R}^{N \times N}$ is a given SPD matrix and $b \in \mathbb{R}^N$ is a given vector. We will now derive the CG method directly from the minimisation of the quadratic functional (2.5.33).

Since A is SPD, the functional $\mathcal{F}(z)$ has a uniquely determined minimiser $z_* \in \mathbb{R}^N$, which is a *stationary point* of $\mathcal{F}(z)$, i.e. $\nabla \mathcal{F}(z_*) = 0$. From

$$\nabla \mathcal{F}(z) = Az - b$$

we see that the minimiser z_* is equal to the uniquely determined solution x of the linear algebraic system $Ax = b$.

---

4. Ritz proposed this approach in 1908 [527], and the English physicist Lord Rayleigh (John William Strutt, 1842–1919), Physics Nobel Prize winner in 1904, wrote in 1911 [518] that a similar method had been used earlier by him. An evaluation of their respective contributions is given in [416]; see also [212].

Now suppose that x_n is an approximation to the minimiser of $\mathcal{F}(z)$, or equivalently, to the solution of $Ax = b$. Then

$$\mathcal{F}(x_n) = \frac{1}{2}(x - x_n)^* A(x - x_n) - \frac{1}{2} x^* A x = \frac{1}{2} \|x - x_n\|_A^2 - \frac{1}{2} \|x\|_A^2.$$

Note that the term $\|x\|_A$ does not depend on the choice of the approximation x_n. Thus, the minimisation of the functional $\mathcal{F}(z)$ over some (affine) subspace of $\mathbb{R}^N$ is mathematically equivalent to the minimisation of $\|x - z\|_A$ over the same (affine) subspace. It is therefore natural to measure the distance of the approximate solution x_k to x in the energy norm.

Let x_0 be an initial approximation and let the sequence of approximations x_n to the solution x be constructed by the recurrence

$$x_n = x_{n-1} + \alpha_{n-1} p_{n-1}, \quad n = 1, 2, \ldots, \tag{2.5.34}$$

where α_{n-1} is a certain scalar coefficient and p_{n-1} is a certain *direction vector*.

For the moment suppose that the vector p_{n-1} is given. Then our goal is to choose the scalar α_{n-1} so that the new approximation x_n minimises $\|x - z\|_A$ or, equivalently, $\mathcal{F}(z)$ over the line $x_{n-1} + \alpha\, p_{n-1}$. A simple calculation shows that

$$\|x - x_n\|_A^2 = \|x - x_{n-1}\|_A^2 - 2\alpha_{n-1} p_{n-1}^* r_{n-1} + \alpha_{n-1}^2 p_{n-1}^* A p_{n-1}.$$

Hence the minimum is attained for

$$\alpha_{n-1} = \frac{p_{n-1}^* r_{n-1}}{p_{n-1}^* A p_{n-1}}. \tag{2.5.35}$$

As an immediate consequence we get

$$p_{n-1}^* r_n = p_{n-1}^* (b - A x_n) = p_{n-1}^* (r_{n-1} - \alpha_{n-1} A p_{n-1}) = 0,$$

which means that the residual r_n (corresponding to the approximation x_n) is orthogonal to the direction vector p_{n-1}. Geometrically, this means that the gradient $\nabla \mathcal{F}(x_n)$ at x_n is orthogonal to the equipotential surface determined by the equation $\mathcal{F}(y) = \mathcal{F}(x_n)$.

It remains to make a suitable choice for the direction vectors. The simplest choice is to start with $p_0 \equiv r_0 = b - A x_0$. If we take

$$p_n \equiv r_n = b - A x_n = -\nabla \mathcal{F}(x_n)$$

also in the subsequent steps, $n = 1, 2, \ldots$, we get the method of steepest descent. The method's name comes from the fact that the search direction in step n is equal to the negative gradient of the given functional, evaluated at the current approximation x_n. Geometrically, the search direction is the direction where the given functional *locally* decreases most quickly. This approach exhibits a very slow convergence behaviour when the functional describes a long and narrow valley. Such a situation is typical when A is ill conditioned, which is the case in many applications. Further mathematical details are given in [250, Section 10.2.1] and [272, Chapter 2].

In 1958 Stiefel described the essential idea (already used in the original derivation of CG in [313]) to obtain a better, even optimal, method based on the iteration (2.5.34) as follows [581, p. 6]:

> In order to secure the convergence of the iteration some *tactic* is needed. Normally, it is recommended that p_i be chosen in such a way that an appropriate error measure is minimized going from x_i to x_{i+1}. This tactical rule concerns only a single step. However, like a good chess player, one should not be guided by the greatest advantage to be gained in a single move, but should consider the widest over-all strategy for winning the game and choose the set of moves that best promote that strategy. Therefore, the following *strategy* problem must be solved. Among all paths having a given number of segments n, it is necessary to find the path yielding an endpoint x_n with least error measure.

We will show how Stiefel's strategy problem can be solved. The choice of α_{n-1} in (2.5.35) gives a one-dimensional minimisation at each step. The choice $p_n = r_n$ for $n = 1, 2, \ldots$ does not guarantee any minimisation property over higher-dimensional subspaces. In order to obtain such property, the vector p_n must combine information from several iteration steps. The simplest possibility is to generate the new search direction p_n as a linear combination of the previous direction vector and the (new) residual, i.e.

$$p_n = r_n + \omega_n p_{n-1}, \qquad (2.5.36)$$

where ω_n is a certain scalar. Independently of the choice of this scalar, the minimisation along the line $x_{n-1} + \alpha \, p_{n-1}$ (see (2.5.34)–(2.5.35)) implies that

$$p_n^* r_n = r_n^* r_n + \omega_n p_{n-1}^* r_n = r_n^* r_n = \|r_n\|^2. \qquad (2.5.37)$$

As a consequence, the resulting iteration process can stop only at the solution x. Indeed, the process stops if either $p_n = 0$ or $\alpha_n = 0$. In both cases

$$0 = p_n^* r_n = \|r_n\|^2,$$

i.e. $r_n = 0$ and $Ax_n = b$.

In order to motivate the choice of ω_n below, we first notice that the change of the error from the step $n - 1$ to the step n,

$$x - x_n = (x - x_{n-1}) - \alpha_{n-1} p_{n-1},$$

with the value

$$\alpha_{n-1} = \frac{p_{n-1}^* r_{n-1}}{p_{n-1}^* A p_{n-1}} = \frac{p_{n-1}^* A (x - x_{n-1})}{p_{n-1}^* A p_{n-1}},$$

can be interpreted as the A-orthogonalisation of the error $x - x_{n-1}$ against the vector p_{n-1}. Equivalently,

$$x - x_{n-1} = \alpha_{n-1} p_{n-1} + (x - x_n)$$

can be viewed as the orthogonal decomposition of $x - x_{n-1}$ with respect to the A-inner product, $(y, z)_A = z^* A y$ for all $y, z \in \mathbb{R}^N$. This means that $x - x_{n-1}$ is decomposed into its A-orthogonal components $\alpha_{n-1} p_{n-1}$ and $x - x_n$. The Pythagorean Theorem then gives

$$\|x - x_{n-1}\|_A^2 = |\alpha_{n-1}|^2 \|p_{n-1}\|_A^2 + \|x - x_n\|_A^2. \qquad (2.5.38)$$

The recursive application of this leads to

$$x - x_0 = \sum_{j=1}^{n} \alpha_{j-1} p_{j-1} + (x - x_n) \qquad (2.5.39)$$

and

$$\|x - x_0\|_A^2 = \sum_{j=1}^{n} |\alpha_{j-1}|^2 \|p_{j-1}\|_A^2 + \|x - x_n\|_A^2. \qquad (2.5.40)$$

It should be emphasised that the expansion (2.5.39) and the identity (2.5.40) hold for *any* choice of the direction vectors. It follows from the fact that the scalar α_{j-1} gives the minimum of $\|x - x_j\|_A$ along the line determined by the direction vector $p_{j-1}, j = 1, \ldots, n$. The idea of obtaining at the nth step a minimisation over the n-dimensional *subspace* generated by *all* direction vectors $p_0, \ldots, p_{n-1}$ leads in a natural way to the mutual A-orthogonality of these vectors.

Indeed, *assume* that all direction vectors $p_0, p_1, \ldots$ are orthogonal with respect to the A-inner product, i.e. *assume* that $p_i^* A p_j = 0$ for $i \neq j$. Then

$$x - x_n = (x - x_0) - \sum_{j=1}^{n} \alpha_{j-1} p_{j-1}$$

represents the A-orthogonal decomposition of $x - x_0$, and, as a consequence, $\|x - x_n\|_A$ is minimal over all possible approximations in the subspace generated by the search directions $p_0, \ldots, p_{n-1}$,

$$\|x - x_n\|_A = \min_{y \in x_0 + \text{span}\{p_0, \ldots, p_{n-1}\}} \|x - y\|_A. \qquad (2.5.41)$$

Moreover, the assumed mutual A-orthogonality of the vectors p_j, $j = 0, 1, \ldots$, implies that $p_N = 0$. This means that (in exact arithmetic) the algorithm reaches the true solution x in at most N steps.

In order to get as close as we can to the desired A-orthogonality of the search directions, take the value of the last undetermined scalar coefficient ω_n such that the two *subsequent* search directions are A-orthogonal, i.e.

$$p_{n-1}^* A p_n = 0, \quad \text{which gives} \quad \omega_n = -\frac{p_{n-1}^* A r_n}{p_{n-1}^* A p_{n-1}}.$$

With ω_n given, the algorithm is fully determined.

For the quantities x_n, p_n, α_{n-1}, and ω_n one can show inductively that

$$r_i^* r_j = 0 \quad \text{and} \quad p_i^* A p_j = 0, \quad i \neq j, \tag{2.5.42}$$

where obviously $r_n = b - A x_n = r_{n-1} - \alpha_{n-1} A p_{n-1}$. Consequently, the described choice of ω_n, motivated by the *local A-orthogonality* of the directions vectors p_n and p_{n-1} in two consecutive steps, guarantees the *global orthogonality* of all residuals with respect to the standard Euclidean inner product and the *global A-orthogonality* of all direction vectors. With this choice of ω_n, the orthogonality assumption mentioned prior to the minimisation property (2.5.41) is satisfied.

It is worth noticing that the global orthogonality of the residuals implies that the formula for the search direction

$$p_n = r_n - \frac{p_{n-1}^* A r_n}{p_{n-1}^* A p_{n-1}} p_{n-1}$$

can be interpreted as the A-orthogonalisation of the residual r_n against the search directions $p_{n-1}, p_{n-2}, \ldots, p_0$ (here the terms corresponding to $p_{n-2}, \ldots, p_0$ vanish due to the orthogonality of the residuals). Finally, using

$$-A p_{n-1} = \frac{1}{\alpha_{n-1}} (r_n - r_{n-1}) = \frac{p_{n-1}^* A p_{n-1}}{p_{n-1}^* r_{n-1}} (r_n - r_{n-1})$$

and $p_{n-1}^* r_{n-1} = \|r_{n-1}\|^2$, we get

$$\alpha_{n-1} = \frac{\|r_{n-1}\|^2}{p_{n-1}^* A p_{n-1}} \quad \text{and} \quad \omega_n = \frac{\|r_n\|^2}{\|r_{n-1}\|^2}. \tag{2.5.43}$$

Combining equations (2.5.34), (2.5.36), and (2.5.43) gives the CG method precisely as stated in Algorithm 2.5.1 above. Because of its relationship with the Rayleigh–Ritz method, the CG method has also been considered a member of the family of variational methods; see [549].

Finally, we point out that (2.5.38) and $|\alpha_{n-1}| \, \|p_{n-1}\|_A > 0$ imply that

$$\|x - x_{n-1}\|_A > \|x - x_n\|_A,$$

i.e. the A-norm of the error in the CG method is strictly monotonically decreasing. Further theoretical properties of the CG method will be studied in detail later in this book; see in particular Sections 3.5 and 5.6.

2.5.4 Hermitian Indefinite Matrices and the SYMMLQ Method

In Sections 2.5.1 and 2.5.3 we derived the CG method in two different ways. The first derivation was based on the Lanczos algorithm applied to the HPD matrix A, and it used the LDL^T factorisation of the Jacobi matrix T_n; see (2.5.4). Positive definiteness of A guarantees the same property for T_n, and thus the existence of its LDL^T factorisation. If A is Hermitian *indefinite*, then this factorisation no longer exists. In this case one can factorise the matrix $T_n = L_n D_n L_n^T$, where, in general, D_n is block-diagonal with 1×1 or 2×2 blocks on its diagonal. This leads to the SYMMBK method, which was originally devised by Chandra [103]. An algorithmic variant of this method based on a different pivoting strategy for the 2×2 blocks was presented in [442]. Although the factorisation of T_n with a block-diagonal matrix D_n exists, the approximation x_n may still be undefined since T_n may be singular when A is indefinite. If this happens, the method of [442] skips x_n and computes the subsequent approximation x_{n+1} using the previous approximation x_{n-1}. Note that if T_n is singular, then both T_{n-1} and T_{n+1} must be nonsingular, so that x_{n-1} and x_{n+1} are both guaranteed to exist. This fact follows from the strict interlacing property of the eigenvalues of Jacobi matrices, which we will show in Chapter 3; see Theorem 3.3.1 and Remark 3.3.2.

The second derivation of the CG method assumed that A is a real SPD matrix, and was based on minimisation of the quadratic functional (2.5.33). Positive definiteness of A guarantees that the functional has a unique minimiser and that the minimum occurs at a stationary point. This no longer holds when A is indefinite. In addition, an indefinite matrix A does not define a norm.

We stress that the reason why the two derivations of the CG method cannot be generalised (in a straightforward way) to indefinite matrices A is not algorithmical, but mathematical: if A is indefinite, then the projected system corresponding to the projection process in case (1) of Theorem 2.3.1 may be singular, and hence the approximation x_n may not be well defined. On the other hand, as mentioned above, the CG approximation to the solution x must exist at least for every second iteration. It can not, however, be computed using Algorithm 2.5.1 or by any other mathematically equivalent formulation.

In [496] Paige and Saunders proposed the SYMMLQ method, which is well defined at every step for symmetric or Hermitian indefinite matrices $A \in \mathbb{F}^{N \times N}$. Like the CG method in Section 2.5.1, the derivation of this method starts with the Hermitian Lanczos algorithm applied to A and the initial vector $v = r_0$. After n steps we get the relation

$$AV_n = V_n T_n + \delta_{n+1} v_{n+1} e_n^T = V_{n+1} \underline{T}_n, \qquad \underline{T}_n \equiv \begin{bmatrix} T_n \\ \delta_{n+1} e_n^T \end{bmatrix}. \qquad (2.5.44)$$

Here it is assumed that n is less than the grade of v with respect to A, and the notation $\underline{T}_n$ indicates that the Jacobi matrix T_n is extended by the additional row $\delta_{n+1} e_n^T$.

As mentioned above, with A indefinite the matrix T_n may not be positive definite, and hence the factorisation (2.5.4) may not exist. To overcome this problem, Paige and Saunders suggested using the LQ factorisation of $\underline{T}_n^T$, which motivated the method's name.

Following the description of SYMMLQ in [574, Section 3],

$$\underline{T}_n^T = [T_n \mid \delta_{n+1} e_n] = [L_n \mid 0] Q_{n+1}^T, \qquad (2.5.45)$$

where L_n is an $n \times n$ lower triangular matrix and Q_{n+1} is an $(n+1) \times (n+1)$ orthogonal matrix. Recall that $\underline{T}_n$ is real (even when A is not) and thus L_n and Q_{n+1} are both real.

In practice the matrix Q_{n+1} is a product of n Givens rotation matrices,

$$Q_{n+1} = Q_{1,2} Q_{2,3} \cdots Q_{n,n+1}, \quad Q_{j,j+1} = \begin{bmatrix} I_{j-1} & & \\ & c_j & s_j \\ & -s_j & c_j \\ & & & I_{n-j} \end{bmatrix},$$

where $c_j^2 + s_j^2 = 1$. Multiplied successively from the right to the matrix $\underline{T}_n^T$, the rotations $Q_{j,j+1}$ annihilate the n superdiagonal entries of $\underline{T}_n^T$. From the structure of the rotations it is easy to see that the matrix L_n has nonzero entries only on its diagonal and its first two subdiagonals. Since $\underline{T}_{n+1}$ arises from $\underline{T}_n$ by appending one more row and column, simple formulas exist for updating L_{n+1} and Q_{n+2} from L_n and Q_{n+1}, respectively. The LQ factorisation (2.5.45) can always be computed in a numerically stable way. It exists even when T_n is singular.

Based on this LQ factorisation, Paige and Saunders constructed an algorithm that in effect implements the projection process given in case (2) of Theorem 2.3.1, i.e.

$$x_n \in x_0 + A\mathcal{K}_n(A, r_0) \quad \text{and} \quad x - x_n \perp A\mathcal{K}_n(A, r_0). \qquad (2.5.46)$$

The fact that SYMMLQ implements this process and thus minimises the Euclidean norm of the error over the space $x_0 + A\mathcal{K}_n(A, r_0)$ is present (though somewhat hidden) in their paper. Using their notation, the minimisation property of SYMMLQ was also shown, e.g. in [608, Theorem 3.1]. A method that is mathematically equivalent to SYMMLQ, but numerically unstable, was derived by Fridman in 1963 [207]; see also [183, 589, 590].

Next we will show how the Lanczos relation (2.5.44) and the LQ factorisation (2.5.45) can be used to implement the projection process (2.5.46). Using the matrix V_n, whose columns form an orthonormal basis of $\mathcal{K}_n(A, r_0)$, we can write the nth approximation as $x_n = x_0 + AV_n t_n$ for some coefficient vector $t_n \in \mathbb{F}^n$. This vector is determined by the orthogonality condition, which can be written as

$$0 = (AV_n)^*(x - x_n) = V_n^* r_0 - V_n^* A^2 V_n t_n = \|r_0\| e_1 - V_n^* A^2 V_n t_n,$$

or, equivalently,

$$V_n^* A^2 V_n t_n = \|r_0\| e_1,$$

which is the projected system corresponding to the projection process (2.5.46). Note that the matrix $V_n^* A^2 V_n$ is nonsingular when A is nonsingular.

From (2.5.44) and (2.5.45) we obtain

$$V_n^* A^2 V_n = \underline{T}_n^T V_{n+1}^* V_{n+1} \underline{T}_n = L_n L_n^T,$$

so that $t_n = L_n^{-T} L_n^{-1} \|r_0\| e_1$. The nth approximation is of the form

$$x_n = x_0 + A V_n t_n = x_0 + V_{n+1} \underline{T}_n t_n$$

$$= x_0 + V_{n+1} Q_{n+1} \begin{bmatrix} L_n^T \\ 0 \end{bmatrix} L_n^{-T} L_n^{-1} \|r_0\| e_1$$

$$= x_0 + \left(V_{n+1} Q_{n+1} \begin{bmatrix} I_n \\ 0 \end{bmatrix} \right) (\|r_0\| L_n^{-1} e_1)$$

$$\equiv x_0 + W_n \widehat{c}_n. \tag{2.5.47}$$

Here the matrix $W_n = [w_1, \ldots, w_n] \in \mathbb{F}^{N \times n}$ consists of the first n columns of $V_{n+1} Q_{n+1}$, and $\widehat{c}_n$ satisfies $L_n \widehat{c}_n = \|r_0\| e_1$.

Since the matrices L_n and Q_{n+1} can be updated easily, it is again possible to derive a simple update formula for x_n. By construction, only the last column of W_n and the last entry of $\widehat{c}_n$ change from step to step, so that

$$x_n = x_{n-1} + \zeta_n w_n \tag{2.5.48}$$

for some scalar ζ_n. Formulas for ζ_n and w_n can be found in [496, Section 5]; see also [442, p. 457] for a compact presentation. A MATLAB implementation of SYMMLQ is given in [183, pp. 183–184], and a detailed description of the algorithm is presented in [574, Fig. 3].

If T_n is nonsingular, then, considering the factorisation $T_n = \widetilde{L}_n Q_n^T$ (see (2.5.45)), the nth CG approximation x_n^{CG} is of the form

$$x_n^{CG} = x_0 + V_n Q_n (\|r_0\| \widetilde{L}_n^{-1} e_1) = x_0 + \widetilde{W}_n \widetilde{c}_n, \tag{2.5.49}$$

where W_n and $\widetilde{W}_n, \widehat{c}_n$ and $\widetilde{c}_n$ differ only in the last column and element, respectively. Here it is worth noticing that while the vectors $w_1, \ldots, w_{n-1}, \widetilde{w}_n$ form a basis of $\mathcal{K}_n(A, r_0)$, the vector w_n belongs, in general, to $\mathcal{K}_{n+1}(A, r_0)$. Therefore it is clear that whenever T_n is nonsingular, the nth CG approximation can be determined as a simple update of the $(n-1)$th SYMMLQ approximation,

$$x_n^{CG} = x_{n-1} + \widetilde{\zeta}_n \widetilde{w}_n. \tag{2.5.50}$$

Hence the SYMMLQ method can be viewed as the numerically stable way (in terms of solving the projected problem determined by the potentially unstable Lanczos algorithm) for computing the CG approximation for symmetric indefinite problems; see [496] together with references on the early attempt at the same, due to Luenberger [434, 436].

2.5.5 Minimising the Residual Norm: MINRES and GMRES

In addition to the SYMMLQ method, Paige and Saunders derived in [496] the MINRES method. For symmetric (or Hermitian) matrices A this method implements the projection process given in case (3) of Theorem 2.3.1,

$$x_n \in x_0 + \mathcal{K}_n(A, r_0) \quad \text{and} \quad r_n \perp A\mathcal{K}_n(A, r_0). \tag{2.5.51}$$

The equivalent optimality property is

$$\|r_n\| = \min_{z \in x_0 + \mathcal{K}_n(A, r_0)} \|b - Az\|, \tag{2.5.52}$$

i.e. the MINRES method minimises the Euclidean norm of the residual over the affine Krylov subspace $x_0 + \mathcal{K}_n(A, r_0)$.

As shown in case (3) of Theorem 2.3.1, the projection process (2.5.51) is well defined for general nonsingular matrices $A \in \mathbb{F}^{N \times N}$. It can be implemented using the Arnoldi algorithm studied in Section 2.4.1. This leads to the GMRES method proposed by Saad and Schultz in 1986 [544]. After our description of GMRES below, we will indicate the possible simplification of the method that arises when A is Hermitian. Such simplification is essential, because it allows us to update the approximate solution recursively instead of storing all computed basis vectors of the Krylov subspace. This is exploited in the MINRES method developed by Paige and Saunders in 1975 [496]. The GMRES method, on the other hand, uses long recurrences. The same is true for the methods that are mathematically equivalent to GMRES, including the generalised conjugate residual (GCR) method [157, 156, 642] and the orthogonal direction (ORTHODIR) method [685], which both are older than GMRES; see [589] for an early survey of methods.

Our approach in this section is analogous to Section 2.4, where we first introduced the Arnoldi algorithm and then derived the Hermitian Lanczos algorithm as a special case. Note that in both cases we proceed contrary to the historical development: the Hermitian Lanczos algorithm is older than the Arnoldi algorithm, and MINRES was published more than a decade earlier than GMRES. In fact, Saad and Schultz presented their GMRES method as a generalisation of the MINRES method to nonsymmetric linear algebraic systems [544, p. 856].

An application of the Arnoldi algorithm to a (general, nonsingular) matrix $A \in \mathbb{F}^{N \times N}$ and the initial vector $v = r_0$ yields after n steps a matrix relation of the form

$$AV_n = V_{n+1}\underline{H}_{n,n}, \quad \underline{H}_{n,n} \equiv \begin{bmatrix} H_{n,n} \\ h_{n+1,n}e_n^T \end{bmatrix}; \tag{2.5.53}$$

see (2.4.3). Here it is assumed that n is less than the grade of v with respect to A. The columns of V_n form an orthonormal basis of $\mathcal{K}_n(A, r_0)$, and $H_{n,n}$ is an unreduced upper Hessenberg matrix.

Writing $x_n = x_0 + V_n t_n$ for some $t_n \in \mathbb{F}^n$, the orthogonality condition in (2.5.51) implies

$$0 = (AV_n)^*(r_0 - AV_n t_n) = V_n^* A^* r_0 - V_n^* A^* AV_n t_n$$
$$= \underline{H}_{n,n}^* (\|r_0\|e_1) - \underline{H}_{n,n}^* \underline{H}_{n,n} t_n$$

or, equivalently,

$$\underline{H}_{n,n}^* \underline{H}_{n,n} t_n = \underline{H}_{n,n}^* (\|r_0\| e_1),$$

which is the projected system corresponding to the projection process (2.5.51). The projected matrix is guaranteed to be nonsingular, and hence the uniquely determined solution is given by

$$t_n = (\underline{H}_{n,n}^* \underline{H}_{n,n})^{-1} \underline{H}_{n,n}^* (\|r_0\| e_1). \tag{2.5.54}$$

The matrix $(\underline{H}_{n,n}^* \underline{H}_{n,n})^{-1} \underline{H}_{n,n}^*$ is the *Moore–Penrose pseudoinverse* of $\underline{H}_{n,n}$.

A mathematically equivalent approach is based on the optimality property (2.5.52). We then get

$$\|r_n\| = \min_{t \in \mathbb{F}^n} \|r_0 - AV_n t\| = \min_{t \in \mathbb{F}^n} \|V_{n+1}(\|r_0\| e_1 - \underline{H}_{n,n} t)\|$$
$$= \min_{t \in \mathbb{F}^n} \| \|r_0\| e_1 - \underline{H}_{n,n} t \|, \tag{2.5.55}$$

where we have used (2.5.53) and the fact that the columns of V_{n+1} are orthonormal. The minimisation problem (2.5.55) is a least-squares problem that can be written as

$$\underline{H}_{n,n} t \approx \|r_0\| e_1. \tag{2.5.56}$$

The matrix $\underline{H}_{n,n} \in \mathbb{F}^{(n+1) \times n}$ has full rank n, and hence the uniquely determined solution $t = t_n$ again is given by (2.5.54).

A numerically stable way to solve least-squares problems is to use the QR factorisation of the given (rectangular) matrix. In the case of the upper Hessenberg matrix $\underline{H}_{n,n}$ this factorisation can be computed very efficiently. In their derivation of the GMRES method in [544], which uses the modified Gram–Schmidt version of the Arnoldi algorithm (Algorithm 2.4.2), Saad and Schultz proposed using a sequence of n Givens rotation matrices to annihilate the n subdiagonal entries of $\underline{H}_{n,n}$, i.e.

$$Q_{n+1} \underline{H}_{n,n} = \begin{bmatrix} R_{n,n} \\ 0 \end{bmatrix}, \tag{2.5.57}$$

$$Q_{n+1} = Q_{n+1,n} \cdots Q_{3,2} Q_{2,1}, \tag{2.5.58}$$

$$Q_{j+1,j} = \begin{bmatrix} I_{j-1} & & \\ & c_j & s_j \\ & -\bar{s}_j & c_j \\ & & & I_{n-j} \end{bmatrix}, \quad j = 1, \ldots, n, \tag{2.5.59}$$

where $c_j^2 + |s_j|^2 = 1$. The matrix $Q_{n+1} \in \mathbb{F}^{(n+1) \times (n+1)}$ is orthogonal (if $\mathbb{F} = \mathbb{R}$) or unitary (if $\mathbb{F} = \mathbb{C}$), and $R_{n,n} \in \mathbb{F}^{n \times n}$ is upper triangular and invertible. Note that in each step only one additional Givens rotation has to be computed. In particular,

$$Q_{n+1} = Q_{n+1,n} \begin{bmatrix} Q_n & 0 \\ 0 & 1 \end{bmatrix}, \tag{2.5.60}$$

where $Q_{n+1,n} \in \mathbb{F}^{(n+1)\times(n+1)}$ is chosen to annihilate the entry in the position $(n+1, n)$ of the matrix

$$\begin{bmatrix} Q_n & 0 \\ 0 & 1 \end{bmatrix} \underline{H}_{n,n} = \begin{bmatrix} Q_n \underline{H}_{n-1,n-1} & Q_n h^{(n)} \\ 0 & h_{n+1,n} \end{bmatrix}. \quad (2.5.61)$$

Here $h^{(n)} \equiv [h_{1,n}, \ldots, h_{n,n}]^T$ denotes the last column of $H_{n,n}$.

Using the QR factorisation (2.5.57)–(2.5.59) we get

$$(\underline{H}_{n,n}^* \underline{H}_{n,n})^{-1} \underline{H}_{n,n}^* = \begin{bmatrix} R_{n,n}^{-1} \mid 0 \end{bmatrix} Q_{n+1}.$$

Then the least-squares solution t_n in (2.5.54) is of the form

$$t_n = \begin{bmatrix} R_{n,n}^{-1} \mid 0 \end{bmatrix} (\|r_0\| Q_{n+1} e_1), \quad (2.5.62)$$

and the vector t_n can be computed by solving the $n \times n$ upper triangular system

$$R_{n,n} t_n = \|r_0\| [Q_{n+1} e_1]_{1:n},$$

where the right-hand side vector consists of the first n elements of $\|r_0\| Q_{n+1} e_1$.

Moreover, a simple calculation shows that the least-squares residual norm in (2.5.55) is given by

$$\|r_n\| = \min_{t \in \mathbb{F}^n} \| \|r_0\| e_1 - \underline{H}_{n,n} t \| = \|r_0\| \left| e_{n+1}^T Q_{n+1} e_1 \right|. \quad (2.5.63)$$

In words, the Euclidean norm of the nth residual is equal to the product of the Euclidean norm of the initial residual and the absolute value of the last entry in the first column of Q_{n+1}.

Algorithm 2.5.2 shows the main steps of the GMRES method as derived above; see [272, p. 41] and [636, Figure 6.1, p. 67] for a more detailed statement. The convergence criterion in Algorithm 2.5.2 is a user-specified tolerance for the residual norm. Since the residual norm can be computed via (2.5.63), we only need to compute t_n and the approximation x_n when the tolerance is satisfied. After determining x_n it is advisable to compute the norm of the residual using the explicit formula $\|b - Ax_n\|$. This verifies whether the iteratively updated residual norm has been computed accurately. Other useful convergence criteria in the GMRES context, in particular the normwise backward error, will be discussed in Chapter 5.

As mentioned above, Saad and Schultz [544] used the modified Gram–Schmidt version of the Arnoldi algorithm to derive the GMRES method. Other possibilities include the classical Gram–Schmidt version (Algorithm 2.4.1) and the orthogonalisation of the Krylov sequence using Householder transformations [654]. All these methods are mathematically equivalent but have different numerical properties. Details will be discussed in Chapter 5; see in particular Section 5.10.

In spite of the simple formula (2.5.60)–(2.5.61) for updating the QR factorisation of $\underline{H}_{n,n}$, work and storage requirements in the Arnoldi recurrence and thus

Algorithm 2.5.2 The GMRES method

Input: nonsingular matrix $A \in \mathbb{F}^{N \times N}$, right-hand side vector $b \in \mathbb{F}^N$, initial approximation $x_0 \in \mathbb{F}^N$, tolerance τ for the residual norm, maximal number of iterations $n_{\max}$.
Output: approximation x_n, after the algorithm has stopped.

Initialise: $r_0 = b - Ax_0$.

For $n = 1, 2, \ldots, n_{\max}$

Compute step n of the Arnoldi algorithm for A and $v = r_0$ to obtain $AV_n = V_{n+1}\underline{H}_{n,n}$; cf. (2.5.53).

Update the QR factorisation of $\underline{H}_{n,n}$ as in (2.5.57)–(2.5.59). Use the update formula for Q as in (2.5.60)–(2.5.61) and the update formula for $\|r_n\|$; see, e.g. [544] or [636, p. 67], [272, p. 41].

Use (2.5.63) to test if $\|r_n\| \leq \tau$. If yes, then compute t_n as in (2.5.62) and return the approximate solution $x_n = x_0 + V_n t_n$.

End

---

the GMRES method itself grows linearly with n. The situation is different when A is Hermitian. In this case, the upper Hessenberg matrix $H_{n,n}$ in the Arnoldi algorithm must be Hermitian as well. Hence the full Arnoldi recurrence reduces to a three-term recurrence, and the matrix relation (2.5.53) can be written as

$$AV_n = V_{n+1}\underline{T}_n, \quad \underline{T}_n \equiv \begin{bmatrix} T_n \\ \delta_{n+1}e_n^T \end{bmatrix}, \qquad (2.5.64)$$

where T_n is real, symmetric, and tridiagonal; see (2.5.44) and the description of the Hermitian Lanczos algorithm in Section 2.4. Implementing the projection process (2.5.51) or, equivalently, the minimisation property (2.5.52) using (2.5.64) leads to the least-squares problem (2.5.56) with $\underline{T}_n$ replacing $\underline{H}_{n,n}$. The SYMMLQ method of the previous section uses the LQ factorisation of $\underline{T}_n^T$; see (2.5.45). Using this factorisation here again leads to

$$t_n = \left(\underline{T}_n^T \underline{T}_n\right)^{-1} \underline{T}_n^T (\|r_0\| e_1) = \begin{bmatrix} L_n^{-T} \mid 0 \end{bmatrix} \left(\|r_0\| Q_{n+1}^T e_1\right).$$

The matrix L_n^{-T} is upper triangular, and thus the formula for t_n is analogous to (2.5.62).

Since the Lanczos basis vectors satisfy a three-term recurrence, it is possible to derive a short recurrence relation for the approximation x_n as well. To obtain the MINRES method of Paige and Saunders, one can consider

$$x_n = x_0 + V_n t_n = \left(V_n L_n^{-T}\right) \left(\|r_0\| [I_n \mid 0] Q_{n+1}^T e_1\right);$$

see [496, equations (6.6)–(6.9)]. As in the SYMMLQ method (cf. (2.5.47) and (2.5.48)), an update formula for x_n can be developed. A MATLAB implementation of the MINRES method is given in [183, p. 187], and details of the algorithm are given in [272, Algorithm 4, p. 44], [574, Fig. 1], and [636, Figure 6.9, p. 86].

2.5.6 Further Krylov Subspace Methods

Using the notation of the previous sections we will now briefly summarise the main ideas of three further Krylov subspace methods: FOM, BiCG, and QMR. There exist numerous other methods that are not mentioned in this section; see, e.g. the books [272, 452, 543, 636] or the survey papers [203, 545, 570]. The three methods discussed in this section have in common that for HPD or Hermitian matrices they are mathematically equivalent to CG or MINRES. They are typically used for solving linear algebraic systems with non-Hermitian matrices $A \in \mathbb{F}^{N \times N}$. An interesting comparison of the residual norms of all three methods with those of GMRES is given in [327].

FOM (FULL ORTHOGONALISATION METHOD)

This method was derived by Saad in 1981 [541] and stated again in the 1986 paper of Saad and Schultz [544, Algorithm 2] to motivate the development of GMRES. The method is sometimes referred to as Arnoldi's method for linear algebraic systems, and it is mathematically equivalent to the orthogonal residual (ORTHORES) method [685]. It uses the Arnoldi algorithm to implement the projection process of case (1) in Theorem 2.3.1. Hence FOM is mathematically equivalent to the CG method when A is HPD. Using the matrix formulation of the Arnoldi algorithm (see (2.4.3)), the projected system is given by

$$H_{n,n} t_n = \|r_0\| e_1.$$

If A is not HPD, then $H_{n,n}$ can be singular and the FOM method can break down.

The relation between FOM and GMRES has been explored in [85, 121]. It can be shown, for example, that $H_{n,n}$ is singular if and only if the two consecutive approximations x_{n-1} and x_n computed by GMRES are identical [85, Theorem 3.1]. In other words, the exact stagnation of GMRES is equivalent to a breakdown of FOM. Further details on this are given in Section 5.7.1.

BiCG (BI-CONJUGATE GRADIENT) METHOD

This method was proposed by Fletcher in 1975 [190] as a generalisation of CG to symmetric indefinite matrices. Fletcher's original derivation is based on the real version of the non-Hermitian Lanczos algorithm based on coupled two-term recurrences, which Lanczos had presented in 1950 [405]. Because of its 'biconjugacy condition' and its formal resemblance with the 'conjugate gradient algorithm' for solving linear algebraic systems with HPD matrices, Fletcher referred to Lanczos' algorithm as the 'biconjugate gradient algorithm'; see [190, p. 80].

Mathematically, the BiCG method implements the projection process

$$x_n \in x_0 + \mathcal{K}_n(A, r_0) \quad \text{and} \quad r_n \perp \mathcal{K}_n(A^*, \tilde{r}_0),$$

where $\tilde{r}_0$ is an arbitrary (nonzero) vector. If A is HPD and $\tilde{r}_0 = r_0$, then BiCG is mathematically equivalent to CG. On the other hand, if A is Hermitian and $\tilde{r}_0 = Ar_0$, then BiCG is mathematically equivalent to MINRES. Both equivalences were shown by Fletcher in [190].

Using the matrix form of the non-Hermitian Lanczos algorithm (see (2.4.8)), the projected system of the BiCG method is given by

$$T_{n,n} t_n = \|r_0\| e_1.$$

Using the LU factorisation $T_{n,n} = L_n U_n$, the nth approximation can be written as

$$x_n = x_0 + V_n t_n = x_0 + V_n T_{n,n}^{-1}(\|r_0\| e_1) = x_0 + (V_n U_n^{-1})(\|r_0\| L_n^{-1} e_1).$$

Analogously to the CG method (cf. (2.5.5)) this formula allows us to work out a simple update formula for x_n.

If A is not HPD, then $T_{n,n}$ may be singular and hence BiCG can break down. Yet, the method is popular in practical applications because it is based on three-term recurrences rather than the full Arnoldi recurrence used in the GMRES method. Detailed statements of the BiCG method can be found in Fletcher's original paper [190, equations (5.1a)–(5.1f) and (5.8)] and in [203, Algorithm 3.2], [206, Algorithm 5.1], [272, p. 80], [543, Algorithm 7.3, p. 223], and [636, Figure 7.1, p. 97]. The relationship of Fletcher's results to other works on Hermitian (symmetric) indefinite problems until the early 1980s has been studied, for example, in [589, 590].

QMR (QUASI-MINIMAL RESIDUAL) METHOD

This method was proposed by Freund and Nachtigal in 1991 [206]. Its nth approximation satisfies $x_n \in x_0 + \mathcal{K}_n(A, r_0)$, but in case of a non-Hermitian matrix A the method is *not* based on a projection process. It rather combines ideas from BiCG and GMRES to tackle the problem of breakdowns in BiCG. More precisely, Freund and Nachtigal use the matrix form of the non-Hermitian Lanczos algorithm (see (2.4.8)) and write the nth residual as

$$r_n = b - Ax_n = r_0 - AV_n t_n = r_0 - V_{n+1} \underline{T}_{n,n} t_n = V_{n+1} (\|r_0\| e_1 - \underline{T}_{n,n} t_n).$$

If V_{n+1} has orthonormal columns, then $\|r_n\| = \|\|r_0\| e_1 - \underline{T}_{n,n} t_n\|$, which is reminiscent of the GMRES minimisation problem (2.5.55). In the non-Hermitian Lanczos algorithm, however, the columns of V_{n+1} are not mutually orthogonal but orthogonal to the columns of W_{n+1}. Nevertheless, Freund and Nachtigal skip the matrix V_{n+1} and compute t_n by solving the minimisation problem

$$\min_{t \in \mathbb{F}^n} \| \|r_0\|e_1 - \underline{T}_{n,n} t\|.$$

This motivates the name *quasi-minimal residual method*. The QMR minimisation problem is a least-squares problem, and the matrix $\underline{T}_{n,n}$ has rank n. Thus, the uniquely determined solution $t = t_n$ is given by

$$t_n = (\underline{T}_{n,n}^* \underline{T}_{n,n})^{-1} \underline{T}_{n,n}^* (\|r_0\|e_1).$$

Similarly to the GMRES method, the QMR least-squares problem is solved by a QR factorisation of $\underline{T}_{n,n}$.

If the non-Hermitian Lanczos algorithm does not break down at step n, an upper bound on the nth QMR residual norm is given by

$$\|r_n^{QMR}\| \leq \sigma_{\max}(V_{n+1}) \, \| \|r_0\|e_1 - \underline{T}_{n,n} t_n\|,$$

where $\sigma_{\max}(V_{n+1})$ denotes the largest singular value of V_{n+1}. On the other hand, since the GMRES residual can also be expressed as a linear combination of the columns of V_{n+1}, we may write, for some $\widehat{t}_n \in \mathbb{F}^n$,

$$\begin{aligned}\|r_n^{GMRES}\| &= \|V_{n+1}(\|r_0\|e_1 - \underline{T}_{n,n} \widehat{t}_n)\| \\ &\geq \sigma_{\min}(V_{n+1}) \, \| \|r_0\|e_1 - \underline{T}_{n,n} \widehat{t}_n\| \\ &\geq \sigma_{\min}(V_{n+1}) \min_{t \in \mathbb{F}^n} \| \|r_0\|e_1 - \underline{T}_{n,n} t\| \\ &\geq \sigma_{\min}(V_{n+1}) \, \| \|r_0\|e_1 - \underline{T}_{n,n} t_n\|.\end{aligned}$$

Putting the two bounds together shows that the Euclidean norms of the QMR and GMRES residuals are related by

$$\|r_n^{QMR}\| \leq \kappa(V_{n+1}) \|r_n^{GMRES}\|, \tag{2.5.65}$$

where $\kappa(V_{n+1}) \equiv \sigma_{\max}(V_{n+1})/\sigma_{\min}(V_{n+1})$ is the condition number of V_{n+1}. This result is due to Nachtigal and was first published in his PhD thesis [468]. Its proof can also be found, for example, in Greenbaum's book [272, Theorem 5.3.1].

If A is Hermitian and $v_1 = w_1$, then the non-Hermitian Lanczos algorithm is mathematically equivalent to the Hermitian Lanczos algorithm, $V_{n+1} = W_{n+1}$ has orthonormal columns, QMR is mathematically equivalent to MINRES (or GMRES), and equality holds in (2.5.65). Relations between the residual norms of QMR and BiCG are studied in [121, Section 4] and [206, Section 5]. Algorithmic details of the QMR method are given in [543, Algorithm 7.4, p. 225] and [636, Figure 7.2, p. 101], [272, p. 83].

While the QMR method avoids one type of the BiCG-like breakdowns due to singularity of $T_{n,n}$, it still requires that the underlying non-Hermitian Lanczos

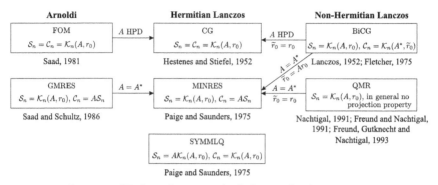

Figure 2.3 Overview of Krylov subspace methods discussed in Section 2.5. An arrow indicates that two methods are mathematically equivalent when the respective conditions are satisfied.

algorithm does not break down. As mentioned in Section 2.4.2, look-ahead versions of the non-Hermitian Lanczos algorithm have been introduced to improve this algorithm's numerical stability. The look-ahead idea has also been employed in the context of the QMR method; see [206] or [203, Section 3] for a survey. The role of the look-ahead technique is often misinterpreted as though it offered a tool to cure the general breakdown situation. There is no doubt that the technique is useful in many practical situations, where it avoids locally (i.e. for small values of n) very large values of the condition number $\kappa(V_{n+1})$ in (2.5.65). Nevertheless, the impact of look-ahead is limited. This objection is not based on the existence of the (rather rare) incurable breakdowns (see the survey of the history and the remarkable links in Parlett's paper [506]), but on the indisputable fact that in finite precision arithmetic short recurrences cannot preserve biorthogonality or even linear independence of the computed basis vectors. This problem is well demonstrated by the numerical behaviour of the Hermitian Lanczos algorithm; see Section 5.9.

An overview of Krylov subspace methods discussed in this section is shown in Figure 2.3. The methods are ordered in three groups according to the three different algorithms that are used in their respective derivations: Arnoldi, Hermitian Lanczos, and non-Hermitian Lanczos. The boxes show the projection properties of the different methods, and arrows indicate that two methods are mathematically equivalent when the respective condition is satisfied. The names and dates below each box show that method's original developer(s) and time of development. It is clear from this overview that, historically, the algorithmic development in the field of Krylov subspace methods took place in a rather complicated way.

2.5.7 Historical Note: A. N. Krylov and the Early History of Krylov Subspace Methods

Krylov subspace methods are named after the Russian scientist and navy general Aleksei Nikolaevich Krylov. Krylov was born on August 15, 1863, in a small village in Siberia, Russia, which today is named Krylovo in his honour. From September

1878 to May 1884 he attended the Maritime High School in St. Petersburg. He left the school as a midshipman in the Russian Navy. After four years in the compass division of the Russian Naval Hydrographic Department, Krylov entered the shipbuilding department of the St. Petersburg Naval Academy in October 1888. There he received his mathematical education from Aleksandr Nikolaevich Korkin, a student of Chebyshev. When Krylov graduated in October 1890, Korkin convinced him to stay on the staff of the Naval Academy. During the following fifty years Krylov taught numerous courses in engineering and mathematics at the Naval Academy.

Beside these teaching activities, Krylov had a successful and illustrious career in science, military, and politics. His major accomplishments include the Gold Medal of the (British) Royal Institutiton of Naval Architects (1898), full membership of the Russian Academy of Sciences (from 1916), and directorship of the Physics and Mathematics Institute of the USSR Academy of Sciences (1927–1932). The following quote from a book by Vucinich gives an indication of the wide range of Krylov's scientific interests [652, p. 358]:

> He was rightfully called an encyclopedist of naval arts and sciences: he was a mathematician, a shipbuilding engineer, a theoretician, an artillery expert, and a historian of science. In all these specialties, Krylov was first and foremost a practical seaman; however, one can also assert with some truth that in all his studies he wrote primarily as a mathematician.

Krylov travelled throughout Europe and maintained many contacts to leading contemporary scientists. He was a member of Russian (later Soviet), English, and French maritime societies. According to Vucinich [652, p. 359], under Krylov's influence the Royal Naval College at Greenwich, England, decided to instruct its students in the application of Chebyshev's quadrature rule to the structural mechanics of ships. Krylov was extremely well read, in particular in the classical works on mechanics and mathematics of the 17th to 19th centuries. One of his outstanding scientific achievements is his Russian translation (from the Latin original) of and extensive commentary on Newton's *Principia Mathematica*, which he published in 1915. An interesting account of Krylov's role (around 1921) in the publication of Euler's collected works is given in [488]. In 1936, Krylov published Chebyshev's lectures on *The Theory of Probability*, which Liapunov had taken down in 1879–1880. (These lectures were translated into English in 1999 [110].) Altogether Krylov wrote more than 500 articles and scientific papers, and several books. His collected works have been published in 12 volumes (1936–1956) by the Academy of Sciences of the USSR. Krylov died in St. Petersburg (then Leningrad) on October 26, 1945. More on his life and work can be found in [1], [652] (in particular pp. 358–361), and in his (Russian) autobiography [399].

Typical of Krylov's work was the application of mathematical methods in the solution of practical problems. His foremost object of study was the oscillations of ships. In this context he wrote his 1931 paper [398], which would later attach his name to the field of Krylov subspace methods. In the paper he derived a new method for solving the secular equation that determines the frequency of small vibrations in mechanical systems. He formulated this method in the context of linear differential equations with constant coefficients, and he referred to previous works of Laplace, Laguerre, and Leverrier inspired by celestial mechanics; see Remark 4.2.12 for mathematical details of Krylov's method.

An algebraic reformulation of the method was given by Luzin in 1931 [438] and by Gantmacher in 1934 [215] (see footnote 1 is Section 2.2). These papers were initially unknown to Lanczos, Hestenes, Stiefel and other scientists in the West who developed the first Krylov subspace methods in the early 1950s. According to a footnote in his paper of 1950, Lanczos learned from Ostrowski that his method 'parallels' earlier work of Krylov [398] and Luzin [438]. Lanczos added [405, p. 255]:

> On the basis of the reviews of these papers in the Zentralblatt, the author believes that the two methods coincide only in the point of departure. The author has not, however, read these Russian papers.

By the time he wrote this footnote, Lanczos was on the research staff at the Institute of Numerical Analysis (INA), which had been envisioned in Curtiss' 'Prospectus' (reprinted in 1989 [123]) and founded by the US National Bureau of Standards on the campus of the University of California at Los Angeles (UCLA) in July 1947. From February 1949 Lanczos headed the two research projects 'Determination of characteristic values of matrices' and 'Approximate solution of sets of arbitrary simultaneous algebraic equations' [473]. Later, in 1949, the two projects were revised and renamed 'Calculation of eigenvalues, eigenvectors, and eigenfunctions of linear operators' and 'Solution of sets of simultaneous algebraic equations and techniques for the inversion and iteration of matrices', respectively. Several researchers, including Rosser, Blanch, and Forsythe joined Lanczos in his investigations [474]. The projects were immensely successful; they not only led to the two papers of Lanczos in 1950 [405] (his method for solving eigenvalue problems) and 1952 [406] (his method for solving linear algebraic systems), but also to the 1952 paper of Hestenes and Stiefel [313] on the CG method. It is interesting to note that Lanczos also presented in 1952 another paper [407] (published in 1953 and sponsored in part by the Office of Naval Research), which was devoted to the use of Chebyshev polynomials in the solution of large-scale linear algebraic systems, and which in many aspects was truly visionary; see the historical note in Section 5.5.3.

The UCLA professor, Hestenes, was the 'UCLA representative' at INA. In June or early July 1951 he developed an iterative method for solving symmetric positive definite linear algebraic systems, which he called the 'conjugate gradient method'. Stiefel independently discovered the same method at the ETH in Zürich, Switzerland. He came to the INA in August 1951 to participate in a symposium on 'Simultaneous Linear Equations and the Determination of Eigenvalues' (the proceedings of this symposium have been published in [503]). Hestenes and Todd recall the following anecdote [314, p. 29]:

> When Stiefel arrived at INA from Switzerland, the librarian gave him a paper describing this [Hestenes'] routine. Shortly thereafter Stiefel came to Hestenes' office with this paper in hand and said, 'Look! This is my talk.'.

Stiefel remained at the INA for six months after the symposium, so that he and Hestenes could write a joint paper on their method. It was finished in May 1952. The history of the INA has been described in detail in [314]; see also [247, 614]. An interesting historical coincidence with Krylov's occupation is that Lanczos, Hestenes, and Stiefel acknowledged sponsorship (in part) by the Office of Naval Research of the United States Navy; see the footnotes on the first pages of [313, 405, 406].

A direct link between Krylov's method of 1931 and a number of more widely known methods (mainly for computing eigenvalues) was made in 1959, when Householder and Bauer [347] reviewed methods for expanding the characteristic polynomial of a (general) matrix A. They found that several methods, including the ones of Lanczos, Hestenes, and Stiefel, are closely related to Krylov's method in the sense that they all are based on the sequence of vectors

$$v_1, \quad v_2 = Av_2, \quad v_3 = Av_2, \quad \ldots$$

The paper of Householder and Bauer probably contains the first use of the term *Krylov sequence*; see [347, p. 33]. (It also contains an early use of the term *Hessenberg form* and a citation of Hessenberg's dissertation of 1941; cf. [311] and footnote 2 in Section 2.4.1 in this book.)

The attractive features of methods based on Krylov sequences, and in particular CG, were not fully realised in the two decades following the first developments. The roundoff properties of CG were poorly understood, it was unclear how to measure the method's convergence, and its termination was considered a major problem. Because of its finite termination property, CG was often considered a *direct method*, which for an $N \times N$ matrix should deliver the exact solution, or an acceptable approximation of it, after iterating exactly N steps. This point of view was underlined, for example, by Householder when he wrote in 1953 that CG 'terminates in n steps to yield the exact solution apart from round-off'. He added that 'the process can be continued beyond n steps for reduction of the round-off error' [345, p. 73]. It should be emphasised that Hestenes and Stiefel *did* understand CG as an iterative method. This will be discussed in detail in the historical note in Section 5.2.1.

Failures of CG were reported, for example, by Ginsburg in 1959 [163]. Similarly, in structural analysis computations performed in the late 1950s Livesley found that in the CG method

> ... rounding errors show a tendency to build up to such an extent that the solution after N steps is often a worse approximation to the correct solution than the starting point. This build-up appears to be due to the fact that some of the displacement components in a structure are often closely coupled, so that the matrix Y is rarely well-conditioned. It could probably be cured by double-length working in certain parts of the program, but this would still further increase the computing time. The method was therefore abandoned in favor of an elimination process. [430, p. 37]

At about the same time, Forsythe and Wasow pointed out in their book on finite-difference methods that the CG method 'has not been widely adopted for partial differential equations because of the relatively high storage requirements and the relatively complex structure of each iterative step' [194, p. 215]. Nevertheless, the references listed by Golub and O'Leary [247] show that CG was used quite frequently in the 1960s in the engineering community. This is also confirmed by the exposition on the CG method in the monograph of Saulyev [549]. Interest in the theory of the CG method among Russian mathematicians is best documented in the remarkable work of Vorobyev [651], which, paradoxically, was later almost forgotten; see Section 3.7 in this book.

Until the early 1970s, CG was treated with suspicion, particularly by mathematicians. Wilkinson found in 1971 that it 'cannot be too strongly emphasised that the procedure *cg* is seldom, if ever, to be preferred to Cholesky type algorithms on the grounds of accuracy'. However, he also remarked (in contrast to Forsythe and Wasow) that the method is optimal in terms of storage requirements, because 'the matrix A is, in general, not stored at all. All that is needed is an auxiliary procedure for determining Ax from a given vector x' [672, p. 4]. In modern large-scale applications this still is one of the main advantages of Krylov subspace methods. At the SIGNUM-SIAM Panel on the *25th Anniversary of Modern Numerical Analysis* held on 17 October 1972, (celebrating the 25th anniversary of the founding of the INA) Todd mentioned that the methods of Hestenes, Stiefel, and Lanczos had 'not been as successful as originally thought for the linear equation problem' [614, p. 364].

A turning point in the history of Krylov subspace methods was Reid's paper of 1971, in which he pointed out that the CG method 'has several very pleasant features when regarded not a direct method for the solution of full systems of equations but as an iterative method for the solution of large and sparse systems' [521, p. 231]. Reid gave convincing numerical examples in which sufficiently accurate solutions were obtained in much fewer than N iterations. Furthermore, he showed examples in which the CG method outmatched Chebyshev Iteration and Successive Overrelaxation (SOR), which by that time were considered state-of-the-art iterative methods (cf. the historical note in Section 5.5.3). He also studied different implementations of the algorithm, the use of 'recursive' versus. true residuals, and the effect of roundoff on the finite termination property. All of these topics would become active areas of research in the next decades.

Reid presented his results at the *Oxford Conference of the Institute of Mathematics and Its Applications* in April 1970. In the discussion session following Reid's talk, which is appended to the paper [521], Wilkinson asked: 'Engeli and Stiefel considered using n iterations plus one or two more, but you are advocating many less and this may explain the different conclusions. Have you tried very ill-conditioned matrices, such as Hilbert matrices?' This question as well as others (e.g. those by Fletcher, Curtiss, and Jennings) indicate how unfamiliar leading scientists of the time were with the original paper of Hestenes and Stiefel and with Reid's conclusions about the true iterative nature of CG.

Also in 1971, the PhD thesis of Paige [489] led to the rediscovery of Lanczos' method for large and sparse (Hermitian or symmetric) eigenvalue problems. Paige built up a groundbreaking theory that links convergence of the *computed* eigenvalue approximations to the loss of orthogonality of the Lanczos vectors. He proved that the accuracy of computed eigenvalue approximations can be guaranteed using easily computable *a posteriori* error bounds. Compared with the discussion following Reid's talk at the Oxford conference in 1970, Paige's PhD thesis and the subsequent papers [490, 491, 492] turned the state-of-the-art of the rounding error analysis of the Lanczos method, and, as shown later by Greenbaum [269] and others (see [457] and Chapter 5 of this book) also of the CG method, upside down. The finite precision behaviour of the Hermitian Lanczos and CG methods can not only be understood, it can actually be described by an elegant mathematical theory.

From the late 1970s, Krylov subspace methods for linear algebraic systems and eigenvalue problems became widely used. Their increasing popularity can be linked also to the pioneering work on preconditioning by Concus, Golub, O'Leary, van der Vorst, Axelsson, and others. The renewed interest from the mathematical community in the 1980s led to the methods' analysis in terms of their projection properties and the underlying subspaces. Apparently, the term *Krylov subspace* was introduced only then. One of the earliest formal definitions of a Krylov subspace was stated by Parlett in 1980 [508, Chapter 12].

2.6 SUMMARY AND OUTLOOK: LINEAR PROJECTIONS AND NONLINEAR BEHAVIOUR

As described in Section 2.1, the main idea of projection processes in general, and hence of Krylov subspace methods in particular, is to find an *approximate solution* of a potentially very large linear algebraic system, $Ax = b$, by solving a system of *much smaller dimensionality*. This smaller system is obtained by projecting the original system onto a suitable subspace. Intuitively, if the projection captures a sufficient part of the information contained in the original data (corresponding to the underlying application, its mathematical model, and the discretisation), then one can expect a sufficiently accurate approximate solution.

Krylov subspace methods use a *sequence of nested subspaces*, thus giving a sequence of approximate solutions. If some subspace eventually captures all information needed to solve the problem, and the method is well defined, then the projection process terminates with a solution of $Ax = b$. Hence well-defined Krylov subspace methods are *finite processes*.

From a computational point of view the finite character of Krylov subspace methods is, however, unimportant: If the original linear algebraic system has a very large size N, then the goal usually is to obtain a sufficiently accurate approximate solution x_n for $n \ll N$. That is why from a practical point of view Krylov subspace methods should be considered *iterative*. Lanczos nicely put it this way in 1952 [406, p. 40]:

> Even if theoretically speaking the last vector vanishes exactly only after N iterations, it is quite possible that it may drop *practically* below negligible bounds after a relatively few iterations.

In a similar way, Hestenes and Stiefel discussed stopping the CG method [313] (also see the historical note in Section 5.2.1).

On the other hand, from a mathematical point of view, the finite termination property is substantial. Krylov subspace methods use at any given iteration the information computed in all previous iterations. In order to achieve the finite termination property using the minimal number of steps (which is in most cases still very large and not used in practice), they anticipate, in some sense, the behaviour in the future iterations. Therefore the *theoretical* finite termination property affects the *practical* behaviour of the iterates from the very first iteration.

The speed of approximating the exact solution crucially depends on the amount of information transferred from the original problem into the projected systems. It therefore depends on the construction of the sequence of nested subspaces used in the projection process. Krylov subspace methods build up these subspaces using increasing powers of the operator (matrix) with respect to the given initial vector. The hope is that powering the operator tends to transfer dominant information as quickly as possible into the projected problem. If a sufficiently accurate approximation to the exact solution is determined by such dominant information present in the data, then appropriately implemented (in order to minimise effects of rounding errors) Krylov subspace methods can find it very efficiently. When no such dominance exists, Krylov subspace methods typically perform poorly. Since the operator powers represent nonlinearity, which in fact is increasing with the iteration number, Krylov subspace methods for solving *linear* algebraic systems via *linear* projections are by their nature *highly nonlinear*.

A closer look at the operator powers leads naturally to the idea of *moments* and the question of whether Krylov subspace methods have *moment matching properties*. This topic and several related issues are investigated in the next chapter.

3

Matching Moments and Model Reduction View

The essential point of the projection process (2.1.1)–(2.1.4) studied in Chapter 2 is that an approximation x_n of the solution x of the linear algebraic system $Ax = b$ of order N can be obtained from the *projected system*

$$C_n^* A S_n t_n = C_n^* r_0,$$

which is of order n. When we think of the data A, b, x_0 as a given and possibly large *model*, and $n < N$, the data $C_n^* A S_n, C_n^* r_0$ of the projected system can be interpreted as a *reduced model*, and the projection process itself as a special type of *model reduction*. The goal is to obtain an x_n 'close to' x for $n \ll N$.

The Krylov subspace idea brings in another aspect. Here the search and constraint spaces $\mathcal{S}_n$ and $\mathcal{C}_n$, whose bases form the columns of S_n and C_n, respectively, are built using powers of the matrix A. As an example, consider an HPD matrix A and the CG method, where $\mathcal{S}_n = \mathcal{C}_n = \mathcal{K}_n(A, r_0)$; cf. case (1) in Theorem 2.3.1. Then the jth column of the matrix $S_n = C_n$ is of the form $p_{j-1}(A) r_0$, where $p_{j-1}(\lambda)$ is a certain polynomial of degree $j - 1$, $j = 1, \ldots, n$. Therefore the entries of the projected matrix $C_n^* A S_n$ can be written as

$$c_i^* A s_j \;=\; \sum_{k=1}^{i+j-1} \alpha_k^{(ij)} r_0^* A^k r_0, \quad i, j = 1, \ldots, n,$$

where $\alpha_1^{(ij)}, \alpha_2^{(ij)}, \ldots, \alpha_{i+j-1}^{(ij)}$ are certain coefficients. It needs to be pointed out that the projected system does not occur explicitly in the CG method. Nevertheless, it implicitly determines the underlying mathematics.

The term $r_0^* A^k r_0$ is called the *kth moment* of the matrix A with respect to the initial residual r_0. Since the projected system is of order n, the projected matrix involves the first $2n - 1$ of these moments,

$$r_0^* A^k r_0, \quad k = 1, \ldots, 2n - 1.$$

The number $2n - 1$ (or $2n$ if we also count the zero moment $r_0^* r_0$) is of particular importance; it will occur frequently in this chapter.

As shown in Theorem 2.3.1, Krylov subspace methods are not 'just' projection methods, but they are additionally characterised by certain *optimality properties*. This naturally leads to the question whether these properties can be linked to *moment matching*. In this chapter we will explore this link in detail, and we will characterise some Krylov subspace methods in terms of *moment matching model reduction*, which explains the chapter's title.

Our characterisation will not be purely algebraic, i.e. in terms of matrices. Rather than that, we will explain deep links between Krylov subspace methods and classical topics of analysis. Working on this chapter (and possibly reading it, too) felt like a long ascent, starting from a simple version of the Stieltjes moment problem of 1894, and peaking with a unified overview of numerous relations, particularly between the Lanczos algorithm and CG on the one hand, and moments, quadrature, orthogonal polynomials, as well as continued fractions on the other. Many results on the latter topics (e.g. the algebraic degree of exactness of the Gauss quadrature and the interlacing property of the roots of orthogonal polynomials) are classical. In order to underline a unified point of view we have chosen to derive everything 'from scratch', with the goal of making the chapter as self-contained as possible.

Before we start we would like to mention that theoretical relationships as well as various applications of matrices, moments, and quadrature are the subject of a recent monograph by Golub and Meurant [246], which can be used as a complementary reference for a part of the material of this chapter. Thorough presentations of the correspondence between descriptions based on moments, polynomials, continued fractions, and matrices can be found also, e.g. in [256] and in the monograph [87]. Our exposition and focus is, however, different from [246, 256, 87]. Most of all, this chapter is influenced by the approach presented by Vorobyev in his monograph [651] from 1958 (original Russian edition) and 1965 (English translation). The essence is given in the following quote from the Preface of both editions:

> Thus, in recent years, a vast number of diverse iterative methods have been developed. This book will present but one class of methods based on a variational principle and closely related to the Chebyshev-Markov classical problem of moments. However, these methods are distinguished by the broad group of problems to which they may be applied and by the rapidity of convergence of the successive approximations.
>
> A general formulation of the problem and the use of functional analysis has enabled all these methods to be consolidated into a single method of moments.

Understanding the moment matching properties of individual methods, including the number of moments that are matched and the way the moments are defined (using either orthogonal projections or oblique projections linked with some auxiliary subspaces), may help in understanding the performance of the methods when applied to particular groups of problems. Concerning the applicability of the method of moments Vorobyev pointed out in [651, p. 113]:

> ... the fact that the method of moments is invariant to the choice of coordinate system permits its use in constructing an approximate solution when the number of degrees of freedom is very great and solving the equations exactly is difficult.

Some biographical information about Vorobyev is given in the introductory part of Section 3.7, which contains a detailed study of Vorobyev's formulation of the moment problem and its application to Krylov subspace methods.

3.1 STIELTJES' MOMENT PROBLEM

In his classical paper of 1894 [588], Stieltjes formulated the following *problem of moments*.

Determine a (real) bounded and non-decreasing function $\omega(\lambda)$ defined on the half-line $[0, \infty)$, such that its *moments*

$$\int_0^\infty \lambda^k d\omega(\lambda), \quad k = 0, 1, 2, \ldots$$

are equal to a set of prescribed numbers $\xi_k, k = 0, 1, 2, \ldots$, i.e. such that

$$\int_0^\infty \lambda^k d\omega(\lambda) = \xi_k, \quad k = 0, 1, 2, \ldots; \qquad (3.1.1)$$

see [588, Sections 24 and 51–54]. The integral with respect to $\omega(\lambda)$ is a *Riemann–Stieltjes integral*; see, e.g. [129, Section 1.6.5] or [391, Section 36] for modern definitions and basic properties.

In [588, Sections 24, 38, and 50], Stieltjes interpreted the non-decreasing function $\omega(\lambda)$ as a *mass distribution* on $[0, \infty)$. Then for any given $a, b \in \mathbb{R}, 0 \leq a < b < \infty$, the integral

$$\int_a^b d\omega(\lambda) = \omega(b) - \omega(a) \qquad (3.1.2)$$

represents the mass contained in the interval $[a, b]$. As pointed out by Shohat and Tamarkin [561, p. vii], this is why $\omega(\lambda)$ was (and still is) referred to as a *distribution function*. The first and the second moment of $\omega(\lambda)$, given by

$$\int_0^\infty \lambda \, d\omega(\lambda) \quad \text{and} \quad \int_0^\infty \lambda^2 d\omega(\lambda),$$

can be interpreted as the mechanical moments with respect to zero of the total mass

$$\int_0^\infty d\omega(\lambda)$$

distributed over the semi-axis $[0, +\infty)$. The first moment divided by the total mass gives the centre of mass, the second gives the centre of inertia. In this interpretation the goal of the Stieltjes moment problem is to find the distribution of positive mass on the half-line $\lambda \geq 0$, given the generalised mechanical moments of this distribution with respect to zero. Further details on the history and motivation of the moment problem are given in the historical note in Section 3.3.5 below.

In the following we will consider a simplified version of the Stieltjes moment problem. In this version we are given a finite real interval $[a, b]$ and a non-decreasing (not necessarily continuous) real function $\omega(\lambda)$ defined on $[a, b]$. We assume that $\omega(a) = 0$ and $\omega(b) = 1$, so that the integral (3.1.2) is equal to one. This means that the given distribution function $\omega(\lambda)$ is *normalised*. The number $\lambda_* \in (a, b]$ is called a *point of increase* of $\omega(\lambda)$, when $\omega(\lambda_*) > \omega(\lambda)$ holds for all $\lambda \in (\lambda_* - \epsilon, \lambda_*)$ and some $\epsilon > 0$. (In a more general way, a point of increase λ_* can be defined as a point in the neighborhood of which the function $\omega(\lambda)$ is not constant; cf. [588, Chapter VI, Sections 37 and 38], [607, p. 22] and [227, p. 3].) We denote the number of points of increase of $\omega(\lambda)$ by $n(\omega)$. Note that $n(\omega)$ may be infinite (even uncountable).

Problem 3.1.1 (**the simplified Stieltjes' moment problem**) *Let $\omega(\lambda)$ be a non-decreasing real function defined on a given finite real interval $[a, b]$ with $\omega(a) = 0$ and $\omega(b) = 1$, and let the moments of $\omega(\lambda)$ be denoted by*

$$\xi_k \equiv \int_a^b \lambda^k d\omega(\lambda), \quad k = 0, 1, 2, \ldots. \qquad (3.1.3)$$

For a given positive integer n we want to determine a non-decreasing real function $\omega^{(n)}(\lambda)$ on $[a, b]$ with n points of increase $\lambda_1^{(n)}, \ldots, \lambda_n^{(n)}$, where

$$a < \lambda_1^{(n)} < \lambda_2^{(n)} < \cdots < \lambda_n^{(n)} \leq b,$$

and associated positive weights $\omega_1^{(n)}, \ldots, \omega_n^{(n)}$, where $\sum_{j=1}^n \omega_j^{(n)} \equiv 1$, i.e.

$$\omega^{(n)}(\lambda) = \begin{cases} 0 & \text{if } a \leq \lambda < \lambda_1^{(n)}, \\ \sum_{j=1}^i \omega_j^{(n)} & \text{if } \lambda_i^{(n)} \leq \lambda < \lambda_{i+1}^{(n)}, \ i = 1, \ldots, n-1, \\ \sum_{j=1}^n \omega_j^{(n)} = 1 & \text{if } \lambda_n^{(n)} \leq \lambda \leq b, \end{cases} \qquad (3.1.4)$$

such that the first $2n$ moments of $\omega^{(n)}(\lambda)$ match the first $2n$ moments of the given function $\omega(\lambda)$, i.e. such that

$$\int_a^b \lambda^k d\omega^{(n)}(\lambda) = \xi_k, \quad k = 0, 1, \ldots, 2n-1. \qquad (3.1.5)$$

Remark 3.1.2

Our text makes links between different areas of mathematics, which all have their standard and established notation. Therefore in some cases a conflict of notation cannot be avoided. In particular, we use $[a,b]$ to denote the given finite real interval (as in the standard references [607, 217, 227]) as well as b to denote the right-hand side of the given linear algebraic system. Whenever the distribution function $\omega(\lambda)$ arises from such a system, it is defined on a finite real interval and the notation in (3.1.5) can be simplified to

$$\xi_k \equiv \int_{-\infty}^{+\infty} \lambda^k d\omega(\lambda) \equiv \int \lambda^k d\omega(\lambda).$$

We will use this simplification where appropriate.

We first note that if $\omega(\lambda)$ has $N < \infty$ points of increase,

$$a < \lambda_1 < \lambda_2 < \cdots < \lambda_N \leq b,$$

and the associated positive weights are $\omega_1, \ldots, \omega_N$, with $\sum_{j=1}^{N} \omega_j = 1$, i.e. if

$$\omega(\lambda) = \begin{cases} 0 & \text{if } a \leq \lambda < \lambda_1, \\ \sum_{j=1}^{i} \omega_j & \text{if } \lambda_i \leq \lambda < \lambda_{i+1}, \ i = 1, \ldots, N-1, \\ \sum_{j=1}^{N} \omega_j = 1 & \text{if } \lambda_N \leq \lambda \leq b, \end{cases} \quad (3.1.6)$$

then the distribution function $\omega(\lambda)$ itself trivially solves the simplified Stieltjes problem with $n = N$. If $\mathfrak{n}(\omega) = N < \infty$, we will therefore always consider $n \leq N$, where for $n = N$ a solution is already known.

For a closer look at Problem 3.1.1 we start with a brief summary of the most important definitions related to *mechanical quadrature*. The concepts in this context are standard and can be found in many books; see, e.g. [355, Section 7.0] or [607, Section 15.1]. Consider n distinct numbers $\mu_1, \ldots, \mu_n \in (a,b)$ and n real numbers $\zeta_1, \ldots, \zeta_n$. If $f(\lambda)$ is a Riemann–Stieltjes integrable function on $[a,b]$, we can write its integral as

$$I_\omega(f) \equiv \int_a^b f(\lambda) d\omega(\lambda) = \sum_{j=1}^{n} \zeta_j f(\mu_j) + E_\omega^n(f), \quad (3.1.7)$$

where $E_\omega^n(f)$ represents some error (or remainder). The term

$$I_\omega^n(f) \equiv \sum_{j=1}^{n} \zeta_j f(\mu_j) \quad (3.1.8)$$

is called an *n-node mechanical quadrature* for approximating the integral $I_\omega(f)$. The numbers $\mu_1, \ldots, \mu_n$ and $\zeta_1, \ldots, \zeta_n$ are the *nodes* and *weights* of the quadrature, respectively. Of course, the general goal is to choose nodes and weights that give a small error $E_\omega^n(f)$. When the error is equal to zero, the quadrature (3.1.8) is called *exact*. A quadrature that is exact for all polynomials of degree k, but not for all polynomials of larger degree, is said to have the *algebraic degree of exactness k*.

Let us come back to Problem 3.1.1. Since n is assumed finite, the Riemann–Stieltjes integral on the left-hand side of (3.1.5) is given by

$$\int_a^b \lambda^k d\omega^{(n)}(\lambda) = \sum_{j=1}^n \omega_j^{(n)} \left\{\lambda_j^{(n)}\right\}^k.$$

If the n distinct numbers $\lambda_1^{(n)}, \ldots, \lambda_n^{(n)} \in (a, b]$ and n positive numbers $\omega_1^{(n)}, \ldots, \omega_n^{(n)}$ (summing up to 1) solve the simplified Stieltjes' moment problem, then

$$\sum_{j=1}^n \omega_j^{(n)} \left\{\lambda_j^{(n)}\right\}^k = \int_a^b \lambda^k d\omega(\lambda), \quad k = 0, 1, \ldots, 2n - 1.$$

Since any polynomial of degree at most $2n - 1$ is a linear combination of the monomials $\lambda^k, k = 0, 1, \ldots, 2n - 1$, the last equation implies that

$$\sum_{j=1}^n \omega_j^{(n)} p\left(\lambda_j^{(n)}\right) = \int_a^b p(\lambda) \, d\omega(\lambda) \quad \text{(3.1.9)}$$

holds for any polynomial $p(\lambda)$ of degree at most $2n - 1$.

The previous considerations can be summarised as follows:

A solution of the simplified Stieltjes' moment problem yields an n-node mechanical quadrature for the given Riemann–Stieltjes integral with the distribution function $\omega(\lambda)$. This quadrature has positive weights and the algebraic degree of exactness $2n - 1$.

Conversely, if an n-node mechanical quadrature for the Riemann–Stieltjes integral with the distribution function $\omega(\lambda)$ has positive weights and the algebraic degree of exactness $2n - 1$, then it gives a solution of the simplified Stieltjes moment problem.

3.2 MODEL REDUCTION VIA ORTHOGONAL POLYNOMIALS

The main goal in this section is to derive the *Gauss–Christoffel quadrature*, and to show that it gives a solution of the simplified Stieltjes moment problem.

The theory of the Gauss quadrature for approximating the Riemann integral is classical and has been covered in many books; see, e.g. [111, Chapter I, Section 6], [225, Chapter 3.2], [355, Chapter 7], [393, Chapter VII, Sections A–C], [408, Chapter VI, Section 10], and [607, Chapters III and XV]. Our exposition follows some of these sources. It is, however, based on the Riemann–Stieltjes integral, and in agreement with Gautschi [221, 222, 227], we use the name *Gauss–Christoffel quadrature* to underline this generality.

The standard approach to the Gauss quadrature for the Riemann integral, and to the Gauss–Christoffel quadrature for the Riemann–Stieltjes integral is based on *Hermite interpolation*, and is attributed to Markov; see his original paper [445] and, e.g. [222, p. 82]. Instead of following this approach we will use a description based on *Lagrange interpolation*. This allows us to retain some free parameters in the error term of the quadrature rule, which will later be convenient in the evaluation of quadrature differences; see Chapter 5.

3.2.1 Derivation of the Gauss–Christoffel Quadrature

As in Section 3.1 we consider a non-decreasing distribution function $\omega(\lambda)$ on a finite real interval $[a,b]$, with $\omega(a) = 0$, $\omega(b) = 1$, and with $\mathfrak{n}(\omega)$ points of increase, where $\mathfrak{n}(\omega)$ may be infinite. Since $\omega(\lambda)$ is of bounded variation, the Riemann–Stieltjes integral

$$I_\omega(f) = \int_a^b f(\lambda)\,d\omega(\lambda) \qquad (3.2.1)$$

exists for any given continuous function $f(\lambda)$; see, e.g. [129, Section 1.6.5] or [391, Section 36].

We will now derive an *interpolatory quadrature* for approximating the integral (3.2.1). For a given positive integer $n \leq \mathfrak{n}(\omega)$ we consider n distinct nodes $\lambda_1^{(n)}, \ldots, \lambda_n^{(n)} \in (a,b]$, and we define the polynomial

$$q_n(\lambda) \equiv \left(\lambda - \lambda_1^{(n)}\right) \cdots \left(\lambda - \lambda_n^{(n)}\right).$$

Then the *Newton–Lagrange polynomial* interpolating the function $f(\lambda)$ at the points $\lambda_1^{(n)}, \ldots, \lambda_n^{(n)}$ can be written as

$$\mathcal{L}_n(\lambda) = \sum_{j=1}^n f\left(\lambda_j^{(n)}\right) \frac{q_n(\lambda)}{q_n'\left(\lambda_j^{(n)}\right)\left(\lambda - \lambda_j^{(n)}\right)}. \qquad (3.2.2)$$

By construction, $\mathcal{L}_n(\lambda)$ is of degree at most $n-1$. Note that

$$\frac{q_n\left(\lambda_i^{(n)}\right)}{q_n'\left(\lambda_j^{(n)}\right)\left(\lambda_i^{(n)} - \lambda_j^{(n)}\right)} = \delta_{ij}, \quad \text{(Kronecker's } \delta),$$

and thus indeed

$$\mathcal{L}_n\left(\lambda_i^{(n)}\right) = f\left(\lambda_i^{(n)}\right), \quad \text{for } i = 1, \ldots, n.$$

Moreover, we can write

$$f(\lambda) = \mathcal{L}_n(\lambda) + q_n(\lambda) f\left[\lambda_1^{(n)}, \ldots, \lambda_n^{(n)}, \lambda\right],$$

where $f[\lambda_1^{(n)}, \ldots, \lambda_n^{(n)}, \lambda]$ is the *nth divided difference* of $f(\lambda)$ with respect to the $n+1$ arguments $\lambda_1^{(n)}, \ldots, \lambda_n^{(n)}, \lambda$. If $f(\lambda)$ is continuous in $[a,b]$, then the divided difference $f[\lambda_1^{(n)}, \ldots, \lambda_n^{(n)}, \lambda]$ is well defined for each λ distinct from $\lambda_1^{(n)}, \ldots, \lambda_n^{(n)}$. If additionally $f'(\lambda)$ is continuous in $[a,b]$, then there exists a (unique) extension so that $f[\lambda_1^{(n)}, \ldots, \lambda_n^{(n)}, \lambda]$ becomes a continuous function of $\lambda \in [a,b]$. This generalises to higher-order derivatives of $f(\lambda)$. These and further results from the theory of divided differences can be found in [355, Section 6.1] and [461, Chapter 1].

Integrating both sides of the last equation, and using (3.2.2) gives

$$I_\omega(f) = \int_a^b \mathcal{L}_n(\lambda)\, d\omega(\lambda) + \int_a^b q_n(\lambda) f\left[\lambda_1^{(n)}, \ldots, \lambda_n^{(n)}, \lambda\right] d\omega(\lambda)$$

$$\equiv \sum_{j=1}^n \omega_j^{(n)} f\left(\lambda_j^{(n)}\right) + E_\omega^n(f). \qquad (3.2.3)$$

The *n-node interpolatory quadrature* for approximating the Riemann–Stieltjes integral (3.2.1) is then defined by

$$I_\omega^n(f) \equiv \sum_{j=1}^n \omega_j^{(n)} f\left(\lambda_j^{(n)}\right), \quad \text{where} \qquad (3.2.4)$$

$$\omega_j^{(n)} \equiv \int_a^b \frac{q_n(\lambda)}{q_n'\left(\lambda_j^{(n)}\right)\left(\lambda - \lambda_j^{(n)}\right)} d\omega(\lambda), \quad \text{for } j = 1, \ldots, n. \qquad (3.2.5)$$

We stress that the weights $\omega_1^{(n)}, \ldots, \omega_n^{(n)}$ in the interpolatory quadrature are uniquely determined by the choice of the nodes $\lambda_1^{(n)}, \ldots, \lambda_n^{(n)}$. The error is given by

$$E_\omega^n(f) = I_\omega(f) - I_\omega^n(f) = \int_a^b q_n(\lambda) f\left[\lambda_1^{(n)}, \ldots, \lambda_n^{(n)}, \lambda\right] d\omega(\lambda). \qquad (3.2.6)$$

If the function $f(\lambda)$ is n-times continuously differentiable in $[a, b]$, then

$$f\left[\lambda_1^{(n)}, \ldots, \lambda_n^{(n)}, \lambda\right] = \frac{f^{(n)}(\zeta(\lambda))}{n!},$$

where $\zeta(\lambda)$ is some number in the interior of the smallest closed interval containing the points $\lambda_1^{(n)}, \ldots, \lambda_n^{(n)}, \lambda$; see, e.g. [355, Section 6.1, Corollary 2, p. 252] or [461, p. 5]. In particular, if $f(\lambda)$ is a polynomial of degree at most $n-1$, then its nth derivative is the zero polynomial. Hence for *any* choice of n distinct nodes $\lambda_1^{(n)}, \ldots, \lambda_n^{(n)} \in (a, b]$, the algebraic degree of exactness of the n-node interpolatory quadrature (3.2.4)–(3.2.5) is at least $n-1$. This can also be concluded from (3.2.3) by noticing that for any polynomial of degree at most $n-1$, the interpolating polynomial $\mathcal{L}_n(\lambda)$ is equal to the polynomial itself.

On the other hand, if an n-node mechanical quadrature (3.1.8) has an algebraic degree of exactness of at least $n-1$, then this quadrature must be an interpolatory quadrature of the form (3.2.4)–(3.2.5). This can easily be seen as follows. Let $\mu_1, \ldots, \mu_n \in (a, b]$ be the nodes and $\zeta_1, \ldots, \zeta_n$ be the weights of an n-node mechanical quadrature, and let $q_n(\lambda) \equiv (\lambda - \mu_1) \cdots (\lambda - \mu_n)$. If the mechanical quadrature is exact for polynomials of degree $n-1$, then

$$\int_a^b \frac{q_n(\lambda)}{q_n'(\mu_j)(\lambda - \mu_j)} d\omega(\lambda) = \sum_{k=1}^n \zeta_k \frac{q_n(\mu_k)}{q_n'(\mu_j)(\mu_k - \mu_j)} = \zeta_j, \quad j = 1, \ldots, n,$$

so that the weights of the quadrature indeed are of the form (3.2.5). We summarise these considerations in a lemma.

Lemma 3.2.1
An n-node mechanical quadrature of the form (3.1.8) has the algebraic degree of exactness at least $n - 1$ if and only if it is an n-node interpolatory quadrature of the form (3.2.4)–(3.2.5).

The question asked and answered by Gauss in his fundamental paper of 1814 [219] was: can we increase the algebraic degree of exactness of an interpolatory quadrature by a special choice of the nodes? Note that for *any* choice of $n < \mathfrak{n}(\omega)$ distinct nodes we have

$$I_\omega(q_n^2) = \int_a^b q_n^2(\lambda)\, d\omega(\lambda) > 0, \quad \text{but} \quad I_\omega^n(q_n^2) = \sum_{j=1}^n \omega_j^{(n)} q_n^2\left(\lambda_j^{(n)}\right) = 0.$$

Thus, we cannot expect that the algebraic degree of exactness is $2n$ or larger. In the following, however, we will show that for each $n < \mathfrak{n}(\omega)$ there exists a choice of n nodes so that the resulting interpolatory quadrature has the algebraic degree of exactness $2n - 1$.

For a given positive integer $n < \mathfrak{n}(\omega)$, where $\mathfrak{n}(\omega)$ may be finite or infinite, let

$$\psi_0(\lambda) \equiv 1, \psi_1(\lambda), \ldots, \psi_n(\lambda) \tag{3.2.7}$$

be the first $n + 1$ *monic orthogonal polynomials* corresponding to the inner product (on the space of polynomials of degree at most n)

$$(\phi, \psi) \equiv \int_a^b \phi(\lambda)\psi(\lambda)\, d\omega(\lambda), \tag{3.2.8}$$

with the associated norm given by

$$\|\psi\|^2 \equiv (\psi, \psi) \equiv \int_a^b \psi^2(\lambda)\, d\omega(\lambda). \tag{3.2.9}$$

The uniquely defined polynomials (3.2.7) can be generated by orthogonalising the sequence of monomials $1, \lambda, \ldots, \lambda^n$ with respect to the inner product (3.2.8) using the Gram–Schmidt method, i.e.

$$\psi_0(\lambda) \equiv 1, \quad \text{and} \quad \psi_i(\lambda) = \lambda^i - \sum_{j=0}^{i-1} \frac{(\lambda^i, \psi_j)}{\|\psi_j(\lambda)\|^2} \psi_j(\lambda), \quad i = 1, 2, \ldots, n.$$

Note that if the distribution function $\omega(\lambda)$ has $N < \infty$ points of increase, i.e. if it is given by (3.1.6), then

$$(\phi, \psi) = \sum_{j=1}^N \omega_j \phi(\lambda_j)\psi(\lambda_j),$$

so that the inner product (3.2.8) is *discrete*.

By construction, $\psi_n(\lambda)$ is orthogonal to any polynomial of degree at most $n-1$. If $\phi(\lambda)$ is any monic polynomial of degree n, then $\widehat{\phi}(\lambda) \equiv \phi(\lambda) - \psi_n(\lambda)$ is of degree at most $n-1$, and the orthogonality of $\psi_n(\lambda)$ to $\widehat{\phi}(\lambda)$ gives

$$\|\phi(\lambda)\|^2 = \|\psi_n(\lambda) + \widehat{\phi}(\lambda)\|^2 = \|\psi_n(\lambda)\|^2 + \|\widehat{\phi}(\lambda)\|^2 \geq \|\psi_n(\lambda)\|^2. \tag{3.2.10}$$

Equality holds in (3.2.10) if and only if $\|\widehat{\phi}(\lambda)\| = 0$, or, equivalently, $\phi(\lambda) = \psi_n(\lambda)$. We thus have proved the following minimisation property of the monic orthogonal polynomials.

Lemma 3.2.2
In the previous notation, the polynomial $\psi_n(\lambda)$ is the unique monic polynomial of degree n that minimises the norm (3.2.9).

(Note that this result trivially holds for $n = 0$.)

Remark 3.2.3
The assertion of Lemma 3.2.2 easily generalises from the special inner product (3.2.8) and the associated norm (3.2.9) to any inner product on a space of polynomials. More precisely, whenever a sequence of monic polynomials is orthogonal with respect to some given inner product, then these polynomials minimise the corresponding norm in the sense of Lemma 3.2.2. This is another instance of the equivalence of *orthogonality* and *optimality*, which we have met previously in Theorems 2.3.1 and 2.3.2.

Now consider the subtle special case when $\omega(\lambda)$ has $N < \infty$ points of increase $a < \lambda_1 < \lambda_2 < \cdots < \lambda_N \leq b$, and $n = N$. We write $\psi_N(\lambda)$ as the result of orthogonalising λ^N with respect to the monic polynomials $\psi_0(\lambda), \ldots, \psi_{N-1}(\lambda)$, i.e.

$$\psi_N(\lambda) = \lambda^N - \sum_{j=0}^{N-1} \frac{(\lambda^N, \psi_j)}{\|\psi_j(\lambda)\|^2} \psi_j(\lambda), \tag{3.2.11}$$

so that

$$(\psi_N, \psi_i) = 0, \quad i = 0, 1, \ldots, N-1. \tag{3.2.12}$$

We will show that, although (3.2.8) is not definite on the space of polynomials of degree at most N (and hence does not define an inner product on this space), $\psi_N(\lambda)$ is uniquely determined by (3.2.12). Indeed, let $\widetilde{\psi}(\lambda)$ be a monic polynomial of degree N satisfying (3.2.12). Then

$$(\psi_N - \widetilde{\psi}, \psi_i) = 0, \quad i = 0, 1, \ldots, N-1.$$

Since $\psi_N(\lambda) - \widetilde{\psi}(\lambda)$ is a polynomial of degree at most $N-1$, the definiteness of the inner product (3.2.8) on the (sub)space of polynomials of degree at most $N-1$ implies $\psi_N(\lambda) - \widetilde{\psi}(\lambda) = 0$, and therefore $\psi_N(\lambda) = \widetilde{\psi}(\lambda)$. With the choice

$$\psi_N(\lambda) = (\lambda - \lambda_1) \cdots (\lambda - \lambda_N),$$

the condition (3.2.12) is clearly satisfied. Therefore, the roots of the unique monic orthogonal polynomial determined by (3.2.12) are equal to $\lambda_1, \ldots, \lambda_N$.

The following lemma states one of the basic properties of the roots of the monic orthogonal polynomials; cf. [607, Theorem 3.3.1]. The proof given here follows an analogous proof in [111, Chapter I, Section 5, Theorem 5.2].

Lemma 3.2.4
In the previous notation, consider a positive integer $n \leq \mathrm{n}(\omega)$ (where $\mathrm{n}(\omega)$ may be finite or infinite). Then the n roots $\lambda_1^{(n)}, \ldots, \lambda_n^{(n)}$ of the monic orthogonal polynomial $\psi_n(\lambda)$ are distinct and located in the interval $(a, b]$. If $n < \mathrm{n}(\omega)$, then the roots of $\psi_n(\lambda)$ are located in the open interval (a, b).

Proof
If $n = \mathrm{n}(\omega) \equiv N < \infty$, then, as shown above, the N roots of $\psi_N(\lambda)$ are equal to the N distinct points of increase of $\omega(\lambda)$, which are located in $(a, b]$.

Otherwise (if $n < \mathrm{n}(\omega)$) the equation

$$0 = (\psi_n, \psi_0) = (\psi_n, 1) = \int_a^b \psi_n(\lambda)\, d\omega(\lambda)$$

implies that $\psi_n(\lambda)$ must have at least one root in the open interval (a, b). Denote by $\zeta_1, \ldots, \zeta_k$ the $k \leq n$ distinct roots of $\psi_n(\lambda)$ of odd multiplicity that are located in the open interval (a, b). Then the polynomial

$$(\lambda - \zeta_1) \cdots (\lambda - \zeta_k)\, \psi_n(\lambda)$$

has no roots of odd multiplicity in (a, b), and therefore it does not change its sign in (a, b). Since $n < \mathrm{n}(\omega)$, we must have

$$\int_a^b (\lambda - \zeta_1) \cdots (\lambda - \zeta_k)\, \psi_n(\lambda)\, d\omega(\lambda) \neq 0. \qquad (3.2.13)$$

If $k < n$, the orthogonality of $\psi_n(\lambda)$ to $(\lambda - \zeta_1) \cdots (\lambda - \zeta_k)$ implies that the integral (3.2.13) is equal to zero, which is a contradiction. Therefore $k = n$, so that $\psi_n(\lambda)$ has n distinct roots in the open interval (a, b). □

If $n < \mathrm{n}(\omega)$, then the n roots $\lambda_1^{(n)}, \ldots, \lambda_n^{(n)}$ of $\psi_n(\lambda)$ are distinct and located in (a, b). Hence these roots may be taken as the n distinct nodes in the interpolatory quadrature (3.2.4)–(3.2.5). As shown next, this yields the *uniquely defined n-node mechanical quadrature that has the algebraic degree of exactness* $2n - 1$.

Lemma 3.2.5
In the previous notation, consider a positive integer $n < \mathrm{n}(\omega)$ (where $\mathrm{n}(\omega)$ may be finite or infinite). An n-node mechanical quadrature has the algebraic degree of exactness $2n - 1$ if and only if it is an interpolatory quadrature of the form (3.2.4)–(3.2.5) with the n nodes given by the roots of $\psi_n(\lambda)$.

Proof
If an n-node mechanical quadrature has the algebraic degree of exactness $2n - 1$, it must be an interpolatory quadrature (cf. Lemma 3.2.1). Let $\lambda_1^{(n)}, \ldots, \lambda_n^{(n)} \in (a, b)$ be the n nodes of this interpolatory quadrature. We will show that the polynomial

$$q_n(\lambda) \equiv \left(\lambda - \lambda_1^{(n)}\right) \cdots \left(\lambda - \lambda_n^{(n)}\right)$$

is equal to $\psi_n(\lambda)$. Let $p(\lambda)$ be any polynomial of degree at most $n-1$. Then the product $q_n(\lambda)p(\lambda)$ is of degree at most $2n-1$, and since the quadrature has the algebraic degree of exactness $2n-1$, we get

$$(q_n, p) = \int_a^b q_n(\lambda)p(\lambda)\, d\omega(\lambda) = \sum_{j=1}^n \omega_j^{(n)} \underbrace{q_n\left(\lambda_j^{(n)}\right)}_{=0} p\left(\lambda_j^{(n)}\right) = 0,$$

so that $q_n(\lambda)$ is orthogonal to all polynomials of degree at most $n-1$ and indeed $q_n(\lambda) = \psi_n(\lambda)$.

Conversely, let $f(\lambda)$ be an arbitrary polynomial of degree $2n-1$. By the division theorem for polynomials,[1] there exist uniquely defined polynomials $p(\lambda)$ and $r(\lambda)$, both of degree at most $n-1$, such that

$$f(\lambda) = \psi_n(\lambda)p(\lambda) + r(\lambda).$$

Note that $(\psi_n, p) = 0$ and $f(\lambda_j^{(n)}) = r(\lambda_j^{(n)}), j = 1, \ldots, n$. Now consider the interpolatory quadrature (3.2.4)–(3.2.5) with nodes given by the n roots of $\psi_n(\lambda)$. We know from Lemma 3.2.1 that its algebraic degree of exactness is at least $n-1$, and thus

$$\int_a^b f(\lambda)\, d\omega(\lambda) = \int_a^b r(\lambda)\, d\omega(\lambda) = \sum_{j=1}^n \omega_j^{(n)} r\left(\lambda_j^{(n)}\right) = \sum_{j=1}^n \omega_j^{(n)} f\left(\lambda_j^{(n)}\right).$$

Hence the algebraic degree of exactness is in fact $2n-1$. □

We now come to the definition of the *Gauss–Christoffel quadrature*.

Definition 3.2.6: *In the previous notation, the n-node Gauss–Christoffel quadrature is the uniquely defined interpolatory quadrature (3.2.4)–(3.2.5) that has the algebraic degree of exactness $2n-1$.*

The 'magic' of the Gauss–Christoffel quadrature is that it is exact for polynomials of degree $2n-1$, although it is based on n nodes only. We can give an alternative characterisation of this fact. The n-node quadrature based on the roots $\lambda_1^{(n)}, \ldots, \lambda_n^{(n)}$ of $\psi_n(\lambda)$ is as accurate as if it had been based on the $2n$ nodes $\lambda_1^{(n)}, \ldots, \lambda_n^{(n)}, \mu_1, \ldots, \mu_n$, where $\mu_1, \ldots, \mu_n$ are n *arbitrary* distinct nodes in (a, b). In other words, one gets n nodes 'for free'. This elegant result is stated in the following theorem (cf. [487, Theorem 1]), which is a slight generalisation of a result found in [355, Section 7.3, Theorem 3, p. 329] (also cf. equation (6) on p. 334 of that book).

---

1. See Theorem 4.2.2 in Chapter 4 of this book for a statement of this theorem in the case of complex polynomials. The theorem in fact holds for polynomials over any field.

Theorem 3.2.7

In the previous notation, consider a positive integer $n < \mathfrak{n}(\omega)$ (where $\mathfrak{n}(\omega)$ may be finite or infinite), let $\psi_n(\lambda) = (\lambda - \lambda_1^{(n)}) \cdots (\lambda - \lambda_n^{(n)})$ be the nth monic orthogonal polynomial with respect to the inner product (3.2.8), and let $\mu_1, \ldots, \mu_n \in (a, b)$ be n arbitrary distinct points.

If the function $f(\lambda)$ is twice continuously differentiable in $[a, b]$, then for any $m = 1, \ldots, n$ the error of the Gauss–Christoffel quadrature based on the n nodes $\lambda_1^{(n)}, \ldots, \lambda_n^{(n)}$ is given by

$$E_\omega^n(f) \equiv I_\omega(f) - I_\omega^n(f) = \int_a^b \psi_n(\lambda) f\left[\lambda_1^{(n)}, \ldots, \lambda_n^{(n)}, \lambda\right] d\omega(\lambda) = \quad (3.2.14)$$

$$\int_a^b \psi_n(\lambda)(\lambda - \mu_1) \cdots (\lambda - \mu_m) f\left[\lambda_1^{(n)}, \ldots, \lambda_n^{(n)}, \mu_1, \ldots, \mu_m, \lambda\right] d\omega(\lambda), \quad (3.2.15)$$

where $f[\lambda_1^{(n)}, \ldots, \lambda_n^{(n)}, \mu_1, \ldots, \mu_m, \lambda]$ is the $(n + m)$th divided difference of the function $f(\lambda)$ with respect to the nodes $\lambda_1^{(n)}, \ldots, \lambda_n^{(n)}, \mu_1, \ldots, \mu_m, \lambda$.

Proof

Recall that the equations in (3.2.14) hold for any n-node interpolatory quadrature; see (3.2.6). The assertion to be shown here is the equality of (3.2.14) with (3.2.15).

First assume that the n distinct points $\mu_1, \ldots, \mu_n$ are distinct from any of the nodes $\lambda_1^{(n)}, \ldots, \lambda_n^{(n)}$. The Gauss–Christoffel quadrature with the n nodes $\lambda_1^{(n)}, \ldots, \lambda_n^{(n)}$ gives

$$\int_a^b f(\lambda) d\omega(\lambda) = \sum_{j=1}^n \omega_j^{(n)} f\left(\lambda_j^{(n)}\right) + E_\omega^n(f), \quad (3.2.16)$$

$$E_\omega^n(f) = \int_a^b \psi_n(\lambda) f\left[\lambda_1^{(n)}, \ldots, \lambda_n^{(n)}, \lambda\right] d\omega(\lambda), \quad (3.2.17)$$

where the continuity of $f'(\lambda)$ guarantees the finiteness of the divided difference as λ varies. On the other hand, for each fixed $m \leq n$, the interpolatory quadrature based on the $n + m$ nodes $\lambda_1^{(n)}, \ldots, \lambda_n^{(n)}, \mu_1, \ldots, \mu_m$ gives

$$\int_a^b f(\lambda) d\omega(\lambda) = \sum_{j=1}^n \widehat{\omega}_j^{(n)} f\left(\lambda_j^{(n)}\right) + \sum_{i=1}^m \zeta_i f(\mu_i) + \widehat{E}_\omega^{n+m}(f), \quad (3.2.18)$$

$$\widehat{E}_\omega^{n+m}(f) = \int_a^b \psi_n(\lambda) p_m(\lambda) f\left[\lambda_1^{(n)}, \ldots, \lambda_n^{(n)}, \mu_1, \ldots, \mu_m, \lambda\right] d\omega(\lambda), \quad (3.2.19)$$

where $p_m(\lambda) \equiv (\lambda - \mu_1) \cdots (\lambda - \mu_m)$. The definition of the weights of an interpolatory quadrature (see (3.2.5)), and the orthogonality of $\psi_n(\lambda)$ to all polynomials of degree at most $n - 1$ shows that

$$\zeta_i = \frac{1}{(\psi_n \cdot p_m)'(\mu_i)} \int_a^b \psi_n(\lambda) \frac{p_m(\lambda)}{\lambda - \mu_i} d\omega(\lambda) = 0, \quad i = 1, \ldots, m.$$

Thus, each additional node μ_i yields a zero weight ζ_i in (3.2.18). In both formulas (3.2.16) and (3.2.18) the error term vanishes when the integrand is a monomial λ^k for $k = 0, 1, \ldots, n-1$ (even up to $k = 2n-1$, but $k \leq n-1$ is sufficient here). This implies that both quadratures have the algebraic degree of exactness at least $n-1$ (in fact $2n-1$). By Lemma 3.2.5 their weights are determined by (3.2.5), so that $\widehat{\omega}_j^{(n)} = \omega_j^{(n)}$ for $j = 1, \ldots, n$, and the assertion is proved.

If some μ_i is equal to some $\lambda_j^{(n)}$, then one needs to consider the Hermite instead of the Lagrange interpolant of $f(\lambda)$. This is somewhat more technical, but the above approach can be applied in an analogous way. We refer to [461, Sections 1.8 and 7.1] and [355, Sections 6.1 and 7.3] for details on divided differences with multiple arguments and corresponding expressions for the error term of the Gauss quadrature. These directly translate to the Gauss–Christoffel case. Note that now the continuity of $f''(\lambda)$ is required to make the divided differences well defined. □

If we take $\mu_j = \lambda_j^{(n)}$, $j = 1, \ldots, n$, then the error term (3.2.15) of the Gauss–Christoffel quadrature becomes

$$E_\omega^n(f) = \int_a^b \psi_n^2(\lambda) f\left[\lambda_1^{(n)}, \lambda_1^{(n)}, \ldots, \lambda_n^{(n)}, \lambda_n^{(n)}, \lambda\right] d\omega(\lambda). \quad (3.2.20)$$

If, moreover, $f(\lambda)$ is $(2n)$-times continuously differentiable on $[a, b]$, then

$$f\left[\lambda_1^{(n)}, \lambda_1^{(n)}, \ldots, \lambda_n^{(n)}, \lambda_n^{(n)}, \lambda\right] = \frac{f^{(2n)}(\zeta(\lambda))}{(2n)!},$$

where $\zeta(\lambda)$ is some number in the interior of the smallest closed interval containing the points $\lambda_1^{(n)}, \ldots, \lambda_n^{(n)}, \lambda$; see, e.g. [355, Section 6.1, Corollary 2, p. 252]. Since $\psi_n^2(\lambda)$ is non-negative and continuous, and $f[\lambda_1^{(n)}, \lambda_1^{(n)}, \ldots, \lambda_n^{(n)}, \lambda_n^{(n)}, \lambda]$ is a continuous function of λ, the mean value theorem for integrals yields

$$E_\omega^n(f) = \frac{f^{(2n)}(\zeta)}{(2n)!} \int_a^b \psi_n^2(\lambda) d\omega(\lambda), \quad (3.2.21)$$

where ζ is some number between a and b. This widely known expression for the error of the Gauss–Christoffel quadrature was originally derived by Markov in 1885 [445]. The derivative of degree $2n$ of any polynomial of degree at most $2n-1$ is equal to zero, which gives another proof that the Gauss–Christoffel quadrature has the algebraic degree of exactness $2n-1$ (recall that only an algebraic degree of exactness of at least $n-1$ was required in the proof of Theorem 3.2.7).

3.2.2 Moment Matching and Convergence Properties of the Gauss–Christoffel Quadrature

Let $\lambda_1^{(n)}, \ldots, \lambda_n^{(n)}$ denote the n distinct roots of the monic orthogonal polynomial $\psi_n(\lambda)$. Then for each $j = 1, \ldots, n$ the corresponding function

$$\left(\frac{\psi_n(\lambda)}{\psi_n'\left(\lambda_j^{(n)}\right)\left(\lambda - \lambda_j^{(n)}\right)}\right)^2$$

is a polynomial of degree $2n - 2$ that is non-negative in $[a, b]$. An application of the Gauss–Christoffel quadrature, which is exact for this polynomial, gives

$$0 < \int_a^b \left(\frac{\psi_n(\lambda)}{\psi_n'\left(\lambda_j^{(n)}\right)\left(\lambda - \lambda_j^{(n)}\right)} \right)^2 d\omega(\lambda)$$

$$= \omega_j^{(n)} + \sum_{\substack{i=1 \\ i \neq j}}^n \omega_i^{(n)} \left(\frac{\psi_n\left(\lambda_i^{(n)}\right)}{\psi_n'\left(\lambda_j^{(n)}\right)\left(\lambda_i^{(n)} - \lambda_j^{(n)}\right)} \right)^2$$

$$= \omega_j^{(n)}.$$

Thus, the weights $\omega_1^{(n)}, \ldots, \omega_n^{(n)}$ of the Gauss–Christoffel quadrature are positive. These weights are sometimes referred to as the *Christoffel numbers*; see, e.g. [607, Section 3.4]. An application of the Gauss–Christoffel quadrature to the constant polynomial $p(\lambda) \equiv 1$ gives

$$\omega_1^{(n)} + \cdots + \omega_n^{(n)} = \int_a^b d\omega(\lambda).$$

The assumption $\omega(a) = 0$ and $\omega(b) = 1$ implies that the integral on the right-hand side is equal to one.

Since the Gauss–Christoffel quadrature has positive weights and the algebraic degree of exactness $2n - 1$, we have arrived at the following essential result.

Theorem 3.2.8
For each positive integer $n < \mathfrak{n}(\omega)$ (where $\mathfrak{n}(\omega)$ may be infinite), the n distinct nodes $\lambda_1^{(n)}, \ldots, \lambda_n^{(n)} \in (a, b)$ (given by the roots of $\psi_n(\lambda)$), and the corresponding n positive weights $\omega_1^{(n)}, \ldots, \omega_n^{(n)}$ (given by (3.2.5) and summing up to 1) of the Gauss–Christoffel quadrature uniquely define a distribution function $\omega^{(n)}(\lambda)$ that solves the simplified Stieltjes' moment problem (i.e. Problem 3.1.1).

A main consequence of Theorem 3.2.8 can be formulated as follows:

The n-node Gauss–Christoffel quadrature represents a *matching moments model reduction*. The original model is given by the distribution function $\omega(\lambda)$ with $\mathfrak{n}(\omega) \leq \infty$ points of increase, and the reduced model by the distribution function $\omega^{(n)}(\lambda)$ with $n < \mathfrak{n}(\omega)$ points of increase given by the roots of the nth monic orthogonal polynomial $\psi_n(\lambda)$. The reduced model matches the first $2n$ moments of the original model.

We will now briefly look at the convergence properties of the Gauss–Christoffel quadrature.

For simplicity, let $\omega(\lambda)$ be a nondecreasing and continuous function on the finite real interval $[a, b]$ with $\omega(a) = 0$ and $\omega(b) = 1$. Then $\mathfrak{n}(\omega)$ is infinite, and the

Gauss–Christoffel quadrature is defined for any integer n. Let $f(\lambda)$ be a continuous function on $[a, b]$. By the classical approximation theorem of Weierstrass [664], $f(\lambda)$ can be approximated arbitrarily closely by polynomials. More precisely, for any $\epsilon > 0$ there exists an integer $n_* = n_*(f, \epsilon)$ and a polynomial $p_*(\lambda)$ of degree at most n_*, such that

$$|f(\lambda) - p_*(\lambda)| < \epsilon, \quad \text{for all } \lambda \in [a, b];$$

see, e.g. [355, Chapter 5, Section 1] for a proof.

For a given $\epsilon > 0$, consider the corresponding integer n_* and the n-node Gauss–Christoffel quadrature, where $2n - 1 > n_*$. Then the error satisfies

$$|E_\omega^n(f)| = |I_\omega(f) - I_\omega^n(f)|$$

$$= \left|(I_\omega(f) - I_\omega(p_*)) + \left(I_\omega(p_*) - I_\omega^{(n)}(p_*)\right) + \left(I_\omega^{(n)}(p_*) - I_\omega^{(n)}(f)\right)\right|$$

$$\leq |I_\omega(f) - I_\omega(p_*)| + \underbrace{\left|I_\omega(p_*) - I_\omega^{(n)}(p_*)\right|}_{=0} + \left|I_\omega^{(n)}(p_*) - I_\omega^{(n)}(f)\right|$$

$$\leq \int_a^b \underbrace{|f(\lambda) - p_*(\lambda)|}_{<\epsilon} d\omega(\lambda) + \sum_{j=1}^n \omega_j^{(n)} \underbrace{\left|p_*\left(\lambda_j^{(n)}\right) - f\left(\lambda_j^{(n)}\right)\right|}_{<\epsilon}$$

$$< 2\epsilon.$$

Note that in the second inequality we require that the weights $\omega_1^{(n)}, \ldots, \omega_n^{(n)}$ be non-negative (they are in fact positive). Since ϵ was arbitrary, we have shown that the Gauss–Christoffel quadrature converges to the exact value of the Riemann–Stieltjes integral, i.e.

$$\lim_{n \to \infty} I_\omega^n(f) = I_\omega(f).$$

This result was first obtained by Stieltjes in 1884 [585].

Remark 3.2.9
The above proof of convergence applies to any interpolatory quadrature (3.2.4)–(3.2.5) that has non-negative weights. More generally, one can show that if a mechanical quadrature (3.1.8) with non-negative weights converges for all polynomials, then it converges for all functions $f(\lambda)$ for which the integral $I_\omega(f)$ exists; see [607, Section 15.2] for details.

Now consider a slight change of notation, and let λ be any fixed (real) number *outside* the interval $[a, b]$. Then the function

$$f(u) \equiv \frac{1}{\lambda - u}$$

is continuous in $[a, b]$, and the convergence result for the Gauss–Christoffel quadrature shows that

$$\lim_{n \to \infty} \sum_{j=1}^n \frac{\omega_j^{(n)}}{\lambda - \lambda_j^{(n)}} = \int_a^b \frac{d\omega(u)}{\lambda - u}, \qquad (3.2.22)$$

where $\lambda_1^{(n)}, \ldots, \lambda_n^{(n)}$ are the roots of $\psi_n(\lambda)$, and $\omega_1^{(n)}, \ldots, \omega_n^{(n)}$ are the corresponding positive weights. Note that if $\omega(\lambda)$ has $\mathfrak{n}(\omega) = N < \infty$ points of increase, then trivially

$$\sum_{j=1}^{N} \frac{\omega_j^{(N)}}{\lambda - \lambda_j^{(N)}} = \int_a^b \frac{d\omega(u)}{\lambda - u}.$$

We will see in Section 3.3 that the expressions in (3.2.22) are closely related to yet another fundamental mathematical concept, namely the *continued fractions*.

Before continuing our mathematical development we will pause for a description of some of the history of the Gauss quadrature.

3.2.3 Historical Note: Gauss' Fundamental Idea, an Early Application, and Later Developments

Mechanical quadratures that are accurate for polynomials up to some given degree are closely linked with polynomial interpolation of functions. The basic idea had already been formulated by Newton. The *n*-node mechanical quadrature (for the Riemann integral) that is exact for polynomials up to degree $2n - 1$ was published by Gauss in 1814 [219]. Gautschi rates Gauss' discovery 'as one of the most significant events of the 19th century in the field of numerical integration and perhaps in all of numerical analysis' [222, p. 78]. In 1826 Jacobi reformulated Gauss' results using a more general mathematical language [356]. In particular, Jacobi explicitly used the idea of *orthogonality*; see the historical note on orthogonality and three-term recurrences in Section 3.3.6, as well as the beautiful exposition in the historical monograph [240], and the comments in [222, 384, 393] and [607, Chapter 3]. Both Gauss in [219] and Jacobi in [356] referred to Newton and Cotes. Jacobi mentioned the 1723 edition of Newton's *Principia Mathematica* [479] and Cotes's *Harmonia Mensurarum* [120] (posthumously published in 1722) as the original sources. The four essential works of Newton, Cotes, Gauss, and Jacobi on interpolation and quadrature were edited and republished together (in German) in 1917 by Kowalewski [394].

The Gauss quadrature had already been used successfully by 19th century astronomers. One of the earliest applications, which is described also in [394, pp. 100–101], was made by Encke, who was a student of Gauss (on and off between 1811 and 1816) and the first Director of the (new) Berlin Observatory, established in 1835. He reported the following in the 1863 edition of the *Berlin Astronomical Yearbook* ('Berliner Astronomisches Jahrbuch') [162, p. 124]:

> During the first five years of the new Berlin Observatory, my then assistant Herr Prof. Galle, now director of the Breslau Observatory, observed the barometer reading and the temperature reading three times daily between sunrise and sundown, at times which were one ninth of a day later than sunrise and earlier than sundown, as well as close to midday. This way he achieved the same accuracy as if he had made five observations daily in equal time intervals,

which in fact was confirmed when the average temperature and barometer readings were derived from his observations and compared with those made for several years.

(Our translation.)

Where is the application of the Gauss quadrature? Consider the linear distribution function $\omega(\lambda) = \lambda$ on the interval $[0, 1]$ and the corresponding inner product

$$(\phi, \psi) \equiv \int_0^1 \phi(\lambda)\psi(\lambda)\,d\lambda.$$

The first four monic orthogonal polynomials with respect to this inner product are given by $\psi_0(\lambda) = 1$, and

$$\psi_1(\lambda) = \lambda - \frac{1}{2}, \quad \psi_2(\lambda) = \lambda^2 - \lambda + \frac{1}{6}, \quad \psi_3(\lambda) = \lambda^3 - \frac{3}{2}\lambda^2 + \frac{3}{5}\lambda - \frac{1}{20}.$$

These polynomials (as well as $\psi_4(\lambda), \ldots, \psi_7(\lambda)$) were explicitly stated by Gauss in Section 22 of his 1814 paper [219]. Gauss also stated that the three roots of $\psi_3(\lambda)$ are

0.1127016653792583, 0.5, 0.8872983346207417.

A computation with MATLAB 7 (executing the command roots([1, -3/2, 3/5, -1/20])) yields

0.112701665379258, 0.5, 0.887298334620742.

Thus, Gauss' computations agree with the MATLAB results up to 16 digits!

In the Berlin Observatory, Encke's assistant Galle used Gauss' data as follows: Supposedly for practical reasons, the roots of $\psi_3(\lambda)$ were approximated by

$$\frac{1}{9}, \quad \frac{1}{2}, \quad \frac{8}{9}.$$

The interval $[0, 1]$ represented the time from sunrise to sundown, and thus measurements were made 'one ninth after sunrise', at midday, and 'one ninth before sundown'. Assume, for simplicity, that the temperature is well approximated by a polynomial of degree $2n - 1 = 5$. Then the Gauss quadrature gives the average temperature for this approximation exactly, and so does an interpolatory quadrature using six arbitrary distinct nodes. Encke and Galle found that the observations at the $n = 3$ times determined by the (approximated) roots of $\psi_3(\lambda)$ led to results as accurate as though they had made five observations at other (for example equally spaced) times.

The theory of the Gauss quadrature was extended by Christoffel in 1858 [113] and 1877 [114]. Christoffel also generalised the Gauss quadrature to weighted integrals with general non-negative and integrable weight functions. For modern surveys of these developments see [129, Section 2.7] and [222]. An interesting contemporary description of the history of mechanical quadrature is contained in the 1881 book [304, I. Theil] of Heine.

3.3 ORTHOGONAL POLYNOMIALS AND CONTINUED FRACTIONS

In this section we continue the development that led us to (3.2.22), and we will link the expressions in this equation to orthogonal polynomials and continued fractions. We start with the derivation of some further classical results on orthogonal polynomials.

3.3.1 Three-term Recurrence and the Interlacing Property

In addition to the monic orthogonal polynomials $\psi_0(\lambda), \psi_1(\lambda), \ldots, \psi_n(\lambda)$ introduced in (3.2.7), we will now consider their *normalised* counterparts

$$\varphi_0(\lambda) \equiv 1, \varphi_1(\lambda), \ldots, \varphi_n(\lambda), \qquad (3.3.1)$$

i.e. $\|\varphi_n\| = 1$. These polynomials can be generated by applying the Gram–Schmidt method to the recursively generated sequence of polynomials

$$\varphi_0(\lambda) \equiv 1, \lambda\varphi_0(\lambda) = \lambda, \lambda\varphi_1(\lambda), \ldots, \lambda\varphi_{n-1}(\lambda).$$

Since the inner product (3.2.8) satisfies

$$(\lambda\phi, \psi) = (\phi, \lambda\psi),$$

the nth step of the Gram–Schmidt method is

$$\begin{aligned}
\delta_{n+1}\varphi_n(\lambda) &= \lambda\varphi_{n-1}(\lambda) - \sum_{j=0}^{n-1}(\lambda\varphi_{n-1}, \varphi_j)\,\varphi_j(\lambda) \\
&= \lambda\varphi_{n-1}(\lambda) - \sum_{j=0}^{n-1}(\varphi_{n-1}, \lambda\varphi_j)\,\varphi_j(\lambda) \\
&= \lambda\varphi_{n-1}(\lambda) - (\varphi_{n-1}, \lambda\varphi_{n-1})\,\varphi_{n-1}(\lambda) - (\varphi_{n-1}, \lambda\varphi_{n-2})\,\varphi_{n-2}(\lambda) \\
&= \lambda\varphi_{n-1}(\lambda) - (\lambda\varphi_{n-1}, \varphi_{n-1})\,\varphi_{n-1}(\lambda) - \delta_n\,\varphi_{n-2}(\lambda), \qquad (3.3.2)
\end{aligned}$$

where the constant δ_{n+1} is used for the normalisation, the inner products $(\varphi_{n-1}, \lambda\varphi_0), \ldots, (\varphi_{n-1}, \lambda\varphi_{n-3})$ vanish since $\varphi_{n-1}(\lambda)$ is orthogonal to all polynomials of degree $n-2$ or less, and $(\varphi_{n-1}, \lambda\varphi_{n-2}) = \delta_n$.

The orthogonalisation step represents a *three-term recurrence*, because the new polynomial $\delta_{n+1}\varphi_n(\lambda)$ is determined using only $\varphi_{n-1}(\lambda)$ and $\varphi_{n-2}(\lambda)$. This recurrence is also known as the *Stieltjes recurrence*. It can be written as

$$\delta_{n+1}\varphi_n(\lambda) = (\lambda - \gamma_n)\varphi_{n-1}(\lambda) - \delta_n\varphi_{n-2}(\lambda), \quad n = 1, 2, \ldots, \qquad (3.3.3)$$

where $\delta_1 \equiv 0, \varphi_{-1}(\lambda) \equiv 0, \varphi_0(\lambda) \equiv 1$,

$$\gamma_n \equiv (\lambda\,\varphi_{n-1}, \varphi_{n-1}), \qquad (3.3.4)$$

and

$$\delta_{n+1} \equiv \|(\lambda - \gamma_n)\varphi_{n-1}(\lambda) - \delta_n\varphi_{n-2}(\lambda)\|. \tag{3.3.5}$$

The orthogonalisation step in (3.3.2) may be rearranged as

$$\begin{aligned}\delta_{n+1}\varphi_n(\lambda) &= (\lambda\varphi_{n-1}(\lambda) - \delta_n\varphi_{n-2}(\lambda)) - (\lambda\varphi_{n-1}, \varphi_{n-1})\varphi_{n-1}(\lambda) \\ &= \lambda\varphi_{n-1}(\lambda) - \delta_n\varphi_{n-2}(\lambda) \\ &\quad - (\lambda\varphi_{n-1} - \delta_n\varphi_{n-2}, \varphi_{n-1})\,\varphi_{n-1}(\lambda),\end{aligned} \tag{3.3.6}$$

where in the second term we have used that $(\lambda\varphi_{n-1} - \delta_n\varphi_{n-2}, \varphi_{n-1}) = (\lambda\varphi_{n-1}, \varphi_{n-1})$ due to the orthogonality of $\varphi_{n-2}(\lambda)$ and $\varphi_{n-1}(\lambda)$. This recurrence corresponds to the *modified Gram–Schmidt orthogonalisation* that should be used in practical implementations; cf. Algorithm 2.4.2 and its discussion. In practical computations, (3.3.3)–(3.3.5) should therefore be replaced by the mathematically equivalent formulas

$$\delta_{n+1}\varphi_n(\lambda) = (\lambda\varphi_{n-1}(\lambda) - \delta_n\varphi_{n-2}(\lambda)) - \gamma_n\varphi_{n-1}(\lambda), \tag{3.3.7}$$

$$\gamma_n = (\lambda\varphi_{n-1} - \delta_n\varphi_{n-2}, \varphi_{n-1}), \tag{3.3.8}$$

$$\delta_{n+1} = \|(\lambda\varphi_{n-1}(\lambda) - \delta_n\varphi_{n-2}(\lambda)) - \gamma_n\varphi_{n-1}(\lambda)\|, \tag{3.3.9}$$

for $n = 1, 2, \ldots$, where $\delta_1 \equiv 0, \varphi_{-1}(\lambda) \equiv 0, \varphi_0(\lambda) \equiv 1$.

Denoting by $\Phi_n(\lambda)$ a column vector having the orthonormal polynomials $\varphi_0(\lambda) \equiv 1, \ldots, \varphi_{n-1}(\lambda)$ as its entries,

$$\Phi_n(\lambda) = [\varphi_0(\lambda), \ldots, \varphi_{n-1}(\lambda)]^T,$$

the cumulative (row by row) matrix form of the Stieltjes recurrence (3.3.3) can be written as

$$\lambda\,\Phi_n(\lambda) = T_n\Phi_n(\lambda) + \delta_{n+1}\varphi_n(\lambda)\,e_n, \tag{3.3.10}$$

where the recurrence coefficients form the *Jacobi matrix*

$$T_n \equiv \begin{bmatrix} \gamma_1 & \delta_2 & & \\ \delta_2 & \gamma_2 & \ddots & \\ & \ddots & \ddots & \delta_n \\ & & \delta_n & \gamma_n \end{bmatrix}, \quad \delta_j > 0, \ j = 2, \ldots, n; \tag{3.3.11}$$

cf. Definition 2.4.1, and see the historical note in Section 3.4.3 for the origin and early applications of Jacobi matrices.

Let $\Psi_n(\lambda)$ be the column vector with the monic orthogonal counterparts to $\varphi_0(\lambda), \ldots, \varphi_{n-1}(\lambda)$ as its entries,

$$\Psi_n(\lambda) = [\psi_0(\lambda), \ldots, \psi_{n-1}(\lambda)]^T.$$

Since the entries of $\Psi_n(\lambda)$ are scalar multiples of the entries of $\Phi_n(\lambda)$, there exists an invertible diagonal matrix $\Upsilon_n \equiv \mathrm{diag}(\varsigma_1, \ldots, \varsigma_n)$ such that

$$\Psi_n(\lambda) = \Upsilon_n \Phi_n(\lambda).$$

Multiplying (3.3.10) with Υ_n gives

$$\lambda \Upsilon_n \Phi_n(\lambda) = \Upsilon_n T_n \Upsilon_n^{-1} \Upsilon_n \Phi_n(\lambda) + \delta_{n+1} \frac{\varsigma_n}{\varsigma_{n+1}} \varsigma_{n+1} \varphi_n(\lambda) e_n,$$

so that the matrix form of the three-term recurrence for the monic orthogonal polynomials can be written as

$$\lambda \Psi_n(\lambda) = \widehat{T}_n \Psi_n(\lambda) + \psi_n(\lambda) e_n. \tag{3.3.12}$$

Here the super-diagonal of $\widehat{T}_n \equiv \Upsilon_n T_n \Upsilon_n^{-1}$ must consist of ones (since the entries of $\Psi_n(\lambda)$ are monic), and

$$\delta_{n+1} \frac{\varsigma_n}{\varsigma_{n+1}} = 1.$$

This immediately gives

$$\varsigma_1 = 1, \; \varsigma_2 = \delta_2, \ldots, \; \varsigma_{n+1} = \prod_{j=2}^{n+1} \delta_j,$$

and hence

$$\widehat{T}_n = \begin{bmatrix} \gamma_1 & 1 & & \\ \delta_2^2 & \gamma_2 & \ddots & \\ & \ddots & \ddots & 1 \\ & & \delta_n^2 & \gamma_n \end{bmatrix}. \tag{3.3.13}$$

The three-term recurrence for the monic orthogonal polynomials is therefore given by

$$\psi_n(\lambda) = (\lambda - \gamma_n)\psi_{n-1}(\lambda) - \delta_n^2 \psi_{n-2}(\lambda), \quad n = 1, 2, \ldots, \tag{3.3.14}$$

with $\delta_1 \equiv 0$, $\psi_{-1}(\lambda) \equiv 0$, $\psi_0(\lambda) \equiv 1$, and

$$\psi_n(\lambda) = \varphi_n(\lambda) \prod_{j=2}^{n+1} \delta_j, \quad \delta_{n+1} = \frac{\|\psi_n\|}{\|\psi_{n-1}\|}. \tag{3.3.15}$$

In particular,

$$\psi_1(\lambda) = \lambda - \gamma_1 = \delta_2 \varphi_1(\lambda), \quad \psi_2(\lambda) = (\lambda - \gamma_2)(\lambda - \gamma_1) - \delta_2^2 = \delta_2 \delta_3 \varphi_2(\lambda).$$

We will next prove the *interlacing property* for the roots of a sequence of orthogonal polynomials.

Theorem 3.3.1
In the previous notation, let

$$\psi_0(\lambda) \equiv 1, \; \psi_1(\lambda), \ldots, \psi_n(\lambda), \ldots$$

be the sequence of the monic orthogonal polynomials corresponding to the inner product (3.2.8). Let $\lambda_1^{(n)} < \lambda_2^{(n)} < \cdots < \lambda_n^{(n)}$ be the ordered roots of $\psi_n(\lambda)$, and suppose that n is smaller than the number of points of increase of $\omega(\lambda)$, i.e. $n < \mathfrak{n}(\omega)$.

Then for any given integer m with $n + 1 \leq m < \mathfrak{n}(\omega)$, there is at least one root of $\psi_m(\lambda)$ contained in any of the $n + 1$ open intervals

$$\left(a, \lambda_1^{(n)}\right), \quad \left(\lambda_j^{(n)}, \lambda_{j+1}^{(n)}\right), \; j = 1, \ldots, n-1, \quad \left(\lambda_n^{(n)}, b\right).$$

If $\omega(\lambda)$ has N points of increase $\lambda_1 < \lambda_2 < \cdots < \lambda_N$ and $m = N$, then the statement remains valid except for $\lambda_N^{(N)} = \lambda_N \in (\lambda_n^{(n)}, b]$.

Proof
Assume that $m < \mathfrak{n}(\omega)$. Using Lemma 3.2.4, the roots of $\psi_n(\lambda)$ are distinct and located in the open interval (a, b). We will prove the rest of the statement by contradiction. First consider, for simplicity of notation, that $n \geq 2$, and suppose that, for some integer j between 1 and $n - 1$, no root of $\psi_m(\lambda)$ is contained in the open interval $(\lambda_j^{(n)}, \lambda_{j+1}^{(n)})$. Since $\psi_n(\lambda)$ is orthogonal to all polynomials of a degree less than n,

$$\int_a^b \psi_n(\lambda) \frac{\psi_n(\lambda)}{\left(\lambda - \lambda_j^{(n)}\right)\left(\lambda - \lambda_{j+1}^{(n)}\right)} \, d\omega(\lambda) = 0.$$

Note that the m-node Gauss–Christoffel quadrature is exact for the integral on the left-hand side. Thus,

$$\sum_{\ell=1}^m \omega_\ell^{(m)} \psi_n\left(\lambda_\ell^{(m)}\right) \frac{\psi_n\left(\lambda_\ell^{(m)}\right)}{\left(\lambda_\ell^{(m)} - \lambda_j^{(n)}\right)\left(\lambda_\ell^{(m)} - \lambda_{j+1}^{(n)}\right)} = 0.$$

Since $\omega_\ell^{(m)} > 0$, and since by assumption no $\lambda_\ell^{(m)}$ lies in the open interval $(\lambda_j^{(n)}, \lambda_{j+1}^{(n)})$, the terms in the sum are non-negative and not all equal to zero. Therefore the sum must be positive, which is a contradiction. Similarly, using the polynomials $\psi_n^2(\lambda)/(\lambda - \lambda_1^{(n)})$ and $\psi_n^2(\lambda)/(\lambda - \lambda_n^{(n)})$ one can prove that there must be a root of $\psi_m(\lambda)$ smaller than $\lambda_1^{(n)} > a$ and another one larger than $\lambda_n^{(n)} < b$. The same polynomial can be considered for $n = 1$, so that the first part of the proof is finished.

If $\omega(\lambda)$ has N points of increase $\lambda_1, \ldots, \lambda_N$ and $m = N$, the proof can be repeated in an analogous way with

$$\sum_{\ell=1}^{N} \omega_\ell \, \psi_n(\lambda_\ell) \, \frac{\psi_n(\lambda_\ell)}{\left(\lambda_\ell - \lambda_j^{(n)}\right)\left(\lambda_\ell - \lambda_{j+1}^{(n)}\right)} = 0,$$

which follows from the fact that $\psi_n(\lambda), n < N$, belongs to the sequence of monic orthogonal polynomials with respect to the inner product (3.2.8). Similarly, $\lambda_1^{(N)} = \lambda_1$ must be smaller than $\lambda_1^{(n)}$ and thus is located in $(a, \lambda_1^{(n)})$, while $\lambda_N^{(N)} = \lambda_N$ must be larger than $\lambda_n^{(n)}$ and thus is located in $(\lambda_n^{(n)}, b]$; cf. Lemma 3.2.4. □

Remark 3.3.2
A related interlacing theorem for the eigenvalues of Hermitian matrices (see, e.g. [322, Section 1]) is attributed to Cauchy [97]; cf. [57, p. 78], and [508, Section 10.1]. This result says that if a row and corresponding column are deleted from an Hermitian matrix of size $(n+1) \times (n+1)$, then the eigenvalues of the resulting Hermitian matrix of size $n \times n$ interlace the ones of the original matrix. In terms of the orthogonal polynomials, this corresponds to an interlacing result for the roots of $\psi_n(\lambda)$ and $\psi_{n+1}(\lambda)$ (see Corollary 3.3.3 below; the relationship between Hermitian matrices and orthogonal polynomials will become apparent in Section 3.5). Note that Theorem 3.3.1 is considerably stronger, since it applies not only to $\psi_n(\lambda)$ and $\psi_{n+1}(\lambda)$, but to $\psi_n(\lambda)$ and *all* subsequent polynomials $\psi_m(\lambda), n+1 \leq m \leq n(\omega)$. Several different formulations of the result for orthogonal polynomials can be found in the literature; see, e.g. [588, Chapter 1, Section 3], [607, Section 3.3, Theorems 3.3.2 and 3.3.3], and [111] (see Chapter I, Section 5, Theorem 5.3, and, in particular, Chapter II, Section 4). The results are closely related to the fact that the orthogonal polynomials form a so-called *Sturm sequence*. For the corresponding definitions and for the related theory we refer to the historical monograph by Khrushchev [384]; see also [607, Section 3.3, p. 44], [217, pp. 68–69, and 105–108], and [5, Chapter 1, Theorem 1.2.2]. The interlacing property has been independently rediscovered and proved many times. In the numerical analysis literature it is often linked to Jacobi matrices, with the proof being attributed to Givens; see [355, Chapter 4, Theorem 4, p. 168], [670, Chapter 2, Section 41 and 47, Chapter 5, Sections 36–38], [235]. The proof given above is based on *orthogonality* (see the historical note on orthogonality and three-term recurrences in Section 3.3.6), similarly to the proof of Lemma 3.2.4; cf. [111, Chapter II, Section 4, Theorem 4.1, pp. 59–60] and [634, Theorem 6.1, pp. 663–664].

The following is an immediate consequence of Lemma 3.2.4 and Theorem 3.3.1.

Corollary 3.3.3
If $n(\omega) = \infty$, *then the roots of two consecutive (monic) orthogonal polynomials strictly interlace, i.e. for* $n = 1, 2, \ldots$

$$a < \lambda_1^{(n+1)} < \lambda_1^{(n)} < \lambda_2^{(n+1)} < \lambda_2^{(n)} < \cdots < \lambda_{n-1}^{(n)} < \lambda_n^{(n+1)} < \lambda_n^{(n)} < \lambda_{n+1}^{(n+1)} < b.$$

(3.3.16)

If $\omega(\lambda)$ has N points of increase $\lambda_1, \ldots, \lambda_N$, then the Nth monic orthogonal polynomial given by the recurrence (3.3.14) is $\psi_N(\lambda) = (\lambda - \lambda_1) \cdots (\lambda - \lambda_N)$, and the statement holds for $n = 1, \ldots, N - 1$.

The next result complements the interlacing of the roots of two consecutive orthogonal polynomials by an analogous property of the sums of weights in the corresponding Gauss–Christoffel quadrature. It can be interpreted as the interlocking property of two consecutive Gauss–Christoffel distribution functions $\omega^{(n)}(\lambda)$ and $\omega^{(n+1)}(\lambda)$; see (3.1.4).

Theorem 3.3.4
In the previous notation, let $\omega^{(n)}(\lambda)$ and $\omega^{(n+1)}(\lambda)$ be two consecutive distribution functions defined by the Gauss–Christoffel quadrature solution of the simplified Stieltjes' moment problem determined by a nondecreasing real function $\omega(\lambda)$ on a given real finite interval $[a, b]$ with $\omega(a) = 0$, $\omega(b) = 1$, $n < \mathfrak{n}(\omega)$; see Problem 3.1.1 in Section 3.1. Then their difference

$$\Delta^{(n)}(\lambda) = \omega^{(n+1)}(\lambda) - \omega^{(n)}(\lambda)$$

has exactly $2n - 1$ sign changes in the interval $[a, b]$.

Proof
Since $\omega^{(n)}(a) = \omega^{(n+1)}(a) = 0$, $\omega^{(n)}(b) = \omega^{(n+1)}(b) = 1$ and $\omega^{(n)}(\lambda)$ has in the interval (a, b) exactly n points of increase which strictly interlace the points of increase of $\omega^{(n+1)}(\lambda)$, the difference $\Delta^{(n)}(\lambda)$ can not have in this interval more than $2n - 1$ sign changes. It remains to prove that it must have at least $2n - 1$ of them. We prove it by contradiction.

Assume that the number of sign changes of $\Delta^{(n)}(\lambda)$ in the interval $[a, b]$ is at most $2n - 2$. Then there exists a polynomial $\sigma_{2n-2}(\lambda)$ of degree at most $2n - 2$ such that the product

$$\sigma_{2n-2}(\lambda) \Delta^{(n)}(\lambda) = \sigma_{2n-2}(\lambda) \left(\omega^{(n+1)}(\lambda) - \omega^{(n)}(\lambda) \right)$$

is non-negative in $[a, b]$. Since it is not identically zero we have

$$\int_a^b \sigma_{2n-2}(\lambda) \Delta^{(n)}(\lambda) \, d\lambda > 0.$$

Let

$$\Omega_{2n-1}(t) = \int_a^t \sigma_{2n-2}(\lambda) \, d\lambda.$$

Since $\sigma_{2n-2}(\lambda)$ is analytic (and therefore continuous) and $\Delta^{(n)}(\lambda)$ is a right continuous piecewise constant function with finitely many points of increase and decrease in $[a, b]$ (and therefore is of bounded variation), we can apply integration by parts for the Riemann–Stieltjes integral (see, e.g. [607, formula (1.4.4)] or [129, formula (1.6.5.12)]), giving

$$\int_a^b \sigma_{2n-2}(\lambda)\Delta^{(n)}(\lambda)\,d\lambda = \Omega_{2n-1}(b)\Delta^{(n)}(b) - \Omega_{2n-1}(a)\Delta^{(n)}(a)$$
$$-\left\{\int_a^b \Omega_{2n-1}(\lambda)\,d\omega^{(n+1)}(\lambda) - \int_a^b \Omega_{2n-1}(\lambda)\,d\omega^{(n)}(\lambda)\right\} > 0.$$

The first two terms on the right-hand side vanish because $\Delta^{(n)}(a) = \Delta^{(n)}(b) = 0$. The difference of the integrals also vanishes because $\Omega_{2n-1}(\lambda)$ is a polynomial of degree at most $2n-1$ and the first $2n$ moments of the distribution functions $\omega^{(n)}(\lambda)$ and $\omega^{(n+1)}(\lambda)$ are equal. Consequently,

$$\int_a^b \sigma_{2n-2}(\lambda)\Delta^{(n)}(\lambda)\,d\lambda = 0$$

gives a contradiction that finishes the proof. □

The interlocking of the weight functions proven in this theorem is visualised in Figure 3.1, and it can also be stated in the following way: For any $k = 1, 2, \ldots, n-1$,

$$\sum_{i=1}^{k} \omega_i^{(n)} < \sum_{j=1}^{k+1} \omega_j^{(n+1)} < \sum_{j=1}^{k+1} \omega_j^{(n)}. \tag{3.3.17}$$

Since for any $k \geq 1$ with $n + k \leq \mathfrak{n}(\omega)$ the distribution function $\omega^{(n+k)}(\lambda)$ is non-decreasing in the interval $[a, b]$, and the same holds for the distribution function $\omega(\lambda)$, the differences

$$\Delta^{(n,k)}(\lambda) = \omega^{(n+k)}(\lambda) - \omega^{(n)}(\lambda) \quad \text{and}$$
$$\Delta^{(n,\infty)}(\lambda) = \omega(\lambda) - \omega^{(n)}(\lambda)$$

have at most $2n - 1$ sign changes in the interval $(a, b]$. Here the notation $\Delta^{(n,\infty)}(\lambda)$ is relevant for the case that $\omega(\lambda)$ has infinitely many points of increase $\mathfrak{n}(\omega)$; the

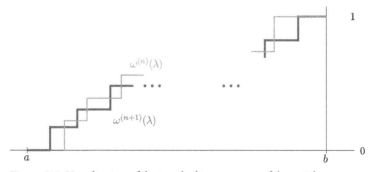

Figure 3.1 Visualisation of the interlocking property of the weight functions $\omega^{(n)}(\lambda)$ and $\omega^{(n+1)}(\lambda)$.

case that $\mathfrak{n}(\omega)$ is finite is included in $\Delta^{(n,k)}(\lambda)$ with $n+k=N=\mathfrak{n}(\omega)$. The rest of the proof of Theorem 3.3.4 relies on the fact that the first $2n$ moments of the distribution functions $\omega^{(n+1)}(\lambda)$ and $\omega^{(n)}(\lambda)$ coincide. But the same is true for $\omega^{(n+k)}(\lambda)$ and $\omega^{(n)}(\lambda)$, and, in particular, for $\omega(\lambda)$ and $\omega^{(n)}(\lambda)$. Therefore the statement of Theorem 3.3.4 can be generalised accordingly, and the differences $\Delta^{(n,k)}(\lambda)$ and $\Delta^{(n,\infty)}(\lambda)$ have both exactly $2n - 1$ sign changes in the interval $(a, b]$.

This is very useful in practical considerations concerning the approximation of $\omega(\lambda)$ by the individual distribution functions $\omega^{(n)}(\lambda), n = 1, 2, \ldots$; see, e.g. [185]. In particular, for any $n < \mathfrak{n}(\omega)$ the first weight $\omega_1^{(n)}$ and the last weight $\omega_n^{(n)}$ give the upper bounds

$$\int_a^{\lambda_1^{(n)}} d\omega(\lambda) < \omega_1^{(n)} \quad \text{and} \quad \int_{\lambda_n^{(n)}}^b d\omega(\lambda) < \omega_n^{(n)}$$

for the integrals representing the cumulated 'mass' distributed in the intervals $[a, \lambda_1^{(n)}]$ and $[\lambda_n^{(n)}, b]$ corresponding to $\omega(\lambda)$. An application of this observation can be found in [325].

The proof of Theorem 3.3.4 follows the proof of Theorem 22.1 in the monograph on moment spaces by Karlin and Shapley [375]. As pointed out by Van Assche [631, formula (4.4)] and Fischer [183, p. 37], the result, however, had already been given by Stieltjes in 1885; see [587] and the paper [582] published posthumously in Stieltjes' collected works. The statement can easily be extended to any two distribution functions which match the first $2n$ moments; see [183, Theorem 2.2.5]. It is rather surprising that the book [375], which contains many results about moments, does not give any reference to the fundamental contributions of Stieltjes and Chebyshev or to any other relevant publication from the 19th century.

3.3.2 Continued Fractions

We will now link the previously discussed orthogonal polynomials to continued fractions. The results we are going to describe are again classical, and they have been described in many books; see, e.g. [5, Chapter 1, Section 4], [111, Chapter III, Section 4], or [432, 657]. A historical perspective on continued fractions, in particular related to the contribution of Euler, can be found in [384]. For a summary and the historical context of the fundamental work of Stieltjes we refer to [631]. A very interesting viewpoint related to the algorithm of Euclid as the underlying principle is presented in the monograph [87], which links linear algebra with rational approximation and theory of linear dynamical systems. We also point out that Hestenes and Stiefel discussed relations between orthogonal polynomials and continued fractions in Sections 14–18 of their founding paper on the CG method [313]. We will give more details on this particular connection in Section 3.5.

As previously, consider a nondecreasing function $\omega(\lambda)$ defined on the finite real interval $[a, b]$, with $\omega(a) = 0$, $\omega(b) = 1$, $\mathfrak{n}(\omega) \leq \infty$, and the associated sequences of orthonormal and monic orthogonal polynomials (3.3.1) and (3.2.7), respectively. These are generated by the respective three-term recurrences (3.3.3) and (3.3.14), which are determined by the unique scalar coefficients γ_n, δ_{n+1}, $n = 1, 2, \ldots$.

Using these coefficients we can define the *n*th convergent $\mathcal{F}_n(\lambda)$ of a *continued fraction*, i.e.

$$\mathcal{F}_n(\lambda) \equiv \cfrac{1}{\lambda - \gamma_1 - \cfrac{\delta_2^2}{\lambda - \gamma_2 - \cfrac{\delta_3^2}{\lambda - \gamma_{n-2} - \cfrac{\delta_n^2}{\lambda - \gamma_{n-1} - \cfrac{\delta_n^2}{\lambda - \gamma_n}}}}} \quad (3.3.18)$$

$$\equiv \frac{\mathcal{R}_n(\lambda)}{\mathcal{P}_n(\lambda)}.$$

Observe that the first two convergents are given by

$$\mathcal{F}_1(\lambda) = \frac{1}{\lambda - \gamma_1} = \frac{1}{\psi_1(\lambda)},$$

$$\mathcal{F}_2(\lambda) = \cfrac{1}{\lambda - \gamma_1 - \cfrac{\delta_2^2}{\lambda - \gamma_2}} = \frac{\lambda - \gamma_2}{(\lambda - \gamma_2)(\lambda - \gamma_1) - \delta_2^2} = \frac{\lambda - \gamma_2}{\psi_2(\lambda)}.$$

If $n(\omega) = N < \infty$, then the sequence of orthogonal polynomials is finite, and also the convergents form the finite sequence $\mathcal{F}_1(\lambda), \ldots, \mathcal{F}_N(\lambda)$. Otherwise the sequence of convergents is formally infinite. The numerator $\mathcal{R}_n(\lambda)$ and the denominator $\mathcal{P}_n(\lambda)$ of $\mathcal{F}_n(\lambda)$ are polynomials of degrees $n - 1$ and n, respectively. The following two theorems (3.3.5 and 3.3.6) contain well-known results about these polynomials.

Theorem 3.3.5
Using the previous notation,

$$\mathcal{P}_n(\lambda) = \psi_n(\lambda), \quad (3.3.19)$$

$$\mathcal{R}_n(\lambda) = \int_a^b \frac{\psi_n(\lambda) - \psi_n(z)}{\lambda - z} \, d\omega(z) = \int_a^b \sum_{j=1}^n \frac{\psi_n^{(j)}(z)}{j!} (\lambda - z)^{j-1} d\omega(z),$$

$$(3.3.20)$$

for $n = 1, 2, \ldots$. *Moreover, the numerators satisfy the three-term recurrence*

$$\mathcal{R}_n(\lambda) = (\lambda - \gamma_n)\mathcal{R}_{n-1}(\lambda) - \delta_n^2 \mathcal{R}_{n-2}(\lambda), \quad n = 2, 3, \ldots, \quad (3.3.21)$$

which starts with $\mathcal{R}_0(\lambda) \equiv 0$ *and* $\mathcal{R}_1(\lambda) \equiv 1$.

(Recall that the recurrence for $\mathcal{P}_n(\lambda) = \psi_n(\lambda)$ starts with $\psi_{-1}(\lambda) \equiv 0$ and $\psi_0(\lambda) \equiv 1$; see (3.3.14)–(3.3.15).)

Proof
The proof is by induction. For $n = 1$ we have

$$\mathcal{R}_1(\lambda) \equiv 1, \quad \mathcal{P}_1(\lambda) = \psi_1(\lambda) = \lambda - \gamma_1, \quad \mathcal{F}_1(\lambda) = \frac{1}{\lambda - \gamma_1}.$$

We will show that the denominator and numerator are for $n = 2, 3, \ldots$ given by the recurrences

$$\mathcal{P}_n(\lambda) = (\lambda - \gamma_n)\mathcal{P}_{n-1}(\lambda) - \delta_n^2 \mathcal{P}_{n-2}(\lambda), \quad \mathcal{P}_0(\lambda) = 1, \; \mathcal{P}_1(\lambda) = \lambda - \gamma_1,$$
$$\mathcal{R}_n(\lambda) = (\lambda - \gamma_n)\mathcal{R}_{n-1}(\lambda) - \delta_n^2 \mathcal{R}_{n-2}(\lambda), \quad \mathcal{R}_0(\lambda) = 0, \; \mathcal{R}_1(\lambda) = 1.$$

Consider formally

$$\mathcal{F}_n(\lambda) \equiv \mathcal{F}_n = \cfrac{1}{a_1 - \cfrac{b_2}{a_2 - \cfrac{b_3}{\ddots a_{n-3} - \cfrac{b_{n-1}}{a_{n-2} - \cfrac{b_n}{a_{n-1} - \cfrac{b_n}{a_n}}}}}}$$

$$= \cfrac{1}{a_1 - \cfrac{b_2}{a_2 - \cfrac{b_3}{\ddots a_{n-3} - \cfrac{b_{n-1}}{a_{n-2} - \cfrac{b_{n-1}}{a_{n-1}^\star}}}}} \equiv \mathcal{F}_{n-1}^\star,$$

where $a_j = \lambda - \gamma_j$, $b_j = \delta_j^2$, and

$$a_{n-1}^\star = \frac{a_{n-1} a_n - b_n}{a_n}.$$

Then $\mathcal{F}_n$ can be viewed as $\mathcal{F}_{n-1}^\star$, with a_{n-1} being replaced by $a_{n-1}^\star$. Note that $\mathcal{F}_{n-1}^\star$ has the form of an $(n-1)$th convergent. The induction assumption applied to $\mathcal{F}_{n-1}^\star$ now gives $\mathcal{F}_{n-1}^\star = \mathcal{R}_{n-1}^\star / \mathcal{P}_{n-1}^\star$, where

$$\mathcal{P}_{n-1}^\star = a_{n-1}^\star \mathcal{P}_{n-2} - b_{n-1} \mathcal{P}_{n-3}, \quad \mathcal{R}_{n-1}^\star = a_{n-1}^\star \mathcal{R}_{n-2} - b_{n-1} \mathcal{R}_{n-3},$$

which leads to

$$a_n \mathcal{P}_{n-1}^\star = a_n(a_{n-1}\mathcal{P}_{n-2} - b_{n-1}\mathcal{P}_{n-3}) - b_n \mathcal{P}_{n-2}$$
$$= a_n \mathcal{P}_{n-1} - b_n \mathcal{P}_{n-2},$$

with the analogous identity for $a_n \mathcal{R}_{n-1}^\star$. Setting $\mathcal{P}_n \equiv a_n \mathcal{P}_{n-1}^\star$ and $\mathcal{R}_n \equiv a_n \mathcal{R}_{n-1}^\star$ then shows

$$\mathcal{F}_n = \mathcal{F}_{n-1}^\star = \frac{\mathcal{R}_{n-1}^\star}{\mathcal{P}_{n-1}^\star} = \frac{a_n \mathcal{R}_{n-1}^\star}{a_n \mathcal{P}_{n-1}^\star} = \frac{\mathcal{R}_n}{\mathcal{P}_n},$$

which proves (3.3.19) and (3.3.21). Now assume that (3.3.20) holds for $n-2$ and $n-1$. Substituting this assumption into (3.3.21) gives

$$\mathcal{R}_n(\lambda) = \int_a^b \frac{(\lambda - \gamma_n)(\psi_{n-1}(\lambda) - \psi_{n-1}(z)) - \delta_n^2(\psi_{n-2}(\lambda) - \psi_{n-2}(z))}{\lambda - z} \, d\omega(z)$$

$$= \int_a^b \frac{\psi_n(\lambda) - \psi_n(z)}{\lambda - z} \, d\omega(z) - \int_a^b \psi_{n-1}(z) \, d\omega(z),$$

where the last term is zero due to the orthogonality of $\psi_{n-1}(\lambda)$ to the constant polynomial $\psi_0(\lambda) \equiv 1$, which finishes the proof. □

The previous Theorem 3.3.5 gives a formula for the continued fraction $\mathcal{F}_n(\lambda)$ in (3.3.18) in terms of the nth monic orthogonal polynomial $\psi_n(\lambda)$ and its derivatives. The next theorem (3.3.6) shows a very important property of the partial fraction decomposition of $\mathcal{F}_n(\lambda)$: it reveals the nodes and weights of the corresponding Gauss–Christoffel quadrature.

Theorem 3.3.6
Using the previous notation, the nth convergent $\mathcal{F}_n(\lambda)$ of the continued fraction corresponding to $\omega(\lambda)$ can be decomposed into the partial fraction

$$\mathcal{F}_n(\lambda) = \frac{\mathcal{R}_n(\lambda)}{\psi_n(\lambda)} = \sum_{j=1}^n \frac{\omega_j^{(n)}}{\lambda - \lambda_j^{(n)}}, \qquad (3.3.22)$$

where $\lambda_1^{(n)}, \ldots, \lambda_n^{(n)}$ and $\omega_1^{(n)}, \ldots, \omega_n^{(n)}$ are the nodes and weights of the n-node Gauss–Christoffel quadrature associated with $\omega(\lambda)$.

Proof
The polynomials $\mathcal{R}_n(\lambda)$ and $\psi_n(\lambda)$ are of degrees $n-1$ and n, respectively. Thus, $\mathcal{R}_n(\lambda)$ can be uniquely expressed as a linear combination of the n linearly independent polynomials of degree $n-1$ given by

$$\frac{\psi_n(\lambda)}{\lambda - \lambda_j^{(n)}} = \prod_{\substack{\ell=1 \\ \ell \neq j}}^n (\lambda - \lambda_\ell^{(n)}), \qquad j = 1, \ldots, n.$$

This means that the partial fraction decomposition

$$\mathcal{F}_n(\lambda) = \frac{\mathcal{R}_n(\lambda)}{\psi_n(\lambda)} = \sum_{j=1}^n \frac{\eta_j}{\lambda - \lambda_j^{(n)}}, \quad \text{where } \eta_j = \frac{\mathcal{R}_n\left(\lambda_j^{(n)}\right)}{\psi_n'\left(\lambda_j^{(n)}\right)}, \quad j = 1, \ldots, n,$$

exists, and by construction it is unique. Substituting for $\mathcal{R}_n(\lambda)$ from (3.3.20), and using the fact that $\psi_n(\lambda_j^{(n)}) = 0, j = 1, \ldots, n,$

$$\eta_j = \int_a^b \frac{\psi_n\left(\lambda_j^{(n)}\right) - \psi_n(z)}{\psi'\left(\lambda_j^{(n)}\right)\left(\lambda_j^{(n)} - z\right)} d\omega(z) = \int_a^b \frac{\psi_n(z)}{\psi'\left(\lambda_j^{(n)}\right)\left(z - \lambda_j^{(n)}\right)} d\omega(z).$$

The right-hand side is equal to the jth weight $\omega_j^{(n)}$ of the n-node Gauss–Christoffel quadrature (cf. (3.2.5)), which finishes the proof. □

The partial fraction decomposition given in the previous Theorem 3.3.6 allows the following development. For $|\lambda|$ sufficiently large we can decompose the jth term of the sum in (3.3.22) as

$$\frac{\omega_j^{(n)}}{\lambda - \lambda_j^{(n)}} = \frac{\omega_j^{(n)}}{\lambda}\left(1 - \frac{\lambda_j^{(n)}}{\lambda}\right)^{-1} = \sum_{\ell=1}^{2n} \omega_j^{(n)} \left\{\lambda_j^{(n)}\right\}^{\ell-1} \frac{1}{\lambda^\ell} + \mathcal{O}\left(\frac{1}{\lambda^{2n+1}}\right),$$

(3.3.23)

and therefore

$$\mathcal{F}_n(\lambda) = \sum_{j=1}^n \frac{\omega_j^{(n)}}{\lambda - \lambda_j^{(n)}} = \sum_{\ell=1}^{2n} \frac{1}{\lambda^\ell} \left(\sum_{j=1}^n \omega_j^{(n)} \left\{\lambda_j^{(n)}\right\}^{\ell-1}\right) + \mathcal{O}\left(\frac{1}{\lambda^{2n+1}}\right).$$

Since $\lambda_1^{(n)}, \ldots, \lambda_n^{(n)}$ and $\omega_1^{(n)}, \ldots, \omega_n^{(n)}$ are the nodes and weights of the n-node Gauss–Christoffel quadrature associated with the distribution function $\omega(\lambda)$, and since this quadrature solves the simplified Stieltjes' moment problem (see Theorem 3.2.8), the first $2n$ coefficients of the expansion of $\mathcal{F}_n(\lambda)$ into the power series around ∞ are equal to the first $2n$ moments of $\omega(\lambda)$, i.e.

$$\mathcal{F}_n(\lambda) = \sum_{\ell=1}^{2n} \frac{\xi_{\ell-1}}{\lambda^\ell} + \mathcal{O}\left(\frac{1}{\lambda^{2n+1}}\right), \qquad (3.3.24)$$

$$\xi_{\ell-1} = \int_a^b \lambda^{\ell-1} d\omega(\lambda) = \sum_{j=1}^n \omega_j^{(n)} \left\{\lambda_j^{(n)}\right\}^{\ell-1}, \quad \ell = 1, \ldots, 2n. \qquad (3.3.25)$$

Now recall that

$$\lim_{n \to \infty} \sum_{j=1}^n \frac{\omega_j^{(n)}}{\lambda - \lambda_j^{(n)}} = \int_a^b \frac{d\omega(u)}{\lambda - u};$$

cf. (3.2.22). As shown in Theorem 3.3.6, the sum on the left-hand side is equal to the nth convergent $\mathcal{F}_n(\lambda)$. The relation between this convergent and the integral on the right-hand side was studied by Chebyshev in 1859 [106, Section IV];

see also [105]. One year before Chebyshev it was published in a remarkable paper by Christoffel [113]. It was also studied by Stieltjes in 1894 [588, Chapter II, Sections 7 and 8; Chapter VIII, Section 51]. They considered expansions of the integral

$$\int_a^b \frac{d\omega(u)}{\lambda - u}.$$

into continued fractions. Using, for $|\lambda|$ sufficiently large, the expansion

$$\frac{1}{\lambda - u} = \sum_{\ell=1}^{2n} \frac{u^{\ell-1}}{\lambda^\ell} + \mathcal{O}\left(\frac{1}{\lambda^{2n+1}}\right),$$

we obtain

$$\int_a^b \frac{d\omega(u)}{\lambda - u} = \sum_{\ell=1}^{2n} \left(\int_a^b u^{\ell-1} d\omega(u)\right) \frac{1}{\lambda^\ell} + \mathcal{O}\left(\frac{1}{\lambda^{2n+1}}\right)$$

$$= \sum_{\ell=1}^{2n} \frac{\xi_{\ell-1}}{\lambda^\ell} + \mathcal{O}\left(\frac{1}{\lambda^{2n+1}}\right),$$

i.e.

$$\int_a^b \frac{d\omega(u)}{\lambda - u} = \mathcal{F}_n(\lambda) + \mathcal{O}\left(\frac{1}{\lambda^{2n+1}}\right). \tag{3.3.26}$$

For important examples and extensions we refer to [105, 588]. Further details on the work of Chebyshev and Stieltjes in this context are described in the historical note in Section 3.3.5.

3.3.3 The Gauss–Christoffel Quadrature for Analytic Functions

For analytic functions it is possible to express the error of the Gauss–Christoffel quadrature without derivatives or divided differences. To show this we will use results of the previous section. With $\psi_n(\lambda)$ as above, the function

$$\rho_n(z) = \int_a^b \frac{\psi_n(\lambda)}{z - \lambda} d\omega(\lambda)$$

is analytic in the complex plane outside the interval $[a, b]$. Using this function in the expression (3.3.20) the numerator $\mathcal{R}_n(\lambda)$ of the nth convergent $\mathcal{F}_n(\lambda)$ can be written as

$$\mathcal{R}_n(z) = \int_a^b \frac{\psi_n(z) - \psi_n(\lambda)}{z - \lambda} d\omega(\lambda) = -\rho_n(z) + \psi_n(z) \int_a^b \frac{d\omega(\lambda)}{z - \lambda}.$$

Suppose that $f(z)$ is analytic in a simply connected domain containing $[a, b]$ in its interior, and let Γ be a simple, closed, and positively oriented curve in that domain encircling $[a, b]$. Then, using the same notation as in Theorem 3.2.7,

$$E_\omega^n(f) \equiv I_\omega(f) - I_\omega^n(f) = \frac{1}{2\pi\iota} \int_\Gamma K_n(z) f(z)\, dz, \qquad (3.3.27)$$

where

$$K_n(z) \equiv \frac{\rho_n(z)}{\Psi_n(z)};$$

see, e.g. [129, p. 303, relation (4.6.18)], or [227, Theorem 2.48]. Indeed, substituting the Cauchy formula

$$f(\lambda) = \frac{1}{2\pi\iota} \int_\Gamma \frac{f(z)}{z - \lambda}\, dz$$

into the quadrature identity

$$\int_a^b f(\lambda)\, d\omega(\lambda) = \sum_{j=1}^n \omega_j^{(n)} f\left(\lambda_j^{(n)}\right) + E_\omega^n(f)$$

gives

$$\int_a^b \left(\frac{1}{2\pi\iota} \int_\Gamma \frac{f(z)}{z - \lambda}\, dz \right) d\omega(\lambda) = \sum_{j=1}^n \frac{\omega_j^{(n)}}{2\pi\iota} \int_\Gamma \frac{f(z)}{z - \lambda_j^{(n)}}\, dz + E_\omega^n(f).$$

Interchanging the order of integration and using the identity (3.3.22) from Theorem 3.3.6, the error term can be written as

$$\begin{aligned}
E_\omega^n(f) &= \frac{1}{2\pi\iota} \int_\Gamma \left(\int_a^b \frac{1}{z - \lambda}\, d\omega(\lambda) - \sum_{j=1}^n \frac{\omega_j^{(n)}}{z - \lambda_j^{(n)}} \right) f(z)\, dz \\
&= \frac{1}{2\pi\iota} \int_\Gamma \left(\int_a^b \frac{1}{z - \lambda}\, d\omega(\lambda) - \frac{R_n(z)}{\Psi_n(z)} \right) f(z)\, dz \\
&= \frac{1}{2\pi\iota} \int_\Gamma \frac{\rho_n(z)}{\Psi_n(z)} f(z)\, dz \\
&= \frac{1}{2\pi\iota} \int_\Gamma K_n(z) f(z)\, dz.
\end{aligned}$$

The kernel $K_n(z)$ depends through $\Psi_n(z)$ and $\rho_n(z)$ on the given distribution function $\omega(\lambda)$. The identity (3.3.27) has been applied to estimate the error and to study its decrease with n for some particular classes of distribution functions $\omega(\lambda)$, see, e.g. [222, 228, 227], or [129, Section 4.6].

3.3.4 Summary of the Previous Mathematical Development

Let us describe the flow of ideas in the preceding sections of this chapter. Given a (normalised) nondecreasing distribution function $\omega(\lambda)$ defined on a finite real interval $[a, b]$, we have considered the sequence of its moments ξ_k, $k = 0, 1, 2, \ldots$; see (3.1.3). The goal of the simplified Stieltjes' moment problem was to determine a distribution function $\omega^{(n)}(\lambda)$ with n points of increase, such that the first $2n$ moments of $\omega^{(n)}(\lambda)$ match the first $2n$ moments $\xi_0, \xi_1 \ldots, \xi_{2n-1}$ of $\omega(\lambda)$; see (3.1.5). We have shown that this problem is solved uniquely by the n-node Gauss–Christoffel quadrature, which has the algebraic degree of exactness $2n - 1$; see Theorem 3.2.8.

The n nodes of the Gauss–Christoffel quadrature are the roots of the nth monic orthogonal polynomial $\psi_n(\lambda)$ with respect to the inner product (3.2.8), which is determined by $\omega(\lambda)$. The polynomial $\psi_n(\lambda)$ represents the denominator of the nth convergent $\mathcal{F}_n(\lambda)$ of the continued fraction associated with $\omega(\lambda)$; see (3.3.18). If this nth convergent is expressed in the form of a partial fraction, then the (scalar) numerators and the scalar terms in the denominators of the individual fractions give the weights and nodes, respectively, of the associated n-node Gauss–Christoffel quadrature; see (3.3.22). When this partial fraction representation of $\mathcal{F}_n(\lambda)$ is expanded into a power series around ∞, the first $2n$ coefficients of the expansion corresponding to the powers $\lambda^{-1}, \ldots, \lambda^{-2n}$ are equal to the first $2n$ moments $\xi_0, \xi_1 \ldots, \xi_{2n-1}$; see (3.3.24).

Moreover, the polynomials $\psi_n(\lambda)$ and their orthonormal counterparts $\varphi_n(\lambda)$, $n = 0, 1, 2, \ldots$, satisfy the three-term recurrences (3.3.14) and (3.3.3), respectively. The numerators $\mathcal{R}_n(\lambda)$ of the continued fractions $\mathcal{F}_n(\lambda)$, $n = 1, 2, \ldots$, satisfy the same recurrence (3.3.14) with different initial functions; see Theorem 3.3.5. The corresponding matrix forms (3.3.12) and (3.3.10) of the recurrences (3.3.14) and (3.3.3) involve real tridiagonal matrices (with positive off-diagonal elements) containing the recurrence coefficients in their rows. In the case of orthonormal polynomials $\varphi_n(\lambda)$ the tridiagonal matrix T_n is symmetric, and thus it is a Jacobi matrix; see (3.3.11).

In summary, given a (nondecreasing and normalised) distribution function $\omega(\lambda)$, we get for $n = 1, 2, \ldots$ the sequences of:

- Distribution functions $\omega^{(n)}(\lambda)$ with n points of increase, which solve the simplified Stieltjes' moment problem and are determined by the n-node Gauss–Christoffel quadrature.
- Monic polynomials $\psi_n(\lambda)$ that are orthogonal with respect to the inner product (3.2.8) determined by $\omega(\lambda)$.
- Polynomials $\varphi_n(\lambda)$ that are orthonormal with respect to the same inner product.
- The nth convergents $\mathcal{F}_n(\lambda)$ of the continued fraction determined by $\omega^{(n)}(\lambda)$.
- The $n \times n$ Jacobi matrices T_n.

All of these mathematical objects are uniquely determined by the distribution function $\omega(\lambda)$, with a one-to-one correspondence between them.

In some way everything is linked to the Jacobi matrix T_n, which from the matrix theory point of view is a purely algebraic object. In the following section we will show that, because of the connections outlined above, T_n has remarkable properties that are highly useful in the analysis of numerical algorithms for solving linear algebraic systems and algebraic eigenvalue problems.

It is worth pointing out that *analytic* objects such as distribution functions $\omega^{(n)}(\lambda)$, monic orthogonal polynomials $\psi_n(\lambda)$, orthonormal polynomials $\varphi_n(\lambda)$, as well as the continued fraction convergents $\mathcal{F}_n(\lambda)$ can be studied using *algebraic* tools applied to Jacobi matrices T_n, and vice versa. We see that here the interplay between analysis and algebra is so strong that the two become one field of mathematics. Before we continue with an investigation of Jacobi matrices, we present two historical sections which document that such a unified view was present in the original developments.

3.3.5 Historical Note: Chebyshev, Markov, Stieltjes, and the Moment Problem

This section briefly recalls, with no claim of completeness, some historical developments of the concept of moments which are relevant to our exposition. For further reading we recommend, besides Stieltjes' paper [588] and other original works referred to in the text, the thorough reviews by Van Assche [631] and Kjeldsen [386].

Since the work of Euler (see [171, Chapter XVII] and [384, Section 4.1, Theorem 4.2]) it has been known how to transform formal infinite series into continued fractions; the application of this transformation to formal power series

$$\sum_{j=1}^{\infty} \frac{\xi_{j-1}}{\lambda^j} \qquad (3.3.28)$$

was straightforward. On the other hand, (3.3.24)–(3.3.25) show how to expand a continued fraction into a power series, with the integral representation given by (3.3.26); see also [432, Sections 3.2 and 3.3]. These results were known to Chebyshev, with the integral in (3.3.26) being replaced by an ordinary Riemann integral with a positive integrable weight $w(u)$.

Inspired by the work of Bienaymé from 1853 (published in 1867 [58]), Chebyshev published in 1874 a short paper [107] in which he considered bounds for the integral

$$\int_{\alpha}^{\beta} w(u)\, du, \qquad (3.3.29)$$

where $w(u)$ is a positive (Riemann integrable) function defined on the interval $[a, b]$, $a < \alpha < \beta < b$, given the values of the integrals over the (larger) interval $[a, b]$,

$$\int_a^b w(u)\, du, \int_a^b u w(u)\, du, \ldots, \int_a^b u^{2n-1} w(u)\, du, \ldots \qquad (3.3.30)$$

Using the formal power series (3.3.28) with the moments defined in (3.3.30), he linked this problem with the continued fraction

$$\mathcal{F}_n(\lambda) = \frac{\mathcal{R}_n(\lambda)}{\psi_n(\lambda)};$$

see (3.3.22). In particular, he stated (without proof) the following inequalities

$$\sum_{\ell=j+1}^{n-1} \frac{\mathcal{R}_n\left(\lambda_\ell^{(n)}\right)}{\psi_n'\left(\lambda_\ell^{(n)}\right)} < \int_{\lambda_j^{(n)}}^{\lambda_n^{(n)}} w(u)\,du < \sum_{\ell=j}^{n} \frac{\mathcal{R}_n\left(\lambda_\ell^{(n)}\right)}{\psi_n'\left(\lambda_\ell^{(n)}\right)}, \qquad (3.3.31)$$

now called the (Bienaymé–) Chebyshev inequalities. Proofs were given independently in 1884 by Markov [444] and Stieltjes [585]. These proofs were similar and Markov claimed the priority, which was acknowledged by Stieltjes in a short note published in 1885 [587]. Another proof given by Stieltjes can be found in his posthumously published paper [582] mentioned above in relation to the interlocking property of two consecutive distribution functions; see Theorem 3.3.4. Chebyshev returned to this topic with a thorough expository paper presented to the St. Petersburg Academy of Sciences in 1885, published in French (with translation by Sophia Kowalevski) in 1886 [109]. For further comments on the contribution of Chebyshev, Markov, Christoffel, Heine, and others we refer, e.g. to the Introduction in [561], [396], [5, Foreword, pp. 22–24, Chapter 3 (Section 3), in particular pp. 112–113], [111, Chapter II, Section 6, Chapter III, Section 4, Theorem 4.3, p. 88], [222], [384], [77], [392, Chapter 1] (written by Akhiezer), and [94].

Although Chebyshev and Markov used moments and the relationship to the integral

$$\int_a^b \frac{w(u)}{\lambda - u}\,du \qquad (3.3.32)$$

and to its continued fraction expansion, they did not formulate a problem of moments. This problem appeared as a by-product of the analytic theory of continued fractions formulated by Stieltjes in his ingenious paper [588]. Stieltjes referred to the work of Chebyshev, Heine, and Darboux, but his motivation and achievement went far beyond all previous related work. In this context he introduced, e.g. what is known today as the *Stieltjes integral*; see [588, Sections 37–38]. The paper was published in 1894, and it was awarded a prize by the Academy of Sciences in Paris. Stieltjes died on December 31 of the same year in Toulouse.

In our notation, Stieltjes used a variable $z \equiv -\lambda$, which gives

$$\int_a^b \frac{d\omega(u)}{\lambda - u} = -\int_a^b \frac{d\omega(u)}{z + u},$$

and he considered a continued fraction in the form

$$\cfrac{1}{a_1 z + \cfrac{1}{a_2 + \cfrac{1}{a_3 z + \cdots + \cfrac{1}{a_{2m} + \cfrac{1}{a_{2m+1} z + \cdots}}}}} \qquad (3.3.33)$$

with $a_1, a_2, \ldots$ being positive real numbers. He described the equivalence of the form (3.3.33) with the form (3.3.18) (as well as with other forms) in the introduction of his paper [588]. It is important to recall that the distribution function $\omega(u)$ is defined on the interval $[0, +\infty)$; see Section 3.1. Consequently, the scalar coefficients $\gamma_1, \delta_2, \ldots$ in the Stieltjes recurrence (3.3.3) are positive (if $\delta_{n+1} = 0$, the recurrence is finite, and the case is trivial). This corresponds to the assumption that $a_1, a_2, \ldots$ are positive; see [588, pp. 610–612]. Some related results and comments can be found, e.g. in [217, Supplement 2], [657, Chapter 1], [5, Appendix], [255, Section III, Appendix A and B]; cf. also [104, 106, 108]. Because of its relation to Jacobi matrices, (3.3.18) is sometimes called a Jacobi type continued fraction [111, p. 85]. Connections of the different forms of continued fractions to the quotient-difference (QD) algorithm are described by Henrici in [310, in particular pp. 36–37].

Stieltjes proved that the continued fraction (3.3.33) is convergent if and only if the series

$$\sum_{j=1}^{\infty} a_j \qquad (3.3.34)$$

is divergent. Then (3.3.33) represents an analytic function of z in the complex plane except for the negative real axis which can be written in the integral form

$$\int_0^{\infty} \frac{d\omega(u)}{z+u}.$$

The coefficients of the continued fraction (3.3.33) can be expressed via moments using the formulas

$$a_{2j} = \frac{A_j^2}{B_j B_{j-1}}, \quad a_{2j+1} = \frac{B_j^2}{A_j A_{j+1}}, \qquad (3.3.35)$$

where $A_0 \equiv B_0 \equiv 1$, and

$$A_j \equiv \begin{vmatrix} \xi_0 & \xi_1 & \cdots & \xi_{j-1} \\ \vdots & \vdots & & \vdots \\ \xi_{j-1} & \xi_j & \cdots & \xi_{2j-2} \end{vmatrix},$$

$$B_j \equiv \begin{vmatrix} \xi_1 & \xi_2 & \cdots & \xi_j \\ \vdots & \vdots & & \vdots \\ \xi_j & \xi_{j+1} & \cdots & \xi_{2j-1} \end{vmatrix}, \quad j = 1, 2, \ldots$$

Analogous formulas were already known to Jacobi and others; see the historical note in Section 3.4.3. For another occurrence of the moment matrices we refer to Section 3.6.

When (3.3.34) converges, the continued fraction (3.3.33) diverges. Stieltjes studied this divergence and posed and answered the question of whether the sequence of moments determines the unique distribution function. This was the origin of the moment problem presented in Section 3.1. In particular, Stieltjes proved that the moment problem has the unique solution (in his terminology, is *determinate*) if and only if

$$A_j > 0, \quad B_j > 0, \quad j = 1, 2, \ldots ;$$

see [588, Chapter II, Section 8, Chapter IV, Section 24 and Chapter VIII]. This paper not only documents the outstanding mathematical creativity of Stieltjes, but it is also an example of extraordinary clarity and thoroughness of presentation.

The classical problem of moments, its history, and results are described by Shohat and Tamarkin [561], and by Akhiezer [5]. For an interesting geometrical view we refer to Karlin and Shapley [375]. As with the method of Krylov [398] and its algebraic interpretation in [215] (cf. the historical note in Section 2.5.7 and Remark 4.2.12 in this book), the results on the problem of moments found immediate applications in investigations of oscillations of mechanical systems. A summary of related achievements, including the relationship to the Sturm–Liouville problem (see, e.g. [514]) is given in the classical monograph by Gantmacher and Krein [217], with Supplement 2 of that monograph devoted to the relationship with Stieltjes' work on continued fractions; see also [5, Appendix, pp. 232–242]. Analytic theory of continued fractions is described, e.g. in the monograph [657].

3.3.6 Historical Note: Orthogonal Polynomials and Three-term Recurrences

Although nowadays orthogonal polynomials and three-term recurrences seem naturally connected, the historical development and understanding of this connection, as well as understanding of the concepts of orthogonality and orthogonalisation, was not straightforward. The history also shows the importance of continued fractions as one of the fundamental concepts in the development of a large part of mathematics.

As pointed out by Brezinski, a specific form of three-term recurrences related to the expansion of a rational number into a continued fraction was presented around 1150 by the Indian mathematician Bhascara II; see [77, Section 1.3, pp. 32–33].

The beginning of the theory of continued fractions is linked with the pioneering work of the 17th century mathematicians Brouncker and Wallis. The latter published in 1655 the book *Arithmetica infinitorium*, in which he exposed the earlier (unpublished) results of Brouncker, and possibly invented the name *continued fraction*; see [77, Section 3.1, in particular pp. 79–81], [384, Section 1.2, in particular Theorem 1.4], and [111, Chapter III, Section 2, pp. 80–81, and Section 4, pp. 85–86, relation (4.3)]. Among other results, Wallis presented the *three-term recurrences* for the numerators and denominators of the convergents of continued fractions, now known as the Brouncker–Wallis formulas. The work of Brouncker and Wallis influenced Euler and many other mathematicians. It indicates an understanding of the relationship between continued fractions and polynomials, which was explicitly described much later; see [384, Section 3.2, p. 143].

In the 18th century, Euler showed (among his many other fundamental contributions) how to expand infinite series into continued fractions; see Section 3.3.5. The idea of how to expand the formal power series into continued fractions was developed further by Chebyshev in the paper [104], published in Russian in 1855, then translated into French and subsequently republished in 1858 by Bienaymé, who significantly influenced, as mentioned in Section 3.3.5, some of Chebyshev's later works. Chebyshev showed that the denominators associated with a continued fraction form a sequence of *orthogonal polynomials*, and presented the three-term recurrence relations for the numerators and denominators of the continued fraction convergents [104, Section 4]; see also the later paper by Christoffel from 1877 [114]. These recurrences for polynomials (cf. Theorem 3.3.5) are formally the same as the Brouncker–Wallis recurrences for numbers. Interesting comments on the starting point of a general theory of orthogonal polynomials in the work of Chebyshev can be found in [403, Section 1] (for a small Corrigendum to this paper see [404]) and also in the commentary by Gautschi in [243, Part V, pp. 347–348].

The orthogonality of polynomials (without giving it a name) had already been used as a basic concept by Jacobi in his 1826 paper [356], which reformulated the new quadrature method invented by Gauss in 1814 [219]; see the historical note in Section 3.2.3. In order to determine quadrature nodes that maximise the quadrature's algebraic degree of exactness, Gauss used in his discovery continued fractions associated with the hypergeometric series. That led him (for the standard Riemann integral without weight) to Legendre polynomials. The approaches of Gauss and Jacobi, as well as further generalisations and developments up to the work of Christoffel, were summarised by Heine in [304, Part I, Mechanische Quadratur, Sections 1–16, pp. 1–31]. Developments up to modern times were beautifully exposed by Gautschi in his survey paper [222]; see also [227, 240].

Orthogonal polynomials were extensively studied under the name *reciprocal functions* by Murphy in the second of his several memoirs published in the Transactions of the Cambridge Philosophical Society in 1835; see [466, Introduction, pp. 113–115, Part IV. Inverse Method for Definite Integrals which vanish; and Theory of Reciprocal Functions, pp. 116–148]. The term *orthogonalisation* which is related, according to [222, Section 1.3, p. 78], to Schmidt's Göttingen dissertation from 1905 (see the historical note in Section 2.4.3), and its use with polynomials, is linked with the early works of Szegö from 1918–1919.

3.4 JACOBI MATRICES

An attentive reader will certainly have noticed that the Jacobi matrix (3.3.11) is related to the matrix (2.4.6), which contains the coefficients of the Lanczos algorithm for computing orthonormal bases of the Krylov subspaces. In fact, we can see the following:

Jacobi Matrices

> Jacobi matrices represent a cornerstone binding two wings of the same building: One is built from moments, continued fractions, and polynomials, with the purpose of approximating functions and integrals. The other is built from vectors, vector spaces, operators, and matrices with the purpose of matrix computations such as solving linear algebraic systems and approximating eigenvalues.

In order to learn more about Jacobi matrices, we now combine views from both sides. Among the sources that inspired our development are [222, Section 5.1], [217, Chapter II, Section 4], [670, Chapter 2, Section 2.4], [5, Chapter 4, Sections 1 and 2], and [508, Chapter 7]. An interesting classical view, which combines the Lanczos algorithm, Jacobi matrices, and orthogonal polynomials can be found in [346, Sections 1.5 and 1.6]. A historical note describing the origin and early use of Jacobi matrices is given in Section 3.4.3.

3.4.1 Algebraic Properties of Jacobi Matrices

As mentioned above, the rows of the Jacobi matrix T_n defined by (3.3.11) contain the coefficients of the three-term recurrence for the orthonormal polynomials $\varphi_0(\lambda), \varphi_1(\lambda), \ldots, \varphi_{n-1}(\lambda)$, and also the coefficients γ_n and δ_n determining the nth (non-normalised) polynomial $\delta_{n+1}\varphi_n(\lambda)$,

$$\delta_{n+1}\varphi_n(\lambda) \equiv \lambda\varphi_{n-1}(\lambda) - \gamma_n\varphi_{n-1}(\lambda) - \delta_n\varphi_{n-2}(\lambda). \tag{3.4.1}$$

If $\omega(\lambda)$ has $N < \infty$ points of increase and if $n = N$, then $\delta_{n+1}\varphi_n(\lambda)$ denotes the polynomial determined by the right-hand side of (3.4.1), without any specific meaning for the individual δ_{n+1} and $\varphi_n(\lambda)$. In the following we can exclude this case and hence assume that $n < \mathfrak{n}(\omega)$.

An easy calculation shows

$$\det(\lambda I - T_n) = (\lambda - \gamma_n)\det(\lambda I - T_{n-1}) - \delta_n^2 \det(\lambda I - T_{n-2}),$$

where $\det(\lambda I - T_2) = (\lambda - \gamma_1)(\lambda - \gamma_2) - \delta_2^2 = \psi_2(\lambda)$, and, if we formally consider $T_1 = \gamma_1$, $\det(\lambda I - T_1) = \lambda - \gamma_1 = \psi_1(\lambda)$. Since $\det(\lambda I - T_n)$ is a monic polynomial of degree n, and the recurrence for this determinant given above is equivalent to the recurrence (3.3.14) for the monic counterparts $\psi_n(\lambda)$ of the normalised polynomials $\varphi_n(\lambda)$, we can conclude

$$\psi_n(\lambda) = \det(\lambda I - T_n). \tag{3.4.2}$$

Consequently, the eigenvalues of the Jacobi matrix T_n are the roots of the polynomials $\psi_n(\lambda)$ and $\varphi_n(\lambda)$.

The polynomial description and the matrix–vector algebraic description can be linked by the following simple observation. *Any* Jacobi matrix T_n can be viewed as the result of n steps of the (Hermitian) Lanczos algorithm (Algorithm 2.4.3) applied to the same T_n with the initial vector e_1,

$$T_n I_n = I_n T_n, \quad I_n = [e_1, \ldots, e_n]; \tag{3.4.3}$$

see (2.4.7); note that the vector e_1 is of grade n with respect to T_n. The orthonormal Lanczos vectors from (3.4.3), given by $e_1, \ldots, e_n$, are generated by the three-term recurrence with the coefficients contained in the *columns* of T_n.

Using Algorithm 2.4.3, the jth Lanczos vector e_j can be viewed as the result of a polynomial of degree $j - 1$ in T_n applied to the initial vector e_1. For $j = 1$ we obtain $e_1 = \varphi_0(T_n) e_1$, and since T_n contains the coefficients of the Stieltjes recurrence (3.3.3), one can prove that

$$e_j = \varphi_{j-1}(T_n) e_1, \quad j = 2, \ldots, n. \tag{3.4.4}$$

Indeed, consider the first columns $\psi_{j-1}(T_n) e_1$ for $j = 2, 3, \ldots, n$. Using the vector $\psi_{j-1}(T_n) e_1$, where the last $n - j$ entries are zero, we can write

$$\psi_{j-1}(T_n) e_1 = \begin{bmatrix} \psi_{j-1}(T_{j-1}) e_1 \\ \prod_{\ell=2}^{j} \delta_\ell \\ 0 \end{bmatrix} = \prod_{\ell=2}^{j} \delta_\ell \, e_j,$$

where we have used $\psi_{j-1}(T_{j-1}) = 0$ (the matrix is annihilated by its characteristic polynomial) and the vector $\psi_{j-1}(T_{j-1}) e_1$ is of length $j - 1$. Consequently, using (3.3.15) we obtain (3.4.4).

The real symmetric matrix T_n can be orthogonally diagonalised, i.e.

$$T_n Z_n = Z_n \operatorname{diag}\left(\theta_1^{(n)}, \ldots, \theta_n^{(n)}\right), \tag{3.4.5}$$

where $\theta_1^{(n)}, \ldots, \theta_n^{(n)}$ are the n eigenvalues of T_n (which must be real), and the eigenvector matrix Z_n satisfies

$$Z_n = \left[z_1^{(n)}, \ldots, z_n^{(n)}\right], \quad Z_n^T Z_n = Z_n Z_n^T = I_n.$$

For simplicity, we denote

$$\operatorname{diag}\left(\theta_j^{(n)}\right) \equiv \operatorname{diag}\left(\theta_1^{(n)}, \ldots, \theta_n^{(n)}\right).$$

The orthonormality of the vectors $e_1, \ldots, e_n$ (with respect to the Euclidean inner product) generated by the Lanczos recurrence (3.4.3) then for $k, \ell = 0, 1, \ldots, n - 1$ gives

$$\begin{aligned}
(e_{k+1}, e_{\ell+1}) &= (\varphi_k(T_n) e_1, \varphi_\ell(T_n) e_1) \\
&= \left(Z_n \operatorname{diag}\left(\varphi_k\left(\theta_j^{(n)}\right)\right) Z_n^T e_1, Z_n \operatorname{diag}\left(\varphi_\ell\left(\theta_j^{(n)}\right)\right) Z_n^T e_1\right) \\
&= \sum_{j=1}^{n} \left(z_j^{(n)}, e_1\right)^2 \varphi_k\left(\theta_j^{(n)}\right) \varphi_\ell\left(\theta_j^{(n)}\right) \\
&= \delta_{k+1, \ell+1} \quad \text{(Kronecker's } \delta\text{)}. \tag{3.4.6}
\end{aligned}$$

Jacobi Matrices

For any two polynomials $\varphi(\lambda)$ and $\psi(\lambda)$ we define

$$(\varphi, \psi) \equiv \left(\varphi(T_n)e_1, \psi(T_n)e_1\right) = \sum_{j=1}^{n} \left(z_j^{(n)}, e_1\right)^2 \varphi\left(\theta_j^{(n)}\right) \psi\left(\theta_j^{(n)}\right). \tag{3.4.7}$$

We will show below that this defines an inner product on the space of the polynomials of degree at most $n - 1$.

Observe from (3.4.6) that

$$(\varphi_k, \varphi_\ell) = (e_{k+1}, e_{\ell+1}) = \delta_{k+1,\ell+1} \quad \text{for } k, \ell = 0, 1, \ldots, n - 1. \tag{3.4.8}$$

This is used in the proof of the following theorem, which links the eigenvalues and eigenvectors of the Jacobi matrix T_n to the n nodes and weights of the Gauss–Christoffel quadrature.

Theorem 3.4.1
As above, let $\varphi_0(\lambda), \varphi_1(\lambda), \ldots, \varphi_n(\lambda)$ be the orthonormal polynomials with respect to the inner product (3.2.8) associated with the distribution function $\omega(\lambda)$ defined on $[a, b]$. Denote by $\theta_1^{(n)}, \ldots, \theta_n^{(n)}$ the eigenvalues and $(z_1^{(n)}, e_1), \ldots, (z_n^{(n)}, e_1)$ the first components of the corresponding eigenvectors of the Jacobi matrix T_n associated with the three-term Stieltjes recurrence (3.3.3).

Then

$$\theta_j^{(n)} = \lambda_j^{(n)}, \quad \text{and} \quad \left(z_j^{(n)}, e_1\right)^2 = \omega_j^{(n)} > 0, \quad j = 1, \ldots, n, \tag{3.4.9}$$

where $\lambda_1^{(n)}, \ldots, \lambda_n^{(n)} \in (a, b)$ are the roots of $\varphi_n(\lambda)$, and $\omega_1^{(n)}, \ldots, \omega_n^{(n)}$ are the weights of the corresponding n-node Gauss–Christoffel quadrature associated with $\omega(\lambda)$.

As a consequence, the eigenvalues of the Jacobi matrix T_n are distinct and the first components of its eigenvectors are nonzero.

Proof
Consider a polynomial $f(\lambda)$ of degree at most $2n - 1$. Using the division theorem for polynomials (cf. footnote 1 in Section 3.2.1) we can write

$$f(\lambda) = \varphi_n(\lambda)p(\lambda) + r(\lambda) = \varphi_n(\lambda)p(\lambda) + \sum_{k=1}^{n-1} v_k \varphi_k(\lambda) + v_0,$$

where $p(\lambda)$ and $r(\lambda)$ are polynomials of degree at most $n - 1$, and $v_0, \ldots, v_{n-1}$ are some scalar coefficients.

Now consider the n-node Gauss–Christoffel quadrature associated with $\omega(\lambda)$. Since it integrates $f(\lambda)$ exactly, the orthogonality of $\varphi_0(\lambda), \varphi_1(\lambda), \ldots, \varphi_n(\lambda)$ with respect to the inner product (3.2.8) and the normalisation of the distribution function $\omega(\lambda)$ give

$$\sum_{j=1}^{n} \omega_j^{(n)} f\left(\lambda_j^{(n)}\right) = I_\omega^n(f) = I_\omega(f) = \int_a^b f(\lambda) \, d\omega(\lambda) = \int_a^b v_0 \, d\omega(\lambda) = v_0.$$

On the other hand, using the form (3.4.7) and the relations (3.4.8), we get

$$v_0 = \left(\varphi_0, v_0 \varphi_0 + \sum_{k=1}^{n-1} v_k \varphi_k\right) = \sum_{j=1}^{n} \left(z_j^{(n)}, e_1\right)^2 \left(v_0 + \sum_{k=1}^{n-1} v_k \varphi_k\left(\theta_j^{(n)}\right)\right)$$
$$= \sum_{j=1}^{n} \left(z_j^{(n)}, e_1\right)^2 f\left(\theta_j^{(n)}\right).$$

The assertion now follows from the fact that the n-node Gauss–Christoffel quadrature is uniquely determined (in fact, we only needed to prove the expression for the weights; the first part of the statement follows immediately from (3.4.2)). □

This theorem has the following immediate consequence.

Corollary 3.4.2
Using the previous notation, (3.4.7) defines an inner product on the space of polynomials of degree at most $n - 1$. The polynomials $\varphi_0(\lambda), \varphi_1(\lambda), \ldots, \varphi_{n-1}(\lambda)$ are not only orthonormal with respect to the inner product (3.2.8), but also with respect to (3.4.7).

Proof
Obviously, (3.4.7) is a symmetric bilinear form on the space of polynomials of degree at most $n - 1$. In addition, for any polynomial $\varphi(\lambda)$ of degree at most $n - 1$,

$$(\varphi, \varphi) = \sum_{j=1}^{n} \left(z_j^{(n)}, e_1\right)^2 \varphi^2\left(\theta_j^{(n)}\right),$$

which by Theorem 3.4.1 is positive, unless $\varphi(\lambda) \equiv 0$. □

Consider the spectral decomposition of the Jacobi matrix T_n as in (3.4.5). Then for any polynomial $p(\lambda)$ we have

$$p(T_n) = Z_n p\left(\text{diag}\left(\theta_j^{(n)}\right)\right) Z_n^T,$$

so that, in particular,

$$e_1^T p(T_n) e_1 = \sum_{j=1}^{n} \left(z_j^{(n)}, e_1\right)^2 p\left(\theta_j^{(n)}\right).$$

If $p(\lambda)$ is of degree at most $2n - 1$, then the Gauss–Christoffel quadrature for the Riemann–Stieltjes integral of $p(\lambda)$ is exact. Using the inner product (3.2.8) and Theorem 3.4.1 we can therefore write

$$(p, 1) = \int_a^b p(\lambda)\, d\omega(\lambda) = \int_a^b p(\lambda)\, d\omega^{(n)}(\lambda) = e_1^T p(T_n) e_1;$$

see [183, p. 32].

We have shown in Theorem 3.4.1 that the eigenvalues of the Jacobi matrix T_n associated with the distribution function $\omega(\lambda)$ are distinct, and that the first components of the corresponding eigenvectors are nonzero. This property in fact holds for any Jacobi matrix.

Theorem 3.4.3
The eigenvalues of a Jacobi matrix are distinct and the first component of each of its eigenvectors is nonzero.

Proof
Let T_M be any $M \times M$ Jacobi matrix, with the orthogonal eigendecomposition

$$T_M Z_M = Z_M \operatorname{diag}\left(\theta_j^{(M)}\right).$$

The equation

$$T_M I_M = I_M T_M \tag{3.4.10}$$

represents, analogously to (3.4.3), M steps of the (Hermitian) Lanczos algorithm applied to the matrix T_M with the initial vector e_1. Then

$$e_j = \varphi_{j-1}^{(M)}(T_M) e_1, \quad j = 2, \ldots, M, \tag{3.4.11}$$

where $\varphi_{j-1}^{(M)}(\lambda)$ is some polynomial of degree $j-1$. Now define, analogously to (3.4.7), the bilinear form

$$(\varphi, \psi)_M \equiv \sum_{j=1}^{M} \left(z_j^{(M)}, e_1\right)^2 \varphi\left(\theta_j^{(M)}\right) \psi\left(\theta_j^{(M)}\right). \tag{3.4.12}$$

The rest of the proof is by contradiction.

Assume that the eigenvalues $\theta_1^{(M)}, \ldots, \theta_M^{(M)}$ of T_M are not all distinct or that $(z_j^{(M)}, e_1) = 0$ for some j. Omitting the zero terms in (3.4.12) and combining the terms with the same eigenvalues then gives

$$(\varphi, \psi)_M = (\varphi, \psi)_K \equiv \sum_{j=1}^{K} \left(\widehat{z}_j^{(M)}, e_1\right)^2 \varphi\left(\widehat{\theta}_j^{(M)}\right) \psi\left(\widehat{\theta}_j^{(M)}\right), \tag{3.4.13}$$

where $K \leq M-1$ and $\widehat{\theta}_1^{(M)}, \ldots, \widehat{\theta}_K^{(M)}$ are the distinct eigenvalues of T_M with $(\widehat{z}_j^{(M)}, e_1) \neq 0$, $j = 1, \ldots, K$. Since the eigenvector matrix Z_M is nonsingular, its first row has at least one nonzero element, and hence $K \geq 1$.

The bilinear form $(\varphi, \psi)_K$ given by (3.4.13) represents an inner product on the set of all polynomials of degree at most $K-1 \leq M-2$. Using (3.4.10)–(3.4.11) we get, analogously to (3.4.8),

$$\left(\varphi_i^{(M)}, \varphi_\ell^{(M)}\right)_K = \left(\varphi_i^{(M)}, \varphi_\ell^{(M)}\right)_M = \delta_{i+1,\ell+1} \quad \text{for } i, \ell = 0, 1, \ldots, M-1. \tag{3.4.14}$$

In particular,

$$\left(\varphi_K^{(M)}, \varphi_K^{(M)}\right)_K = \sum_{j=1}^{K} \left(\widehat{z}_j^{(M)}, e_1\right)^2 \left(\varphi_K\left(\widehat{\theta}_j^{(M)}\right)\right)^2 = 1.$$

Since $\varphi_0^{(M)}(\lambda) = 1, \varphi_1^{(M)}(\lambda), \ldots, \varphi_{K-1}^{(M)}(\lambda)$ are orthonormal polynomials with respect to the inner product $(\varphi, \psi)_K$ defined by (3.4.13), the unique monic orthogonal polynomial corresponding to $\varphi_K^{(M)}(\lambda)$ is determined by

$$\psi_K^{(M)}(\lambda) = \lambda^K - \sum_{j=0}^{K-1} (\lambda^K, \varphi_j)_K \varphi_j(\lambda);$$

see (3.2.11)–(3.2.12). Analogously to Section 3.2.1 we conclude that this polynomial is equal to

$$\psi_K^{(M)}(\lambda) = \left(\lambda - \widehat{\theta}_1^{(M)}\right) \cdots \left(\lambda - \widehat{\theta}_K^{(M)}\right),$$

and therefore $\varphi_K^{(M)}\left(\widehat{\theta}_j^{(M)}\right) = 0$, $j = 1, \ldots, K$, giving $\left(\varphi_K^{(M)}, \varphi_K^{(M)}\right)_K = 0$, a contradiction. □

The proof of Theorem 3.4.3 uses properties of orthogonal polynomials and inner products on the space of polynomials. One can also prove this result purely algebraically as follows.

Let $z_j^{(M)} = \left[z_{j1}^{(M)}, \ldots, z_{jM}^{(M)}\right]^T$ be an eigenvector with the corresponding eigenvalue $\theta_j^{(M)}$ of the Jacobi matrix

$$T_M \equiv \begin{bmatrix} \gamma_1 & \delta_2 & & \\ \delta_2 & \gamma_2 & \ddots & \\ & \ddots & \ddots & \delta_M \\ & & \delta_M & \gamma_M \end{bmatrix}.$$

Suppose that $z_{j1}^{(M)} = 0$. Comparing the first components on both sides of the equation

$$T_M z_j^{(M)} = \theta_j^{(M)} z_j^{(M)}$$

gives

$$\delta_2 z_{j2}^{(M)} = 0.$$

Since $\delta_2 > 0$ we must have $z_{j2}^{(M)} = 0$. Continuing with the next components leads to $z_j^{(M)} = 0$, which is a contradiction since $z_j^{(M)}$ is an eigenvector. Therefore $z_{j1}^{(M)} = (z_j^{(M)}, e_1) \neq 0$, and in the same way we can show that $z_{jM}^{(M)} = (z_j^{(M)}, e_M) \neq 0$.

The distinctness of the eigenvalues of T_M follows algebraically from the fact that $T_M - \lambda I$ always has a minor of order $n - 1$ which is nonzero (just skip the first column and the last row); see [508, Lemma 7.7.1].

Remark 3.4.4
The proof of Theorem 3.4.3 has shown that for any given Jacobi matrix T_M the corresponding bilinear form (3.4.12) is an inner product on the space of polynomials of degree at most $M - 1$. Moreover, with respect to this inner product the polynomials $\varphi_0^{(M)}(\lambda), \ldots, \varphi_{M-1}^{(M)}(\lambda)$ form an orthonormal basis of the space; cf. (3.4.14). The characteristic polynomial of T_M is given by

$$\psi_M^{(M)}(\lambda) = \left(\lambda - \theta_1^{(M)}\right) \cdots \left(\lambda - \theta_M^{(M)}\right)$$

and $\psi_j^{(M)}(\lambda)$, the monic orthogonal counterpart to $\varphi_j^{(M)}(\lambda)$, $j = 1, \ldots, M - 1$, is the characteristic polynomial of the left $j \times j$ principal submatrix of T_M. Therefore Theorem 3.3.1 and Corollary 3.3.3 can be reformulated as the interlacing theorem and corollary for the eigenvalues of the left principal submatrices of a Jacobi matrix; see [355, Chapter 4, Theorem 4, p. 168], [670, Chapter 2, Section 47, Chapter 5, Section 38], or [235]. As pointed out in Remark 3.3.2, the Cauchy interlacing theorem for eigenvalues of Hermitian matrices is considerably weaker than the interlacing theorem for eigenvalues of Jacobi matrices.

The eigenvectors of a Jacobi matrix can be expressed in terms of the associated orthogonal polynomials. Rewriting (3.3.10) as

$$T_n \Phi_n(\lambda) = \lambda \, \Phi_n(\lambda) - \delta_{n+1} \varphi_n(\lambda) \, e_n, \qquad (3.4.15)$$

and considering that $\varphi_n(\theta_j^{(n)}) = 0, j = 1, \ldots, n$, we immediately get

$$T_n \Phi_n\left(\theta_j^{(n)}\right) = \theta_j^{(n)} \Phi_n\left(\theta_j^{(n)}\right), \quad j = 1, \ldots, n, \qquad (3.4.16)$$

where

$$\Phi_n\left(\theta_j^{(n)}\right) = \left[1, \varphi_1\left(\theta_j^{(n)}\right), \ldots, \varphi_{n-1}\left(\theta_j^{(n)}\right)\right]^T.$$

Thus, the normalised eigenvectors of T_n are given by

$$z_j^{(n)} = \Phi_n\left(\theta_j^{(n)}\right) \Big/ \left\|\Phi_n\left(\theta_j^{(n)}\right)\right\|, \quad j = 1, \ldots, n. \qquad (3.4.17)$$

Due to the interlacing property of the roots of orthogonal polynomials (see Theorem 3.3.1), we must have $\varphi_{n-1}(\theta_j^{(n)}) \neq 0, j = 1, \ldots, n$, so that not only the first, but also the last element of each eigenvector of T_n is nonzero.

Writing the spectral decomposition (3.4.5) of a Jacobi matrix T_n in the form

$$\operatorname{diag}\left(\theta_j^{(n)}\right) Z_n^T = Z_n^T T_n, \qquad (3.4.18)$$

we can consider T_n as the result of the Lanczos algorithm applied to the diagonal matrix $\operatorname{diag}(\theta_j^{(n)})$ with the initial vector $Z_n^T e_1$ that is composed of the first elements of the normalised eigenvectors of T_n. The Jacobi matrix T_n is therefore uniquely determined by its eigenvalues and the first components of the corresponding normalised eigenvectors, which means that it can be 'reconstructed' using just this data. We will formulate this result as a proposition.

Proposition 3.4.5
Given $2n$ real numbers $\theta_1^{(n)}, \ldots, \theta_n^{(n)}, \zeta_1, \ldots, \zeta_n$ with

$$\theta_1^{(n)} < \theta_2^{(n)} < \cdots < \theta_n^{(n)}, \quad \zeta_j \neq 0,\ j = 1, \ldots, n, \quad \text{and} \quad \sum_{j=1}^n \zeta_j^2 = 1,$$

the Lanczos algorithm applied to the diagonal matrix $\operatorname{diag}(\theta_j^{(n)})$ with the initial vector $[\zeta_1, \ldots, \zeta_n]^T$ determines the unique Jacobi matrix T_n having the eigenvalues $\theta_j^{(n)}$ such that the corresponding normalised eigenvectors $z_j^{(n)}$ satisfy $(z_j^{(n)}, e_1) = \zeta_j$, $j = 1, \ldots, n$. The Jacobi matrix is independent of the sign change in any of the numbers ζ_j, $j = 1, \ldots, n$.

Proof
It remains to prove the last part. Let the sign changes be realised via multiplication by the diagonal matrix D, which has diagonal entries ± 1,

$$Z_n \longrightarrow Z_n D.$$

Then from (3.4.18)

$$D \operatorname{diag}\left(\theta_\ell^{(n)}\right) Z_n^T = D Z_n^T T_n,$$

which shows the assertion. □

In the context of the Gauss–Christoffel quadrature, the considerations above lead to a solution of an inverse problem: given nodes and weights of the n-node Gauss–Christoffel quadrature, compute the entries of the corresponding Jacobi matrix (which determine the three-term recurrence for the associated orthogonal polynomials). The investigation of the last problem has a remarkably long history; for a brief recollection of the literature see [487, Section 3.1] and also [596, Section 3]. The first and very elegant solution of reconstruction of a Jacobi matrix from spectral data (in a mathematical sense, without considering computational issues) known to us was published by Wendroff in 1961 [665]. In the classical language of orthogonal polynomials, however, the problem is solved by the Stieltjes recurrence (3.3.3). As pointed out by many authors, see, e.g. [130, 520], and as also in our development above, the Stieltjes recurrence (implemented with modified Gram–Schmidt orthogonalisation and normalisation of the orthogonal polynomials as in (3.3.7)–(3.3.9)) is equivalent to the Lanczos algorithm. In finite precision arithmetic this algorithm is numerically unstable in the sense that the computed vectors gradually (and sometimes very quickly) lose orthogonality, and eventually even linear independence. This fact, thoroughly analysed starting from the pioneering PhD thesis of Paige in 1971 [489], has been noted in the orthogonal polynomial literature; see, e.g. [223, 224], [188, Section 2]. Surprisingly, reorthogonalisation has been rejected as too costly [259, p. 325], although the cost is negligible as long as n is small. Moreover, the analysis of the finite precision behaviour of the Lanczos algorithm by Paige, Parlett, Scott, Simon, Greenbaum, and others (reviewed, for example, in [457]) also supplied convincing examples for illustrating and testing numerical instabilities. In spite of the works [253, 130, 378, 68, 242, 226], which emphasise the interplay between the classical polynomial and vector algebraic formulations, the analysis of the finite precision behaviour of the Lanczos algorithm is rarely mentioned in the literature on orthogonal polynomials.

In order to overcome the numerical instability of the Lanczos algorithm, Gragg and Harrod suggested in their beautiful paper of 1984 [259] a new algorithm based on ideas of Rutishauser; for an interesting experimental comparison see [520, Section 2]. An alternative approach from [188] is based on a remarkable result by Nevai on modifications of the recurrence coefficients when adding a single point of increase to the given distribution function; see [476, Section 7, Lemma 15, p. 131], [188, Section 3, Lemma 1, p. 187]. From numerical results Gragg and Harrod spotted a curious phenomenon: close distribution functions can give very different Jacobi matrices. They concluded that the inverse problem of reconstructing a Jacobi matrix from the nodes and weights is ill-conditioned [259, p. 330 and p. 332]. This conclusion has been examined by Laurie [414], who pointed out that the negative statement is linked to the use of the max-norm for vectors. He suggested measuring the perturbation of the weights in the componentwise relative sense; see [414, p. 179] and [415, Section 6]. The main part of [414] is devoted to a constructive proof of the following statement, which is found on p. 168 of that paper.

Theorem 3.4.6
In the previous notation, let the n weights, the first node and the $n - 1$ positive differences between the consecutive nodes of the n-node Gauss–Christoffel quadrature be given. Then the main diagonal entries, shifted by the smallest node, and the off-diagonal entries of the corresponding Jacobi matrix T_n can be computed in $\frac{9}{2}n^2 + O(n)$ arithmetic operations, all of which can involve only addition, multiplication, and division of positive numbers.

Consequently, in finite precision arithmetic, the shifted main diagonal entries as well as the off-diagonal entries of T_n, can be computed to a *relative accuracy* no worse than $\frac{9}{2}n^2\varepsilon + O(n\varepsilon)$, where ε denotes the machine precision. This result also gives a bound on the condition number of the problem. If the weights, the first node, and the $n - 1$ positive differences between the consecutive nodes are perturbed, with the size of the relative perturbations of the individual entries bounded by some small ϵ, then such perturbation can cause a relative change of the individual entries of the shifted main diagonal and of the individual off-diagonal entries of T_n, no larger than $\frac{9}{2}n^2\epsilon + O(n\epsilon)$. The resulting algorithm combines ideas from earlier works from approximation theory, orthogonal polynomials, and numerical linear algebra.

The following proposition shows that any orthonormal basis of a finite-dimensional vector space can be generated by the Hermitian Lanczos algorithm. Note that it is sufficient to formulate this statement for the space $\mathbb{C}^M$.

Proposition 3.4.7
Any orthonormal basis $v_1, \ldots, v_M$ of $\mathbb{C}^M$ can be generated by the Lanczos algorithm applied to a Hermitian matrix having any given M distinct eigenvalues $\theta_1 < \theta_2 < \cdots < \theta_M$ and the initial vector v_1.

Proof
Let T_M be any Jacobi matrix with the eigenvalues $\theta_1, \ldots, \theta_M$ and the spectral decomposition

$$T_M Z = Z \operatorname{diag}(\theta_j), \quad Z^T Z = Z Z^T = I.$$

Choosing the first row of Z, such matrix is determined as in (3.4.18),

$$\operatorname{diag}(\theta_j) Z^T = Z^T T_M.$$

Let $V = [v_1, \ldots, v_M]$. Multiplying the previous equality from the left by VZ and substituting $(VZ)^*(VZ) = I$ gives

$$(VZ) \operatorname{diag}(\theta_j) (VZ)^* (VZ) Z^T = (VZ) Z^T T_M.$$

Denoting $A \equiv (VZ) \operatorname{diag}(\theta_j) (VZ)^*$, we finally get

$$AV = V T_M, \qquad (3.4.19)$$

which finishes the proof. □

The relationship between the matrix of the normalised eigenvectors Y of A, the matrix of the Lanczos vectors V, and the matrix of the normalised eigenvectors Z of the Jacobi matrix T_M in (3.4.19) is then given by

$$Y = VZ. \qquad (3.4.20)$$

In the rest of this section we present some further (and mostly well-known) results connecting Jacobi matrices, the Gauss–Christoffel quadrature and continued fractions. In the proof of Theorem 3.3.6 we have observed that the weights in the n-node Gauss–Christoffel quadrature can be expressed in terms of the numerator $\mathcal{R}_n(\lambda)$ of the nth convergent $\mathcal{F}_n(\lambda)$ of the associated continued fraction (see (3.3.18)) as

$$\omega_j^{(n)} = \frac{\mathcal{R}_n\left(\lambda_j^{(n)}\right)}{\psi_n'\left(\lambda_j^{(n)}\right)}.$$

Jacobi Matrices

With $\lambda_j^{(n)} = \theta_j^{(n)}$, $j = 1, \ldots, n$, and noticing that the recurrence (3.3.21) can be considered as the recurrence giving the polynomial $\psi_{2,n}(\lambda)$, which is the characteristic polynomial of the matrix $T_{2,n}$ obtained from T_n by deleting the first row and column,

$$\omega_j^{(n)} = \left(z_j^{(n)}, e_1\right)^2 = \frac{\psi_{2,n}\left(\theta_j^{(n)}\right)}{\psi_n'\left(\theta_j^{(n)}\right)}, \quad j = 1, \ldots, n. \tag{3.4.21}$$

Another expression for the weights $\omega_j^{(n)}$ can be obtained in the following way. For any $j = 1, \ldots, n$ we can write

$$\psi_{n-1}(\lambda) = \frac{\psi_n(\lambda)}{\lambda - \theta_j^{(n)}} + \widetilde{\psi}_{n-2,j}(\lambda),$$

where $\widetilde{\psi}_{n-2,j}(\lambda)$ is some polynomial of degree at most $n-2$. Using this and applying the n-node Gauss–Christoffel quadrature gives

$$\prod_{i=2}^{n} \delta_i^2 = \int_a^b \psi_{n-1}^2(\lambda) \, d\omega(\lambda) = \int_a^b \psi_{n-1}(\lambda) \frac{\psi_n(\lambda)}{\lambda - \theta_j^{(n)}} \, d\omega(\lambda)$$

$$= \sum_{\ell=1}^{n} \omega_\ell^{(n)} \left[\psi_{n-1}(\lambda) \frac{\psi_n(\lambda)}{\lambda - \theta_j^{(n)}}\right]_{\lambda = \theta_\ell^{(n)}} = \omega_j^{(n)} \psi_{n-1}\left(\theta_j^{(n)}\right) \psi_n'\left(\theta_j^{(n)}\right),$$

which yields

$$\omega_j^{(n)} = \frac{\prod_{i=2}^{n} \delta_i^2}{\psi_{n-1}\left(\theta_j^{(n)}\right) \psi_n'\left(\theta_j^{(n)}\right)}, \quad j = 1, \ldots, n. \tag{3.4.22}$$

Moreover, using (3.4.17),

$$\omega_j^{(n)} = \left(z_j^{(n)}, e_1\right)^2 = \frac{1}{\left\|\Phi_n\left(\theta_j^{(n)}\right)\right\|^2} = \left(\sum_{k=0}^{n-1} \varphi_k^2\left(\theta_j^{(n)}\right)\right)^{-1}, \quad j = 1, \ldots, n,$$

which has been proven by various authors using the so-called *Christoffel–Darboux identity*; see, e.g. [222, Section 1.4] and [667] (particularly Sections 2.5 and 2.9; relations (35) and (69)).

We will finish this section with two important observations. First, let

$$\Pi_n \equiv \begin{bmatrix} 0 & \cdots & \cdots & 0 & 1 \\ & & & 1 & 0 \\ & & \cdot\cdot\cdot & & \\ 0 & 1 & & & \\ 1 & 0 & \cdots & \cdots & 0 \end{bmatrix}, \quad \Pi_n^2 = I_n,$$

be the permutation matrix corresponding to the reordering $\{1, 2, \ldots, n\} \longrightarrow \{n, n-1, \ldots, 1\}$ (sometimes called the *reverse identity matrix*). Then (3.4.3) can be written as

$$(\Pi_n T_n \Pi_n) I_n = I_n (\Pi_n T_n \Pi_n).$$

Denoting

$$T_n^\Pi \equiv \Pi_n T_n \Pi_n = \begin{bmatrix} \gamma_n & \delta_n & & \\ \delta_n & \gamma_{n-1} & \ddots & \\ & \ddots & \ddots & \delta_2 \\ & & \delta_2 & \gamma_1 \end{bmatrix}, \qquad (3.4.23)$$

we get the permuted counterpart of (3.4.3),

$$T_n^\Pi I_n = I_n T_n^\Pi, \qquad (3.4.24)$$

with the spectral decomposition of T_n^Π given by

$$T_n^\Pi Z_n^\Pi = Z_n^\Pi \operatorname{diag}\left(\theta_j^{(n)}\right), \quad Z_n^\Pi = \Pi_n Z_n. \qquad (3.4.25)$$

As a consequence, the results developed throughout this subsection hold for T_n^Π and Z_n^Π. The individual eigenvectors of T_n^Π correspond to the eigenvectors of T_n, with the order of the entries being reversed, and the characteristic polynomial of the matrix $T_{2,n}^\Pi$ is proportional to $\psi_{n-1}(\lambda)$. Therefore we get, in particular, a counterpart of (3.4.21),

$$\left(z_j^{(n)}, e_n\right)^2 = \frac{\psi_{n-1}\left(\theta_j^{(n)}\right)}{\psi_n'\left(\theta_j^{(n)}\right)}, \quad j = 1, \ldots, n. \qquad (3.4.26)$$

Analogously to (3.4.22),

$$\left(z_j^{(n)}, e_n\right)^2 = \frac{\prod_{i=2}^n \delta_i^2}{\psi_{2,n}\left(\theta_j^{(n)}\right) \psi_n'\left(\theta_j^{(n)}\right)}, \quad j = 1, \ldots, n. \qquad (3.4.27)$$

Combining this with (3.4.21) gives

$$\left(z_j^{(n)}, e_1\right)^2 \left(z_j^{(n)}, e_n\right)^2 = \frac{\prod_{i=2}^n \delta_i^2}{\left(\psi_n'\left(\theta_j^{(n)}\right)\right)^2}, \quad j = 1, \ldots, n.$$

Second, the previous results on Jacobi matrices can be generalised to any real tridiagonal matrix of the form

$$\widetilde{T}_n = \begin{bmatrix} \gamma_1 & \delta_2^U & & \\ \delta_2^L & \gamma_2 & \ddots & \\ & \ddots & \ddots & \delta_n^U \\ & & \delta_n^L & \gamma_n \end{bmatrix}, \quad \text{with } \delta_j^L \delta_j^U > 0, j = 2, \ldots, n. \quad (3.4.28)$$

Indeed, the polynomial

$$\widetilde{\psi}_j(\lambda) = \det(\lambda I - \widetilde{T}_j),$$

i.e. the characteristic polynomial of the jth leading principal submatrix of $\widetilde{T}_n$, is determined by the recurrence

$$\widetilde{\psi}_j(\lambda) = (\lambda - \gamma_j)\widetilde{\psi}_{j-1}(\lambda) - \left(\delta_j^L \delta_j^U\right)\widetilde{\psi}_{j-2}(\lambda), \quad j = 1, \ldots, n,$$

where $\delta_1^L \equiv \delta_1^U = 0$, $\widetilde{\psi}_{-1}(\lambda) \equiv 0$, $\widetilde{\psi}_0(\lambda) \equiv 1$, $j = 2, \ldots, n$. The off-diagonal entries enter the recurrence only in the form of products $\delta_j^L \delta_j^U$, $j = 2, \ldots, n$, which suggests that replacing in $\widetilde{T}_n$ the numbers δ_j^L and δ_j^U by $\sqrt{\delta_j^L \delta_j^U}$, $j = 2, \ldots, n$, symmetrises the matrix without changing the characteristic polynomials of any of its left principal submatrices. The symmetrisation can be achieved by the following diagonal similarity transformation

$$T_n \equiv D_n \widetilde{T}_n D_n^{-1}, \quad D_n^2 = \mathrm{diag}\left(1, \frac{\delta_2^U}{\delta_2^L}, \ldots, \prod_{j=2}^n \frac{\delta_j^U}{\delta_j^L}\right), \quad (3.4.29)$$

so that the spectral decomposition of $\widetilde{T}_n$ is given by

$$\widetilde{T}_n = D_n^{-1} T_n D_n = D_n^{-1} Z_n \,\mathrm{diag}\left(\theta_j^{(n)}\right) Z_n^T D_n = \widetilde{Z}_n \,\mathrm{diag}\left(\theta_j^{(n)}\right) \widetilde{Z}_n^T,$$

where $\widetilde{Z}_n \equiv D_n^{-1} Z_n$. For further details and for a description of the oscillatory properties of eigenvectors and the relationship with the problem of oscillations of mechanical systems we refer to [217]; see particularly the Introduction and Chapter II of that book.

3.4.2 The Persistence Theorem, Stabilisation of Nodes and Weights in the Gauss–Christoffel Quadrature

Given a distribution function $\omega(\lambda)$, one can ask how the nodes and weights of the n-node Gauss–Christoffel quadrature evolve with increasing n. In Theorem 3.4.1 we have shown that the nodes of the nth Gauss–Christoffel quadrature are equal to the eigenvalues of the associated $n \times n$ Jacobi matrix, and that the weights are given by the squared first elements of the corresponding normalised eigenvectors. Given an $n \times n$ Jacobi matrix T_n, one can therefore reformulate the question and ask what can be said about location of eigenvalues and the first components of the corresponding eigenvectors of the Jacobi matrix that contains T_n as its left principal submatrix.

In this section we will consider a non-decreasing distribution function $\omega(\lambda)$ defined on a real finite interval $[a, b]$, with $\omega(a) = 0$, $\omega(b) = 1$, and with M points of increase $a < \lambda_1 < \lambda_2 < \cdots < \lambda_M \leq b$, and corresponding positive weights $\omega_1, \omega_2, \ldots, \omega_M$ (summing up to 1). Associated with the M nodes and weights is an $M \times M$ Jacobi matrix T_M; see Proposition 3.4.5. We denote its spectral decomposition by

$$T_M Z = Z \operatorname{diag}(\lambda_\ell), \quad Z^T Z = Z Z^T = I_M, \quad Z = [z_1, \ldots, z_M],$$

where we write Z instead of Z_M for simplicity.

The nth left principal submatrix of T_M will be denoted by T_n, with its spectral decomposition denoted by

$$T_n Z_n = Z_n \operatorname{diag}\left(\theta_j^{(n)}\right), \quad Z_n^T Z_n = Z_n Z_n^T = I_n, \quad Z_n = \left[z_1^{(n)}, \ldots, z_n^{(n)}\right].$$

The following proposition shows how each eigenvalue $\theta_j^{(n)}$ of T_n can be written as a convex combination of the eigenvalues $\lambda_1, \ldots, \lambda_M$ of T_M.

Proposition 3.4.8
In the previous notation, let $n < M$ and suppose that $V_n = [e_1, \ldots, e_n]$ is the $M \times n$ matrix generated by n steps of the Lanczos algorithm applied to T_M and the initial vector e_1; see (3.4.10). Define the $M \times n$ matrix

$$W_n \equiv Z^T V_n Z_n \equiv \left[w_1^{(n)}, \ldots, w_n^{(n)}\right],$$

then

$$\theta_j^{(n)} = \sum_{\ell=1}^M \left(w_j^{(n)}, e_\ell\right)^2 \lambda_\ell, \quad \sum_{\ell=1}^M \left(w_j^{(n)}, e_\ell\right)^2 = 1, \quad j = 1, \ldots, n. \quad (3.4.30)$$

Proof
The first n steps of the Lanczos algorithm applied to T_M with the initial vector e_1 produce the left principal submatrix T_n of T_M. Using the definition of V_n we can write

$$T_n = V_n^T T_M V_n = V_n^T Z \operatorname{diag}(\lambda_\ell) Z^T V_n.$$

The spectral decomposition of T_n then gives

$$\operatorname{diag}(\theta_j^{(n)}) = Z_n^T V_n^T Z \operatorname{diag}(\lambda_\ell) Z^T V_n Z_n = W_n^T \operatorname{diag}(\lambda_\ell) W_n,$$

which, together with $\|w_j^{(n)}\| = 1, j = 1, \ldots, n$, finishes the proof. □

Note that
$$W_n = Z^T V_n Z_n = Z^T \begin{bmatrix} Z_n \\ 0 \end{bmatrix}.$$

Consequently, the coefficients of the convex combination (3.4.30) depend only on the first n elements of the eigenvectors of T_M.

Theorem 3.3.1 and Corollary 3.3.3 on interlacing show that for any Jacobi matrix T_k that contains T_n as its left principal submatrix, $n < k \leq M$, there is at least one eigenvalue of T_k in each open interval determined by the $n+2$ ordered points $a, \theta_1^{(n)}, \ldots, \theta_n^{(n)}, b$; see Remark 3.4.4. We may ask when the information obtained from T_n guarantees that for a given eigenvalue $\theta_j^{(n)}$ there is an eigenvalue of T_k close to it. This must certainly take place, as a trivial consequence of interlacing, when

$$\text{either} \quad \theta_j^{(n)} - \theta_{j-1}^{(n)} \quad \text{or} \quad \theta_{j+1}^{(n)} - \theta_j^{(n)} \quad \text{is small.}$$

In general, considering

$$T_n z_j^{(n)} = \theta_j^{(n)} z_j^{(n)}, \quad \left\| z_j^{(n)} \right\| = 1,$$

we denote, for any $k, n < k \leq M$,

$$\widehat{z}_j^{(n,k)} = \begin{bmatrix} z_j^{(n)} \\ 0 \end{bmatrix} \in \mathbb{R}^k.$$

Then

$$T_k \widehat{z}_j^{(n,k)} = \begin{bmatrix} T_n z_j^{(n)} \\ 0 \end{bmatrix} + \begin{bmatrix} 0 \\ \delta_{n+1} \left(z_j^{(n)}, e_n \right) e_1 \end{bmatrix}$$

$$= \theta_j^{(n)} \widehat{z}_j^{(n,k)} + \begin{bmatrix} 0 \\ \delta_{n+1} \left(z_j^{(n)}, e_n \right) e_1 \end{bmatrix},$$

which gives

$$\left\| T_k \widehat{z}_j^{(n,k)} - \theta_j^{(n)} \widehat{z}_j^{(n,k)} \right\| = \delta_{n+1} \left| \left(z_j^{(n)}, e_n \right) \right|. \tag{3.4.31}$$

This leads to the *persistence theorem*, originally proved by Paige in his PhD thesis [489]; cf. [492, p. 241] and [670, p. 171].

Theorem 3.4.9
Using the previous notation, $n < k \leq M$, we get for any $j = 1, \ldots, n$,

$$\min_\ell \left| \theta_j^{(n)} - \theta_\ell^{(k)} \right| \leq \delta_{n+1} \left| \left(z_j^{(n)}, e_n \right) \right| \equiv \delta(j).$$

In words, any Jacobi matrix T_k which contains T_n as its left principal submatrix has for any eigenvalue $\theta_j^{(n)}$ of T_n an eigenvalue within $\delta(j)$ of $\theta_j^{(n)}$. If $\delta(j)$ is small, then T_k has an eigenvalue close to the eigenvalue $\theta_j^{(n)}$ of T_n.

Proof
Using (3.4.31) and the spectral decomposition of T_k,

$$\delta_{n+1}\left|\left(z_j^{(n)}, e_n\right)\right| = \left\|Z_k \operatorname{diag}\left(\theta_\ell^{(k)}\right) Z_k^T \widehat{z}_j^{(n,k)} - \theta_j^{(n)} \widehat{z}_j^{(n,k)}\right\|$$

$$= \left\|\operatorname{diag}\left(\theta_\ell^{(k)}\right)\left(Z_k^T \widehat{z}_j^{(n,k)}\right) - \theta_j^{(n)}\left(Z_k^T \widehat{z}_j^{(n,k)}\right)\right\|$$

$$= \left\|\operatorname{diag}\left(\theta_\ell^{(k)} - \theta_j^{(n)}\right)\left(Z_k^T \widehat{z}_j^{(n,k)}\right)\right\|$$

$$= \left(\sum_{\ell=1}^k \left(\theta_\ell^{(k)} - \theta_j^{(n)}\right)^2 \left(\widehat{z}_j^{(n,k)}, z_\ell^{(k)}\right)^2\right)^{\frac{1}{2}}$$

$$\geq \min_\ell \left|\theta_\ell^{(k)} - \theta_j^{(n)}\right| \left(\sum_{\ell=1}^k \left(\widehat{z}_j^{(n,k)}, z_\ell^{(k)}\right)^2\right)^{\frac{1}{2}}$$

$$= \min_\ell \left|\theta_\ell^{(k)} - \theta_j^{(n)}\right|,$$

which completes the proof. □

The persistence theorem motivates the following terminology introduced by Paige.

Definition 3.4.10: Consider the notation of Theorem 3.4.9, with T_n being the $n \times n$ left principal submatrix of a $k \times k$ Jacobi matrix T_k, $n < k \leq M$. An eigenvalue $\theta_j^{(n)}$ of T_n is called stabilised to within $\delta_{n+1}|(z_j^{(n)}, e_n)|$, and T_k has an eigenvalue within $\delta_{n+1}|(z_j^{(n)}, e_n)|$ of $\theta_j^{(n)}$.

This concept of stabilisation plays an important role in the theory of the finite precision behaviour of the Lanczos algorithm developed by Paige [489, 490, 491, 492], as well as in the subsequent work of Parlett [508, 507] and his students, see [557, 558, 559, 565, 567, 566], Greenbaum [269, 270], and others; see also the review of the results in [457, Sections 4–5]. In this section we will not consider effects of rounding errors. Instead, we will examine stabilisation of the eigenvalues and the behaviour of the first elements of the associated normalised eigenvectors of Jacobi matrices. Translated into the language of orthogonal polynomials and quadrature, we will examine stabilisation of nodes and weights in the Gauss–Christoffel quadrature.

As for the nodes, the interlacing property given in Theorem 3.3.1 guarantees that if the n-node Gauss–Christoffel quadrature has two nodes, say $\lambda_j^{(n)}$ and $\lambda_{j+1}^{(n)}$, close to each other, then for any k-node Gauss–Christoffel quadrature with $k > n$ there must be at least one node between $\lambda_j^{(n)}$ and $\lambda_{j+1}^{(n)}$. One can ask whether this means that at least one of the nodes $\lambda_j^{(n)}$ or $\lambda_{j+1}^{(n)}$ is stabilised to within some small number δ, i.e. whether

$$\delta_{n+1}\left|(z_j^{(n)},e_n)\right| \quad \text{or} \quad \delta_{n+1}\left|(z_{j+1}^{(n)},e_n)\right| \quad \text{is small.}$$

This question was posed in 1992 in the unpublished report [596], and the *negative* answer was given by Wüling in 2005 [678, Section 3.1]. However, if the nodes of the nth Gauss–Christoffel quadrature form a tight cluster (which can be formed of a single node) that is well separated from the other nodes, and if at the $(n-1)$th or $(n+1)$th Gauss–Christoffel quadrature there is a tight well separated cluster with the *same number of nodes* close to the original cluster, then all nodes in the cluster must be stabilised to within small δ. For details we refer to [678, Section 4]. The proofs in Wüling's paper are based on the following very clever observation.

Let Γ be a simple, closed, and positively oriented curve encircling the Gauss–Christoffel quadrature nodes $\theta_m^{(n)}, \ldots, \theta_{m+c-1}^{(n)}$, but no others. Using (3.4.26), an application of the residue theorem (see, e.g. [2]) gives

$$\sum_{\ell=1}^{c}\left(z_{m+\ell-1}^{(n)},e_n\right)^2 = \sum_{\ell=1}^{c}\frac{\psi_{n-1}\left(\theta_{m+\ell-1}^{(n)}\right)}{\psi_n'\left(\theta_{m+\ell-1}^{(n)}\right)} = \frac{1}{2\pi\iota}\int_\Gamma \frac{\psi_{n-1}(z)}{\psi_n(z)}\,dz, \quad (3.4.32)$$

which can easily be seen by considering the partial fraction expansion of the integrand $\psi_{n-1}(z)/\psi_n(z)$ analogously to the proof of Theorem 3.3.6. In this way the problem is transformed to estimating the contour integral in the complex plane on the right-hand side of (3.4.32). The equality (3.4.32) gives an example of a useful application of complex analysis in the context of real symmetric matrices and their spectral decompositions (see Section 3.3.3 for similar applications).

Not only the nodes, but also the weights of the n-node and k-node Gauss–Christoffel quadratures can be quantitatively related, as indicated above by Theorem 3.3.4 and shown in the following result; see [596, Theorem 3.1], or [634, Lemma 5.9]. The instructive paper [634] by van der Sluis and van der Vorst contains many further interesting results and is highly recommended.

Theorem 3.4.11
Using the previous notation, if T_n is the $n \times n$ left principal submatrix of the $k \times k$ Jacobi matrix T_k, $n < k \leq M$, then

$$\omega_j^{(n)} = \left(z_j^{(n)}, e_1\right)^2 = \frac{1}{\psi_n'\left(\theta_j^{(n)}\right)}\sum_{\ell=1}^{k}\omega_\ell^{(k)}\frac{\psi_n\left(\theta_\ell^{(k)}\right)}{\theta_\ell^{(k)}-\theta_j^{(n)}}$$

$$= \frac{1}{\left(\psi_n'\left(\theta_j^{(n)}\right)\right)^2}\sum_{\ell=1}^{k}\omega_\ell^{(k)}\left(\frac{\psi_n\left(\theta_\ell^{(k)}\right)}{\theta_\ell^{(k)}-\theta_j^{(n)}}\right)^2, \quad j=1,\ldots,n,$$

where $\omega_\ell^{(k)} = \left(z_\ell^{(k)},e_1\right)^2$, $\ell=1,\ldots,k$.

Proof
Let $p(\lambda)$ be any polynomial of degree at most $2n - 1$. Then the remainder $\widehat{p}(\lambda)$ of the division of $p(\lambda)$ by $\psi_n(\lambda)$,

$$p(\lambda) = \widetilde{p}(\lambda)\psi_n(\lambda) + \widehat{p}(\lambda), \qquad (3.4.33)$$

is of degree at most $n - 1$ (see footnote 1 in Section 3.2.1), and it can therefore be expressed as its (Newton–) Lagrange interpolant on the n points $\theta_1^{(n)}, \ldots, \theta_n^{(n)}$, i.e.

$$\widehat{p}(\lambda) = \sum_{j=1}^{n} \widehat{p}\left(\theta_j^{(n)}\right) \frac{\psi_n(\lambda)}{\psi_n'\left(\theta_j^{(n)}\right)\left(\lambda - \theta_j^{(n)}\right)} = \sum_{j=1}^{n} p\left(\theta_j^{(n)}\right) \frac{\psi_n(\lambda)}{\psi_n'\left(\theta_j^{(n)}\right)\left(\lambda - \theta_j^{(n)}\right)},$$

where we have used that $\widehat{p}(\theta_j^{(n)}) = p(\theta_j^{(n)})$, $j = 1, \ldots, n$. Since $\psi_n(\lambda)$ represents also the nth monic orthogonal polynomial with respect to the inner product based on the nodes $\theta_\ell^{(k)}$ and weights $\omega_\ell^{(k)}$, $\ell = 1, \ldots, k$ (see Corollary 3.4.2), its orthogonality to $\widetilde{p}(\lambda)$ gives

$$\sum_{\ell=1}^{k} \omega_\ell^{(k)} \widetilde{p}\left(\theta_\ell^{(k)}\right) \psi_n\left(\theta_\ell^{(k)}\right) = 0.$$

Consequently, using (3.4.33),

$$\sum_{\ell=1}^{k} \omega_\ell^{(k)} p\left(\theta_\ell^{(k)}\right) = \sum_{\ell=1}^{k} \omega_\ell^{(k)} \widehat{p}\left(\theta_\ell^{(k)}\right).$$

Substituting for $\widehat{p}(\theta_\ell^{(k)})$ and interchanging the order of summation gives

$$\sum_{\ell=1}^{k} \omega_\ell^{(k)} p\left(\theta_\ell^{(k)}\right) = \sum_{\ell=1}^{k} \omega_\ell^{(k)} \sum_{j=1}^{n} p\left(\theta_j^{(n)}\right) \frac{\psi_n\left(\theta_\ell^{(k)}\right)}{\psi_n'\left(\theta_j^{(n)}\right)\left(\theta_\ell^{(k)} - \theta_j^{(n)}\right)}$$

$$= \sum_{j=1}^{n} \left(\frac{1}{\psi_n'\left(\theta_j^{(n)}\right)} \sum_{\ell=1}^{k} \omega_\ell^{(k)} \frac{\psi_n\left(\theta_\ell^{(k)}\right)}{\theta_\ell^{(k)} - \theta_j^{(n)}} \right) p\left(\theta_j^{(n)}\right). \qquad (3.4.34)$$

There exists a monic polynomial $\chi_{n-2}(\lambda)$ of degree $n - 2$ such that

$$\frac{\psi_n(\lambda)}{\lambda - \theta_j^{(n)}} = \left(\lambda - \theta_j^{(n)}\right) \chi_{n-2}(\lambda) + \psi_n'\left(\theta_j^{(n)}\right).$$

Using this equality we get

$$\sum_{\ell=1}^{k} \omega_\ell^{(k)} \left(\frac{\psi_n\left(\theta_\ell^{(k)}\right)}{\theta_\ell^{(k)} - \theta_j^{(n)}} \right)^2 = \sum_{\ell=1}^{k} \omega_\ell^{(k)} \psi_n\left(\theta_\ell^{(k)}\right) \chi_{n-2}\left(\theta_\ell^{(k)}\right)$$

$$+ \psi_n'\left(\theta_j^{(n)}\right) \sum_{\ell=1}^{k} \omega_\ell^{(k)} \frac{\psi_n\left(\theta_\ell^{(k)}\right)}{\theta_\ell^{(k)} - \theta_j^{(n)}}$$

$$= \psi_n'\left(\theta_j^{(n)}\right) \sum_{\ell=1}^{k} \omega_\ell^{(k)} \frac{\psi_n\left(\theta_\ell^{(k)}\right)}{\theta_\ell^{(k)} - \theta_j^{(n)}}, \qquad (3.4.35)$$

because the first term on the right-hand side vanishes due to the orthogonality of $\psi_n(\lambda)$ to $\chi_{n-2}(\lambda)$. Combining (3.4.34) and (3.4.35) finally gives

$$\sum_{\ell=1}^{k} \omega_\ell^{(k)} p\left(\theta_\ell^{(k)}\right) = \sum_{j=1}^{n} \widehat{\omega}_j^{(n)} p\left(\theta_j^{(n)}\right), \qquad (3.4.36)$$

where

$$\widehat{\omega}_j^{(n)} = \frac{1}{\psi_n'\left(\theta_j^{(n)}\right)} \sum_{\ell=1}^{k} \omega_\ell^{(k)} \frac{\psi_n\left(\theta_\ell^{(k)}\right)}{\theta_\ell^{(k)} - \theta_j^{(n)}}$$

$$= \frac{1}{\left(\psi_n'\left(\theta_j^{(n)}\right)\right)^2} \sum_{\ell=1}^{k} \omega_\ell^{(k)} \left(\frac{\psi_n\left(\theta_\ell^{(k)}\right)}{\theta_\ell^{(k)} - \theta_j^{(n)}} \right)^2 > 0.$$

Moreover,

$$\sum_{j=1}^{n} \widehat{\omega}_j^{(n)} = \sum_{j=1}^{n} \widehat{\omega}_j^{(n)} p_0(\lambda) = \sum_{\ell=1}^{k} \omega_\ell^{(k)} p_0(\lambda) = \sum_{\ell=1}^{k} \omega_\ell^{(k)} = 1,$$

where we have used $p_0(\lambda) = 1$ and the equality (3.4.36). Since the right-hand side of (3.4.36) represents the (uniquely determined) n-node Gauss–Christoffel quadrature, we must have $\widehat{\omega}_j^{(n)} = \omega_j^{(n)}$, and the proof is finished. □

Theorem 3.4.11 describes modifications of weights in the Gauss–Christoffel quadrature when the number of the quadrature nodes changes. As for any mathematical result, it is important to ask how much insight this theorem offers into the problem. Here the answer is pessimistic. Although the result gives (in a mathematical sense) complete information, it is not very useful for obtaining a deeper general understanding. Orthogonal polynomials typically oscillate, and a small change in their argument λ can cause a large change in their values. Therefore the quantitative information given by Theorem 3.4.11 is not easy to use. Further investigations must be based on different tools.

As an example, one can ask whether the accumulated weight of a tight, well separated cluster of nodes in the Gauss–Christoffel quadrature is stabilised. A similar 'stabilisation property' is sometimes used, based on experimental observations, as a known fact without giving a firm mathematical justification; see, e.g. [458, Section II, p. 6197], [95, Section 3]. The problem can be specified as follows.

Let $\lambda_m^{(n)}, \ldots, \lambda_{m+c-1}^{(n)}$ be a tight, well separated cluster of $c \geq 1$ nodes of the n-node Gauss–Christoffel quadrature, and let $\lambda_j^{(n+1)}, \ldots, \lambda_{j+c-1}^{(n+1)}$ be the corresponding tight, well separated cluster (with the same number of nodes) of the $(n+1)$-node Gauss–Christoffel quadrature approximating the same distribution function $\omega(\lambda)$. From the interlacing property

$$\lambda_j^{(n+1)} < \lambda_m^{(n)} < \lambda_{j+1}^{(n+1)} < \cdots < \lambda_{j+c-1}^{(n+1)} < \lambda_{m+c-1}^{(n)}$$

when $j = m$, or,

$$\lambda_m^{(n)} < \lambda_j^{(n+1)} < \lambda_{m+1}^{(n)} < \cdots < \lambda_{m+c-1}^{(n)} < \lambda_{j+c-1}^{(n+1)}$$

when $j = m+1$. We will further consider the second case; for the first case the approach is analogous. Using (3.4.22) with $\lambda_j^{(n)} = \theta_j^{(n)}, j = 1, \ldots, n$, and $\lambda_j^{(n+1)} = \theta_j^{(n+1)}, j = 1, \ldots, n+1$,

$$\sum_{\ell=1}^{c} \omega_{m+\ell}^{(n+1)} = \prod_{j=2}^{n+1} \delta_j^2 \sum_{\ell=1}^{c} \frac{1}{\psi_n\left(\theta_{m+\ell}^{(n+1)}\right) \psi_{n+1}'\left(\theta_{m+\ell}^{(n+1)}\right)}, \quad (3.4.37)$$

and analogously

$$\sum_{\ell=1}^{c} \omega_{m+\ell-1}^{(n)} = \prod_{j=2}^{n} \delta_j^2 \sum_{\ell=1}^{c} \frac{1}{\psi_{n-1}\left(\theta_{m+\ell-1}^{(n)}\right) \psi_n'\left(\theta_{m+\ell-1}^{(n)}\right)}. \quad (3.4.38)$$

Since

$$\psi_{n+1}\left(\theta_{m+\ell-1}^{(n)}\right) = \left(\theta_{m+\ell-1}^{(n)} - \gamma_n\right) \psi_n\left(\theta_{m+\ell-1}^{(n)}\right) - \delta_{n+1}^2 \psi_{n-1}\left(\theta_{m+\ell-1}^{(n)}\right)$$
$$= -\delta_{n+1}^2 \psi_{n-1}\left(\theta_{m+\ell-1}^{(n)}\right),$$

the difference between the cumulated weights of the clusters in the $(n+1)$-node and the n-node Gauss–Christoffel quadrature is given by

$$\sum_{\ell=1}^{c} \left(\omega_{m+\ell}^{(n+1)} - \omega_{m+\ell-1}^{(n)}\right) = \prod_{j=2}^{n+1} \delta_j^2 \sum_{\ell=1}^{c} \left(\frac{1}{\psi_n\left(\theta_{m+\ell}^{(n+1)}\right) \psi_{n+1}'\left(\theta_{m+\ell}^{(n+1)}\right)} \right.$$
$$\left. + \frac{1}{\psi_{n+1}\left(\theta_{m+\ell-1}^{(n)}\right) \psi_n'\left(\theta_{m+\ell-1}^{(n)}\right)} \right).$$
$$(3.4.39)$$

Now let Γ be a simple, closed, and positively oriented curve encircling the nodes

$$\theta_m^{(n)}, \theta_{m+1}^{(n+1)}, \theta_{m+1}^{(n)}, \ldots, \theta_{m+c-1}^{(n)}, \theta_{m+c}^{(n+1)},$$

but no others. Since the function $1/(\psi_n(z)\psi_{n+1}(z))$ has simple poles at $\theta_j^{(n+1)}$ and $\theta_i^{(n)}$, $j = 1, \ldots, n+1$ and $i = 1, \ldots, n$, respectively, its decomposition into partial fractions reveals that (3.4.39) contains nothing but the sum of its residues inside the curve Γ. Using the residue theorem,

$$\sum_{\ell=1}^{c}\left(\omega_{m+\ell}^{(n+1)} - \omega_{m+\ell-1}^{(n)}\right) = \frac{1}{2\pi i}\prod_{i=2}^{n+1}\delta_i^2\int_\Gamma \frac{1}{\psi_n(z)\psi_{n+1}(z)}\,dz. \quad (3.4.40)$$

Consequently, the problem of stabilisation of cumulated weight of a tight, well separated cluster of nodes in the Gauss–Christoffel quadrature can be resolved via estimating the contour integral given on the right-hand side of (3.4.40). Wülling showed in [679] that the cumulated weight indeed stabilises, and the result remains valid when the number of nodes in the associated cluster in two consecutive (n-node and $(n+1)$-node) Gauss–Christoffel quadratures decreases or increases by one (note that any other option is excluded by the interlacing property).

Finally, Theorem 3.4.9 and the reconstruction of a Jacobi matrix from the given spectral data (see Proposition 3.4.5), have the following interesting consequence, which was first formulated by Scott [558].

Theorem 3.4.12
Using the previous notation (see, in particular, (3.4.31) and Definition 3.4.10) let M arbitrary distinct real numbers $\lambda_1 < \lambda_2 < \cdots < \lambda_M$ be given, and let T_M be an $M \times M$ Jacobi matrix with the eigenvalues $\lambda_1, \ldots, \lambda_M$, such that the eigenvalues of its $(M-1)\times(M-1)$ left principal submatrix T_{M-1} satisfy

$$\theta_j^{(M-1)} = (\lambda_{j+1} + \lambda_j)/2, \quad j = 1, \ldots, M-1, \quad (3.4.41)$$

i.e. the eigenvalues of T_{M-1} lie in the centres of the intervals determined by the eigenvalues of T_M. Let δ_T denote the minimal distance between two eigenvalues of T_M, i.e.

$$\delta_T \equiv \min_{i\neq\ell}|\lambda_i - \lambda_\ell|.$$

Then for any $n \in \{1, \ldots, M-1\}$,

$$\left\|T_M\widehat{z}_j^{(n,M)} - \theta_j^{(n)}\widehat{z}_j^{(n,M)}\right\| = \delta_{n+1}\left|\left(z_j^{(n)}, e_n\right)\right| \geq \delta_T/4, \quad j = 1, \ldots, n,$$

which means that no eigenvalue of a left principal submatrix of T_M is stabilised to within a value less than $\delta_T/4$.

Proof
Given $\lambda_1, \ldots, \lambda_M$ and $\theta_1^{(M-1)}, \ldots, \theta_{M-1}^{(M-1)}$, the Jacobi matrix T_M is uniquely determined. Indeed, consider

$$\psi_M(\lambda) \equiv (\lambda - \lambda_1)\cdots(\lambda - \lambda_M), \quad \psi_{M-1}(\lambda) \equiv \left(\lambda - \theta_1^{(M-1)}\right)\cdots\left(\lambda - \theta_1^{(M-1)}\right),$$

together with

$$\zeta_j^2 \equiv (z_j, e_1)^2 = \frac{\vartheta}{\psi_{M-1}(\lambda_j)\psi'_M(\lambda_j)}, \quad j = 1, \ldots, M,$$

where ϑ is determined from the normalisation condition

$$\sum_{j=1}^{M} \zeta_j^2 = 1.$$

Proposition 3.4.5 proves (by construction) the existence of T_M, which finishes the first part of the proof.

The rest will be proved by contradiction. Suppose some eigenvalue $\theta_j^{(n)}$ of a left principal submatrix T_n of T_M satisfies

$$\delta_{n+1}\left|\left(z_j^{(n)}, e_n\right)\right| < \delta_T/4.$$

Then, using Theorem 3.4.9, for any k with $n < k \leq M$ there must be an eigenvalue $\theta_\ell^{(k)}$ of the Jacobi matrix T_k that contains T_n as its left principal submatrix and that satisfies

$$\left|\theta_\ell^{(k)} - \theta_j^{(n)}\right| < \delta_T/4. \tag{3.4.42}$$

This must hold, in particular, for $k = M - 1$. For some index ℓ,

$$\lambda_\ell < \theta_j^{(n)} < \lambda_{\ell+1},$$

and, using (3.4.42) with $k = M - 1$, the distance of $\theta_j^{(n)}$ from the centre of the interval $[\lambda_\ell, \lambda_{\ell+1}]$ is less than $\delta_T/4$,

$$\left|\theta_j^{(n)} - (\lambda_\ell + \lambda_{\ell+1})/2\right| < \delta_T/4.$$

Considering the definition of $\delta_T/4$ this gives

$$\theta_j^{(n)} - \lambda_\ell > \delta_T/4 \quad \text{and} \quad \lambda_{\ell+1} - \theta_j^{(n)} > \delta_T/4,$$

which is a contradiction, because for $k = M$ there is no eigenvalue of T_M within $\delta_T/4$ from $\theta_j^{(n)}$. □

Theorem 3.4.12 has very interesting consequences for the Lanczos and the CG method, including their behaviour in finite precision arithmetic. Details are given in [457]; see also Section 5.9.1.

3.4.3 Historical Note: The Origin and Early Applications of Jacobi Matrices

In the previous sections we emphasised the importance of Jacobi matrices. As are numerous other mathematical objects and methods, they were named after Carl Gustav Jacob Jacobi (1804–1851). Jacobi is considered one of the most productive and versatile mathematicians of the 19th century. In his career of just 25 years he wrote about 170 articles (more than 30 of them published after his death). These articles were concerned with such diverse mathematical topics as number theory, differential equations, algebra, algebraic and differential geometry, calculus of variations, and the theory of functions. He also contributed to analytical mechanics, celestial mechanics, mathematical physics, and last but not least, the history of science. Jacobi's collected works fill seven volumes [361]. In his memorial speech on Jacobi, held on 1 July 1852, in the Academy of Sciences in Berlin, Dirichlet praised Jacobi as a man 'who with a strong hand influenced almost all areas of a science that has grown to an overwhelming extent through two thousand years of work, and who, wherever he directed his creative mind, unveiled important and often deeply hidden truths' [139, p. 193] (our translation).

When speaking of a *Jacobi matrix* we mean a real symmetric tridiagonal matrix with positive off-diagonal elements; see Definition 2.4.1. In this book we usually denote such matrices by

$$T_n = \begin{bmatrix} \gamma_1 & \delta_2 & & & \\ \delta_2 & \gamma_2 & \delta_3 & & \\ & \ddots & \ddots & \ddots & \\ & & \delta_{n-1} & \gamma_{n-1} & \delta_n \\ & & & \delta_n & \gamma_n \end{bmatrix}. \qquad (3.4.43)$$

This Jacobi matrix should not be confused with the *Jacobian matrix* (shortly the *Jacobian*). The latter object is the matrix of first-order partial derivatives of a differentiable function $f : \mathbb{R}^k \to \mathbb{R}^\ell$. If $f = (f_1, f_2, \ldots, f_\ell)$ and $x = (x_1, x_2, \ldots, x_k)$, then the Jacobian of f is defined as

$$J_f \equiv \frac{\partial f}{\partial x} \equiv \begin{bmatrix} \frac{\partial f_1}{\partial x_1} & \frac{\partial f_1}{\partial x_2} & \cdots & \frac{\partial f_1}{\partial x_k} \\ \vdots & \vdots & & \vdots \\ \frac{\partial f_\ell}{\partial x_1} & \frac{\partial f_\ell}{\partial x_2} & \cdots & \frac{\partial f_\ell}{\partial x_k} \end{bmatrix}.$$

The origin of the Jacobi matrices (3.4.43) can be traced back to the winter of 1848. According to Koenigsberger's biography of Jacobi [390, pp. 456–457], Jacobi's plan for the winter semester 1848/1849 had been to lecture on 'Differential calculus with applications' (our translation) at the University of Berlin. But when he started on October 30, only seven students signed up. Ill-tempered Jacobi abandoned his plan and decided to lecture on elliptic functions instead.

This change of plans may also have motivated him to read the note 'On quadratic forms and hyperelliptic functions' (our translation) in the Berlin Academy of Sciences on 9 November 1848. A part of this note was published (in a preliminary version) in 1848 [358] and (in a more refined version) in 1850 [359]. In the published paper Jacobi did not consider hyperelliptic functions, but explained how any quadratic form with n variables (and integer coefficients) can be reduced by a linear transformation that has a determinant ± 1 (i.e. a unimodular transformation) into a quadratic form which he wrote as

$$a_0 w_0 w_0 + a_1 w_0 w_1 + a_2 w_1 w_1 + a_3 w_1 w_2 + a_4 w_2 w_2$$
$$+ a_5 w_2 w_3 + a_6 w_3 w_3 + \ldots + a_{2n-3} w_{n-2} w_{n-1} + a_{2n-2} w_{n-1} w_{n-1}; \quad (3.4.44)$$

see [359, p. 291]. Jacobi described this fact and his formula (3.4.44), in which he wrote a and w instead of a_0 and w_0, as follows [359, pp. 291–292]:

> For any given quadratic form of n variables one can find an *equivalent* form containing, except for the variables' squares, only $n - 1$ products, which from an assumed ordering of the variables are obtained by multiplication of each variable into the *immediately following one*.

(Our translation; words emphasised as in Jacobi's text.)

In modern matrix notation, we write a quadratic form as $v^T A v$, where A is a symmetric matrix. Jacobi's expression (3.4.44) is of this form, where

$$A = \begin{bmatrix} a_0 & \tfrac{1}{2}a_1 & & & & \\ \tfrac{1}{2}a_1 & a_2 & \tfrac{1}{2}a_3 & & & \\ & \ddots & \ddots & \ddots & & \\ & & \tfrac{1}{2}a_{2n-5} & a_{2n-4} & \tfrac{1}{2}a_{2n-3} \\ & & & \tfrac{1}{2}a_{2n-3} & a_{2n-2} \end{bmatrix}, \text{ and } v = \begin{bmatrix} w_0 \\ w_1 \\ \vdots \\ w_{n-1} \end{bmatrix},$$

so that A is symmetric and tridiagonal.

As indicated by the title of the paper [359], 'On the reduction of the quadratic forms to the minimal number of terms' (our translation), Jacobi's goal was to have, for a given quadratic form, an equivalent form containing a smaller, at best the minimal, number of terms. Today it is well known that any quadratic form can be orthogonally reduced to a sum of squares. In matrix notation, a real symmetric (complex Hermitian) matrix can be orthogonally (unitarily) diagonalised, and its eigenvalues are real. (Already, in 1829 [97], Cauchy had shown that (in modern terms) real symmetric matrices have real eigenvalues.) Hence despite the paper's title and explicitly stated goal, Jacobi did not give the reduction of a quadratic form to the minimal possible number of terms in [359].

In 1857, Jacobi's student Borchardt published a paper of Jacobi that gave a reduction of a general quadratic form to a sum of squares, i.e. its diagonalisation, as well as some more general results [360]. In particular, given a quadratic form with $n + 1$ variables,

$$\sum_{i,k=0}^{n} a_{ik} x_i x_k \quad \text{with} \quad a_{ik} = a_{ki}, \qquad (3.4.45)$$

Jacobi expressed its diagonalised form as

$$\frac{u_0^2}{\rho_0} + \frac{u_1^2}{\rho_0 \rho_1} + \cdots + \frac{u_n^2}{\rho_{n-1} \rho_n}; \qquad (3.4.46)$$

cf. [360, p. 270]. He also gave formulas for the new variables $u_0, u_1, \ldots, u_n$ and the new coefficients $\rho_0, \rho_1, \ldots, \rho_n$ in terms of determinants of matrices formed from the given coefficients a_{ik} and the partial derivatives of the given form with respect to the original variables. Koenigsberger [390, p. 457–458] reports that the posthumously published paper [360] was written at about the same time as [358].

As pointed out by Dieudonné (see [137, Chapter VII, p. 154] and [138, pp. 571–572]), the formula (3.4.46) reveals the roots of the denominator of the following rational function $\mathcal{F}_n(\lambda)$ given by the partial sum

$$\mathcal{F}_n(\lambda) = \sum_{j=0}^{n} \frac{u_j^2}{\rho_{j-1}\rho_j - \lambda}$$

(here $\rho_{-1} \equiv 1$ and, for simplicity, $\sum_{j=0}^{n} u_j^2 = 1$). It can equivalently be expressed as a finite continued fraction (cf. Theorem 3.3.6 in this book), which has been a common tool in number theory as well as approximation theory. In the modern language of matrices,

$$\mathcal{F}_n(\lambda) = u^T (D - \lambda I)^{-1} u,$$

where $u^T = [u_0, u_1, \ldots, u_n]$ and $D = \text{diag}(\rho_0, \rho_0\rho_1, \ldots, \rho_{n-1}\rho_n)$. Therefore the transformation of a quadratic form (3.4.45) into (3.4.46) 'reveals the eigenvalues of the form', or, in a modern language, reveals the eigenvalues of the corresponding symmetric matrix. When applied to the already reduced form (3.4.44), this transformation would give the eigenvalues of the corresponding symmetric tridiagonal matrix. Jacobi, however, did not formulate his results in terms of matrices, and he did not explicitly state results about eigenvalues.

In this context it is interesting to note that in another seminal paper [357] on the iterative diagonalisation of a homogeneous symmetric system of linear algebraic equations Jacobi practically worked with matrices, although he always included the unknown variables and wrote the matrix elements as coefficients of systems of equations. The method described in that paper is now called the *Jacobi method*, and it gained a lot of attention for its natural parallelism and for remarkable numerical properties; see, e.g. [662, Section 7.2], [147, 148]. In the paper Jacobi defined what is now called a *Givens rotation* [357, Section 6, pp. 64–71].

Most probably the first appearance of a tridiagonal matrix representation (written in the form of a determinant) can be found, in a slightly different context, in the 1858 paper [504] of Painvin. Heine picked up this representaion 20 years later in the chapter on infinite continued fractions of his book [303, p. 262]. Following Painvin, he expressed related three-term recurrences for the numerators and denominators of the nth convergent of an infinite continued fraction as determinants of the real tridiagonal matrix, where the products of the corresponding off-diagonal entries are positive; see (3.3.18)–(3.3.20) in our book. Analogous determinants were applied to general quadratic forms in the posthumously published paper of Jacobi [360].

It should also be pointed out that by that time it was already known, due to the fundamental work of Chebyshev from 1855 [104], that the denominators of the convergents of the continued fraction $\mathcal{F}_n(\lambda)$ form a sequence of orthogonal polynomials, and that they can be determined by a three-term recurrence (cf. the historical note in Section 3.3.6). A complete understanding of the advantages of the matrix representation of Jacobi's results came, however, much later.

In the early 20th century, Jacobi's reductions published in 1850 [359] and 1857 [360] turned out to be important ingredients in Toeplitz' work on bilinear forms with infinitely many variables. After obtaining his PhD in Breslau in 1905 with a thesis on algebraic geometry entitled 'On systems of forms whose functional determinant identically vanishes' (our translation), Toeplitz spent seven years in Göttingen (afterwards he went on to professorships in Kiel (1913) and Bonn (1928)). By the time Toeplitz arrived in Göttingen, Hilbert was in the middle of his fundamental work on the theory of integral equations and systems of equations with infinitely many unknowns (this work also played a role in the development of the Gram–Schmidt method; see the historical note in Section 2.4.3). Toeplitz became deeply involved in this effort, and focused himself on bilinear and quadratic forms with infinitely many unknowns. Already in February 1907 Hilbert had presented a paper of Toeplitz to the Academy of Sciences in Göttingen [615]. In this paper Toeplitz used Jacobi's reduction of 1857 [360] to obtain shorter and algebraic proofs of some of Hilbert's analytic results.

Toeplitz was an algebraist, and similar to Schmidt he strived for elegant and elementary derivations. He clearly described these intentions in the introduction of a paper in 1910 [616, p. 489]:

> ... the final goal is to complement the theory of Hilbert and Hellinger by another independent, one could say algebraic, component that replaces the differential-analytic tools ... by a development of elementary character.

(Our translation.)

Hilbert's student, Hellinger mentioned in this quote, had considered orthogonal transformations of bounded quadratic forms in his PhD dissertation of 1907 entitled 'The orthogonal invariants of quadratic forms of infinitely many variables' (our translation). In [616], Toeplitz applied Jacobi's reduction of 1850 [359] in the context of infinitely many unknowns. He referred to a quadratic form of the type

$$a_1 x_1^2 + a_2 x_2^2 + \cdots + 2b_1 x_1 x_2 + 2b_2 x_2 x_3 + \cdots \tag{3.4.47}$$

as a *Jacobi form* ('Jacobische Form') [616, p. 492].

Toeplitz wrote in [616] that the transformation reducing a given finite quadratic form into a Jacobi form is orthogonal, and he briefly stated an algorithm for the reduction. In modern matrix notation and in the case of a finite matrix this algorithm may be described as follows. Consider a symmetric matrix

$$B = [b_{ij}] \in \mathbb{R}^{n \times n}, \quad n > 2, \quad \text{with} \quad b \equiv [b_{12}, \ldots, b_{1n}]^T \neq 0.$$

Let $W_1 \in \mathbb{R}^{(n-1) \times (n-2)}$ be defined by $W_1^T W_1 = I_{n-2}$ and $W^T b = 0$. This means that the $n - 2$ columns of W_1 form an orthonormal basis of the orthogonal complement of span$\{b\}$. Then the matrix

$$W \equiv \begin{bmatrix} 1 & 0 & 0 \\ 0 & b/\|b\| & W_1 \end{bmatrix} \in \mathbb{R}^{n \times n}$$

is orthogonal, $W^T W = W W^T = I_n$, and

$$W^T B W = \begin{bmatrix} b_{11} & \|b\| & 0 \\ \|b\| & * & * \\ 0 & * & * \end{bmatrix}.$$

The reduction can now be continued using the $(n-1) \times (n-1)$ matrix with the entries symbolically indicated by '$*$'. One can show that this reduction is mathematically equivalent to the application of the (Hermitian) Lanczos algorithm (Algorithm 2.4.3) to the matrix B and the initial vector $e_1 = [1, 0, \ldots, 0]^T$. If e_1 is of grade n with respect to B, then the result after n steps of the reduction is an orthogonal transformation of B to a symmetric tridiagonal form with positive off-diagonal elements. As mentioned above, Toeplitz mainly studied this reduction in the case of infinitely many unknowns, which is significantly more complicated since convergence issues have to be considered.

While in Göttingen, Toeplitz became friends with Hellinger. Jointly they worked on Hilbert's theory and the vastly developing field of functional analysis. In particular, in an elegant paper of 1914 [306] they explained the links between quadratic forms having infinitely many unknowns and the analytic theory of continued fractions developed by Stieltjes in his fundamental treatise of 1894 [588] (for more details on that work see the historical note in Section 3.3.5). In [306] Hellinger and Toeplitz referred to quadratic forms of the type (3.4.47) as *J-forms* ('J-Formen'). Page 202 of their paper shows the corresponding symmetric tridiagonal coefficient matrix,

$$\begin{bmatrix} a_1 & b_1 & & & \\ b_1 & a_2 & b_2 & & \\ & b_2 & a_3 & b_3 & \\ & & \ddots & \ddots & \ddots \end{bmatrix},$$

and this possibly is the first *Jacobi matrix* that appeared in print (though Hellinger and Toeplitz assumed $b_j \neq 0$ rather than $b_j > 0$). Their joint work culminated more than a decade later in a 200-page review article in Teubner's multi-volume 'Encyklopädie der mathematischen Wissenschaften' [307].

Let us mention that Hilbert praised Stieltjes' work on continued fractions in [319, p. 12]. Hilbert's paper [319] of 1906 is the fourth in his sequence of six 'Mitteilungen' between 1904 and 1910 that laid the foundations of functional analysis. The complete collection was published as a book in 1912 [320]. The representation theorem of Riesz, proven in 1909 [525] and Hilbert's *resolution of unity*, led to the integral representation of self-adjoint operators in Hilbert spaces; see Remark 3.5.1 below for some mathematical details. These, in turn, played a fundamental role in the development of the mathematical foundations of quantum mechanics. A course given by Hilbert in Göttingen in the winter semester 1926/27 resulted in his joint paper with von Neumann and Nordheim [321], and is viewed today as the starting point for von Neumann's own investigations leading to the fundamental publications [647, 648]. Historical details can be found, e.g. in [548] (see in particular the Introduction to Chapter 6) and in the paper [579]. In Weyl's obituary for Hilbert, written in 1944, one finds the following statement concerning this development [666, p. 651]:

> But then a sort of miracle happened: the spectrum theory in Hilbert space was discovered to be the adequate mathematical instrument of the new quantum physics inaugurated by Heisenberg and Schrödinger in 1923.

Further seminal contributions linking Jacobi matrices with representation of self-adjoint operators can be found in the monograph of Stone [592], originally published in 1932. A review of the general theory of the unitary analogue of Jacobi matrices (the so-called CVM matrices) together with detailed historical comments is given in [564]. In this way, the work on analytic theory of continued fractions by Stieltjes and on transformations of quadratic forms started by Jacobi affected developments in vast areas of mathematics and mathematical physics.

3.5 MODEL REDUCTION VIA MATRICES: HERMITIAN LANCZOS AND CG

In Section 3.2.2 we have shown that the n-node Gauss–Christoffel quadrature can be considered a matching moments model reduction. In particular, the distribution function $\omega^{(n)}(\lambda)$ determined by the Gauss–Christoffel quadrature matches the first $2n$ moments of the original model given by the distribution function $\omega(\lambda)$.

We now formulate the same matching moments model reduction in the language of matrices. This will bring us back to the topic of Krylov subspace methods. Consider the linear algebraic system $Ax = b$ with an $N \times N$ Hermitian matrix A. Denote the spectral decomposition of A by

$$AY = Y \operatorname{diag}(\lambda_j), \quad \text{where} \quad Y = [y_1, \ldots, y_N], \quad Y^*Y = YY^* = I_N. \quad (3.5.1)$$

For a given initial vector x_0 consider the initial residual $r_0 = b - Ax_0$ and the Lanczos initial vector $v_1 = r_0/\|r_0\|$. Then we can define a distribution function $\omega(\lambda)$ as in (3.1.6), with N points of increase equal to the eigenvalues $\lambda_1, \ldots, \lambda_N$ of A, and the associated weights $\omega_1, \ldots, \omega_N$ equal to the squared size of the components of v_1 in the corresponding invariant subspaces, i.e.

$$\omega_j = |(v_1, y_j)|^2, \quad j = 1, \ldots, N. \quad (3.5.2)$$

We assume that the eigenvalues of A are distinct and that $\omega_j > 0, j = 1, \ldots, N$. This assumption is used for convenience of notation only, and it does not represent any factual restriction.

The moments of $\omega(\lambda)$ can be expressed as (see Remark 3.1.2)

$$\int \lambda^k \, d\omega(\lambda) = \sum_{j=1}^{N} \omega_j \{\lambda_j\}^k = v_1^* A^k v_1, \quad k = 0, 1, 2, \ldots. \quad (3.5.3)$$

Now consider the Jacobi matrix T_n that results from n steps of the Stieltjes recurrence determined by the distribution function $\omega(\lambda)$ (see (3.3.3)) or, equivalently, from n steps of the (Hermitian) Lanczos algorithm applied to A and v_1,

$$AV_n = V_n T_n + \delta_{n+1} v_{n+1} e_n^T, \qquad (3.5.4)$$

$V_n^* V_n = I_n$, $V_n^* v_{n+1} = 0$; cf. (2.4.5). Then, analogously to (3.5.3), the n-node Gauss–Christoffel quadrature of the monomials (i.e. the moments) can be expressed using the spectral decomposition of T_n as

$$\int \lambda^k d\omega^{(n)}(\lambda) = \sum_{j=1}^{n} \omega_j^{(n)} \left\{ \lambda_j^{(n)} \right\}^k = \sum_{j=1}^{n} \omega_j^{(n)} \left\{ \theta_j^{(n)} \right\}^k$$

$$= e_1^T T_n^k e_1, \quad k = 0, 1, 2, \ldots. \qquad (3.5.5)$$

Since the Gauss–Christoffel quadrature is exact for polynomials up to degree $2n - 1$, the expressions (3.5.3) and (3.5.5) are equal for $k = 0, 1, \ldots, 2n - 1$, and thus

$$v_1^* A^k v_1 = e_1^T T_n^k e_1, \quad k = 0, 1, \ldots, 2n - 1. \qquad (3.5.6)$$

In the Lanczos *method* (see [405] or [508, Chapter 11]) the eigenvalues of T_n give approximations of the eigenvalues of A. The main consequence of (3.5.6) for the (Hermitian) Lanczos *algorithm* can be formulated as follows:

The result of the Lanczos algorithm for a Hermitian matrix A and an initial vector v_1 can be considered a model reduction from A and v_1 to T_n and e_1, such that the first $2n$ moments of the reduced model match those of the original model.

In our book, we deal primarily with linear algebraic systems $Ax = b$, where A can be viewed as a linear operator on the vector space $\mathbb{C}^N$. When A is HPD, we can extend the above considerations to the CG method [313] (Algorithm 2.5.1 in Section 2.5.1). If the columns of V_n form an orthonormal basis of the Krylov subspace $\mathcal{K}_n(A, r_0)$ determined by n steps of the Lanczos algorithm, then the nth approximation $x_n = x_0 + V_n t_n$ generated by the CG method is characterised by

$$x_n = x_0 + V_n t_n, \quad T_n t_n = \|r_0\| e_1, \quad T_n = V_n^* A V_n; \qquad (3.5.7)$$

see (2.5.5) in Section 2.5.1. Analogously to the model reduction properties of the Hermitian Lanczos algorithm or, equivalently, the n-node Gauss–Christoffel quadrature, we observe the following:

The nth CG approximation x_n can be considered as a result of the model reduction from $Ax = b$ to $T_n t_n = \|r_0\| e_1$, such that the first $2n$ moments of the reduced model match those of the original model.

Clearly, there is more behind the Lanczos algorithm and the CG method than 'just' the projection idea based on Krylov subspaces that was outlined in Chapter 2. Through their link to moment matching they are connected to topics in other areas of mathematics we have used in this chapter, among them orthogonal polynomials, Gauss–Christoffel quadrature, and continued fractions.

It is interesting to note that many of these connections of their newly proposed method were already mentioned by Hestenes and Stiefel in their seminal paper of 1952 [313]. A look at the titles of Sections 14–18:

14. A duality between orthogonal polynomials and n-dimensional geometry
15. An algorithm for orthogonalization
16. A new approach to the cg-method, eigenvalues
17. Example, Legendre polynomials
18. Continued fractions

indicates the extraordinary depth and thoroughness of Hestenes' and Stiefel's work.[2] Unfortunately, Sections 14–18 of [313] have rarely been quoted in later publications on CG or in numerical linear algebra textbooks. This may be attributed, in part, to an explosive *algorithmic* development of the whole field stimulated by the development of computers and applications. In the spirit of [313], we have established here the link between the polynomial description of the moment problem from Section 3.2, and the matrix description represented by the Hermitian Lanczos algorithm and the CG method.

Let us recap. A given Hermitian $N \times N$ matrix A and a normalised initial vector $v_1 = r_0/\|r_0\|$ determine the distribution function $\omega(\lambda)$ with N points of increase equal to the N distinct eigenvalues λ_j of A, and the corresponding weights equal to $|(v_1, y_j)|^2, j = 1, \ldots, N$. On the other hand, given $\|r_0\|$ and a unitary matrix Y, $\omega(\lambda)$ determines A and $r_0 = \|r_0\|v_1$. Indeed, using the results from Section 3.4.1, in particular Proposition 3.4.5, $\omega(\lambda)$ determines a Jacobi matrix T_N with the (real, symmetric) matrix Z containing its normalised eigenvectors. Then $Y = VZ$ (see (3.4.20)), and

$$A = V T_N V^* = Y \operatorname{diag}(\lambda_j) Y^*.$$

Given $\omega(\lambda)$, Y and $\|r_0\|$, the matrices A and V are uniquely determined except for the signs of the components of v_1 in the individual invariant subspaces.

Analogously, T_n and e_1 determine the distribution function $\omega^{(n)}(\lambda)$, and $\omega^{(n)}(\lambda)$ determines uniquely the Jacobi matrix T_n, which can be viewed as a result of the model reduction that matches the first $2n$ moments (3.5.6). Similarly,

---

2. Among the early works related to CG we also want to point out the papers by Stiefel [581], Henrici [310], and Rutishauser [539], all published in the 1958 volume of the Applied Mathematics Series, as well as the work by Ljusternik [431]; see also interesting historical comments given by Forsythe in [194, pp. 25–26]. A nice survey of the approximation theory issues and algorithmic developments related to the Lanczos algorithm was given by Gutknecht at the Lanczos Centenary Conference [285].

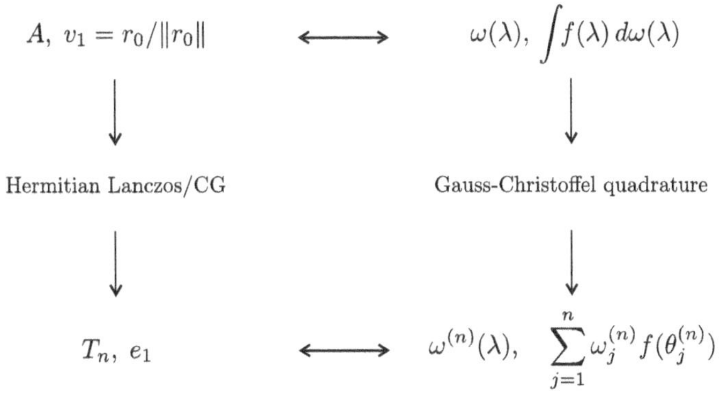

Model Reduction via Matrices: Hermitian Lanczos and CG

$$A, \ v_1 = r_0/\|r_0\| \quad \longleftrightarrow \quad \omega(\lambda), \ \int f(\lambda) \, d\omega(\lambda)$$

$$\downarrow \qquad\qquad\qquad\qquad\qquad \downarrow$$

Hermitian Lanczos/CG Gauss–Christoffel quadrature

$$\downarrow \qquad\qquad\qquad\qquad\qquad \downarrow$$

$$T_n, \ e_1 \quad \longleftrightarrow \quad \omega^{(n)}(\lambda), \ \sum_{j=1}^{n} \omega_j^{(n)} f(\theta_j^{(n)})$$

Figure 3.2 Visualisation of the fundamental relationships between the Hermitian Lanczos and CG methods on the one hand, and the Gauss–Christoffel quadrature on the other.

$\omega^{(n)}(\lambda)$ can be viewed as the result of the model reduction that solves the simplified Stieltjes moment problem; cf. Section 3.2.2. Finally, $\omega^{(n)}(\lambda)$ represents the distribution function determined by the n-node Gauss–Christoffel quadrature approximation of the Riemann–Stieltjes integral with the distribution function $\omega(\lambda)$. These fundamental relationships are visualised in Figure 3.2, and in words they may be expressed as follows:

> The Hermitian Lanczos algorithm and the CG method are matrix formulations of the Gauss–Christoffel quadrature. Vice versa, the Gauss–Christoffel quadrature is the essence of the Hermitian Lanczos algorithm and the CG method.

The fact that the nth step of Lanczos or CG gives an optimal approximation $\omega^{(n)}(\lambda)$ (in the sense of the Gauss–Christoffel quadrature) of the distribution function $\omega(\lambda)$ explains the *highly nonlinear behaviour* of Lanczos and CG that has already been indicated in Section 2.6.

Remark 3.5.1
Representations of matrices and operators using the Riemann–Stieltjes integral similar to (3.5.3) and (3.5.5) have been used in the development of the mathematical foundations of quantum mechanics; cf. also the end of the historical note on Jacobi matrices in Section 3.4.3. We will briefly outline the approach used in that field. Considering the spectral decomposition of a Hermitian matrix A,

$$A = \sum_{j=1}^{N} \lambda_j y_j y_j^*,$$

we can write symbolically

$$v_1^* A v_1 = v_1^* \left(\sum_{j=1}^{N} \lambda_j y_j y_j^* \right) v_1 \equiv v_1^* \left(\int \lambda \, dE(\lambda) \right) v_1$$

$$= \sum_{j=1}^{N} \lambda_j v_1^* y_j y_j^* v_1 = \sum_{j=1}^{N} \lambda_j \omega_j = \int \lambda \, d\omega(\lambda).$$

This leads to the representation

$$A = \int \lambda \, dE(\lambda), \quad y_j y_j^* = \int_{\Delta_j} dE(\lambda), \quad j = 1, \ldots, N,$$

where $\Delta_1, \ldots, \Delta_N$ are nonintersecting intervals that contain the respective eigenvalues $\lambda_1, \ldots, \lambda_N$ in their interiors (possibly as their respective centres), and

$$I_N = \sum_{j=1}^{N} y_j y_j^* \equiv \int dE(\lambda)$$

represents the so-called *resolution of unity*. In this way the Riemann–Stieltjes integral representation of operators (matrices) and of their functions was used in the fundamental monograph on quantum mechanics by von Neumann originally published in 1932 [648, Chapter II, Section 7, starting from p. 112]. The main ideas had however already been formulated by von Neumann in the paper [647], with the Riemann–Stieltjes representation of operators outlined on pp. 31–33, and the resolution of unity attributed to Hilbert; see also Section 3.3.5. As pointed out to us by Michele Benzi, another fundamental work on the mathematical foundation of quantum mechanics was published by Wintner in 1929 [674]. It developed a representation based on infinite matrices which was, three years later, overshadowed by the concept of von Neumann. The matrix representation using the Riemann–Stieltjes integral is there linked to the Riesz representation theorem for functionals, see [674, Chapter II, Section 49, pp. 106–108]. For a comprehensive description of the integral representation of operators used in quantum mechanics we refer, e.g. to [371, Chapters 11–15], and for basic mathematical theory of operator representations via the Riemann–Stieltjes integral to, e.g. [83, Section 4] and [6, Chapter VI, Sections 63–68]. Although the motivations of von Neumann and Wintner were theoretical rather than computational, their work illustrates an ingenious combination of functional analytic and algebraic principles, that later found a computational expression in a different way in Krylov subspace methods.

Considering the equivalence (in the previously described sense) of the CG method and the Gauss–Christoffel quadrature, one can ask whether there is also a relationship between *quantitative measures* of the error in the CG method and the approximation error in the Gauss–Christoffel quadrature. To describe this relationship we observe that

$$\frac{\|x - x_0\|_A^2}{\|r_0\|^2} = \frac{(r_0, A^{-1} r_0)}{\|r_0\|^2} = (v_1, A^{-1} v_1) = (e_1, T_N^{-1} e_1) = \int \lambda^{-1} \, d\omega(\lambda).$$

Using (3.5.7) and the fact that the vector $t_n = \|r_0\| T_n^{-1} e_1$ is real, the A-norm of the CG error at step n can be written as

$$\frac{\|x - x_n\|_A^2}{\|r_0\|^2} = \frac{\|x - x_0 - V_n t_n\|_A^2}{\|r_0\|^2}$$

$$= \frac{\|x - x_0\|_A^2}{\|r_0\|^2} - \frac{2\Re((r_0, V_n t_n)) - (AV_n t_n, V_n t_n)}{\|r_0\|^2}$$

$$= \frac{\|x - x_0\|_A^2}{\|r_0\|^2} - \frac{2\Re((V_n^* r_0, t_n)) - (V_n^* A V_n t_n, t_n)}{\|r_0\|^2}$$

$$= \frac{\|x - x_0\|_A^2}{\|r_0\|^2} - \frac{2(\|r_0\|e_1, \|r_0\|T_n^{-1}e_1)) - (\|r_0\|e_1, \|r_0\|T_n^{-1}e_1)}{\|r_0\|^2}$$

$$= \frac{\|x - x_0\|_A^2}{\|r_0\|^2} - (e_1, T_n^{-1}e_1)$$

$$= \frac{\|x - x_0\|_A^2}{\|r_0\|^2} - \sum_{i=1}^{n} \omega_i^{(n)} \left\{\theta_i^{(n)}\right\}^{-1}.$$

Summarising, in terms of the A-norm of the CG error,

$$\frac{\|x - x_0\|_A^2}{\|r_0\|^2} = \sum_{j=1}^{n} \omega_j^{(n)} \left\{\theta_j^{(n)}\right\}^{-1} + \frac{\|x - x_n\|_A^2}{\|r_0\|^2}, \qquad (3.5.8)$$

or, in terms of the n-node Gauss–Christoffel quadrature for $f(\lambda) = \lambda^{-1}$,

$$\int \lambda^{-1} d\omega(\lambda) = \sum_{j=1}^{n} \omega_j^{(n)} \left\{\theta_j^{(n)}\right\}^{-1} + \frac{\|x - x_n\|_A^2}{\|r_0\|^2}. \qquad (3.5.9)$$

We have thus proved the following theorem.

Theorem 3.5.2
Using the previous notation, the squared A-norm of the nth error in the CG method, scaled by the squared Euclidean norm of the initial residual, is equal to the approximation error of the n-node Gauss–Christoffel quadrature for $f(\lambda) = \lambda^{-1}$ with the distribution function $\omega(\lambda)$ defined by the matrix A and the normalised initial residual.

In terms of the quadratic forms and the $(1,1)$-element of the inverse of the associated Jacobi matrix T_n, the relations (3.5.8) and (3.5.9) can be equivalently written as

$$\frac{(r_0, A^{-1} r_0)}{\|r_0\|^2} = (e_1, T_n^{-1} e_1) + \frac{(r_n, A^{-1} r_n)}{\|r_0\|^2}. \qquad (3.5.10)$$

Using the partial fraction decomposition (3.3.22),

$$\sum_{j=1}^{n} \omega_j^{(n)} \left\{ \theta_j^{(n)} \right\}^{-1} = (e_1, T_n^{-1} e_1) = -\mathcal{F}_n(0), \qquad (3.5.11)$$

where

$$-\mathcal{F}_n(0) \equiv \cfrac{1}{\gamma_1 - \cfrac{\delta_2^2}{\gamma_2 - \cfrac{\delta_3^2}{\ddots \gamma_{n-2} - \cfrac{\delta_n^2}{\gamma_{n-1} - \cfrac{\delta_n^2}{\gamma_n}}}}}. \qquad (3.5.12)$$

More details, mathematically equivalent formulations of the terms in (3.5.8)–(3.5.12) suitable for numerical calculations, and applications to error estimates and stopping criteria for the CG method can be found in [244, 245, 457] and [246].

Before we proceed with generalisations to other Krylov subspace methods, we briefly discuss in the following section some results related to the explicit use of the matrix of moments.

3.6 FACTORISATION OF THE MATRIX OF MOMENTS

Connections between the simplified Stieltjes' moment problem (Problem 3.1.1) and methods for solving $Ax = b$, where A is HPD, were strongly promoted by Golub. His views have influenced and inspired many of his collaborators and thus have led to a great variety of results in theory and also in applications; see, e.g. [226, 243] for reviews of his (selected) important contributions. In this section we will discuss one of these contributions, which will reveal an interesting link between the moments of a distribution function and the projected matrix $C_n^* A S_n$ studied in Chapter 2; see (2.1.9).

Based on the results of Wilf [667] and Mysovskih [467], Golub and Welsch described in [253] (among other topics) the relationship between the nodes and weights of the Gauss–Christoffel quadrature, Jacobi matrices and *matrices of moments*. We will present their elegant argument related to the last topic (see [513] for an analogous idea). The nth matrix of moments corresponding to a given nondecreasing distribution function $\omega(\lambda)$ with $\mathfrak{n}(\omega) \geq n$ points of increase is the Hankel matrix[3]

$$M_n \equiv \begin{bmatrix} \xi_0 & \xi_1 & \xi_2 & \cdots & \xi_n \\ \xi_1 & \xi_2 & & & \\ \xi_2 & & \ddots & & \vdots \\ \vdots & & & & \\ \xi_n & & \cdots & & \xi_{2n} \end{bmatrix} \in \mathbb{R}^{(n+1)\times(n+1)}, \qquad (3.6.1)$$

---

3. Hankel matrices, named after the German mathematician Hermann Hankel (1839–1873), are square matrices with constant skew-diagonals.

where
$$\xi_k = \int \lambda^k d\omega(\lambda), \quad k = 0, 1, \ldots 2n;$$

cf. (3.1.1) and Remark 3.1.2. With the inner product (3.2.8), this matrix can alternatively be written as

$$M_n = \begin{bmatrix} (1,1) & (1,\lambda) & (1,\lambda^2) & \cdots & (1,\lambda^n) \\ (\lambda,1) & (\lambda,\lambda) & & & \\ (\lambda^2,1) & & \ddots & & \vdots \\ \vdots & & & & \\ (\lambda^n,1) & & \cdots & & (\lambda^n,\lambda^n) \end{bmatrix}, \quad (3.6.2)$$

so that M_n may also be considered the *Gram matrix* (cf. (2.4.9)) for the monomials $1, \lambda, \ldots, \lambda^n$. Given any nonzero vector $z = [\eta_0, \ldots, \eta_n]^T$,

$$\begin{aligned} z^T M_n z &= \sum_{i,j=0}^{n} \eta_i \eta_j (\lambda^i, \lambda^j) \\ &= \left(\eta_0 + \eta_1 \lambda + \cdots + \eta_n \lambda^n, \eta_0 + \eta_1 \lambda + \cdots + \eta_n \lambda^n \right) \\ &= \left\| \eta_0 + \eta_1 \lambda + \cdots + \eta_n \lambda^n \right\|^2 \\ &> 0, \end{aligned}$$

i.e. the matrix of moments M_n is symmetric positive definite.

The results presented in this section are mainly of theoretical interest. Matrices of moments are notoriously ill-conditioned. For example, Tyrtyshnikov proved in [628] that the condition number of any real positive definite Hankel matrix of order n is larger than 2^{n-6}. Further results can be found, e.g. in [179, 48, 49]; see also [246, Section 5.2] and the survey in Section 3 of the paper on sensitivity of the Gauss–Christoffel quadrature [487].

Let $\omega(\lambda)$ be the distribution function associated with an HPD matrix A and a normalised initial vector v_1 in the Lanczos method, or with a linear algebraic system $Ax = b$ with the HPD matrix A, an initial approximation x_0, and the initial residual $r_0 = b - Ax_0$. Then, taking $C_n = S_n \equiv [v_1, Av_1, \ldots, A^{n-1}v_1]$, the projected matrix in (2.1.9), which by case (1) in Theorem 2.3.1 is nonsingular, is given by

$$C_n^* A S_n = \begin{bmatrix} v_1^* A v_1 & \cdots & v_1^* A^n v_1 \\ \vdots & & \vdots \\ v_1^* A^n v_1 & \cdots & v_1^* A^{2n-1} v_1 \end{bmatrix} = \begin{bmatrix} (\lambda, 1) & \cdots & (\lambda, \lambda^{n-1}) \\ \vdots & & \vdots \\ (\lambda^n, 1) & \cdots & (\lambda^n, \lambda^{n-1}) \end{bmatrix},$$

cf. (3.5.3) and the introduction to this chapter. Thus, the projected matrix is obtained from the matrix of moments M_n by deleting the first row and the last column.[4]

---

4. It is interesting to note that the determinant of this matrix was used by Stieltjes in his solution of the moment problem; see [588], [386, pp. 34–35].

Now consider the *Cholesky factorisation* (cf. [250, Theorem 4.2.5]) of the moment matrix,

$$M_n = L_n L_n^T, \qquad (3.6.3)$$

and denote

$$L_n^{-1} = \begin{bmatrix} \zeta_{11} & & \\ \vdots & \ddots & \\ \zeta_{n+1,1} & \cdots & \zeta_{n+1,n+1} \end{bmatrix}. \qquad (3.6.4)$$

Then we observe that

$$L_n^{-1} M_n L_n^{-T} = \begin{bmatrix} (g_0, g_0) & (g_0, g_1) & \cdots & (g_0, g_n) \\ (g_1, g_0) & (g_1, g_1) & & \\ \vdots & & \ddots & \vdots \\ (g_n, g_0) & & \cdots & (g_n, g_n) \end{bmatrix} = I_{n+1},$$

where

$$g_0(\lambda) \equiv \zeta_{11} = 1, \quad g_j(\lambda) \equiv \zeta_{j+1,1} + \zeta_{j+2,2}\lambda + \cdots + \zeta_{j+1,j+1}\lambda^j, \quad j = 1,\ldots,n,$$

are polynomials in λ. Since

$$L_n^{-1} M_n L_n^{-T} = I_{n+1},$$

$g_0(\lambda) = \varphi_0(\lambda) \equiv 1$, $g_1(\lambda) = \varphi_1(\lambda), \ldots, g_n(\lambda) = \varphi_n(\lambda)$ are the first $n+1$ orthonormal polynomials associated with the distribution function $\omega(\lambda)$. We can formulate this result as a theorem; cf. [253, Section 4].

Theorem 3.6.1
Using the previous notation, for each $j = 0, 1, \ldots, n$ the entries of the $(j+1)$th row of the matrix L_n^{-1} in (3.6.4) are equal to the $j+1$ coefficients of the $(j+1)$th orthonormal polynomial $\varphi_j(\lambda)$ associated with the given distribution function $\omega(\lambda)$.

Remarkable related developments in the fields of computational physics and chemistry started in the late 1930s (see [208]), with a milestone represented by Gordon's paper of 1968 [255]. That paper also affected further developments in computational mathematics. It is interesting to note, as pointed out by Gautschi in a commentary of Golub's Selected Works [243, p. 346], that the papers of Golub and Welsch [253] and Gordon [255] were independently submitted for publication within one month. Both papers present similar ideas and refer to the pioneering work on the QR algorithm by Francis [198] (see also [197]), with the latter referring also to the equally important and almost simultaneously published paper by Kublanovskaya [401] (see also [400]). Another seminal paper written from the computational physics perspective is due to Reinhardt [522]. Further references from the area of computational physics and chemistry can be found in Section 3.9; see also [249].

In computational mathematics the development continued with the work of Sack and Donovan [547] on modified moments. As pointed out by Gautschi in a commentary of Golub's Selected Works [243, pp. 347–348], the idea as well as the corresponding algorithm based on modified moments are in fact present in the work of Chebyshev from 1859 [106]; see [223]. A survey of related developments can also be found in [487, Section 3]. In 1979 Dahlquist, Golub, and Nash [126] laid a foundation for error estimation in the iterative solution of linear algebraic systems; see also the paper of Dahlquist, Golub, and Eisenstat published in 1972 [125]. For further information and references see the papers [226, 242, 244, 245, 249, 457] and Golub's *Selected Works* [243].

Up to now we have assumed that A is HPD, or at least Hermitian. The connections between the matching moments model reduction, the Lanczos method, and the CG method have been explained through the underlying relationship with orthogonal polynomials and the Gauss–Christoffel quadrature. An extension to a *general* matrix A can therefore possibly also be based on an appropriate generalisation of the concept of the Gauss–Christoffel quadrature. An alternative approach that we prefer to follow in this book is based on the following observation:

If we consider an $N \times N$ Hermitian matrix A as an operator on $\mathbb{F}^N$, we can restrict it to the subspace $\mathcal{K}_n(A, v_1)$ using the orthogonal projector $V_n V_n^*$, where the columns of V_n form the orthonormal Lanczos basis. Then the matrix representation of the orthogonally projected restriction of A with respect to the basis V_n is given by

$$V_n^*(V_n V_n^* A) V_n = V_n^* A V_n = T_n. \quad (3.6.5)$$

This motivates our further exposition, in which we will interpret the operator restriction (3.6.5) as a result of the moment matching model reduction. We will then extend this approach to a general matrix A. Later we will show the relationship of this approach to the projection process (2.1.8)–(2.1.9) introduced in Chapter 2. Our main tool in the following will be Vorobyev's formulation of the moment problem.

3.7 VOROBYEV'S MOMENT PROBLEM

In his book [651], Vorobyev suggested an operator formulation of the problem of moments. In the finite-dimensional and Hermitian case this formulation is equivalent to (3.1.5) and (3.5.6). In this section we will first consider the Hermitian and the HPD cases and their relationship to Lanczos and CG, respectively. We then will describe non-Hermitian generalisations and their relationship to the non-Hermitian Lanczos algorithm and the Arnoldi algorithm.

We know very little about the person Yu V. (Yuri Vasilevich) Vorobyev. His book of 1965 [651] is an English translation of the Russian original published in 1958 (another English translation published in 1962 is not very well done). A German translation appeared in 1961. Apart from the three versions of his book, the MathSciNet database of the AMS lists only ten further publications of Vorobyev, dated between 1949 and 1968. On the papers we checked the only affiliation is

'Leningrad'; no institution is given. Nine of the ten papers are singly authored. The only co-author is named V. N. Drozdovič, who seems to have just this one publication. In the preface of his book, Vorobyev acknowledges discussions on the material with Lyusternik. We are not aware of any further interactions by him with his contemporaries.

Yet, Vorobyev must have been very knowledgeable. In Chapter 3 of his book he linked the moment problem for self-adjoined operators (Hermitian matrices) to orthogonal polynomials and continued fractions, citing works of Chebyshev, Markov, and Stieltjes. In addition, he was very up-to-date with the 'Western' literature. Already in his paper [650], which he submitted on June 11, 1953, he had cited the works of Lanczos [405] (published October 4, 1950) as well as Hestenes and Stiefel [313] (published December 6, 1952). On the other hand, the 'Westerners' were aware of Vorobyev. Householder reviewed the Russian original of his book in 1960 for the *Mathematical Reviews* of the AMS and concluded (see http://www.ams.org/mathscinet): 'The exposition is good, and the method seems to be of real value.' A few years later he cited the Russian version in [346, p. 28], but he did not give any details; see also the list of references in the 1960 paper on moments and characteristic roots by Bauer and Householder [44]. In 1968, the Ukranian mathematician Petryshyn (by then at the University of Chicago) published a one-page review of the English translation in *Mathematics of Computation*, in which he described the book as 'clearly written and well motivated'.

In the 1960s and 1970s, Vorobyev's work was cited quite frequently in particular in the physics literature, where moment methods were very popular; see, e.g. [169, 298, 562]. In the context of numerical mathematics his work was rediscovered in Kent's PhD thesis [380], completed in 1989 under the supervision of Golub. Almost a decade later Brezinski presented and generalised Vorobyev's method in [78]; also see Brezinski's book of 1997 [79, Section 4.2]. Later he applied Vorobyev's method of moments for interpretation of the Quadratic Extrapolation; see [80].

3.7.1 Application to the Hermitian Lanczos Algorithm and the CG Method

In this section we will mainly translate results of [651, Chapter 3] into our setting and notation. The main result (see Theorem 3.7.1 below) is equivalent to the moment matching properties of the Hermitian Lanczos algorithm and the CG method that we have already seen in Section 3.5.

As we focus on solving finite-dimensional linear algebraic systems $Ax = b$, we deal with operators on finite-dimensional Hilbert spaces. Vorobyev considered the more general case of bounded self-adjoint operators on infinite-dimensional Hilbert spaces, and he applied his method to various problems in applied mathematics. In this respect it is interesting to compare [651, Chapter 3] with the one of Vorobyev's Russian countryman Akhiezer [5, Chapter 4]. Although both books were published at about the same time and some views and approaches are closely related (see, in particular, the relationship between the spectral theory of operators and the problem of moments), the authors did not refer to each other.

Consider a linear algebraic system $Ax = b$, where $A \in \mathbb{F}^{N \times N}$ is HPD, and $r_0 = b - Ax_0 \equiv \|r_0\| v_1$. In Section 3.5 we linked the CG method to matching moments; cf. (3.1.5) for the moment problem and (3.5.6) for the corresponding matrix formulation. We have seen in (3.6.5) that the Jacobi matrix T_n can be considered as the *matrix representation* of the orthogonally projected restriction of the operator A defined on $\mathbb{F}^N$ to the n-dimensional subspace $\mathcal{K}_n(A, v_1)$ with respect to the orthonormal basis defined by the columns of V_n. The *operator* that is represented in matrix form by T_n is given by

$$A_n \equiv V_n V_n^* A V_n V_n^* = V_n T_n V_n^*. \qquad (3.7.1)$$

We assume that the dimension of $\mathcal{K}_n(A, v_1)$ is equal to n, because otherwise the CG approximation x_n is equal to the exact solution on or before the step $n-1$; cf. case (1) in Theorem 2.3.1.

The action of the operator A_n defined on $\mathcal{K}_n(A, v_1)$ is identical to the action of the operator A resticted to $\mathcal{K}_n(A, v_1)$. This means that A_n can be constructed using Vorobyev's formulation of the moment problem, which can be written as follows (cf. [651, Chapter III, Section 2, p. 54]):

Given A and v_1, determine A_n such that

$$\begin{aligned} A_n v_1 &= A v_1 \\ A_n(Av_1) &= A^2 v_1 \\ &\vdots \\ A_n(A^{n-2}v_1) &= A^{n-1}v_1 \\ A_n(A^{n-1}v_1) &= V_n V_n^*(A^n v_1), \end{aligned} \qquad (3.7.2)$$

or, equivalently,

$$\begin{aligned} A_n v_1 &= A v_1, \\ A_n^2 v_1 &= A^2 v_1, \\ &\vdots \\ A_n^{n-1} v_1 &= A^{n-1} v_1, \\ A_n^n v_1 &= V_n V_n^*(A^n v_1). \end{aligned} \qquad (3.7.3)$$

Since an operator on an n-dimensional subspace is uniquely determined by its action on a basis of the subspace, A_n is uniquely determined by (3.7.2) or (3.7.3). The construction above does *not* require that A is positive definite or even nonsingular. We will formulate the main result as a theorem.

Theorem 3.7.1
Using the previous notation, let A be a Hermitian matrix, and let $\mathcal{K}_n(A, v_1)$ be the n-dimensional Krylov subspace determined by A and an initial vector v_1. Then the operator A_n determined by the orthogonal restriction of A onto $\mathcal{K}_n(A, v_1)$ (see (3.7.2) or (3.7.3)) satisfies

$$v_1^* A^k v_1 = v_1^* A_n^k v_1, \quad k = 0, 1, \ldots, 2n - 1, \tag{3.7.4}$$

which can be reformulated using the matrix $T_n = V_n^* A V_n$ of A_n in the orthonormal basis represented by V_n as

$$v_1^* A^k v_1 = e_1^T T_n^k e_1, \quad k = 0, 1, \ldots, 2n - 1; \tag{3.7.5}$$

cf. (3.5.6).

Proof
Both (3.7.4) and (3.7.5) are trivial for $k = 0$. By construction,

$$v_1^* A^k v_1 = v_1^* A_n (A^{k-1}) v_1 = v_1^* A_n^k v_1, \quad k = 1, \ldots, n - 1. \tag{3.7.6}$$

We will prove that this also holds for $k = n, \ldots, 2n - 1$. Since $(V_n V_n^*)^2 = V_n V_n^*$, a multiplication of the last row of (3.7.3) implies that

$$V_n V_n^* (A^n v_1 - A_n^n v_1) = 0.$$

But since $V_n V_n^*$ is the orthogonal projector onto $\mathcal{K}_n(A, v_1)$, we see that the vector $A^n v_1 - A_n^n v_1$ is orthogonal to all basis vectors $v_1, A v_1 = A_n v_1, \ldots, A^{n-1} v_1 = A_n^{n-1} v_1$, i.e.

$$v_1^* A^j (A^n v_1 - A_n^n v_1) = 0, \quad j = 0, 1, \ldots, n - 1,$$

so that (3.7.4) indeed holds. Here we have used the assumption that A is Hermitian, so that $A = A^*$ and $A_n = A_n^*$. Combining (3.7.4) and (3.7.1) gives (3.7.5). $\square$

Summarising, the moment problems (3.1.5) and (3.5.6) can be equivalently represented by the construction of the operator A_n on the n-dimensional subspace $\mathcal{K}_n(A, v_1)$, given by (3.7.2) or (3.7.3). Note that, apart from the relationship to the CG method, the positive definiteness of A has not been used. The Vorobyev moment problems (3.7.2) and (3.7.3), as well as Theorem 3.7.1 are valid for any Hermitian matrix A.

3.7.2 Application to the Non-Hermitian Lanczos Algorithm

Given a matrix $A \in \mathbb{F}^{N \times N}$ and two starting vectors $v_1, w_1 \in \mathbb{F}^N$ with $\|v_1\| = 1$ and $w_1^* v_1 = 1$, the matrix form of the non-Hermitian Lanczos algorithm is given by

$$AV_n = V_n T_{n,n} + \delta_{n+1} v_{n+1} e_n^T,$$

$$A^* W_n = W_n T_{n,n}^* + \beta_{n+1}^* w_{n+1} e_n^T, \tag{3.7.7}$$

where $W_n^* V_n = I_n$, $\|v_j\| = 1, j = 1, \ldots, n + 1$, $w_{n+1}^* v_{n+1} = 1$, and

$$T_{n,n} = W_n^* A V_n = \begin{bmatrix} \gamma_1 & \beta_2 & & \\ \delta_2 & \gamma_2 & \ddots & \\ & \ddots & \ddots & \beta_n \\ & & \delta_n & \gamma_n \end{bmatrix}, \quad \delta_j > 0, \; \beta_j \neq 0, \; j = 2, \ldots, n;$$

(3.7.8)

cf. (2.4.8). Note that $T_{n,n}$ is in general *not* a Jacobi matrix. We stress that here we have assumed that the algorithm does not break down in steps 1 through n. Then the columns of V_n form a basis of $\mathcal{K}_n(A, v_1)$, and the columns of W_n form a basis of $\mathcal{K}_n(A^*, w_1)$. Because of the biorthogonality $W_n^* V_n = I_n$, the oblique projector onto $\mathcal{K}_n(A, v_1)$ orthogonally to $\mathcal{K}_n(A^*, w_1)$ is given by $V_n W_n^*$. We will formulate the main result in the following theorem.

Theorem 3.7.2
Using the above assumptions and notation, the non-Hermitian Lanczos algorithm applied to the matrix A and the initial vectors v_1 and w_1 represents a model reduction that matches the first 2n moments, i.e.

$$w_1^* A^k v_1 = e_1^T T_{n,n}^k e_1, \quad k = 0, 1, \ldots, 2n - 1, \tag{3.7.9}$$

where $T_{n,n}$ is given by (3.7.8).

Proof
The restriction of A to $\mathcal{K}_n(A, v_1)$ projected orthogonally to $\mathcal{K}_n(A^*, w_1)$ is given by

$$A_n = V_n W_n^* A V_n W_n^* = V_n T_{n,n} W_n^*. \tag{3.7.10}$$

It can be determined by the following generalisation of Vorobyev's moment problem (3.7.2)–(3.7.3):
Given A, v_1, and w_1, determine A_n such that

$$\begin{aligned}
A_n v_1 &= A v_1, \\
A_n (A v_1) &= A^2 v_1, \\
&\vdots \\
A_n (A^{n-2} v_1) &= A^{n-1} v_1, \\
A_n (A^{n-1} v_1) &= V_n W_n^* A^n v_1,
\end{aligned} \tag{3.7.11}$$

or, equivalently,

$$\begin{aligned}
A_n v_1 &= A v_1, \\
A_n^2 v_1 &= A^2 v_1, \\
&\vdots \\
A_n^{n-1} v_1 &= A^{n-1} v_1, \\
A_n^n v_1 &= V_n W_n^* A^n v_1.
\end{aligned} \tag{3.7.12}$$

First note that from $W_n^* V_n = I_n$ and (3.7.10) it easily follows that

$$A_n^k = V_n T_{n,n}^k W_n^*$$

for all $k \geq 1$. Thus, the first $n-1$ rows of (3.7.12) yield

$$w_1^* A^k v_1 = w_1^* A_n^k v_1 = e_1^T T_{n,n}^k e_1, \quad k = 0, 1, \ldots, n-1.$$

It remains to show the equality in (3.7.9) for $k = n, \ldots, 2n-1$. Since $(V_n W_n^*)^2 = V_n W_n^*$, the last row of (3.7.12) implies that

$$0 = V_n W_n^* (A^n v_1 - A_n^n v_1),$$

i.e. the vector $A^n v_1 - A_n^n v_1$ is orthogonal to the basis vectors $w_1, A^* w_1, \ldots, (A^*)^{n-1} w_1$ of the Krylov subspace $\mathcal{K}_n(A^*, w_1)$. We therefore get

$$((A^*)^j w_1)^* (A^n v_1 - A_n^n v_1) = 0, \quad j = 0, 1, \ldots, n-1.$$

A simple rearrangement leads with (3.7.10) to

$$w_1^* A^{j+n} v_1 = w_1^* A^j A_n^n v_1 = w_1^* A^j V_n T_{n,n}^n e_1, \quad j = 0, 1, \ldots, n-1. \quad (3.7.13)$$

Since $T_{n,n}$ is tridiagonal, we have

$$e_n^T (T_{n,n}^*)^j e_1 = 0, \quad j = 0, 1, \ldots, n-2,$$

and thus

$$\begin{aligned}
(A^*)^{n-1} w_1 &= (A^*)^{n-2} (A^* w_1) \\
&= (A^*)^{n-2} (A^* W_n e_1) \\
&= (A^*)^{n-2} \left(W_n T_{n,n}^* + \beta_{n+1}^* w_{n+1} e_n^T \right) e_1 \\
&= (A^*)^{n-2} W_n T_{n,n}^* e_1 \\
&= (A^*)^{n-3} (A^* W_n) T_{n,n}^* e_1 \\
&= (A^*)^{n-3} \left(W_n (T_{n,n}^*)^2 e_1 + \beta_{n+1}^* w_{n+1} e_n^T T_{n,n}^* e_1 \right) \\
&= (A^*)^{n-3} W_n (T_{n,n}^*)^2 e_1 \\
&= \cdots \\
&= W_n (T_{n,n}^*)^{n-1} e_1 + \beta_{n+1}^* w_{n+1} e_n^T (T_{n,n}^*)^{n-2} e_1 \\
&= W_n (T_{n,n}^*)^{n-1} e_1.
\end{aligned}$$

An analogous result obviously holds for the powers less than $n-1$, so that

$$(A^*)^j w_1 = W_n (T_{n,n}^*)^j e_1, \quad j = 0, 1, \ldots, n-1. \quad (3.7.14)$$

Using these expressions for $(A^*)^j w_1$ in (3.7.13) gives

$$w_1^* A^{j+n} v_1 = e_1^T T_{n,n}^{j+n} e_1, \quad j = 0, 1, \ldots, n-1,$$

which completes the proof. □

Remark 3.7.3
The previous proof is based on Vorobyev's moment problem. An alternative proof of an analogous result in [7, Chapter 11] uses a factorisation of the matrix of moments. Yet another proof in [34] requires the additional assumption that the non-Hermitian Lanczos algorithm does not break down until step N (when A is $N \times N$). Note that this assumption means a loss of generality, because it rules out a possible incurable breakdown in steps $n+1, n+2, \ldots, N$ of the non-Hermitian Lanczos method. Mentioning a look-ahead scheme proposed in [204] as a possible general cure (see [34, p. 16]) is mathematically incorrect. Look-ahead techniques are useful in practical computations, but the context of their use is different. They can *not* cure incurable breakdowns.

Analogously to the Hermitian case, the non-Hermitian Lanczos algorithm and the corresponding method for approximating the eigenvalues of A can be considered as a model reduction that matches at its nth step the first $2n$ moments; see (3.7.9). It is worth noticing that here we have *two* vectors v_1 and w_1, which represent the starting vectors for the two coupled recurrences with A and A^*, respectively. In general, v_1 can be different from w_1.

We finish this section with sketching an alternative proof of Theorem 3.7.2 that was suggested to us by Tichý. The term $w_1^* A^j V_n$ in (3.7.13) can be expressed by considering the following *dual* version of Vorobyev's moment problem for $A_n^* = W_n T_n^* V_n^*$, which represents the restriction of A^* onto $\mathcal{K}_n(A^*, w_1)$ projected orthogonally to $\mathcal{K}_n(A, v_1)$:

$$\begin{align} A_n^* w_1 &= A^* w_1, \\ (A_n^*)^2 w_1 &= (A^*)^2 w_1, \\ &\vdots \\ (A_n^*)^{n-1} w_1 &= (A^*)^{n-1} w_1, \\ (A_n^*)^n w_1 &= W_n V_n^* (A^*)^n w_1. \end{align} \quad (3.7.15)$$

Taking the Hermitian transpose of the identities in (3.7.15) gives

$$w_1^* A^j V_n = w_1^* A^j V_n W_n^* V_n = w_1^* A_n^j V_n = e_1^T W_n^* A_n^j V_n = e_1^T T_{n,n}^j, \quad j = 0, 1, \ldots, n,$$

and substituting this in (3.7.13) proves the assertion of Theorem 3.7.2.

3.7.3 Application to the Arnoldi Algorithm

Given a matrix $A \in \mathbb{F}^{N \times N}$ and an initial vector $v_1 \in \mathbb{F}^N$, the matrix form of the Arnoldi algorithm can be written as

$$AV_n = V_n H_{n,n} + h_{n+1,n} v_{n+1} e_n^T, \quad (3.7.16)$$

where $V_n^* V_n = I_n$, $V_n^* v_{n+1} = 0$, and $H_{n,n} = V_n^* A V_n$ is an upper Hessenberg matrix with positive entries on the first subdiagonal; cf. (2.4.3). For this algorithm we can formulate the following moment matching result.

Theorem 3.7.4
Consider the nth step of the Arnoldi algorithm applied to the matrix A and the initial vector v_1. For any additional vector u_1, the model represented by A, v_1 and u_1 is reduced to the model represented by $H_{n,n}, e_1$ and $t_n \equiv V_n^* u_1$, so that the first n moments are matched, i.e.

$$u_1^* A^k v_1 = t_n^* H_{n,n}^k e_1, \quad k = 0, 1, \ldots, n - 1. \tag{3.7.17}$$

If $u_1 = v_1$, then $t_n = e_1$, and the model reduction matches $n + 1$ moments, i.e. the equality in (3.7.17) holds for $k = n$ as well.

Proof
The n steps of the Arnoldi algorithm applied to the matrix A with the initial vector v_1 determine the restriction of A to $\mathcal{K}_n(A, v_1)$ projected orthogonally to $\mathcal{K}_n(A, v_1)$, i.e.

$$A_n = V_n V_n^* A V_n V_n^* = V_n H_{n,n} V_n^*. \tag{3.7.18}$$

Expressing A_n through the Vorobyev moment problem gives

$$\begin{aligned} A_n v_1 &= A v_1, \\ A_n(A v_1) &= A^2 v_1, \\ &\vdots \\ A_n(A^{n-2} v_1) &= A^{n-1} v_1, \\ A_n(A^{n-1} v_1) &= V_n V_n^*(A^n v_1), \end{aligned} \tag{3.7.19}$$

or, equivalently,

$$\begin{aligned} A_n v_1 &= A v_1, \\ A_n^2 v_1 &= A^2 v_1, \\ &\vdots \\ A_n^{n-1} v_1 &= A^{n-1} v_1, \\ A_n^n v_1 &= V_n V_n^* A^n v_1. \end{aligned} \tag{3.7.20}$$

Given an arbitrary vector u_1, a multiplication of the first $n - 1$ rows in (3.7.20) from the left by u_1^* results in

$$u_1^* A^k v_1 = u_1^* A_n^k v_1, \quad k = 0, 1, \ldots, n - 1. \tag{3.7.21}$$

With (3.7.18) we get

$$A_n^k = V_n H_{n,n}^k V_n^*, \quad k = 1, \ldots, n - 1.$$

We orthogonally decompose u_1 as

$$u_1 = V_n V_n^* u_1 + (I_N - V_n V_n^*) u_1.$$

With $t_n \equiv V_n^* u_1$ and (3.7.21) this yields

$$u_1^* A^k v_1 = u_1^* V_n H_{n,n}^k e_1 = t_n^* H_{n,n}^k e_1, \quad k = 0, 1, \ldots, n-1. \tag{3.7.22}$$

A multiplication of the last row in (3.7.20) from the left by u_1^* gives

$$u_1^* A_n^n v_1 = u_1^* V_n V_n^* A^n v_1 = (V_n^* u_1)^* (V_n^* A^n V_n) e_1. \tag{3.7.23}$$

Using $u_1 = v_1$ therefore results in the additional equation

$$v_1^* A^n v_1 = e_1^T H_{n,n}^n e_1, \tag{3.7.24}$$

which completes the proof. □

In a nutshell, the reduced model generated by the Arnoldi algorithm only matches half as many moments as the reduced model generated by the non-Hermitian Lanczos algorithm. A closer look at the proofs of Theorems 3.7.2 and 3.7.4 may give some insight into why this is so. The key for obtaining the 'additional moments' in the proof of the former result is that the vector $A^n v_1 - A_n^n v_1$ is orthogonal to the Krylov subspace generated by A^* and w_1. Forming the inner product with the vector $(A^*)^j w_1$ means to multiply with its Hermitian transpose $w_1^* A^j$. Consequently, (3.7.13) contains only powers of A and A_n, and not of their Hermitian transposes. On the other hand, an analogous argument in the proof of Theorem 3.7.4 would lead to a vector $A^n v_1 - A_n^n v_1$ that is orthogonal to the Krylov subspace generated by A and v_1. The equality analogous to (3.7.13) then is

$$v_1^* (A^*)^j A^n v_1 = v_1^* (A^*)^j A_n^n v_1, \quad j = 0, 1, \ldots, n-1.$$

In the Arnoldi context, where only a Krylov subspace with the matrix A is generated, the terms containing powers of both A and A^* cannot be linked to moment matching properties for A. Hence we have to settle for n (or $n+1$) instead of $2n-1$ moments that are matched.

3.7.4 Matching Moments and Generalisations of the Gauss–Christoffel Quadrature

Motivated by the insightful work on the HPD case by Fischer and Freund [185], Hochbruck and Freund presented in [205] a generalisation of the Gauss–Christoffel quadratures related to the non-Hermitian Lanczos algorithm and to the Arnoldi algorithm. We will briefly describe their approach.

Consider a nonsingular matrix $A \in \mathbb{F}^{N \times N}$, and initial vectors $v_1, w_1 \in \mathbb{F}^N$, $\|v_1\| = 1$, $w_1^* v_1 = 1$, for the non-Hermitian Lanczos algorithm. Define a bilinear form $[\cdot, \cdot]$ on the set of polynomials by

$$[\varphi, \psi] \equiv w_1^* \psi(A) \varphi(A) v_1. \tag{3.7.25}$$

Assume that the non-Hermitian Lanczos algorithm does not break down in steps 1 to n. Then the generated vectors can be expressed in terms of polynomials as

$$v_j = \varphi_{j-1}(A)\, v_1, \quad j = 1, \ldots, n+1,$$

where each polynomial $\varphi_{j-1}(\lambda)$ is of exact degree $j - 1$; cf. the discussion of Algorithm 2.4.4 in Chapter 2. Since the vector v_{n+1} is orthogonal to $\mathcal{K}_n(A^*, w_1)$, it must be orthogonal to the basis vectors $w_1, \overline{\varphi}_1(A^*)w_1, \ldots, \overline{\varphi}_{n-1}(A^*)w_1$, where $\overline{\varphi}_\ell$ denotes the polynomial complex conjugate to $\varphi_\ell, \ell = 1, \ldots, n - 1$. Consequently,

$$w_1^* \,\varphi_j(A)\, \varphi_n(A)\, v_1 = [\varphi_n, \varphi_j] = 0, \quad j = 0, 1, \ldots, n - 1,$$

and the polynomials $\varphi_0 = 1, \varphi_1, \ldots, \varphi_n$ are orthogonal with respect to the bilinear form (3.7.25). The polynomials in the sequence $\varphi_0(\lambda), \varphi_1(\lambda), \ldots$ are called *formal orthogonal polynomials*; see [144], [506, Section 12], [286], [79].

Now consider an approximation to (3.7.25) determined by the matrix $T_{n,n}$,

$$[\varphi, \psi]_n = e_1^T \, \psi(T_{n,n})\, \varphi(T_{n,n})\, e_1. \tag{3.7.26}$$

Then the moment matching property (3.7.9) implies that

$$[\varphi, 1] = [\varphi, 1]_n, \quad \text{for all } \varphi(\lambda) \text{ of degree at most } 2n - 1.$$

Therefore (3.7.26) can be considered as a generalised Gauss–Christoffel quadrature approximation to (3.7.25) associated with the non-Hermitian Lanczos algorithm; see [205, Theorem 2]. The outlined proof of Freund and Hochbruck is formulated in terms of polynomials, and it is essentially identical to our proof based on Vorobyev's method of moments. The generalised Gauss–Christoffel quadrature (3.7.26) associated with the non-Hermitian Lanczos algorithm represents an alternative formulation of the matching moments model reduction described in Section 3.7.2.

The situation for the Arnoldi method is a little bit different. Given a normalised starting vector v_1, the orthonormal Arnoldi vectors can be expressed as

$$v_{j+1} = \varphi_j(A)\, v_1, \quad j = 1, \ldots, n,$$

where each polynomial $\varphi_j(\lambda)$ is of exact degree j, and it is assumed that the algorithm does not stop on or before step n. The polynomials $\varphi_0(\lambda) \equiv 1$, $\varphi_1(\lambda), \ldots, \varphi_n(\lambda)$ are orthonormal with respect to the inner product on a set of polynomials defined by

$$\langle \varphi, \psi \rangle \equiv v_1^* \, (\psi(A))^* \, \varphi(A)\, v_1. \tag{3.7.27}$$

Using the upper Hessenberg matrix $H_{n,n}$ generated in n steps of the Arnoldi algorithm, we can consider an approximation to (3.7.27) defined by

$$\langle \varphi, \psi \rangle_n = e_1^T \, (\psi(H_{n,n}))^* \, \varphi(H_{n,n})\, e_1. \tag{3.7.28}$$

Analogously to (3.7.14) in the non-Hermitian Lanczos algorithm, it can be shown that in the Arnoldi algorithm

$$A^j v_1 = V_n H_{n,n}^j e_1, \quad j = 0, 1, \ldots, n-1,$$

and

$$A^n v_1 = V_n H_{n,n}^n e_1 + h_{n+1,n} v_{n+1} e_n^T H_{n,n}^{n-1} e_1.$$

With $V_n^* V_n = I_n$, $V_n^* v_{n+1} = 0$, we therefore obtain for any polynomials $\varphi(\lambda)$ and $\psi(\lambda)$ of degrees at most n and $n-1$, respectively,

$$\langle \varphi, \psi \rangle = (\psi(A) v_1)^* (\varphi(A) v_1) = \langle \varphi, \psi \rangle_n. \qquad (3.7.29)$$

Using that $\langle \varphi, \psi \rangle = \overline{\langle \psi, \varphi \rangle}$ and $\langle \varphi, \psi \rangle_n = \overline{\langle \psi, \varphi \rangle}_n$, the relation (3.7.29) is valid also for any polynomials $\varphi(\lambda)$ and $\psi(\lambda)$ of degrees at most $n-1$ and n, respectively. Therefore, following [205], we can consider (3.7.28) as a generalised Gauss–Christoffel quadrature approximation to (3.7.27) associated with the Arnoldi algorithm.

It is worth pointing out that here the construction is not based merely on matching the first $n + 1$ moments, i.e.

$$v_1^* A^j v_1 = e_1^T H_{n,n}^j e_1, \quad j = 0, 1, \ldots, n,$$

or, equivalently, on

$$\langle \varphi, 1 \rangle = \langle \varphi, 1 \rangle_n, \quad \text{for all } \varphi(\lambda) \text{ of degree at most } n,$$

(cf. Theorem 3.7.4), but on the stronger property

$$v_1^* (A^*)^\ell A^j v_1 = e_1^T (H_{n,n}^*)^\ell H_{n,n}^j e_1, \quad j = 0, 1, \ldots, n, \quad \ell = 0, 1, \ldots, n-1,$$

that is equivalent to (3.7.29). Therefore we can not put this into the framework of the model reduction described in Theorem 3.7.4 (cf. also the discussion following this result).

The generalised Gauss–Christoffel quadrature interpretations of the non-Hermitian Lanczos algorithm and of the Arnoldi algorithm in [205] focus on the essence of the given relationships. They do not consider a possible formal polynomial expression of the generalised Gauss–Christoffel quadrature in the complex plane using nodes and weights. That was given under various assumptions later by several authors; see, in particular, the papers by Saylor and Smolarski [551, 550], with references to computation of the scattering amplitude [658] (also see [96, 248, 600]). While such expressions can be useful in some special cases, they do not extend fully the deep quantitative relationships of the real Gauss–Christoffel quadrature for the HPD case that we have studied on numerous occasions in this chapter.

3.8 MATCHING MOMENTS AND PROJECTION PROCESSES

In Chapter 2 we presented Krylov subspace methods within the framework of projection processes. In the current Chapter 3 we have exploited further the particular operator powering structure of Krylov subspaces. We have linked Krylov subspace methods with moments and matching moments model reduction; with the CG method, the Hermitian and non-Hermitian Lanczos algorithms, and the Arnoldi algorithm as examples. Vorobyev's moment problem for the Hermitian case was conveniently used for the matrix–vector representation of the scalar moment problem, with subsequent generalisations to the non-Hermitian case. In this section we will investigate in more detail the relationship between Vorobyev's moment problem and the projection processes of Chapter 2.

First we present Vorobyev's moment problem in a slightly generalised form. Consider $n+1$ linearly independent vectors $z_0, z_1, \ldots, z_n \in \mathbb{F}^N$ and denote $\mathcal{S}_n \equiv \operatorname{span}\{z_0, \ldots, z_{n-1}\}$. Let $\mathcal{C}_n$ be a given n-dimensional subspace of $\mathbb{F}^N$, such that the direct sum of its orthogonal complement $\mathcal{C}_n^\perp$ with $\mathcal{S}_n$ gives the whole space $\mathbb{F}^N$,

$$\mathbb{F}^N = \mathcal{S}_n \oplus \mathcal{C}_n^\perp. \tag{3.8.1}$$

Then there exists a unique projector Q_n onto $\mathcal{S}_n$ and orthogonal to $\mathcal{C}_n$; see, Section 2.1. Using this notation and assumptions we can formulate the following vector moment problem.

Determine a linear operator A_n on $\mathcal{S}_n$ such that

$$\begin{aligned} A_n z_0 &= z_1, \\ A_n z_1 &= z_2, \\ &\vdots \\ A_n z_{n-2} &= z_{n-1}, \\ A_n z_{n-1} &= Q_n z_n. \end{aligned} \tag{3.8.2}$$

Since Q_n projects onto $\mathcal{S}_n$,

$$Q_n z_n = -\sum_{j=0}^{n-1} \alpha_j z_j, \tag{3.8.3}$$

for some coefficients $\alpha_0, \alpha_1, \ldots, \alpha_{n-1}$. For clarity we will consider in the following only the case

$$\alpha_0 \neq 0.$$

If $\alpha_0 = 0$, then the vectors on the right-hand side of (3.8.2) would be linearly dependent and the operator A_n defined on $\mathcal{S}_n$ would be singular (that can happen, e.g. in the Lanczos algorithm). Using (3.8.3), the operator A_n satisfies

$$(A_n^n + \alpha_{n-1} A_n^{n-1} + \cdots + \alpha_0 I) z_0 = 0$$

and its matrix representation with respect to the basis $z_0, \ldots, z_{n-1}$ is given by

$$\begin{bmatrix} 0 & & & & -\alpha_0 \\ 1 & 0 & & & \\ & 1 & \ddots & & \vdots \\ & & \ddots & 0 & \\ & & & 1 & -\alpha_{n-1} \end{bmatrix},$$

which is the companion matrix of the monic polynomial

$$\psi_n(\lambda) = \lambda^n + \alpha_{n-1}\lambda^{n-1} + \cdots + \alpha_0. \tag{3.8.4}$$

Consequently, $\psi_n(\lambda)$ is the characteristic polynomial of A_n.

Using the CG method as an example, we will now describe the relationship between the moment problem (3.8.2) and the projection processes of Chapter 2. Consider a linear algebraic system $Ax = b$, where A is HPD, an initial approximation x_0 and the initial residual $r_0 = b - Ax_0$. We assume, without loss of generality, that $\mathcal{K}_{n+1}(A, r_0)$ has dimension $n + 1$. Furthermore, for later convenience, we will not normalise the initial residual as the starting vector of the Krylov sequence. We set

$$z_j \equiv A^j r_0, \quad j = 0, 1, \ldots, n.$$

Substituting in (3.8.2) for z_j, we obtain

$$\begin{aligned} A_n r_0 &= A r_0, \\ A_n(A r_0) &= A^2 r_0, \\ &\vdots \\ A_n(A^{n-2} r_0) &= A^{n-1} r_0, \\ A_n(A^{n-1} r_0) &= Q_n(A^n r_0), \end{aligned} \tag{3.8.5}$$

with (3.8.3) giving

$$Q_n(A^n r_0) = -\sum_{j=0}^{n-1} \alpha_j A^j r_0, \quad \alpha_0 \neq 0. \tag{3.8.6}$$

The condition $\alpha_0 \neq 0$ in (3.8.6) is for the CG method satisfied by construction. Indeed, in the CG method, the matrix representation of A_n in the orthonormal basis V_n is the Jacobi matrix T_n, which is nonsingular due to the interlacing property; see Section 3.4.1, in particular Remark 3.4.4.

Up to now, no result from Section 2.1 has been used in the Vorobyev moment problem associated with the CG method. In particular, it should be pointed out that the nonsingularity of the matrix (operator) A_n has been shown independently of the nonsingularity of the projected matrix from Section 2.1.

Since Q_n is a projector onto $\mathcal{S}_n \equiv \mathcal{K}_n(A, r_0)$ and orthogonal to $\mathcal{C}_n$, where, for the CG method, $\mathcal{C}_n \equiv \mathcal{S}_n$ (cf. case (1) in Theorem 2.3.1),

$$A^n r_0 - Q_n A^n r_0 \perp \mathcal{C}_n. \tag{3.8.7}$$

i.e.

$$A^n r_0 - Q_n A^n r_0 \perp \mathcal{S}_n. \tag{3.8.8}$$

It should be noted that while in (2.1.7) the initial residual r_0 is uniquely decomposed into parts from $A\mathcal{S}_n$ and $\mathcal{C}_n^\perp$, the relation (3.8.7) corresponds to the decomposition of the vector $A^n r_0$ into parts from $\mathcal{S}_n = \mathcal{K}_n(A, r_0)$ and $\mathcal{C}_n^\perp = \mathcal{S}_n^\perp = \mathcal{K}_n(A, r_0)^\perp$,

$$A^n r_0 = A^n r_0 \Big|_{\mathcal{K}_n(A,r_0)} + A^n r_0 \Big|_{\mathcal{K}_n(A,r_0)^\perp}. \tag{3.8.9}$$

Using (3.8.3),

$$A^n r_0 \Big|_{\mathcal{K}_n(A,r_0)^\perp} = A^n r_0 - Q_n A^n r_0 = A^n r_0 + \sum_{j=0}^{n-1} \alpha_j A^j r_0.$$

This gives

$$r_0 = -\frac{1}{\alpha_0}(\alpha_1 A r_0 + \cdots + \alpha_{n-1} A^{n-1} r_0 + A^n r_0) + \frac{1}{\alpha_0} A^n r_0 \Big|_{\mathcal{K}_n(A,r_0)^\perp},$$

which is another expression for the decomposition of the initial residual r_0 into parts in $A\mathcal{S}_n = A\mathcal{K}_n(A, r_0)$ and $\mathcal{C}_n^\perp$. We summarise these relations in the following theorem.

Theorem 3.8.1
Suppose that the CG method applied to $Ax = b$, where A is HPD, does not converge to the true solution on or before the step n. Then the decomposition (3.8.9) of $A^n r_0$ determined by the Vorobyev moment problem (3.8.5) is equivalent to the decomposition (2.1.7) of r_0 determined by the Petrov–Galerkin projection. In particular,

$$r_0 = r_0 \Big|_{A\mathcal{K}_n(A,r_0)} + r_0 \Big|_{\mathcal{K}_n(A,r_0)^\perp},$$

where

$$r_0 \Big|_{A\mathcal{K}_n(A,r_0)} = -\frac{1}{\alpha_0}(\alpha_1 A r_0 + \cdots + \alpha_{n-1} A^{n-1} r_0 + A^n r_0),$$

$$r_0 \Big|_{\mathcal{K}_n(A,r_0)^\perp} = \frac{1}{\alpha_0} A^n r_0 \Big|_{\mathcal{K}_n(A,r_0)^\perp} = r_n.$$

Remark 3.8.2
We point out the difference between the roles of Q_n in (3.8.5)–(3.8.7) and P_n in (2.1)–(2.1.12). While Q_n projects $A^n r_0$ onto $\mathcal{K}_n(A, r_0)$ and orthogonal to $\mathcal{K}_n(A, r_0)$, P_n

projects the initial residual r_0 onto $A\mathcal{K}_n(A, r_0)$ and orthogonal to $\mathcal{K}_n(A, r_0)$. With the moment problem of Vorobyev, the frequently used terminology explained in Remark 2.1.5 would be consistent, because Q_n is an orthogonal projector.

Using, in addition, case (1) in Theorem 2.3.1, the nth approximation x_n of the CG method is determined by any of the three equivalent statements:

1. The nth error $x - x_n$ results from the A-orthogonal projection of the initial error $x - x_0$ onto $\mathcal{K}_n(A, r_0)$;
2. The nth residual r_n results from the oblique projection of the initial residual r_0 onto $A\mathcal{K}_n(A, r_0)$ and orthogonal to $\mathcal{K}_n(A, r_0)$;
3. The nth residual r_n results from the orthogonal projection of the nth Krylov vector $A^n r_0$ onto $\mathcal{K}_n(A, r_0)$.

The operator A_n from (3.8.5) is defined on the subspace $\mathcal{S}_n = \mathcal{K}_n(A, r_0)$ and it is injective, $A_n \mathcal{S}_n \subset \mathcal{S}_n$. Since A in HPD, it is epimorphic, $A_n \mathcal{S}_n = \mathcal{S}_n$, i.e. the matrix A_n is nonsingular. The operator A_n can be viewed as the approximation to A restricted to $\mathcal{S}_n$ that matches the first $2n$ moments

$$r_0^* A_n^k r_0 = r_0^* A^k r_0, \quad k = 0, 1, \ldots, 2n - 1;$$

see (3.7.4). It is therefore natural to investigate its relationship to the projected matrix $C_n^* A S_n$ in (2.1.9). In order to indicate that the ideas can be easily extended to the case $\mathcal{S}_n \neq \mathcal{C}_n$, i.e. beyond the CG method, we use the general notation.

As previously, let $S_n = [z_0, \ldots, z_{n-1}]$ and C_n be matrices with their column vectors forming bases of $\mathcal{S}_n$ and $\mathcal{C}_n$, respectively. Then the projector onto $\mathcal{S}_n$ and orthogonal to $\mathcal{C}_n$ is given by

$$Q_n = S_n (C_n^* S_n)^{-1} C_n^*.$$

With $z_j = A^j r_0$ we have $Q_n z_j = z_j, j = 0, 1, \ldots, n - 1$. Then the moment problem (3.8.5) immediately gives

$$A_n z_j = Q_n A z_j, \quad j = 0, 1, \ldots, n - 1.$$

The image of the basis determines the operator uniquely. Therefore, with restriction to $\mathcal{S}_n$,

$$A_n = Q_n A = S_n (C_n^* S_n)^{-1} C_n^* A.$$

The matrix of A_n in the basis S_n is given by $S_n^\dagger A_n S_n$, where $S_n^\dagger$ denotes the Moore–Penrose pseudoinverse of S_n. Multiplying the previous equality by $S_n^\dagger$ from the left and by S_n from the right gives

$$S_n^\dagger A_n S_n = (C_n^* S_n)^{-1} C_n^* A S_n. \tag{3.8.10}$$

We see that, unless the bases of $\mathcal{S}_n$ and $\mathcal{C}_n$ are biorthonormal, i.e. unless $C_n^* S_n = I_n$, the matrix (3.8.10) is in general *different* from the projected matrix $C_n^* A S_n$ in (2.1.9). This conclusion can be extended analogously to the non-Hermitian Lanczos and Arnoldi algorithms.

It should be emphasised that in the Vorobyev moment problem (3.8.5) we require, in general, $\mathbb{F}^N = \mathcal{S}_n \oplus \mathcal{C}_n^\perp$; see (3.8.1). This is not equivalent to the condition $\mathbb{F}^N = A\mathcal{S}_n \oplus \mathcal{C}_n^\perp$; see (2.1.6). An an example, consider

$$A = \begin{bmatrix} 0 & & & & & 1 \\ 1 & 0 & & & & 0 \\ & 1 & \ddots & & & \vdots \\ & & \ddots & & & \\ & & & 1 & 0 & \\ & & & & 1 & 0 \end{bmatrix} \in \mathbb{R}^{N \times N}, \quad r_0 = e_1,$$

where for $n < N$ we have $\mathcal{S}_n = \mathrm{span}\{e_1, \ldots, e_n\}$, $A\mathcal{S}_n = \mathrm{span}\{e_2, \ldots, e_{n+1}\}$. With $\mathcal{C}_n = A\mathcal{S}_n$ we have

$$A\mathcal{S}_n \oplus \mathcal{C}_n^\perp = \mathbb{R}^N$$

(here $\mathbb{F} = \mathbb{R}$), while

$$\mathcal{S}_n \oplus \mathcal{C}_n^\perp \neq \mathbb{R}^N.$$

Therefore the Vorobyev moment problem gives for $\mathcal{S}_n = \mathcal{C}_n = \mathcal{K}_n(A, r_0)$ the Arnoldi algorithm (see Section 3.7.3), but it is not directly applicable to the GMRES method with $\mathcal{S}_n = \mathcal{K}_n(A, r_0)$ and $\mathcal{C}_n = A\mathcal{K}_n(A, r_0)$; see case (3) in Theorem 2.3.1. However, GMRES iterations are closely related to the so-called FOM iterations (see [543, Section 6.4], and, in particular, [85]), and hence GMRES iterations can be characterised by the Vorobyev moment problem via this relationship.

In the next section we will study the relationship of the previous results on model reduction to the modern concept of model reduction of large-scale dynamical systems.

3.9 MODEL REDUCTION OF LARGE-SCALE DYNAMICAL SYSTEMS

Projections in combination with Krylov subspaces have been used successfully for many years in the approximation of large-scale dynamical systems. In that area a reduction of dimensionality of a given problem is a necessary part of any efficient computational modelling, and therefore model reduction represents a standard technique.

In this section we will relate the modern concept of matching moments model reduction developed in the area of dynamical systems with its roots in the 19th century work of Christoffel, Chebyshev, Stieltjes, Heine, Markov, and even Jacobi and Gauss, as well as with the 20th century work of Lanczos, Hestenes, and Stiefel, and, in particular, the largely unknown work of Vorobyev [651]. Our brief exposition greatly benefits from the comprehensive description in the short, yet remarkably thorough and informative paper [211], from the PhD thesis of Grimme [283], which also includes historical perspectives on model reduction of dynamical systems and references to earlier works, and from the paper by Parlett [506], which presents the algebraic viewpoint of minimal realisations. Moreover, we point out the monograph by Antoulas [7] and the survey papers of Bai [34] and Freund [202]. We are very much aware that our brief description does not include fundamental contributions of many other authors, among them of Mehrmann and Xu; see, e.g. [447]. While studying the link between the model reduction of linear dynamical systems and the classical 19th century results on orthogonal polynomials, approximation by rational functions, and continued fractions, one should not miss the pioneering papers of Gragg [256], and Gragg and Lindquist [260].

For connections between Krylov subspace techniques and model reduction, the literature on linear dynamical systems refers frequently to the paper [260] (see also [256]), and for the partial realisation to the paper of Kalman [373]. Other early model reduction papers are referred to in [211] and in [510, p. 16]; see also the papers of Moore [463], and Laub and Linnemann [413], as well as the work of Villemagne and Skelton [641]. That Gragg and Lindquist, as well as many others, have made fundamental contributions in the field of model reduction is an indisputable fact. For the sake of fairness, however, the description of the history given in many publications on model reduction should be revisited. For example, Grimme recalls that prior to the application of the Lanczos algorithm to the matching moments model reduction in dynamical systems, this algorithm was utilised in structural dynamics for model reduction based on eigenvalue analysis; see [283, p. 28]. He refers to works published in 1970, 1982, and 1984. This information should be complemented by a reference to the book of Vorobyev [651]. This book, of which the Russian original was published in 1958 (see the introductory part in Section 3.7) contains a very clear formulation of the matching moments model reduction ideas together with links to their foundations in the classical works on the moment problem from the 19th century, as well as the works of Krylov, Lanczos, Hestenes, and Stiefel. The book also contains applications of matching moments model reduction ideas to various practical problems. In addition to Vorobyev, one should not omit the extensive literature on moment matching in the fields of computational physics and chemistry, which started in the 1960s; cf. [255, 522, 552] and the works referred to there.

3.9.1 Approximation of the Transfer Function

Let us now consider the linear dynamical system

$$E \frac{dz}{dt} = A z(t) + b u(t),$$
$$y(t) = c^* z(t) + d u(t), \qquad (3.9.1)$$

where $A \in \mathbb{R}^{N \times N}, E \in \mathbb{R}^{N \times N}, b \in \mathbb{R}^N, c \in \mathbb{R}^N, d \in \mathbb{R}$, are given, $z(t) \in \mathbb{R}^N$ represents the inner variables of the system, $u(t) \in \mathbb{R}$ the one-dimensional input (control), and $y(t) \in \mathbb{R}$ the one-dimensional output; see, e.g. [211]. Typically N is very large, and the basic idea of model reduction is to approximate the given system (3.9.1) by a system of the same type, but with the state–space dimension N reduced to a much smaller n. This approximation should be carried out efficiently (at a low cost) and to a sufficient accuracy. Applying the Laplace transform (see, e.g. [7, Section 4.1], [506, Section 4], or [34, Section 2]), the description (3.9.1) is replaced by

$$\mathbf{T}(\lambda) = d + c^*(\lambda E - A)^{-1} b, \quad \lambda \in \mathbb{C}, \tag{3.9.2}$$

where $\mathbf{T}(\lambda)$ is called the *transfer function*. In brief, the model reduction problem (in the transfer function formulation) is to find the reduced-order matrices and vectors $A_n, E_n, b_n, c_n,$ and d_n, such that

$$\mathbf{T}_n(\lambda) = d_n + c_n^*(\lambda E_n - A_n)^{-1} b_n, \quad \lambda \in \mathbb{C},$$

closely approximates (in some sense) $\mathbf{T}(\lambda)$ *within a given frequency range* $\lambda \in \mathbb{C}_A \subset \mathbb{C}$. For simplicity, set $d = 0$ and $E = I$, then

$$\mathbf{T}(\lambda) = c^*(\lambda I - A)^{-1} b. \tag{3.9.3}$$

The problem of finding efficient numerical approximation to (3.9.3) arises in many applications beyond linear dynamical systems (3.9.1). A more general case can be written as

$$c^* F(A) b,$$

where F is a given function so that the matrix function $F(A)$ is defined; see [242, 244, 249]. The particular case $c = b$ and $F(A) = (\lambda I - A)^{-1}$, i.e. $F(A)$ being equal to the matrix resolvent with λ being outside of the spectrum of A, is of great importance in physical chemistry and solid state physics computations starting from the 1960s; see, e.g. [255, 552], the beautiful expository paper [522] containing valuable historical comments and references, and, among many other papers, [480, 464, 338, 339, 340, 337, 151, 481, 458] (ordered chronologically).

Model reduction of linear dynamical systems based on projections onto Krylov subspaces is linked with *local approximations* of the transfer function $\mathbf{T}(\lambda)$. First consider the expansion of (3.9.3) around infinity, i.e.

$$\begin{aligned} \mathbf{T}(\lambda) &= \lambda^{-1} c^* (I - \lambda^{-1} A)^{-1} b \\ &= \lambda^{-1}(c^* b) + \lambda^{-2}(c^* A b) + \cdots + \lambda^{-2n}(c^* A^{2n-1} b) + \cdots ; \end{aligned} \tag{3.9.4}$$

cf. [211], [283, Section 2.2], [7, Section 11.1]. A reduced model of order n that matches the first $2n$ terms in the expansion (3.9.4) is known as the *minimal partial realisation*; see the instrumental papers by Kalman [373], Gragg and Lindquist [260], and also [256] and [641]. The paper [256] describes the connection with the problem of moments, Gauss–Christoffel quadrature, continued fractions, the Padé table, etc., and it refers to many relevant founding papers published during over more than one hundred years, including the work of Chebyshev and Stieltjes.[5]

In particular, with the restriction to $c = b$ and A being HPD, the expansion (3.9.4) is nothing but the expansion (3.3.24)–(3.3.25) used by Chebyshev in 1859 [104, 105] and Christoffel in 1858 [113]; cf. the comment in [87, p. 23]. It was described in an instructive way by Stieljes in 1894 [588] (cf. Chapter II, Sections 7–8 and Chapter VIII, Section 51 in that paper). In order to see this link, we consider the distribution function $\omega(\lambda)$ with N points of increase associated with the HPD matrix A and the initial vector b; see Section 3.5. Then

$$b^*(\lambda I - A)^{-1}b = \sum_{j=1}^{N} \frac{\omega_j}{\lambda - \lambda_j} = \mathcal{F}_N(\lambda),$$

cf. Theorem 3.3.6, where the continued fraction on the right-hand side can be expanded, for any $n < N$, analogously to (3.3.24)–(3.3.25),

$$\mathcal{F}_N(\lambda) = \sum_{\ell=1}^{2n} \frac{\xi_{\ell-1}}{\lambda^\ell} + \mathcal{O}\left(\frac{1}{\lambda^{2n+1}}\right) = \mathcal{F}_n(\lambda) + \mathcal{O}\left(\frac{1}{\lambda^{2n+1}}\right), \quad (3.9.5)$$

$$\xi_{\ell-1} = \int \lambda^{\ell-1} \, d\omega(\lambda) = \sum_{j=1}^{n} \omega_j^{(n)} \left\{\lambda_j^{(n)}\right\}^{\ell-1}, \quad \ell = 1, \ldots, 2n,$$

or, in matrix form (cf. (3.5.6)),

$$\xi_{\ell-1} = b^* A^{\ell-1} b = e_1^T T_n^{\ell-1} e_1.$$

Clearly, $\mathcal{F}_n(\lambda)$ approximates $\mathcal{F}_N(\lambda)$ with the error being of the order $1/|\lambda|^{2n+1}$; cf. [7, Section 11.2, pp. 351–352].

The minimal partial realisation in the model reduction of linear dynamical systems (see [373, 256, 506, 211, 7]) matches the first $2n$ moments

$$\mu_{-\ell} = c^* A^{\ell-1} b, \quad \ell = 1, \ldots, 2n. \quad (3.9.6)$$

In the dynamical systems literature these moments are often called *Markov parameters*. (Note that Markov was a student of Chebyshev.)

Because of the disadvantages of partial realisation based on matching the Markov parameters in (3.9.4), i.e. the approximation of the transfer function $\mathbf{T}(\lambda)$ valid for large $|\lambda|$ (approximation at infinity), the dynamical systems model reduction literature typically prefers an expansion of $\mathbf{T}(\lambda)$ in the neighbourhood of some (finite) $\lambda_0 \in \mathbb{C}$; see, e.g. [211], [283, Chapters 2 and 4], [7, Sections 11.1 and 11.3]. Here

---

5. Euler studied in 1737 the number e using its continued fraction expansion. He introduced a variable, which led him to a Riccati equation. The English translation [170] of Euler's paper was published in 1985, and the editors (B. F. Wyman and M. F. Wyman) pointed out the relationship of Euler's early result to another formulation of partial realisations in systems theory.

the model reduction is achieved by matching the 'moments' (we use the simplified case (3.9.3) and $\lambda_0 = 0$)

$$\mu_\ell = c^*(A^{-1})^\ell b, \quad \ell = 1, 2, \ldots \quad (3.9.7)$$

of the expansion

$$\begin{aligned} -\mathbf{T}(\lambda) &= c^* A^{-1} (I - \lambda A^{-1})^{-1} b \\ &= c^* A^{-1} b + \lambda \, (c^* A^{-2} b) + \cdots + \lambda^{2n-1} (c^* A^{-2n} b) + \cdots \end{aligned} \quad (3.9.8)$$

Remark 3.9.1
Note that the terminology used in the literature on moments is not consistent. In the dynamical systems literature, as mentioned above, $\mu_{-\ell}$ in (3.9.6) is called a Markov parameter and μ_ℓ in (3.9.7) is called a moment. This terminology is different from the classical literature on the problem of moments, Gauss–Christoffel quadrature and orthogonal polynomials (cf. the extensive literature review in [487, Section 3]), as well as the linear algebra literature; see [246].

It is worth pointing out that the expansions (3.9.4) and (3.9.8) correspond to two different expansions of the same continued fraction $\mathcal{F}_N(\lambda)$. The first one is valid for $|\lambda|$ sufficiently large, the second for $|\lambda|$ sufficiently small. For illustrative examples we refer to [432, Chapter I, Sections 3.2–3.4].

In order to efficiently apply Krylov subspace methods for the model reduction around λ_0, Freund and his co-workers suggested transforming the representation of the transfer function; see [202, Section 3.2], [34, Section 2.1, relations (5)–(6)]. The Fourier model reduction by Willcox and Megretski [673] uses another transformation which allows interpretation of the matching of the first r moments μ_ℓ of $\mathbf{T}(\lambda)$ at λ_0 as matching the first r Markov parameters $\mu_{-\ell}$ of the transformed transfer function. This leads to construction of a robust criterion for determining an appropriate value of r, see [284].

It is important to note that using (3.9.7)–(3.9.8) with the transfer function approximated in the neighbourhood of the origin, associated numerical methods that compute the model reduction are based on matching the moments with the powers of A^{-1}. The computation therefore involves the solution of linear algebraic systems with the matrix A. In comparison, in the model reduction methods based on (3.9.4)–(3.9.6), the computation of the approximation needs only much cheaper matrix–vector multiplications with the matrix A. In both cases the practical computation uses Krylov subspace methods; see [7, Part IV].

We will finish this subsection with a brief discussion of an issue that is important in model reduction of dynamical systems, as well as in some physical disciplines. In linear dynamical systems described by first-order differential equations (see (3.9.1)) the autonomous system with $E = I$ and $u(t) = 0$ is called *asymptotically stable* if and only if all eigenvalues of A have negative real parts; for a detailed description we refer to [7, Section 5.8]. If the original system is stable with the eigenvalues of A located in the left-half plane, it is important to preserve this stability in reduced-order modelling, i.e. to ensure that a projected matrix of the form $T_n = V_n^* A V_n$ (corresponding to the Hermitian Lanczos algorithm), $H_{n,n} = V_n^* A V_n$ (Arnoldi algorithm), or $T_{n,n} = W_n^* A V_n$ (non-Hermitian Lanczos algorithm), also have their eigenvalues in the left-half plane. In matching moments model reduction using Kylov subspace methods this is, unfortunately, in general not possible when A is non-normal, i.e. when $AA^* \neq A^*A$.

If A is Hermitian negative definite, then all is fine due to the interlacing property of the roots of orthogonal polynomials (the eigenvalues of the associated Jacobi matrices) described in Section 3.3.1; see also Remark 3.4.4. If A is normal, then the field of values is equal to the convex hull of its spectrum, i.e. a polygon whose vertices are eigenvalues of A [342, Section 1.2, Property 1.2.9]. For any matrix V_n having normalised columns, the field of values of $V_n^* A V_n$ is contained in the field of values of A. Therefore, for a stable and *normal* A, the matrix $V_n^* A V_n$, where V_n is obtained from the (non-Hermitian) Lanczos or Arnoldi algorithm, can never be unstable.

If A is non-normal, however, the field of values can be much larger than the convex hull of the spectrum, and the stability of $V_n^* A V_n$ can not be assured (nor can it be assured for $W_n^* A V_n$, $W_n \neq V_n$, even if A is normal). This sometimes causes significant trouble because it can lead to nonphysical solutions. For an example of propagation of electromagnetic waves we refer to [309, Sections 3.4, and 4] (this paper deals with second-order differential equations and therefore the stability is related to the positive (non-negative) *imaginary* parts of the eigenvalues). On the other hand, one may argue that the lack of stability in model reduction is not necessarily a bad thing, cf. the remark in [7, Section 11.2, pp. 350–351].

3.9.2 Estimates in Quadratic Forms

The following application from signal processing illustrates some of the main ideas of this book. The transfer function (3.9.3) gives, apart from the sign, for *the particular value* $\lambda = 0$ the quantity

$$c^* A^{-1} b, \tag{3.9.9}$$

which in signal processing is called the *scattering amplitude*. Many applications require the approximation of this quantity; see [244, 249, 658, 550, 248]. We will recall some of the results in order to demonstrate that the requirements of efficiency and numerical stability of (finite precision) computations must always be taken into account. In particular, an approximation to the scalar value (3.9.9) can not be efficiently computed by an application of the methods developed for approximation of the transfer function (3.9.3). In spite of existing links, the task of computing an efficient and numerically stable approximation to (3.9.9) represents a different mathematical problem, and it requires a different approach from approximating the transfer function (3.9.3).

Considering an HPD matrix A, and using (3.5.10)–(3.5.12), we can relate the (squared) initial and nth error norms of the CG method to the scattering amplitude (3.9.9) with $b = c = r_0$,

$$r_0^* A^{-1} r_0 = \|x - x_0\|_A^2 = - \|r_0\|^2 \mathcal{F}_n(0) + \|x - x_n\|_A^2. \tag{3.9.10}$$

Here the quantity $-\|r_0\|^2 \mathcal{F}_n(0)$ is an approximation of

$$r_0^* A^{-1} r_0 = -\|r_0\|^2 \mathcal{F}_N(0) = \|r_0\|^2 \sum_{j=1}^{N} \omega_j \lambda_j^{-1},$$

with the *approximation error* given by

$$\|x - x_n\|_A^2.$$

Thus, information generated by the CG method can be used to compute approximations to the scattering amplitude. This fact, and related approaches for approximations of $c^* A^{-1} b$ based on the Gauss–Christoffel quadrature are described, e.g. in [244, 242, 249, 183, 245]. Also see the survey paper [457] as well as Golub's *Selected Works* [243, Parts II and IV] (with commentaries by Greenbaum and Gautschi), and [246, Chapter 12]. The paper [598] describes the relationship of the Gauss–Christoffel quadrature and the scattering amplitude approximations with the results present in the original paper on CG by Hestenes and Stiefel [313]. The estimates suggested in [598] were shown to be *numerically stable*, and they have been further used in the construction of stopping criteria for iterative solvers; see [13, 8, 12, 368], and also [210], the book [453] and the survey paper [457]. For non-Hermitian extensions we refer to, e.g. [248] and [600].

One can also suggest, in a straightforward way, that $c^* A^{-1} b = c^* x$, where x solves the linear system $Ax = b$. Therefore x can, in principle, be approximated by computing $c^* x_n$, where x_n is the nth approximation generated by a Krylov subspace method. This approach is used under the name 'computation of the scattering amplitude' in signal processing, see, e.g. [658, 551, 550], and under the name 'the primal linear output' in the context of optimisation [433]; see also the nice exposition in [248] and in the PhD thesis [591], as well as the survey and comparison of various approaches in [600]. A potential user of this approach should, however, be aware that the approximation of $c^* x$ by the explicit numerical computation of $c^* x_n$ can be highly inefficient even in the HPD case [598, 599]. The efficient estimation of $c^* A^{-1} b = c^* x$ is described and justified in [598, 600]; see also [457, Theorem 3.4].

Numerical approximation of the scalar value $c^* A^{-1} b$ is based, *for a good reason*, on matching the moments $c^* A^{j-1} b, j = 1, 2, \ldots$ The cost of the whole computation must be substantially smaller than the cost of computation of a (highly accurate) numerical approximation to the solution of a linear system with the matrix A. In fact, an approximate solution is not even required to obtain an acceptable approximation to $c^* A^{-1} b$.

In the approximation of the transfer function $\mathbf{T}(\lambda)$, matching the moments $c^* A^{j-1} b, j = 1, 2, \ldots$, is used in the (minimal) partial realisation (3.9.4). A numerical approximation of the scalar $c^* A^{-1} b$ may therefore look to be a simple particular case of the partial realisation for $\lambda = 0$. This interpretation is, however, mathematically erroneous. The partial realisation (3.9.4) uses the expansion of $\mathbf{T}(\lambda)$ around $\lambda = \infty$, and it is therefore valid for $|\lambda|$ large, which prohibits its use for $\lambda = 0$, i.e. for (3.9.9). Note that here we deal with the *formulation of the problem and its mathematical interpretation*, not with computational issues.

The matching moments model reduction of the transfer function $\mathbf{T}(\lambda)$ around $\lambda = 0$ is based, on the other hand, on matching the moments $c^* A^{-\ell} b$, $\ell = 1, 2, \ldots$; see (3.9.8). Because, as explained above, an approximation to the solution of $Ax = b$ is in principle to be avoided, the results developed for the transfer function can not be used in computing an efficient *approximation* of the single value $-\mathbf{T}(0) = c^* A^{-1} b$. Unlike in the previous paragraph, here the argument is based on evaluation of the *computational inefficiency*.

In summary, for both mathematical and computational reasons one can not interpret the matching moments approximation of the *single scalar value* $-\mathbf{T}(0) = c^* A^{-1} b$ as a special case of the approximation of the *function* $\mathbf{T}(\lambda)$ with the subsequent substitution $\lambda = 0$.

4

Short Recurrences for Generating Orthogonal Krylov Subspace Bases

Krylov subspace methods are based on subspaces spanned by an initial vector r_0 and vectors formed by repeated multiplication of r_0 by the given square matrix A. As described in Sections 2.3–2.4, the basis of the Krylov subspace $\mathcal{K}_n(A, r_0)$ given by the vectors $r_0, Ar_0 \ldots, A^{n-1}r_0$ is typically very ill conditioned (recall the power method), and hence useless for actual computations. A well-conditioned, at best *orthogonal* basis should be used to prevent loss of information due to the repeated matrix–vector multiplications.

For efficiency reasons it is desirable to generate such a basis with a *short recurrence*, meaning that in each iteration step only a few of the latest basis vectors are required to generate the new basis vector. In this chapter we discuss when orthogonal Krylov subspace bases can be generated by short recurrences. More precisely, given a Hermitian positive definite (HPD) matrix B and the corresponding B-inner product $(v, w)_B \equiv w^*Bv$, we analyse necessary and sufficient conditions on a nonsingular matrix A, such that for *any* initial vector r_0, a B-orthogonal basis of $\mathcal{K}_n(A, r_0)$, $n = 1, 2, \ldots,$ is generated by an *optimal short recurrence* (the precise definition of this term is given in Section 4.3; see Definition 4.3.3). It is important to emphasise that all derivations in this chapter assume *exact arithmetic*. Finite precision arithmetic aspects of short recurrences, which are of principal importance for the efficiency of practical computations, will be discussed in Chapter 5; see in particular Section 5.9.

We start this chapter by showing that the existence of optimal short recurrences is closely related to the existence of efficient Krylov subspace methods for linear algebraic systems that minimise the error in the B-norm, $\|v\|_B \equiv (v, v)_B^{1/2}$. We thereby also illustrate the historical background of this challenging research topic.

4.1 THE EXISTENCE OF CONJUGATE GRADIENT LIKE DESCENT METHODS

The classical CG method of Hestenes and Stiefel from 1952 [313], which is well defined for symmetric (or Hermitian) positive definite matrices A, is characterised by two main properties. First, at every step n the method minimises the A-norm of the error $x - x_n$ over the affine Krylov subspace $x_0 + \mathcal{K}_n(A, r_0)$; cf. item (1) in Theorem 2.3.1. Second, the two bases of $\mathcal{K}_n(A, r_0)$ employed in this minimisation process (consisting of the residual vectors $r_0, \ldots, r_{n-1}$ and the direction vectors $p_0, \ldots, p_{n-1}$) are orthogonalised by recurrences that contain only three terms. This means that only two previous basis vectors are required to generate a new basis vector that is orthogonal with respect to *all* previously generated basis vectors. In the case of residual vectors the orthogonality is with respect to the Euclidean inner product, while the direction vectors are A-orthogonal; see Algorithm 2.5.1.

More than two decades after Hestenes and Stiefel, the work of Paige and Saunders in 1975 showed [496] that Krylov subspace methods minimising the error in a fixed norm and based on orthogonal bases generated by three-term recurrences also exist for symmetric (or Hermitian) indefinite matrices. They constructed the methods SYMMLQ and MINRES, which are characterised mathematically in items (2) and (3) of Theorem 2.3.1, respectively; see Sections 2.5.4 and 2.5.5 for derivation of these methods.

It was unknown at that time whether an analogous method could also be constructed for nonsymmetric (or non-Hermitian) matrices. At the Householder Symposium held in Oxford in June 1981, Golub offered a prize of 500 US-Dollars

> ... for the construction of a 3-term conjugate gradient like descent method for non-symmetric real matrices or a proof that there can be no such method.

A lively report of this meeting, including the above quote, is given in the SIGNUM Newsletter, vol. 16, no. 4, 1981. Faber and Manteuffel answered Golub's question in 1984 [176], and they were awarded the prize money. In a nutshell, their answer is negative. *For general nonsymmetric matrices there can be no three-term conjugate gradient like descent method.* A large part of this chapter is devoted to explaining the underlying concepts and the intricacies of this fundamental theorem of Faber and Manteuffel. Moreover, we will give a complete proof of this result.

Before doing this we will explain what kind of method Golub had in mind, i.e. what is meant by the term *conjugate gradient like descent method*. Our discussion will show how these methods are related to generating an orthogonal Krylov subspace basis. This in turn will lead to a more accessible formulation of Golub's original question.

Suppose that we want to solve a linear system $Ax = b$, where A is a square (nonsingular) matrix, by an iterative method. Starting from an initial guess x_0, consider the nth iterate of the form

$$x_n = x_{n-1} + \alpha_{n-1} p_{n-1}, \quad n = 1, 2, \ldots, \tag{4.1.1}$$

where α_{n-1} is a scalar coefficient, and p_{n-1} is a nonzero vector. This means that at each step the new iterate is obtained by adding a scalar multiple of a certain nonzero *direction vector* to the previous iterate; see (2.5.34) in Section 2.5.3. By construction,

$$x_n \in x_0 + \text{span}\{p_0, \ldots, p_{n-1}\},$$

and thus the nth error satisfies

$$x - x_n \in x - x_0 + \text{span}\{p_0, \ldots, p_{n-1}\}.$$

We speak of a *conjugate gradient like descent method*, when

$$\text{span}\{p_0, \ldots, p_{n-1}\} = \mathcal{K}_n(A, r_0), \quad n = 1, 2, \ldots,$$

and at each step the error is minimised in some inner product norm,

$$\|x - x_n\|_B \equiv (x - x_n, x - x_n)_B^{1/2}$$
$$= \min_{z \in x_0 + \text{span}\{p_0, \ldots, p_{n-1}\}} \|x - z\|_B, \quad (4.1.2)$$

where B is a *fixed and given* HPD matrix that is *independent* of the initial residual.

Numerous different methods of this type exist. They can be classified by their choices of inner product matrix B, scalar coefficients α_j, and direction vectors p_j. The standard CG method of Hestenes and Stiefel has been described in Sections 2.5.1–2.5.3; see also Section 3.5. General frameworks for such methods are given in [17, 643]. A framework for methods whose norm for measuring the error depends on the initial residual can be found in [39]. Conjugate gradient like methods based on a definite bilinear form that is not necessarily symmetric or Hermitian are studied in [177].

We will now show that error minimisation in the B-norm of the iterative method (4.1.1) is essentially equivalent to the direction vectors $p_0, \ldots, p_{n-1}$ being mutually B-orthogonal, i.e. $(p_i, p_j)_B = 0$ for $i \neq j$. By construction, $x_n - x_0 \in \text{span}\{p_0, \ldots, p_{n-1}\}$. If $\text{span}\{p_0, \ldots, p_{n-1}\}$ plays the role of $\mathcal{U}$ in Theorem 2.3.2, we see that

$$\|x - x_n\|_B = \|(x - x_0) - (x_n - x_0)\|_B = \min_{z \in x_0 + \text{span}\{p_0, \ldots, p_{n-1}\}} \|x - z\|_B$$

holds if and only if

$$x - x_n \perp_B \text{span}\{p_0, \ldots, p_{n-1}\}. \quad (4.1.3)$$

The last condition is equivalent with

$$0 = (x - x_n, p_j)_B$$
$$= (x - x_{n-1}, p_j)_B - \alpha_{n-1}(p_{n-1}, p_j)_B, \quad j = 0, 1, \ldots, n-1. \quad (4.1.4)$$

These relations will be useful in the proof of the following result.

Lemma 4.1.1
Consider a linear algebraic system $Ax = b$ with $A \in \mathbb{F}^{N \times N}$ and $b \in \mathbb{F}^N$. Let $B \in \mathbb{F}^{N \times N}$ be a given HPD matrix, and let $x_0 \in \mathbb{F}^N$ be a given initial approximation. Consider an iterative method of the form (4.1.1), with scalars $\alpha_{n-1} \in \mathbb{F}$ and nonzero direction vectors $p_{n-1} \in \mathbb{F}^N$.

(1) If, for an iterative step $n \geq 1$ of this method, we have
$$\alpha_j = \frac{(x - x_j, p_j)_B}{(p_j, p_j)_B} \neq 0, \quad \text{for } j = 0, 1, \ldots, n-1, \text{ and} \tag{4.1.5}$$
$$(p_i, p_j)_B = 0, \quad \text{for } i \neq j, \; i, j = 0, 1, \ldots, n-1,$$

then the errors are minimal in the B-norm,
$$\|x - x_j\|_B = \min_{z \in x_0 + \text{span}\{p_0, \ldots, p_{j-1}\}} \|x - z\|_B, \quad \text{for } j = 1, \ldots, n. \tag{4.1.6}$$

(2) On the other hand, if (4.1.6) holds for some $n \geq 1$, and if $(x - x_j, p_j)_B \neq 0$ for $j = 0, 1, \ldots, n-1$, then (4.1.5) holds.

Proof
We first show item (1). Consider a step $n \geq 1$ of the iterative method (4.1.1), and assume that (4.1.5) holds. From $\alpha_0 = (x - x_0, p_0)_B / (p_0, p_0)_B$ it easily follows that
$$(x - x_1, p_0)_B = (x - x_0, p_0)_B - \alpha_0 (p_0, p_0)_B = 0,$$

and thus (4.1.6) is satisfied for $j = 1$. For $n = 1$ we are done. If $n > 1$, we assume that (4.1.6) is satisfied up to a certain j, $1 \leq j < n$, and we show that $\|x - x_{j+1}\|_B$ is minimal. By construction,
$$(x - x_{j+1}, p_k)_B = (x - x_j, p_k)_B - \alpha_j (p_j, p_k)_B, \quad k = 0, 1, \ldots, j. \tag{4.1.7}$$

Since $\|x - x_j\|_B$ is minimal,
$$(x - x_j, p_k)_B = 0, \quad k = 0, 1, \ldots, j - 1.$$

If the direction vectors are mutually B-orthogonal, then in particular
$$(p_j, p_k)_B = 0, \quad k = 0, 1, \ldots, j - 1, \quad \text{and thus}$$
$$(x - x_{j+1}, p_k)_B = 0, \quad k = 0, 1, \ldots, j - 1.$$

If additionally the scalar α_j is chosen as
$$\alpha_j = \frac{(x - x_j, p_j)_B}{(p_j, p_j)_B},$$

then (4.1.7) becomes $(x - x_{j+1}, p_k)_B = 0$, $k = 0, 1, \ldots, j$. The B-orthogonality of $x - x_{j+1}$ to $\text{span}\{p_0, \ldots, p_j\}$ means that $\|x - x_{j+1}\|_B$ is indeed minimal.

To show item (2), suppose that (4.1.6) holds for some $n \geq 1$. Then for each $j = 1, \ldots, n$,

$$0 = (x - x_j, p_k)_B$$
$$= (x - x_{j-1}, p_k)_B - \alpha_{j-1}(p_{j-1}, p_k)_B, \quad k = 0, 1, \ldots, j-1. \quad (4.1.8)$$

This equality for $k = j - 1$ and the assumption $(x - x_{j-1}, p_{j-1})_B \neq 0$ for $j = 1, \ldots, n$ yield

$$\alpha_{j-1} = \frac{(x - x_{j-1}, p_{j-1})_B}{(p_{j-1}, p_{j-1})_B} \neq 0, \quad j = 1, \ldots, n,$$

so that the scalar coefficients are indeed of the form stated in (4.1.5). If $n = 1$, this is all we need to show. If $n > 1$, consider any j, $1 < j \leq n$. Since $\|x - x_{j-1}\|_B$ is minimal, we know that $(x - x_{j-1}, p_k)_B = 0$ for $k = 0, 1, \ldots, j - 2$. Now (4.1.8) implies that $(p_{j-1}, p_k)_B = 0$ for $k = 0, 1, \ldots, j - 2$. Since this holds for all $j = 1, \ldots, n$, the direction vectors are mutually B-orthogonal. □

The condition in item (2) that $(x - x_j, p_j)_B \neq 0$ for $j = 0, 1 \ldots, n - 1$ is necessary for the B-orthogonality of the direction vectors in item (1). Indeed, if $(x - x_j, p_j)_B = 0$ for some j, then one can easily construct an example where $\|x - x_{j+1}\|_B$ is minimal, but the direction vectors $p_0, \ldots, p_j$ are *not* mutually B-orthogonal. The point is that if $(x - x_j, p_j)_B = 0$, which means that $\alpha_j = 0$, then $x_{j+1} = x_j$ and the direction vector p_j plays no role in the minimisation of the error norm.

The equivalence of orthogonality of the direction vectors and minimality of the errors can be summarised as follows. On the one hand, if the direction vectors form a B-orthogonal set and the scalar coefficients are chosen as in (4.1.5), then the errors are minimal in the B-norm. On the other hand, if the errors are minimal in the B-norm, and all direction vectors are used in the minimisation of the error norm (i.e. all scalars α_j in (4.1.1) are nonzero), then the direction vectors must form a B-orthogonal set.

In particular, a B-orthogonal basis of the Krylov subspace $\mathcal{K}_n(A, r_0)$ can be used to construct a conjugate gradient like descent method, whose iterates minimise the B-norm of the error over the Krylov subspace. Essential for the efficiency of such a method is the length of the recurrence for generating the B-orthogonal basis vectors. Short recurrences are of course highly desirable, since they limit the work and storage requirements of the method. Naturally, we arrive at the following variant of Golub's question:

> What are the conditions on a (square) matrix A, so that there exists an HPD matrix B for which each Krylov sequence $r_0, Ar_0, A^2 r_0, \ldots$ can be B-orthogonalised with a three-term, or, more generally, a short recurrence?

4.2 CYCLIC SUBSPACES AND THE JORDAN CANONICAL FORM

The main goal of this chapter is to give an answer to the question of Golub. The most important result in this context is the Faber–Manteuffel Theorem [176], which we state and prove in Sections 4.3–4.5 below. Our proof of this theorem is based on a classical result of Linear Algebra, namely the decomposition of a finite-dimensional complex vector space into cyclic invariant subspaces with respect to a linear operator **A** on the space. We consider cyclic subspaces important for understanding Krylov subspace methods in general, and short recurrences for generating orthogonal Krylov subspace bases in particular. Nevertheless, they have been rarely used in this context. Using a formulation in terms of linear operators rather than matrices, we derive in this section the main results on cyclic subspaces, starting from very basic definitions. Our model for this development is the elegant treatment of Gantmacher in [216, Vol. 1, Chapter VII]. One can also find some inspiration in the operator-oriented approach by Vorobyev [651]; see Section 3.7 in this book. We will maintain the operator formulation until we have completed the proof of the Faber–Manteuffel Theorem in Section 4.5.

4.2.1 Invariant Subspaces and the Cyclic Decomposition

In the following, let **A** be a linear operator on a finite-dimensional vector space $\mathcal{V}$ over the field of complex numbers $\mathbb{C}$. A subspace $\mathcal{S} \subseteq \mathcal{V}$ is called *invariant with respect to* **A**, or **A**-*invariant*, if $\mathbf{A}\mathcal{S} \subseteq \mathcal{S}$, which means that $\mathbf{A}v \in \mathcal{S}$ for all $v \in \mathcal{S}$. Note that if $\mathcal{S}$ is **A**-invariant, then $\mathbf{A}\mathcal{S}$ may be a proper subset of $\mathcal{S}$. For every linear operator $\mathbf{A} : \mathcal{V} \to \mathcal{V}$ a number of subspaces are always **A**-invariant: $\{0\}$, $\mathcal{V}$, $\text{Kernel}(\mathbf{A}) \equiv \{v \in \mathcal{V} : \mathbf{A}v = 0\}$, and $\text{Range}(\mathbf{A}) \equiv \{\mathbf{A}v : v \in \mathcal{V}\}$.

Now consider a nonzero vector $v \in \mathcal{V}$. Since $\mathcal{V}$ is finite-dimensional, the Krylov sequence $v, \mathbf{A}v, \mathbf{A}^2 v, \ldots$ contains a first vector, say $\mathbf{A}^d v$, which is a linear combination of the preceding ones,

$$\mathbf{A}^d v = \sum_{j=0}^{d-1} \gamma_j \mathbf{A}^j v;$$

cf. Section 2.2 and in particular equation (2.2.1). Therefore, $p_v(\mathbf{A})v = 0$ for the uniquely defined monic polynomial of the minimal degree d,

$$p_v(\lambda) = \lambda^d - \sum_{j=0}^{d-1} \gamma_j \lambda^j.$$

The formal definition is analogous to the notation of Section 2.2.

Definition 4.2.1: Using the previous notation, the polynomial $p_v(\lambda)$ is called the minimal polynomial of v with respect to **A***, and its degree is called the grade of v with respect to* **A**. *We denote the grade of v with respect to* **A** *by $d(\mathbf{A}, v)$. For the zero vector $v = 0$ we set $p_v(\lambda) \equiv 1$ and $d(\mathbf{A}, v) = 0$.*

It is easy to see that $d(\mathbf{A}, v) \leq \dim \mathcal{V}$ for every vector $v \in \mathcal{V}$. Also note that the zero vector $v = 0$ is the only vector of grade zero with respect to **A**. In particular, if v is any nonzero vector in the kernel of **A**, then $p_v(\lambda) = \lambda$ (since $\mathbf{A}v = 0$), so that $d(\mathbf{A}, v) = 1$.

If v is of grade $d = d(\mathbf{A}, v) \geq 1$, then $\mathcal{K}_d(\mathbf{A}, v)$ is a d-dimensional $\mathbf{A}$-invariant subspace of $\mathcal{V}$, i.e. $\mathbf{A}\mathcal{K}_d(\mathbf{A}, v) \subseteq \mathcal{K}_d(\mathbf{A}, v)$. By construction, $\mathbf{A}\mathcal{K}_d(\mathbf{A}, v) = \mathcal{K}_d(\mathbf{A}, v)$ implies that $v \in \text{Range}(\mathbf{A})$; see item (3) of Lemma 2.2.2. Note that a strict inclusion $\mathbf{A}\mathcal{K}_d(\mathbf{A}, v) \subset \mathcal{K}_d(\mathbf{A}, v)$ may indeed happen when $\mathbf{A}$ is not invertible. For example, if v is any nonzero vector in the kernel of $\mathbf{A}$, then $d(\mathbf{A}, v) = 1$ and $\{0\} = \mathcal{K}_1(\mathbf{A}, \mathbf{A}v) = \mathbf{A}\mathcal{K}_1(\mathbf{A}, v) \subset \mathcal{K}_1(\mathbf{A}, v) = \text{span}\{v\}$.

If $v \neq 0$, then the vectors $v, \mathbf{A}v, \ldots, \mathbf{A}^{d-1}v$ are linearly independent (regardless of $\mathbf{A}$ being invertible or not) and form a basis of $\mathcal{K}_d(\mathbf{A}, v)$. Application of the operator $\mathbf{A}$ carries the first vector of this basis into the second, the second into the third, etc., and the last one into a linear combination of the basis vectors. This can be illustrated as follows:

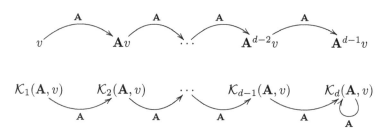

Because of this 'recurring nature' of the operation with $\mathbf{A}$, the $\mathbf{A}$-invariant subspace $\mathcal{K}_d(\mathbf{A}, v)$ is called a *cyclic subspace*. In summary, a cyclic subspace is a Krylov subspace generated by $\mathbf{A}$ and v that is of maximal possible dimension $d = d(\mathbf{A}, v)$.

We will next have a closer look at the minimal polynomials of vectors with respect to $\mathbf{A}$. In order to simplify the notation we will omit the arguments of the polynomials whenever possible. We will also usually skip the phrase 'with respect to $\mathbf{A}$', when it is clear which operator $\mathbf{A}$ is meant. By $\deg(p)$ we will denote the degree of a nonzero polynomial; the degree of the zero polynomial is defined as $-\infty$.

The following well-known division theorem for polynomials, which can be found in most books on (linear) algebra (see, e.g. [335, Section 4.4]), will be applied frequently.

Theorem 4.2.2
If f and $g \neq 0$ are polynomials over $\mathbb{C}$, then there exist uniquely defined polynomials q and r over $\mathbb{C}$, such that

$$f = g \cdot q + r \quad \text{and} \quad \deg(r) < \deg(g).$$

The polynomials q and r in Theorem 4.2.2 are called the *quotient* and the *remainder* of the division of f by g. If $r = 0$, or $f = g \cdot q$, we say that g is a *divisor* of f. Note that this is only defined for $g \neq 0$.

Lemma 4.2.3
Let $\mathbf{A}$ and v be given, and suppose that $\widetilde{p}(\mathbf{A})v = 0$ for a polynomial $\widetilde{p}$. Then p_v, the minimal polynomial of v with respect to $\mathbf{A}$, is a divisor of $\widetilde{p}$.

Proof

Since any polynomial is a divisor of the zero polynomial, the assertion is trivial for $\widetilde{p} = 0$. Hence we may consider $\widetilde{p} \neq 0$. If $v = 0$, then $p_v = 1$, which is a divisor of any polynomial. Now let $v \neq 0$. If $\widetilde{p}(\mathbf{A})v = 0$, then $\deg(\widetilde{p}) \geq \deg(p_v) \geq 1$. By Theorem 4.2.2, there exist uniquely defined polynomials q and r, such that $\widetilde{p} = p_v \cdot q + r$ and $\deg(r) < \deg(p_v)$. Hence

$$0 = \widetilde{p}(\mathbf{A})v = (p_v \cdot q + r)(\mathbf{A})v = q(\mathbf{A})\underbrace{(p_v(\mathbf{A})v)}_{=0} + r(\mathbf{A})v = r(\mathbf{A})v.$$

But then $\deg(r) < \deg(p_v)$ implies that $r = 0$, and thus p_v is a divisor of $\widetilde{p}$. □

We will now use the concept of the minimal polynomial of a vector with respect to $\mathbf{A}$ to derive the minimal polynomial of $\mathbf{A}$ with respect to an $\mathbf{A}$-invariant subspace. To this end, consider a k-dimensional $\mathbf{A}$-invariant subspace $\mathcal{S} \subseteq \mathcal{V}$, $k \geq 1$, and let $v_1, \ldots, v_k$ be a basis of this subspace (here $\mathcal{S}$ is allowed to be the whole space $\mathcal{V}$). Let $p_{v_1}, \ldots, p_{v_k}$ be the minimal polynomials of the basis vectors $v_1, \ldots, v_k$ with respect to $\mathbf{A}$, and let $\varphi_{\mathcal{S}}$ be the least common multiple of the polynomials $p_{v_1}, \ldots, p_{v_k}$. This means that $\varphi_{\mathcal{S}}$ is the uniquely defined monic polynomial of smallest degree that is divided by each minimal polynomial p_{v_j}, $j = 1, \ldots, k$. Clearly, we have $d(\mathbf{A}, v) = \deg(p_v) \leq \deg(\varphi_{\mathcal{S}})$ for all $v \in \mathcal{S}$. Some further important properties of the polynomial $\varphi_{\mathcal{S}}$ are shown in the next result.

Lemma 4.2.4
The polynomial $\varphi_{\mathcal{S}}$ constructed above is the unique monic polynomial of smallest degree that satisfies $\varphi_{\mathcal{S}}(\mathbf{A})v = 0$ for each $v \in \mathcal{S}$, so that $\varphi_{\mathcal{S}}(\mathbf{A}) = \mathbf{0}$ (the zero operator) on $\mathcal{S}$, and p_v is a divisor of $\varphi_{\mathcal{S}}$ for each $v \in \mathcal{S}$. The polynomial $\varphi_{\mathcal{S}}$ is independent of the choice of the basis of $\mathcal{S}$. If a polynomial $\widetilde{\varphi}$ satisfies $\widetilde{\varphi}(\mathbf{A})v = 0$ for all $v \in \mathcal{S}$, then $\varphi_{\mathcal{S}}$ is a divisor of $\widetilde{\varphi}$.

Proof

By construction, $\varphi_{\mathcal{S}}$ is a monic polynomial that satisfies $\varphi_{\mathcal{S}}(\mathbf{A})v_j = 0$ for each basis vector v_j. Thus, $\varphi_{\mathcal{S}}(\mathbf{A})v = 0$ holds for each $v \in \mathcal{S}$. Hence $\varphi_{\mathcal{S}}(\mathbf{A}) = \mathbf{0}$ on $\mathcal{S}$, and Lemma 4.2.3 implies that p_v divides $\varphi_{\mathcal{S}}$ for each $v \in \mathcal{S}$.

Now let $\widetilde{\varphi}$ be another polynomial with $\widetilde{\varphi}(\mathbf{A})v = 0$ for all $v \in \mathcal{S}$. Then in particular $\widetilde{\varphi}(\mathbf{A})v_j = 0$ for $j = 1, \ldots, k$, and Lemma 4.2.3 implies that p_{v_j}, $j = 1, \ldots, k$, is a divisor of $\widetilde{\varphi}$. But since $\varphi_{\mathcal{S}}$ is the least common multiple of $p_{v_1}, \ldots, p_{v_k}$, the polynomial $\varphi_{\mathcal{S}}$ is a divisor of $\widetilde{\varphi}$. If $\widetilde{\varphi}$ has the same degree as $\varphi_{\mathcal{S}}$, then the two polynomials must be identical, since they are both monic. Thus, $\varphi_{\mathcal{S}}$ is the uniquely defined monic polynomial of smallest degree with $\varphi_{\mathcal{S}}(\mathbf{A})v = 0$ for all $v \in \mathcal{S}$, and $\varphi_{\mathcal{S}}$ is independent of the choice of the basis of $\mathcal{S}$. □

Lemma 4.2.4 justifies the following definition.

Definition 4.2.5:
The polynomial $\varphi_{\mathcal{S}}$ constructed above is called the *minimal polynomial of $\mathbf{A}$ with respect to the subspace $\mathcal{S}$*. If $\mathcal{S} = \mathcal{V}$, then $\varphi_{\mathcal{S}} = \varphi_{\mathcal{V}}$ is called the *minimal polynomial of $\mathbf{A}$* (here we maintain the subscript $\mathcal{V}$ for consistency of notation), and we denote the degree of $\varphi_{\mathcal{V}}$ by $d_{\min}(\mathbf{A})$.

If ψ denotes the characteristic polynomial of $\mathbf{A}$, then the classical Cayley–Hamilton Theorem (for historical notes see Section 4.2.3) says that $\psi(\mathbf{A}) = \mathbf{0}$, i.e. that $\psi(\mathbf{A})$ is the zero operator on $\mathcal{V}$. From Lemma 4.2.4 we see that also the minimal polynomial of $\mathbf{A}$ satisfies $\varphi_{\mathcal{V}}(\mathbf{A}) = \mathbf{0}$. Moreover, $\varphi_{\mathcal{V}}$ is a divisor of ψ. In Section 4.2.2 we will prove the well-known fact that $\varphi_{\mathcal{V}} = \psi$ holds if and only if each of the distinct eigenvalues of $\mathbf{A}$ has geometric multiplicity one and thus occurs in exactly one Jordan block of $\mathbf{A}$. Matrices and linear operators having this property are called called *nonderogatory*; see, e.g. [341, p. 58].

It is useful to know that the grade of vector as well as the degree of the minimal polynomial is invariant under shifts of the operator $\mathbf{A}$.

Lemma 4.2.6
If $\mathbf{A}$ is a linear operator on a finite-dimensional vector space $\mathcal{V}$ over $\mathbb{C}$ and $\mu \in \mathbb{C}$ is any given scalar, then $d(\mathbf{A}, v) = d(\mathbf{A} - \mu\mathbf{I}, v)$ for all $v \in \mathcal{V}$, so that in particular $d_{\min}(\mathbf{A}) = d_{\min}(\mathbf{A} - \mu\mathbf{I})$.

Proof
Let $v \in \mathcal{V}$ be given. If p_v is the minimal polynomial of v with respect to $\mathbf{A}$, then the polynomial $q_v(\lambda) \equiv p_v(\lambda + \mu)$ has the same degree as p_v and it satisfies

$$q_v(\mathbf{A} - \mu\mathbf{I})v = p_v(\mathbf{A})v = 0,$$

so that $d(\mathbf{A} - \mu\mathbf{I}, v) \leq d(\mathbf{A}, v)$.

On the other hand, if r_v is the minimal polynomial of v with respect to $\mathbf{A} - \mu\mathbf{I}$, then the polynomial $q_v(\lambda) \equiv r_v(\lambda - \mu)$ has the same degree as r_v and it satisfies

$$q_v(\mathbf{A})v = r_v(\mathbf{A} - \mu\mathbf{I})v = 0,$$

so that $d(\mathbf{A}, v) \leq d(\mathbf{A} - \mu\mathbf{I}, v)$. Thus, $d(\mathbf{A}, v) = d(\mathbf{A} - \mu\mathbf{I}, v)$. □

It is worth mentioning the particular case where μ is an eigenvalue of $\mathbf{A}$ and v is the corresponding eigenvector. Then

$$\mathbf{A}v = \mu v \quad \text{and} \quad (\mathbf{A} - \mu\mathbf{I})v = 0.$$

Hence the minimal polynomials of v with respect to $\mathbf{A}$ and to $\mathbf{A} - \mu\mathbf{I}$ are $\lambda - \mu$ and λ, respectively, giving $d(\mathbf{A}, v) = d(\mathbf{A} - \mu\mathbf{I}, v) = 1$. For $\mathbf{A} = \mathbf{I}$ and $\mu = 1$ we get the trivial case $\mathbf{A} - \mathbf{I} = \mathbf{0}$.

We continue with a result for the minimal polynomial of a linear operator $\mathbf{A}$ with respect to a cyclic subspace.

Lemma 4.2.7
If v is a vector of grade d with respect to $\mathbf{A}$, then the minimal polynomial of v with respect to $\mathbf{A}$ is equal to the minimal polynomial of $\mathbf{A}$ with respect to $\mathcal{K}_d(\mathbf{A}, v)$, i.e. $p_v = \varphi_{\mathcal{K}_d(\mathbf{A},v)}$. Moreover, for any $w \in \mathcal{K}_d(\mathbf{A}, v)$ its minimal polynomial p_w with respect to $\mathbf{A}$ is a divisor of p_v.

Proof

For each $j = 0, 1, \ldots, d-1$,

$$p_v(\mathbf{A})(\mathbf{A}^j v) = \mathbf{A}^j p_v(\mathbf{A}) v = 0,$$

and hence $p_v(\mathbf{A}) w = 0$ for each $w \in \mathcal{K}_d(\mathbf{A}, v)$. Lemma 4.2.4 then implies that $\varphi_{\mathcal{K}_d(\mathbf{A},v)}$ is a divisor of p_v. On the other hand, p_v is a divisor of $\varphi_{\mathcal{K}_d(\mathbf{A},v)}$, since by definition $\varphi_{\mathcal{K}_d(\mathbf{A},v)}$ is the least common multiple of $p_v, p_{\mathbf{A}v}, \ldots, p_{\mathbf{A}^{d-1}v}$. Since the two monic polynomials p_v and $\varphi_{\mathcal{K}_d(\mathbf{A},v)}$ are divisors of each other, they must be equal.

For the last part, again observe that $p_v(\mathbf{A})w = 0$ for each $w \in \mathcal{K}_d(\mathbf{A}, v)$ and apply Lemma 4.2.3. $\square$

We now prove the main result of this section, namely that the whole vector space $\mathcal{V}$ can be decomposed into the direct sum of cyclic subspaces, where the minimal polynomial of $\mathbf{A}$ with respect to each of these cyclic subspaces is a power of an *irreducible polynomial*. This means that each such polynomial is of the form $(\lambda - \lambda_j)^{d_j}$ for some $\lambda_j \in \mathbb{C}$ and integer $d_j \geq 1$. There are several different proofs of this important result, and all of them are quite technical. Our proof is adapted to our notation and context from the one given in [238]. Later in this chapter we will show that the decomposition of $\mathcal{V}$ into cyclic subspaces is mathematically equivalent to constructing the Jordan canonical form of $\mathbf{A}$.

Theorem 4.2.8
Let $\mathbf{A}$ be a linear operator on a finite-dimensional vector space $\mathcal{V}$ over $\mathbb{C}$. Then there exist vectors $v_1, \ldots, v_m \in \mathcal{V}$ of respective grades $d_1 \geq \cdots \geq d_m \geq 1$, such that

$$\mathcal{V} = \bigoplus_{j=1}^{m} \mathcal{K}_{d_j}(\mathbf{A}, v_j), \tag{4.2.1}$$

where $p_{v_j} = (\lambda - \lambda_j)^{d_j}$ and $\lambda_j \in \mathbb{C}, j = 1, \ldots, m$.

Proof

The proof is by induction on the dimension of $\mathcal{V}$. The assertion is trivial when $\mathcal{V}$ is one-dimensional. Consider $\dim \mathcal{V} = n$ with $n \geq 2$, and suppose that the assertion is true for any vector space of dimension at most $n - 1$. Let $\mathbf{A} : \mathcal{V} \to \mathcal{V}$ be a singular linear operator. We show how to deal with an invertible $\mathbf{A}$ at the end of the proof.

Since $\mathbf{A}$ is singular, $\text{Range}(\mathbf{A})$ has dimension at most $n - 1$. Let $\mathcal{W} \subset \mathcal{V}$ be any subspace of dimension $n - 1$ that contains $\text{Range}(\mathbf{A})$. Then

$$\mathbf{A}\mathcal{W} \subseteq \text{Range}(\mathbf{A}) \subseteq \mathcal{W},$$

and hence $\mathbf{A} : \mathcal{W} \to \mathcal{W}$, where $\mathcal{W}$ has dimension $n - 1$. By the induction hypothesis, there are vectors $w_1, \ldots, w_k \in \mathcal{W}$ of the respective grades $d_1 \geq \cdots \geq d_k \geq 1$, such that

$$\mathcal{W} = \bigoplus_{j=1}^{k} \mathcal{K}_{d_j}(\mathbf{A}, w_j), \tag{4.2.2}$$

where $p_{w_j} = (\lambda - \lambda_j)^{d_j}$ for some $d_j \geq 1$ and $\lambda_j \in \mathbb{C}$, $j = 1, \ldots, k$. Our task now is to extend this cyclic decomposition of $\mathcal{W}$ to a cyclic decomposition of the whole space $\mathcal{V}$.

Let $\mathbf{I}_\mathcal{W}$ denote the identity operator on $\mathcal{W}$. Then for each $j = 1, \ldots, k$,

$$0 = p_{w_j}(\mathbf{A})w_j = (\mathbf{A} - \lambda_j \mathbf{I}_\mathcal{W})^{d_j} w_j = \left(\sum_{i=0}^{d_j} \binom{d_j}{i} \lambda_j^{d_j - i} \mathbf{A}^i \right) w_j$$

$$= \lambda_j^{d_j} w_j + \sum_{i=1}^{d_j} \binom{d_j}{i} \lambda_j^{d_j - i} \mathbf{A}^i w_j.$$

Note that

$$\text{if } \lambda_j \neq 0, \text{ then } w_j \in \text{span}\{\mathbf{A}w_j, \ldots, \mathbf{A}^{d_j} w_j\}. \tag{4.2.3}$$

Now define the set $\mathcal{N} \equiv \{j \mid \lambda_j = 0\}$, i.e. the set of all indices j, $1 \leq j \leq k$, for which $\lambda_j = 0$. In other words, the set $\mathcal{N}$ contains all indices j for which $p_{w_j}(\lambda) = \lambda^{d_j}$.

Consider any vector y in the set $\mathcal{V} \setminus \mathcal{W}$, which does not contain the zero vector and hence is not a subspace. Then $\mathcal{V} = \mathcal{W} \oplus \text{span}\{y\}$. Since $\mathbf{A}y \in \text{Range}(\mathbf{A}) \subseteq \mathcal{W}$, we can write $\mathbf{A}y$ as a linear combination of the basis vectors

$$w_j, \mathbf{A}w_j, \ldots, \mathbf{A}^{d_j - 1} w_j, \quad 1 \leq j \leq k,$$

of $\mathcal{W}$. We have to distinguish two cases.

First, if $\mathbf{A}y$ is a linear combination of these vectors where no w_j corresponding to some $\lambda_j = 0$ occurs, then by (4.2.3) there exists a vector $z \in \mathcal{W}$ such that $\mathbf{A}y = \mathbf{A}z$. Taking $u \equiv y - z$ we get $u \in \mathcal{V} \setminus \mathcal{W}$ and $\mathbf{A}u = 0$. In this case span$\{u\} = \mathcal{K}_1(\mathbf{A}, u)$ is a nontrivial cyclic subspace, $p_u = \lambda$, and

$$\mathcal{V} = \mathcal{W} \oplus \mathcal{K}_1(\mathbf{A}, u),$$

so we are done.

Second, if the expansion of $\mathbf{A}y$ in the basis of $\mathcal{W}$ contains at least one vector w_j corresponding to some $\lambda_j = 0$, then for some scalars $\alpha_j \in \mathbb{C}$ that are not all zero and a vector $z \in \mathcal{W}$,

$$\mathbf{A}y = \left(\sum_{j \in \mathcal{N}} \alpha_j w_j \right) + \mathbf{A}z.$$

Then, for $u \equiv y - z$,

$$\mathbf{A}u = \mathbf{A}y - \mathbf{A}z = \sum_{j \in \mathcal{N}} \alpha_j w_j \neq 0. \tag{4.2.4}$$

Hence $\mathcal{K}_1(\mathbf{A}, u)$ is not a cyclic subspace, so there is more work to do. Let ℓ, $1 \leq \ell \leq k$, be the smallest integer with $\alpha_\ell \neq 0$ in (4.2.4). Such an ℓ must exist since $\mathbf{A}u \neq 0$. Let $\tilde{u} \equiv \alpha_\ell^{-1} u$, so that

$$\mathbf{A}\widetilde{u} = \alpha_\ell^{-1}\mathbf{A}u = w_\ell + \sum_{\substack{j \in \mathcal{N} \\ j > \ell}} \frac{\alpha_j}{\alpha_\ell} w_j \equiv w_\ell + \widehat{w}_\ell.$$

Now consider the **A**-invariant subspace

$$\mathcal{X} \equiv \bigoplus_{\substack{j \in \mathcal{N} \\ j > \ell}} \mathcal{K}_{d_j}(\mathbf{A}, w_j). \qquad (4.2.5)$$

(Note that if $\ell = k$, then $\widehat{w}_\ell = 0$ and $\mathcal{X} = \{0\}$.) By construction, $w_\ell = \mathbf{A}\widetilde{u} - \widehat{w}_\ell$, where $\widehat{w}_\ell \in \mathcal{X}$, so that

$$\begin{aligned}
\mathcal{K}_{d_\ell}(\mathbf{A}, w_\ell) \oplus \mathcal{X} &= \text{span}\left\{w_\ell, \ldots, \mathbf{A}^{d_\ell - 1} w_\ell\right\} \oplus \mathcal{X} \\
&= \text{span}\left\{\mathbf{A}\widetilde{u} - \widehat{w}_\ell, \ldots, \mathbf{A}^{d_\ell - 1}(\mathbf{A}\widetilde{u} - \widehat{w}_\ell)\right\} \oplus \mathcal{X} \\
&= \text{span}\left\{\mathbf{A}\widetilde{u}, \mathbf{A}^2 \widetilde{u}, \ldots, \mathbf{A}^{d_\ell}\widetilde{u}\right\} \oplus \mathcal{X}. \qquad (4.2.6)
\end{aligned}$$

The subspaces on both sides of this equation must have the same dimension, so that in particular

$$d_\ell = \dim \mathcal{K}_{d_\ell}(\mathbf{A}, w_\ell) = \dim\left(\text{span}\left\{\mathbf{A}\widetilde{u}, \mathbf{A}^2 \widetilde{u}, \ldots, \mathbf{A}^{d_\ell}\widetilde{u}\right\}\right).$$

We next use that $\mathbf{A}\widetilde{u} = w_\ell + \widehat{w}_\ell$ to derive

$$\mathbf{A}^{d_\ell+1}\widetilde{u} = \mathbf{A}^{d_\ell}(\mathbf{A}\widetilde{u}) = \mathbf{A}^{d_\ell}(w_\ell + \widehat{w}_\ell) = \mathbf{A}^{d_\ell} w_\ell + \sum_{\substack{j \in \mathcal{N} \\ j > \ell}} \frac{\alpha_j}{\alpha_\ell} \mathbf{A}^{d_\ell} w_j.$$

But by construction, for all $j \in \mathcal{N}$ with $j \geq \ell$ we have $d_\ell \geq d_j$ (the grades are ordered decreasingly) and thus $\mathbf{A}^{d_\ell} w_j = \mathbf{A}^{d_\ell - d_j} \mathbf{A}^{d_j} w_j = 0$. Hence all vectors in the last expression above are equal to zero, giving

$$\mathbf{A}^{d_\ell + 1}\widetilde{u} = 0,$$

which will be important below. Using (4.2.2), (4.2.5), and (4.2.6) we obtain

$$\begin{aligned}
\mathcal{W} &= \left(\bigoplus_{j \notin \mathcal{N}} \mathcal{K}_{d_j}(\mathbf{A}, w_j)\right) \oplus \left(\bigoplus_{j \in \mathcal{N}} \mathcal{K}_{d_j}(\mathbf{A}, w_j)\right) \\
&= \left(\bigoplus_{j \notin \mathcal{N}} \mathcal{K}_{d_j}(\mathbf{A}, w_j)\right) \oplus \left(\bigoplus_{\substack{j \in \mathcal{N} \\ j < \ell}} \mathcal{K}_{d_j}(\mathbf{A}, w_j)\right) \oplus \mathcal{K}_{d_\ell}(\mathbf{A}, w_j) \oplus \mathcal{X} \\
&= \bigoplus_{j \neq \ell} \mathcal{K}_{d_j}(\mathbf{A}, w_j) \oplus \text{span}\left\{\mathbf{A}\widetilde{u}, \mathbf{A}^2\widetilde{u}, \ldots, \mathbf{A}^{d_\ell}\widetilde{u}\right\}. \qquad (4.2.7)
\end{aligned}$$

But since $\widetilde{u} \in \mathcal{V} \setminus \mathcal{W}$ with $\dim \mathcal{W} = n - 1$ and $\dim \mathcal{V} = n$, we see that

$$\mathcal{V} = \mathcal{W} \oplus \text{span}\{\widetilde{u}\}$$
$$= \bigoplus_{j \neq \ell} \mathcal{K}_{d_j}(\mathbf{A}, w_j) \oplus \text{span}\left\{\widetilde{u}, \mathbf{A}\widetilde{u}, \mathbf{A}^2\widetilde{u}, \ldots, \mathbf{A}^{d_\ell}\widetilde{u}\right\}$$
$$= \bigoplus_{j \neq \ell} \mathcal{K}_{d_j}(\mathbf{A}, w_j) \oplus \mathcal{K}_{d_\ell+1}(\mathbf{A}, \widetilde{u}). \tag{4.2.8}$$

Finally, since $\mathbf{A}^{d_\ell+1}\widetilde{u} = 0$, the space $\mathcal{K}_{d_\ell+1}(\mathbf{A}, \widetilde{u})$ is a $(d_\ell + 1)$-dimensional cyclic subspace, and

$$p_{\widetilde{u}} = \lambda^{d_\ell+1}.$$

This completes the proof for singular $\mathbf{A}$.

It remains to consider the case of $\mathbf{A}$ being invertible. Then for any eigenvalue $\mu \in \mathbb{C}$ of $\mathbf{A}$ the linear operator $\mathbf{A} - \mu \mathbf{I}_\mathcal{V}$ is singular. Hence for this operator a cyclic decomposition of $\mathcal{V}$ of the form (4.2.1) exists. The assertion for $\mathbf{A}$ now follows when considering that the cyclic subspaces for $j = 1, \ldots, k$ are shift invariant, i.e.

$$\mathcal{K}_{d_j}(\mathbf{A}, v_j) = \mathcal{K}_{d_j}(\mathbf{A} - \mu \mathbf{I}_\mathcal{V}, v_j), \quad \text{for all } \mu \in \mathbb{C},$$

and the fact that if the minimal polynomial of v_j with respect to $\mathbf{A} - \mu \mathbf{I}_\mathcal{V}$ is given by $(\lambda - \lambda_j)^{d_j}$, then the minimal polynomial of v_j with respect to $\mathbf{A}$ is $(\lambda - (\lambda_j + \mu))^{d_j}$. □

Let us illustrate the decomposition (4.2.1) by an example that is similar to the one given in [174, Section 2]. Suppose that $\mathbf{A}$ is the linear operator on $\mathcal{V} = \mathbb{C}^3$ whose matrix representation in the canonical basis of $\mathbb{C}^3$ is

$$A = \begin{bmatrix} 1 & 2 & -6 \\ 1 & 2 & 3 \\ 1 & -1 & 6 \end{bmatrix} = S \left[\begin{array}{cc|c} 3 & 3 & 0 \\ 0 & 3 & 0 \\ \hline 0 & 0 & 3 \end{array}\right] S^{-1}, \quad \text{where}$$

$$S \equiv \begin{bmatrix} 2 & -1 & -1 \\ -1 & -1 & -1 \\ -1 & -1 & 0 \end{bmatrix}.$$

We see that the characteristic polynomial of A is $(\lambda - 3)^3$, and the minimal polynomial is $\varphi_\mathcal{V} = (\lambda - 3)^2$ so that $d_{\min}(A) = 2$. Any nonzero vector in $\mathbb{C}^3$ is either of grade one (and hence is an eigenvector) or of grade two with respect to A. Obviously, the first canonical basis vector is not an eigenvector of A. Therefore $v_1 \equiv [1, 0, 0]^T$ is of grade $d_1 = 2$ with respect to A, the cyclic subspace $\mathcal{K}_{d_1}(A, v_1)$ has dimension two, and the minimal polynomial of v_1 is equal to $\varphi_\mathcal{V}$. Note that

$$\mathcal{K}_{d_1}(A, v_1) = \text{span}\left\{\begin{bmatrix} 1 \\ 0 \\ 0 \end{bmatrix}, \begin{bmatrix} 1 \\ 1 \\ 1 \end{bmatrix}\right\} = \left\{\begin{bmatrix} \alpha \\ \beta \\ \beta \end{bmatrix} : \alpha, \beta \in \mathbb{C}\right\}.$$

Since $\mathbb{C}^3$ has dimension three, it remains to find a vector $v_2 \notin \mathcal{K}_{d_1}(A, v_1)$ that is of grade $d_2 = 1$ and has the minimal polynomial $p_{v_2} = \lambda - 3$ (since p_{v_2} must divide p_{v_1}). This means that v_2 must be an eigenvector with respect to the eigenvalue 3 of A, which is *not* contained in $\mathcal{K}_{d_1}(A, v_1)$. These requirements are satisfied by the vector $v_2 \equiv [-1, -1, 0]^T$, giving the decomposition

$$\mathbb{C}^3 \;=\; \mathcal{K}_{d_1}(A, v_1) \oplus \mathcal{K}_{d_2}(A, v_2) \;=\; \mathrm{span}\,\{v_1, Av_1\} \oplus \mathrm{span}\,\{v_2\}.$$

The next lemma deals with the composition of cyclic subspaces for two vectors v and w. It will be used below to obtain another decomposition of $\mathcal{V}$ into cyclic subspaces. Our proof of this lemma is adapted from [388, p. 491].

Lemma 4.2.9
Let $v, w \in \mathcal{V}$ be nonzero vectors of respective grades d_v, d_w with respect to $\mathbf{A}$. If their minimal polynomials p_v, p_w have no common zeros (i.e. are relatively prime), then $v + w$ is of grade $d_{v+w} = d_v + d_w$, the minimal polynomial of $v + w$ is given by $p_{v+w} = p_v \cdot p_w$, and

$$\mathcal{K}_{d_v + d_w}(\mathbf{A}, v+w) \;=\; \mathcal{K}_{d_v}(\mathbf{A}, v) \oplus \mathcal{K}_{d_w}(\mathbf{A}, w).$$

Proof
We first show that $\mathcal{K}_{d_v}(\mathbf{A}, v)$ and $\mathcal{K}_{d_w}(\mathbf{A}, w)$ have a trivial intersection. If $u \in \mathcal{K}_{d_v}(\mathbf{A}, v) \cap \mathcal{K}_{d_w}(\mathbf{A}, w)$, then by Lemma 4.2.7 the minimal polynomial p_u divides both p_v and p_w. Since p_v and p_w are relatively prime, we must have $p_u = 1$ and thus $u = 0$, which yields $\mathcal{K}_{d_v}(\mathbf{A}, v) \cap \mathcal{K}_{d_w}(\mathbf{A}, w) = \{0\}$. Therefore we can write

$$\mathcal{K}_{d_v}(\mathbf{A}, v) + \mathcal{K}_{d_w}(\mathbf{A}, w) \;=\; \mathcal{K}_{d_v}(\mathbf{A}, v) \oplus \mathcal{K}_{d_w}(\mathbf{A}, w).$$

Since $v \in \mathcal{K}_{d_v}(\mathbf{A}, v)$ and $w \in \mathcal{K}_{d_w}(\mathbf{A}, w)$, we have

$$v + w \in \mathcal{K}_{d_v}(\mathbf{A}, v) \oplus \mathcal{K}_{d_w}(\mathbf{A}, w),$$

and

$$\mathcal{K}_{d_{v+w}}(\mathbf{A}, v+w) \;\subseteq\; \mathcal{K}_{d_v}(\mathbf{A}, v) \oplus \mathcal{K}_{d_w}(\mathbf{A}, w). \tag{4.2.9}$$

Because p_v and p_w are relatively prime, there exist polynomials f and g so that

$$1 = f \cdot p_v + g \cdot p_w.$$

From this standard result on polynomials (see, e.g. [533, p. 5]) we get

$$v \;=\; f(\mathbf{A})\,p_v(\mathbf{A})v + g(\mathbf{A})\,p_w(\mathbf{A})v \;=\; g(\mathbf{A})\,p_w(\mathbf{A})v \;=\; g(\mathbf{A})\,p_w(\mathbf{A})\,(v+w),$$

and hence $v \in \mathcal{K}_{d_{v+w}}(\mathbf{A}, v+w)$. Similarly, $w \in \mathcal{K}_{d_{v+w}}(\mathbf{A}, v+w)$, and thus, together with (4.2.9),

$$\mathcal{K}_{d_{v+w}}(\mathbf{A}, v+w) \;=\; \mathcal{K}_{d_v}(\mathbf{A}, v) \oplus \mathcal{K}_{d_w}(\mathbf{A}, w).$$

Since $v \in \mathcal{K}_{d_{v+w}}(\mathbf{A}, v + w)$, we must have $p_{v+w}(\mathbf{A})v = 0$, and hence p_v is a divisor of p_{v+w}; cf. Lemma 4.2.3. Similarly, p_w is a divisor of p_{v+w}. On the other hand, $p_v(\mathbf{A})p_w(\mathbf{A})(v + w) = 0$, and hence p_{v+w} is a divisor of $p_v \cdot p_w$. This yields $p_{v+w} = p_v \cdot p_w$, and completes the proof. □

Using the decomposition (4.2.1) we will now derive another decomposition of the complex vector space $\mathcal{V}$ into the direct sum of cyclic subspaces. Suppose that we have decomposed $\mathcal{V}$ as in (4.2.1), where $p_{v_j} = (\lambda - \lambda_j)^{d_j}$ for $j = 1, \ldots, m$. Then the main information in the decomposition (4.2.1) can be written schematically as follows:

$$\begin{array}{llll} v_1 & v_2 \cdots & v_m & \text{(vectors)} \\ d_1 \geq d_2 & \cdots \geq d_m & & \text{(grades)} \\ \lambda_1 & \lambda_2 \cdots & \lambda_m & \text{(roots)} \end{array} \qquad (4.2.10)$$

Let $\widetilde{\lambda}_1, \ldots, \widetilde{\lambda}_k$ be the k *distinct* complex numbers among the numbers $\lambda_1, \ldots, \lambda_m$. Suppose that the number $\widetilde{\lambda}_j$ occurs $\ell_j \geq 1$ times in the sequence $\lambda_1, \ldots, \lambda_m$. We now group the cyclic subspaces corresponding to the same $\widetilde{\lambda}_j$ together. Within each such group we order the subspaces decreasingly by their respective dimensions:

$$\underbrace{\begin{array}{c} \widetilde{v}_1, \ldots, \widetilde{v}_{\ell_1} \\ \widetilde{d}_1 \geq \cdots \geq \widetilde{d}_{\ell_1} \end{array}}_{\widetilde{\lambda}_1} \quad \underbrace{\begin{array}{c} \widetilde{v}_{\ell_1+1}, \ldots, \widetilde{v}_{\ell_1+\ell_2} \\ \widetilde{d}_{\ell_1+1} \geq \cdots \geq \widetilde{d}_{\ell_1+\ell_2} \end{array}}_{\widetilde{\lambda}_2} \quad \cdots \quad \underbrace{\begin{array}{c} \widetilde{v}_{m-\ell_k+1}, \ldots, \widetilde{v}_m \\ \widetilde{d}_{m-\ell_k+1} \geq \cdots \geq \widetilde{d}_m \end{array}}_{\widetilde{\lambda}_k} \qquad (4.2.11)$$

$$(\widetilde{\lambda}_1, \widetilde{\lambda}_2, \ldots, \widetilde{\lambda}_k \text{ are distinct})$$

For each $\widetilde{\lambda}_j$, $j = 1, \ldots, k$, we consider the first vector in its associated group in (4.2.11). Then the corresponding minimal polynomials, i.e. the polynomials

$$p_{\widetilde{v}_1} = (\lambda - \widetilde{\lambda}_1)^{\widetilde{d}_1}, \quad p_{\widetilde{v}_{\ell_1+1}} = (\lambda - \widetilde{\lambda}_2)^{\widetilde{d}_{\ell_1+1}}, \quad \ldots, \quad p_{\widetilde{v}_{m-\ell_k+1}} = (\lambda - \widetilde{\lambda}_k)^{\widetilde{d}_{m-\ell_k+1}},$$

are relatively prime. We define

$$w_1 \equiv \widetilde{v}_1 + \widetilde{v}_{\ell_1+1} + \cdots + \widetilde{v}_{m-\ell_k+1}.$$

Then, by Lemma 4.2.9, the minimal polynomial of the vector w_1 is given by

$$p_{w_1} = p_{\widetilde{v}_1} \cdot p_{\widetilde{v}_{\ell_1+1}} \cdots p_{\widetilde{v}_{m-\ell_k+1}}.$$

Since p_{w_1} is the least common multiple of the polynomials $p_{v_1}, \ldots, p_{v_m}$, it is clear that p_{w_1} is equal to the minimal polynomial of $\mathbf{A}$, with its degree given by $c_1 \equiv \widetilde{d}_1 + \widetilde{d}_{\ell_1+1} + \cdots + \widetilde{d}_{m-\ell_k+1}$. Schematically,

$$\begin{aligned} w_1 &\equiv \widetilde{v}_1 + \widetilde{v}_{\ell_1+1} + \cdots + \widetilde{v}_{m-\ell_k+1}, \\ c_1 &\equiv \widetilde{d}_1 + \widetilde{d}_{\ell_1+1} + \cdots + \widetilde{d}_{m-\ell_k+1}, \\ p_{w_1} &\equiv \varphi_\mathcal{V} \quad \text{(the minimal polynomial of } \mathbf{A}\text{)}. \end{aligned} \qquad (4.2.12)$$

We now proceed sequentially through the k groups in (4.2.11), where at each step we pick from each group one vector (from the remaining ones) corresponding to a cyclic subspace of maximal dimension. In the first step this gives

$$w_2 \equiv \tilde{v}_2 + \tilde{v}_{\ell_1+2} + \cdots + \tilde{v}_{m-\ell_k+2}$$
(some of these terms may not exist),
$$c_2 \equiv \tilde{d}_2 + \tilde{d}_{\ell_1+2} + \cdots + \tilde{d}_{m-\ell_k+2} \leq c_1 \quad (4.2.13)$$
(some of these numbers may be zero),
$$p_{w_2} \equiv p_{\tilde{v}_2} \cdot p_{\tilde{v}_{\ell_1+2}} \cdots p_{\tilde{v}_{m-\ell_k+2}} \text{ divides } p_{w_1}.$$

We continue until we reach $\ell \equiv \max\{\ell_j : j = 1, \ldots, k\}$. The resulting decomposition of the space $\mathcal{V}$ is stated in the following corollary of Theorem 4.2.8.

Corollary 4.2.10
Let $\mathbf{A}$ be a linear operator on a finite-dimensional vector space $\mathcal{V}$ over $\mathbb{C}$. Then there exist vectors $w_1, \ldots, w_\ell \in \mathcal{V}$ of respective grades $c_1 \geq \cdots \geq c_\ell \geq 1$, such that

$$\mathcal{V} = \bigoplus_{j=1}^{\ell} \mathcal{K}_{c_j}(\mathbf{A}, w_j), \quad (4.2.14)$$

where $p_{w_1} = \varphi_\mathcal{V}$ (the minimal polynomial of $\mathbf{A}$), and $p_{w_{j+1}}$ is a divisor of p_{w_j} for $j = 1, \ldots, \ell - 1$.

In some books on (linear) algebra, this result is called the *cyclic decomposition theorem*; see, e.g. [335, Section 7.2]. A short and self-contained proof of the decomposition (4.2.14) can be found in [388].

We will use the decomposition (4.2.14) in the proof of the following lemma to construct a basis of $\mathcal{V}$ consisting of vectors that are all of 'full grade' with respect to $\mathbf{A}$. The existence of such a basis will turn out helpful later in this chapter.

Lemma 4.2.11
Let $\mathbf{A}$ be a linear operator on a finite-dimensional vector space $\mathcal{V}$ over $\mathbb{C}$, and let $d_{\min}(\mathbf{A})$ denote the degree of the minimal polynomial of $\mathbf{A}$. Then there exists a basis of $\mathcal{V}$ consisting of vectors that are all of grade $d_{\min}(\mathbf{A})$.

Proof
From the cyclic decomposition (4.2.14) we see that the vectors

$$w_1, \ldots, \mathbf{A}^{c_1-1} w_1, \quad w_2, \ldots, \mathbf{A}^{c_2-1} w_2, \quad \ldots, \quad w_\ell, \ldots, \mathbf{A}^{c_\ell-1} w_\ell \quad (4.2.15)$$

form a basis of $\mathcal{V}$. Consider any fixed scalar $\mu \in \mathbb{C}$ that is not an eigenvalue of $\mathbf{A}^j$ for $j = 1, \ldots, c_1 - 1$. Then the operators $\mathbf{A}^j - \mu \mathbf{I}$ are invertible for $j = 1, \ldots, c_1 - 1$. For any polynomial q we have

$$q(\mathbf{A})(\mathbf{A}^j - \mu \mathbf{I}) w_1 = (\mathbf{A}^j - \mu \mathbf{I}) q(\mathbf{A}) w_1.$$

Since $\mathbf{A}^j - \mu\mathbf{I}$ is invertible, we have $q(\mathbf{A})(\mathbf{A}^j - \mu\mathbf{I})w_1 = 0$ if and only if $q(\mathbf{A})w_1 = 0$. Consequently, the polynomial p_{w_1} (the minimal polynomial of the vector w_1) is the minimal polynomial of the vector $(\mathbf{A}^j - \mu\mathbf{I})w_1 = \mathbf{A}^j w_1 - \mu w_1$, for $j = 1, \ldots, c_1 - 1$. Hence in particular $d(\mathbf{A}, \mathbf{A}^j w_1 - \mu w_1) = d(\mathbf{A}, w_1) = d_{\min}(\mathbf{A})$, for $j = 1, \ldots, c_1 - 1$.

Now we shift all vectors in (4.2.15) (except the first one) by $-\mu w_1$ and thus obtain the vectors

$$
\begin{array}{cccc}
w_1, & \mathbf{A}w_1 - \mu w_1, & \ldots, & \mathbf{A}^{c_1-1}w_1 - \mu w_1, \\
w_2 - \mu w_1, & \mathbf{A}w_2 - \mu w_1, & \ldots, & \mathbf{A}^{c_2-1}w_2 - \mu w_1, \\
& \cdots & & \\
w_\ell - \mu w_1, & \mathbf{A}w_\ell - \mu w_1, & \ldots, & \mathbf{A}^{c_\ell-1}w_\ell - \mu w_1.
\end{array}
\qquad (4.2.16)
$$

Obviously, the vectors in (4.2.16) are linearly independent, and therefore they form a basis of $\mathcal{V}$.

Next note that for any $j = 1, \ldots, \ell - 1$ the minimal polynomial $p_{w_{j+1}}$ of the vector w_{j+1} in (4.2.15) divides the minimal polynomial p_{w_j}. This means that the minimal polynomial of each vector in the basis (4.2.16) is given by p_{w_1}, so that each vector in this basis is of grade $d_{\min}(\mathbf{A})$. $\square$

The previous lemma represents a generalisation of [174, Lemma 4.2] to the case of a possibly singular operator $\mathbf{A}$.

Remark 4.2.12

In 1931 Krylov [398] published a method for transforming the characteristic (or secular) equation $\det(\lambda\mathbf{I} - \mathbf{A}) = 0$ of a matrix $\mathbf{A} \in \mathbb{C}^{N \times N}$ to a simpler form; see the historical note in Section 2.5.7 for details on Krylov and the early developments in the context of his method. The following brief description of this method is based on Gantmacher's book [216, Vol. 1, Chapter VII, Section 8]; see also his beautiful paper on this subject [215], which is discussed in a footnote in Section 2.2 of this book. Many references and a detailed account of Krylov's paper can also be found in [178, Section 42].

Suppose, for simplicity, that $v \in \mathbb{C}^N$ is a vector of grade N with respect to $\mathbf{A}$. (Gantmacher refers to this as the 'regular case'.) Such a vector v exists if and only if $\mathbf{A}$ is nonderogatory; see Section 4.2.2 below. If $p_v(\lambda) = \lambda^N - \sum_{j=0}^{N-1} \alpha_j \lambda^j$ is the minimal polynomial of v with respect to $\mathbf{A}$, then

$$
\begin{bmatrix} v & \cdots & \mathbf{A}^{N-1}v & \mathbf{A}^N v \\ 1 & \cdots & \lambda^{N-1} & \lambda^N - p_v(\lambda) \end{bmatrix} \begin{bmatrix} -\alpha_0 \\ \vdots \\ -\alpha_{N-1} \\ 1 \end{bmatrix} = \begin{bmatrix} 0 \\ \vdots \\ 0 \\ 0 \end{bmatrix}.
$$

Denote the $((N+1) \times (N+1))$–matrix on the left-hand side by M. Since the homogeneous linear system with M has a nonzero solution, M must be singular. Using the linearity of the determinant in the last row, we get

$$0 = \det(M)$$
$$= \det\left(\begin{bmatrix} v & \cdots & \mathbf{A}^{N-1}v & \mathbf{A}^N v \\ 1 & \cdots & \lambda^{N-1} & \lambda^N \end{bmatrix}\right) + \det\left(\begin{bmatrix} v & \cdots & \mathbf{A}^{N-1}v & \mathbf{A}^N v \\ 0 & \cdots & 0 & -p_v(\lambda) \end{bmatrix}\right)$$
$$= \det\left(\begin{bmatrix} v & \cdots & \mathbf{A}^{N-1}v & \mathbf{A}^N v \\ 1 & \cdots & \lambda^{N-1} & \lambda^N \end{bmatrix}\right) - p_v(\lambda) \cdot \det\left([v, \ldots, \mathbf{A}^{N-1}v]\right),$$

and hence

$$p_v(\lambda) = \det\left(\begin{bmatrix} v & \cdots & \mathbf{A}^{N-1}v & \mathbf{A}^N v \\ 1 & \cdots & \lambda^{N-1} & \lambda^N \end{bmatrix}\right) \Big/ \det\left([v, \ldots, \mathbf{A}^{N-1}v]\right). \quad (4.2.17)$$

Note that the matrix $[v, \ldots, \mathbf{A}^{N-1}v]$ is nonsingular by construction. Since $\deg(p_v) = N$, the polynomial $p_v(\lambda)$ is equal to the characteristic polynomial of $\mathbf{A}$. The advantage of Krylov's formula (4.2.17) in comparison with the standard formula for the characteristic polynomial (i.e. $\det(\lambda I - \mathbf{A})$), is that the unknown λ only occurs in one row of the matrix whose determinant has to be computed. Hence the computation of the coefficients simplifies. For more on Krylov's method, in particular the so-called 'singular case' when $\deg(p_v) < N$, we refer to [216, Vol. 1, Chapter VII, Section 8], [178, Section 42], or [346, Chapter 6.2].

Krylov's method was also known to Vorobyev, who in fact pointed out in [651, p. 113]: 'Hence, it follows that the only distinction between the method of moments and Krylov's method for the case of a finite symmetric matrix (where, as previously noted, it coincides with Lanczos' 'method of minimal iterations') is in their computational schemes.'

4.2.2 The Jordan Canonical Form and the Length of the Krylov Sequences

In this section we use the cyclic decomposition (4.2.1) to derive the Jordan canonical form of the operator $\mathbf{A}$, and we further characterise the grade of the vectors $v \in \mathcal{V}$ with respect to $\mathbf{A}$. We also construct a vector of grade d for any given integer d between zero and $d_{\min}(\mathbf{A})$ (the degree of the minimal polynomial of $\mathbf{A}$).

Let $\mathbf{I}$ denote the identity operator on $\mathcal{V}$. For a given linear operator $\mathbf{A}$, consider a vector v of grade $d \geq 1$ with its minimal polynomial with respect to $\mathbf{A}$ given by $p_v(\lambda) = (\lambda - \widehat{\lambda})^d$ for some $\widehat{\lambda} \in \mathbb{C}$. Then the d vectors

$$v, (\mathbf{A} - \widehat{\lambda}\mathbf{I})v, \ldots, (\mathbf{A} - \widehat{\lambda}\mathbf{I})^{d-1}v \quad (4.2.18)$$

are linearly independent. Indeed, if the vectors were linearly dependent, then for some complex scalars $\alpha_0, \ldots, \alpha_{d-1}$, not all equal to zero,

$$0 = \sum_{j=0}^{d-1} \alpha_j (\mathbf{A} - \widehat{\lambda}\mathbf{I})^j v = q(\mathbf{A})v,$$

where q is a nonzero polynomial of degree at most $d - 1$, contradicting the minimality assumption on d.

Since

$$(\mathbf{A} - \widehat{\lambda}\mathbf{I})^j v \in \mathcal{K}_d(\mathbf{A}, v), \quad \text{for} \quad j = 0, 1, \ldots, d-1,$$

it follows that these d vectors form a basis of $\mathcal{K}_d(\mathbf{A}, v)$. For $j = 0, 1, \ldots, d-1$ we observe

$$\begin{aligned}
\mathbf{A} \cdot (\mathbf{A} - \widehat{\lambda}\mathbf{I})^j v &= \left(\widehat{\lambda}\mathbf{I} + (\mathbf{A} - \widehat{\lambda}\mathbf{I})\right) \cdot (\mathbf{A} - \widehat{\lambda}\mathbf{I})^j v \\
&= \widehat{\lambda}(\mathbf{A} - \widehat{\lambda}\mathbf{I})^j v + (\mathbf{A} - \widehat{\lambda}\mathbf{I})^{j+1} v
\end{aligned} \quad (4.2.19)$$

Since $(\mathbf{A} - \widehat{\lambda}\mathbf{I})^d v = 0$, equation (4.2.19) for $j = d - 1$ reads

$$\mathbf{A} \cdot (\mathbf{A} - \widehat{\lambda}\mathbf{I})^{d-1} v = \widehat{\lambda}(\mathbf{A} - \widehat{\lambda}\mathbf{I})^{d-1} v,$$

which shows that the vector $(\mathbf{A} - \widehat{\lambda}\mathbf{I})^{d-1} v$ is an eigenvector of $\mathbf{A}$ corresponding to the eigenvalue $\widehat{\lambda}$.

Let us reverse the order of the vectors in (4.2.18), i.e. consider the vectors

$$(\mathbf{A} - \widehat{\lambda}\mathbf{I})^{d-1} v, \ldots, (\mathbf{A} - \widehat{\lambda}\mathbf{I}) v, v. \quad (4.2.20)$$

Then (4.2.19) shows that the matrix representation of the linear operator $\mathbf{A}$ restricted to the subspace $\mathcal{K}_d(\mathbf{A}, v)$ and with respect to this basis is given by the *Jordan block*

$$J_d(\widehat{\lambda}) \equiv \begin{bmatrix} \widehat{\lambda} & 1 & & \\ & \ddots & \ddots & \\ & & \ddots & 1 \\ & & & \widehat{\lambda} \end{bmatrix} \in \mathbb{C}^{d \times d}. \quad (4.2.21)$$

Accordingly, the basis (4.2.20) is called a *Jordan basis* of the cyclic subspace $\mathcal{K}_d(\mathbf{A}, v)$. Note that while the vector v is the *initial* vector in the Krylov sequence $v, \mathbf{A}v, \ldots, \mathbf{A}^{d-1}v$, it is the *last* vector in the corresponding Jordan basis (4.2.20). This fact has previously been pointed out in [14, Section 3].

Using the decomposition of $\mathcal{V}$ in Theorem 4.2.8 we can now easily state the following important result.

Theorem 4.2.13
Let $\mathbf{A}$ be a linear operator on a finite-dimensional vector space $\mathcal{V}$ over $\mathbb{C}$. Then there exists a basis of $\mathcal{V}$, such that the matrix representation of $\mathbf{A}$ with respect to this basis is given by

$$\begin{bmatrix} J_{d_1}(\lambda_1) & & \\ & \ddots & \\ & & J_{d_m}(\lambda_m) \end{bmatrix}, \quad (4.2.22)$$

where each diagonal block $J_{d_j}(\lambda_j)$, $j = 1, \ldots, m$, is a Jordan block of the form (4.2.21), and $d_1 \geq d_2 \geq \cdots \geq d_m$.

Proof
By Theorem 4.2.8, the whole space $\mathcal{V}$ can be decomposed into the direct sum of cyclic subspaces, where the minimal polynomial of $\mathbf{A}$ with respect to each of these subspaces is of the form $(\lambda - \widehat{\lambda})^d$ for some $\widehat{\lambda} \in \mathbb{C}$ and integer $d \geq 1$. As shown above, for each of these cyclic subspaces there exists a Jordan basis of the form (4.2.20), which proves the assertion. □

The matrix (4.2.22) is called a *Jordan canonical form* of $\mathbf{A}$; see Section 4.2.3 for historical notes on this term. This form is determined uniquely up to the order of the Jordan blocks on the diagonal; see, e.g. [335, Section 7.3] for a proof of this fact. Any permutation of the order of the blocks gives another Jordan canonical form of $\mathbf{A}$. Permuting the order of the blocks corresponds to a reordering of the cyclic subspaces in the decomposition of the space $\mathcal{V}$. Above we have already seen three different orderings:

- The form of (4.2.22) corresponds to the ordering of the cyclic subspaces as in (4.2.10), where the blocks are ordered decreasingly by their sizes; see the first matrix in Figure 4.1 for an illustration.
- Ordering these spaces as in (4.2.11) leads to a Jordan canonical form where the blocks corresponding to the same eigenvalue are grouped together. Within each of the individual groups the blocks are ordered by decreasing size; see the second matrix in Figure 4.1.
- The construction illustrated in (4.2.12)–(4.2.13), which leads to the cyclic decomposition (4.2.14), groups the largest blocks corresponding to each of the distinct eigenvalues first, the second largest next, and so on; see the third matrix in Figure 4.1.

If $\mathbf{A}$ is a given linear operator on a finite-dimensional vector space $\mathcal{V}$ over $\mathbb{C}$ and $\mathcal{K}_d(\mathbf{A}, v)$ is any cyclic subspace, then $d \leq d_{\min}(\mathbf{A})$, with equality if and only if $p_v = \varphi_\mathcal{V}$, i.e. the minimal polynomial of the vector v is equal to the minimal polynomial of $\mathbf{A}$. By Corollary 4.2.10, such a vector v always exists; it is the vector w_1 in the decomposition (4.2.14). More generally, any vector v having nonzero components in the directions of the vectors $\widetilde{v}_1, \widetilde{v}_{\ell_1+1}, \ldots, \widetilde{v}_{m-\ell_k+1}$ (see (4.2.11)) has

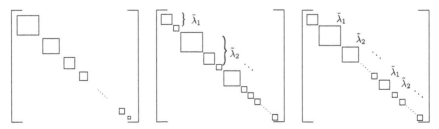

Figure 4.1 An illustration of the three different Jordan forms of $\mathbf{A}$ derived in Section 4.2.

this property. When $d = d_{\min}(\mathbf{A})$, the Krylov sequence $v, \mathbf{A}v, \ldots, \mathbf{A}^{d-1}v$, is called a *Krylov sequence of maximal length*.

As described in Section 2.2 (see in particular Theorem 2.2.3), the length of the Krylov sequence determines how many steps certain well-defined Krylov subspace methods require until termination (in exact arithmetic). We next show that for *any* given integer d, $0 \leq d \leq d_{\min}(\mathbf{A})$, there exists a vector $v \in \mathcal{V}$ that is of grade d. For $d = 0$ we have $v = 0$, and a vector v with $d = d_{\min}(\mathbf{A})$ has already been constructed above. It remains to consider the case $1 \leq d < d_{\min}(\mathbf{A})$.

Let $\widetilde{\lambda}_1, \ldots, \widetilde{\lambda}_k \in \mathbb{C}$ be the distinct eigenvalues of $\mathbf{A}$, and consider the construction (4.2.12)–(4.2.13), where the vector

$$w_1 \equiv \widetilde{v}_1 + \cdots + \widetilde{v}_{m-\ell_k+1}$$

is of grade $d_{\min}(\mathbf{A})$ and

$$\varphi_v(\lambda) = p_{w_1}(\lambda) = (\lambda - \widetilde{\lambda}_1)^{\widetilde{d}_1} \cdots (\lambda - \widetilde{\lambda}_k)^{\widetilde{d}_m - \ell_k + 1}$$

Consider the Jordan basis of $\mathcal{K}_{\widetilde{d}_1}(\mathbf{A}, \widetilde{v}_1)$,

$$(\mathbf{A} - \widetilde{\lambda}_1 \mathbf{I})^{\widetilde{d}_1 - 1} \widetilde{v}_1, \ldots, (\mathbf{A} - \widetilde{\lambda}_1 \mathbf{I}) \widetilde{v}_1, \widetilde{v}_1.$$

By construction, for $\ell = 1, \ldots, \widetilde{d}_1$ the minimal polynomial of the ℓth vector of this basis is given by $(\lambda - \widetilde{\lambda}_1)^\ell$, and thus the ℓth vector is of grade ℓ. The same holds for the other vectors $\widetilde{v}_{\ell_1+1}, \ldots, \widetilde{v}_{m-\ell_k+1}$ that occur in the linear combination of the vector w_1.

Therefore, if d is any given integer, $1 \leq d \leq d_{\min}(\mathbf{A})$, and we want to construct a vector v of grade d, we can choose k integers ξ_j, $j = 1, \ldots, k$, with $1 \leq \xi_1 \leq \widetilde{d}_1$, $1 \leq \xi_2 \leq \widetilde{d}_{\ell_1+1}$, etc., and $\sum_{j=1}^k \xi_j = d$. Then the vector

$$v \equiv (\mathbf{A} - \widetilde{\lambda}_1 \mathbf{I})^{\widetilde{d}_1 - \xi_1} \widetilde{v}_1 + \cdots + (\mathbf{A} - \widetilde{\lambda}_k \mathbf{I})^{\widetilde{d}_m - \ell_k + 1 - \xi_k} \widetilde{v}_{m-\ell_k+1} \qquad (4.2.23)$$

is of grade d since $\widetilde{\lambda}_1, \ldots, \widetilde{\lambda}_k$ are pairwise distinct (cf. Lemma 4.2.9). We can combine the number d from the numbers ξ_j in many different ways, so that the construction is in general by no means unique.

4.2.3 Historical Note: Classical Results of Linear Algebra

Many results on the spectral theory of matrices that are nowadays considered classical were developed in the second half of the 19th century. A first milestone in this development was Cauchy's proof in 1829 that (in modern terminology) real symmetric matrices have real eigenvalues [97]. The general case of the spectral decomposition was solved independently and almost simultaneously by Weierstrass in 1868 [663] and Jordan in 1870 [369]. These two treatments were developed from different mathematical viewpoints and formulated using very different notation (and, in addition, two different languages). A quarrel broke out between Paris and

Berlin mathematicians over priority and the usefulness of the results. Stating three problems in the context of canonical forms, Jordan insisted in 1873 [370] that 'the solutions given by the eminent geometers of Berlin are nevertheless incomplete as some exceptional cases have been left aside despite their significance. [The Berliners'] analysis is also quite difficult to follow — Especially M. Weierstraß' one' (translation taken from [74]). Weierstrass' colleague in Berlin, Kronecker, replied in 1874 [397] that in Jordan's paper 'the solution to the first problem is not original, the solution to the second problem is incorrect, and the solution to the third problem is incomplete. In addition to that, the third problem, which actually includes the first two others as special cases, has been completely solved by M. Weierstraß work of 1868 and by my additional contribution to it' (translation taken from [74]). By the end of 1874 it became clear that the results of Jordan and Weierstrass were mathematically equivalent. For more on this interesting historical episode we refer to [74, 299, 300, 301]. Today, the term *Jordan canonical form* refers to the canonical form for a single matrix or operator (as in Theorem 4.2.13 and Figure 4.1), while the term *Weierstrass canonical form* is typically used in connection with matrix or operator pencils $A - \lambda B$.

The concept of the *minimal polynomial* was introduced by Frobenius in his landmark paper of 1878 [209]. Frobenius had been a student of Weierstrass and Kummer in Berlin, where he received his PhD in 1870. He wrote this paper as a professor at the Eidgenössische Polytechnikum (today ETH) in Zürich. He used the language of bilinear forms and the theory of elementary divisors ('Elementarteiler') introduced by Weierstrass. He applied the minimal polynomial (called by him 'the degree of the equation of smallest degree $\varphi(A) = 0$'; our translation) to give one of the first complete proofs of the Cayley–Hamilton Theorem.

Hamilton had shown this result for quaternions in 1853 [293, p. 567], and Cayley had proven it for matrices of orders $N = 2$ and $N = 3$ in 1858 [98]. In [209], Frobenius proved the Cayley–Hamilton Theorem by showing that the minimal polynomial φ of a matrix (or bilinear form) A is a divisor of the characteristic polynomial ψ. Hence $\varphi(A) = 0$ implies that $\psi(A) = 0$. He also showed that if p is any polynomial with $p(A) = 0$, then φ is a divisor of p; cf. Lemma 4.2.4 in our book. For an interesting discussion of Frobenius' paper from a historical perspective we refer to [302].

4.3 OPTIMAL SHORT RECURRENCES

We now come back to the question of Golub, discussed in Section 4.1, whether for a given matrix (or linear operator) and any starting vector there exists a short recurrence for generating orthogonal Krylov subspace bases. Throughout Sections 4.3–4.5 we will assume that

A is a linear operator on a finite-dimensional vector space $\mathcal{V}$ over $\mathbb{C}$ with a *given* inner product $(\cdot, \cdot)$, and $d_{\min}(\mathbf{A})$ is the degree of the minimal polynomial of **A**.

Our development in this section is based on [426, Section 2]. Let $v \in \mathcal{V}$ be a given vector of grade $d = d(\mathbf{A}, v) \geq 1$. Our goal is to orthogonalise the basis

$v, \mathbf{A}v, \ldots, \mathbf{A}^{d-1}v$ of the cyclic subspace $\mathcal{K}_d(\mathbf{A}, v)$ with respect to the given inner product on $\mathcal{V}$. More precisely, we want to generate vectors $v_1, \ldots, v_d$, such that

$$\text{span}\{v_1, \ldots, v_n\} = \mathcal{K}_n(\mathbf{A}, v), \quad n = 1, \ldots, d, \tag{4.3.1}$$

$$(v_j, v_k) = 0, \quad j \neq k, \; j, k = 1, \ldots, d. \tag{4.3.2}$$

To orthogonalise the Krylov sequence $v, \mathbf{A}v, \ldots, \mathbf{A}^{d-1}v$ we use the Arnoldi algorithm [16] (see Algorithm 2.4.1),

$$v_1 = v,$$

$$v_{n+1} = \mathbf{A}v_n - \sum_{m=1}^{n} h_{m,n} v_m, \quad h_{m,n} = \frac{(\mathbf{A}v_n, v_m)}{(v_m, v_m)}, \tag{4.3.3}$$

$$n = 1, \ldots, d-1, \quad d = d(\mathbf{A}, v).$$

The vectors $v_1, \ldots, v_d$ generated by (4.3.3) form the uniquely defined (up to scaling) basis of $\mathcal{K}_d(\mathbf{A}, v)$ that satisfies (4.3.1) and (4.3.2). Note that in comparison with Algorithm 2.4.1 here we have skipped the normalisation of the basis vectors for notational convenience.

We call v (or v_1) the *initial vector* of the algorithm stated in (4.3.3). This algorithm can formally be written as (cf. (2.4.3))

$$v_1 = v,$$

$$\mathbf{A} \underbrace{[v_1, \ldots, v_{d-1}]}_{\equiv V_{d-1}} = \underbrace{[v_1, \ldots, v_d]}_{\equiv V_d} \underbrace{\begin{bmatrix} h_{1,1} & \cdots & h_{1,d-1} \\ 1 & \ddots & \vdots \\ & \ddots & h_{d-1,d-1} \\ & & 1 \end{bmatrix}}_{\equiv H_{d,d-1}}, \tag{4.3.4}$$

$$(v_i, v_j) = 0 \quad \text{for } i \neq j, \; i, j = 1, \ldots, d.$$

We point out that the whole basis $v_1, \ldots, v_d$ is generated in $d-1$ steps of the Arnoldi algorithm, and hence the matrix $H_{d,d-1} \in \mathbb{C}^{d \times (d-1)}$ is *nonsquare*.

As mentioned above, any other basis $\widehat{v}_1, \ldots, \widehat{v}_d$ satisfying (4.3.1) and (4.3.2) can be obtained by scaling the basis vectors $v_1, \ldots, v_d$. Formally we can write this as

$$\widehat{V}_d \equiv [\widehat{v}_1, \ldots, \widehat{v}_d] = V_d S_d,$$

where $S_d \in \mathbb{C}^{d \times d}$ is a nonsingular diagonal matrix. Then

$$\mathbf{A} \widehat{V}_{d-1} = \widehat{V}_d \widehat{H}_{d,d-1}, \quad \text{where } \widehat{H}_{d,d-1} = S_d^{-1} H_{d,d-1} S_{d-1}, \tag{4.3.5}$$

and S_{d-1} is the $(d-1) \times (d-1)$ leading principal submatrix of S_d. Clearly, with any such scaling the nonzero pattern of $\widehat{H}_{d,d-1}$ is identical to the nonzero pattern of $H_{d,d-1}$. In the following we will mostly be interested in this pattern, and particularly in the upper bandwidth of $H_{d,d-1}$. To analyse this bandwidth we start with the following definition.

Definition 4.3.1: Let s be a given non-negative integer. An unreduced upper Hessenberg matrix is called $(s+2)$-band Hessenberg, when its s-th superdiagonal contains at least one nonzero entry and all its entries above its s-th superdiagonal are zero. Here the 0-th superdiagonal means the diagonal of the matrix.

Let the matrix $H_{d,d-1}$ be $(s+2)$-band Hessenberg; see Figure 4.2. Then for each $n = 1, \ldots, d-1$, (4.3.3) reduces to

$$v_{n+1} = \mathbf{A}v_n - \sum_{m=\max\{n-s,1\}}^{n} h_{m,n} v_m, \quad h_{m,n} = \frac{(\mathbf{A}v_n, v_m)}{(v_m, v_m)}. \quad (4.3.6)$$

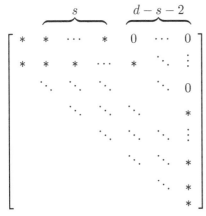

Figure 4.2 The band structure of an $(s+2)$-band Hessenberg matrix of size $d \times (d-1)$. All entries above its s-th superdiagonal are zero, and at least one entry in its s-th superdiagonal is nonzero.

The orthogonal Krylov subspace basis $v_1, \ldots, v_d$ is then generated by an $(s+2)$-term recurrence.

Since precisely the last $s+1$ basis vectors $v_n, \ldots, v_{n-s}$ are required to determine v_{n+1} (and not just any collection of $s+1$ previous basis vectors), and only one operation with $\mathbf{A}$ is performed, we call an $(s+2)$-term recurrence of the form (4.3.6) an *optimal $(s+2)$-term recurrence*. We stress that in the following we will only be concerned with this type of recurrence. A review of results on various other types of recurrences for generating orthogonal Krylov subspace bases that are not of the form (4.3.6) is given in Section 4.8.

Remark 4.3.2
Extending Definition 4.3.1 formally to the case $s = -1$ would include 'down shift' matrices $H_{d,d-1}$ in (4.3.4) that correspond to a recurrence of the form

$$v_{n+1} = \mathbf{A}v_n, \quad n = 1, \ldots, d-1.$$

Equivalently,

$$\mathbf{A}V_d = V_d C_d, \quad (4.3.7)$$

where $C_d = [c_{ij}] \in \mathbb{C}^{d \times d}$ has the form of a companion matrix. Then

$$(\mathbf{A}v_j, v_i) = c_{ij}, \quad \text{and} \quad (v_i, v_j) = \delta_{ij}, \quad i,j = 1, \ldots, d. \tag{4.3.8}$$

Matrices $A \in \mathbb{C}^{N \times N}$ with this property can be easily constructed in the following way. Consider the Euclidean inner product $(\cdot, \cdot)$ on $\mathbb{C}^N$, take any companion matrix $C_d \in \mathbb{C}^{d \times d}$ and any matrix $V_d \in \mathbb{C}^{N \times d}$ with $V_d^* V_d = I_d$. Then

$$AV_d = V_d C_d, \quad \text{where } A \equiv V_d C_d V_d^*. \tag{4.3.9}$$

Clearly, the matrix A (which is of rank d) satisfies (4.3.8). If $d = N$, then (4.3.7)–(4.3.8) and (4.3.9) are equivalent. In any case, here the special form (4.3.7) is obtained for a *particular* initial vector v. In the following we will be interested in optimal short recurrences for a given matrix A or operator $\mathbf{A}$ that hold for *any* initial vector v. Therefore the case $s = -1$ can be excluded from our considerations.

Optimal $(s+2)$-term recurrences of the form (4.3.6) are highly desirable, since they conveniently limit work and storage requirements for orthogonalising the Krylov sequence $v, \mathbf{A}v, \ldots, \mathbf{A}^{d-1}v$ with respect to the given inner product. Given an inner product, and a small integer s, it is therefore essential to understand for which linear operators $\mathbf{A}$ the recurrence (4.3.4) leads for *any* initial vector v of grade $d \geq s + 2$ to a matrix $H_{d,d-1}$ which is at most $(s+2)$-band Hessenberg.

Definition 4.3.3: *Let a non-negative integer s, $s + 2 \leq d_{\min}(\mathbf{A})$, be given.*

(1) If for an initial vector v the matrix $H_{d,d-1}$ in (4.3.4) is $(s+2)$-band Hessenberg, then $\mathbf{A}$ admits for the given inner product and the given vector v an optimal $(s+2)$-term recurrence.

(2) If $\mathbf{A}$ admits for the given inner product and any initial vector v an optimal recurrence of length at most $s + 2$, while it admits for the given inner product and at least one v an optimal $(s+2)$-term recurrence, then $\mathbf{A}$ admits for the given inner product an optimal $(s+2)$-term recurrence.

Let us comment on some subtleties of this definition:

First, the definition intentionally distinguishes a property that holds for $\mathbf{A}$, the given inner product, s and a *particular* v (item (1)) from a property that holds for $\mathbf{A}$, the given inner product, s and *all* v (item (2)). This distinction is made to avoid ambiguities and inaccuracies.

Second, consistently with Definition 4.3.1, s is assumed non-negative and the case $s = -1$ is excluded; cf. Remark 4.3.2.

Third, no recurrence of the form (4.3.3) can produce more than $d_{\min}(\mathbf{A})$ linearly independent vectors. Therefore it is meaningless to consider $s + 2 > d_{\min}(\mathbf{A})$.

We now study what happens with the length of an optimal recurrence for the operator $\mathbf{A}$ and the given vector v when $\mathbf{A}$ is shifted. Let $v \in \mathcal{V}$ be a vector of grade d with respect to $\mathbf{A}$, so that after d steps of the Arnoldi algorithm we obtain the relation

(4.3.4). Suppose that $\mathbf{A}$ admits for the given vector v and the given inner product an optimal $(s+2)$-term recurrence for some $s \geq 1$. Then for any $\mu \in \mathbb{C}$ we have

$$(\mathbf{A} - \mu \mathbf{I})V_{d-1} = \mathbf{A}V_{d-1} - \mu V_{d-1} = V_d H_{d,d-1} - \mu V_{d-1}$$

$$= V_d \begin{bmatrix} h_{1,1} - \mu & h_{1,2} & \cdots & & h_{1,d-1} \\ 1 & \ddots & \ddots & & \vdots \\ & \ddots & \ddots & & h_{d-2,d-1} \\ & & 1 & h_{d-1,d-1} - \mu \\ & & & 1 \end{bmatrix}, \quad (4.3.10)$$

where the last matrix on the right-hand side is $(s+2)$-band Hessenberg. Thus, for $s \geq 1$ the operator $\mathbf{A} - \mu \mathbf{I}$ admits for the vector v and the given inner product an optimal $(s+2)$-term recurrence.

If $\mathbf{A}$ admits for the given vector v and the given inner product an optimal 2-term recurrence (i.e. $s = 0$) and if additionally the diagonal of the matrix $H_{d,d-1}$ is constant, then there exists exactly one $\widehat{\mu} \in \mathbb{C}$ for which the upper Hessenberg matrix on the right-hand side of (4.3.10) is *not* a 2-band Hessenberg matrix. For any $\mu \neq \widehat{\mu}$ the operator $\mathbf{A} - \mu \mathbf{I}$ admits for the given vector v and the given inner product an optimal 2-term recurrence.

The main point is that, apart from some very particular and practically uninteresting cases, results on optimal short recurrences obtained for a nonsingular operator can be transferred to a singular operator by considering shifts of the operator. In the following sections we use this observation and formulate results from [174, 426], which were derived using an invertibility assumption on $\mathbf{A}$, for the case of a general (possibly singular) linear operator $\mathbf{A}$.

In practical applications $d_{\min}(\mathbf{A})$ is usually very large. Given an inner product we are interested in conditions on $\mathbf{A}$, so that it admits an optimal $(s+2)$-term recurrence with $1 \leq s \ll d_{\min}(\mathbf{A})$. Sufficient and necessary conditions will be derived and proved in the following two sections, which are based on [426, Section 2.1] and [174, Sections 3 and 4], respectively.

4.4 SUFFICIENT CONDITIONS

Our goal in this section is to derive sufficient conditions on $\mathbf{A}$ so that it admits for the given inner product an optimal $(s+2)$-term recurrence.

Note that if $s + 2 = d_{\min}(\mathbf{A})$, then $\mathbf{A}$ *can* admit an optimal recurrence of length at most $s + 2$. It *does* admit an optimal $(s+2)$-term recurrence, if there exists an initial vector $v \in \mathcal{V}$ of grade $d = s + 2$, such that the upper right element $h_{1,d-1}$ of $H_{d,d-1}$ is nonzero. As we will see, this property is nontrivial. Therefore the case $s + 2 = d_{\min}(\mathbf{A})$, though practically uninteresting, cannot be discarded right away.

Let $v \in \mathcal{V}$ be any initial vector of grade $d \geq s + 2$. If $\mathbf{A}$ admits for the given inner product and the given initial vector v an optimal $(s+2)$-term recurrence, then the entries $h_{m,n}$ of $H_{d,d-1}$ in (4.3.4) must satisfy (see Figure 4.2)

$$h_{m,n} = 0, \quad \text{for} \quad m+s < n \le d-1, \quad m = 1,\dots,d. \qquad (4.4.1)$$

From (4.3.3) it follows that

$$0 = h_{m,n} = \frac{(\mathbf{A}v_n, v_m)}{(v_m, v_m)},$$

if and only if

$$0 = (\mathbf{A}v_n, v_m) = (v_n, \mathbf{A}^\times v_m), \qquad (4.4.2)$$

where the operator $\mathbf{A}^\times : \mathcal{V} \to \mathcal{V}$ is the uniquely defined *adjoint of* $\mathbf{A}$ with respect to the given inner product; see, e.g. [216, Vol. 1, p. 265] for the definition and properties of the adjoint operator.

Now *assume* that $\mathbf{A}^\times = p_s(\mathbf{A})$, where p_s is a polynomial of degree s, and, for clarity, that no polynomial of smaller degree and with the same property exists. Then

$$\mathbf{A}^\times v_m = p_s(\mathbf{A})v_m \in \mathcal{K}_{m+s}(\mathbf{A}, v_1).$$

For $n > m+s$, the vector v_n is orthogonal to $\mathcal{K}_{m+s}(\mathbf{A}, v_1)$ by construction, so that $(v_n, \mathbf{A}^\times v_m) = 0$, giving $h_{m,n} = 0$. Thus, the matrix $H_{d,d-1}$ is *at most* $(s+2)$-band Hessenberg. In our further analysis of this situation we will use the following formal definition.

Definition 4.4.1: *If, in the previous notation,*

$$\mathbf{A}^\times = p_s(\mathbf{A}), \qquad (4.4.3)$$

where p_s is a polynomial of the smallest possible degree s having this property, then $\mathbf{A}$ is called normal of degree s, or briefly normal(s), with respect to the given inner product.

When it is clear which inner product is meant, we usually skip the term 'with respect to the given inner product'. Nevertheless, we emphasise that the normal(s) property of $\mathbf{A}$ always refers to the given inner product.

Note that in Definition 4.4.1 the integer s is *uniquely determined*. In particular, if $\mathbf{A}$ is normal(s) with respect to the given inner product, then $\mathbf{A}$ is *not* normal(t) with respect to the same inner product for any $t \ne s$. It should also be noted that if $\mathbf{A}$ is normal(s), then

$$\mathbf{A}\mathbf{A}^\times = \mathbf{A}p_s(\mathbf{A}) = p_s(\mathbf{A})\mathbf{A} = \mathbf{A}^\times \mathbf{A}.$$

Since $\mathbf{A}$ and $\mathbf{A}^\times$ commute, the operator $\mathbf{A}$ is normal on $\mathcal{V}$.

We will now show that the property of $\mathbf{A}$ being normal(s) indeed leads to matrices $H_{d,d-1}$ that are $(s+2)$-band Hessenberg.

Lemma 4.4.2

If $\mathbf{A}$ is normal(s) for some non-negative integer s, $s + 2 < d_{\min}(\mathbf{A})$, then for any initial vector v of grade $d \geq s + 2$ the corresponding matrix $H_{d,d-1}$ is $(s + 2)$-band Hessenberg.

Proof

From $\mathbf{A}^\times = p_s(\mathbf{A})$ it follows that for any initial vector v of grade $d \geq s + 2$ the matrix $H_{d,d-1}$ is at most $(s + 2)$-band Hessenberg; cf. the discussion preceding Definition 4.4.1. It remains to show that the s-th superdiagonal of $H_{d,d-1}$ is nonzero.

Since $\mathbf{A}$ is normal(s), any initial vector $v = v_1$ of grade $d \geq s + 2$ satisfies

$$\mathbf{A}^\times v_1 = p_s(\mathbf{A})v_1 \in \mathcal{K}_{s+1}(\mathbf{A}, v_1) = \operatorname{span}\{v_1, \ldots, v_{s+1}\}.$$

Therefore

$$\mathbf{A}^\times v_1 = \zeta v_{s+1} + w$$

for some scalar ζ and a vector $w \in \mathcal{K}_s(\mathbf{A}, v_1)$. The scalar ζ must be nonzero, since the polynomial p_s is of *exact* degree s. Using the orthogonality of v_{s+1} to $\mathcal{K}_s(\mathbf{A}, v_1)$ then yields

$$h_{1,s+1} = \frac{(\mathbf{A}v_{s+1}, v_1)}{(v_1, v_1)} = \frac{(v_{s+1}, \mathbf{A}^\times v_1)}{(v_1, v_1)} = \bar{\zeta}\frac{(v_{s+1}, v_{s+1})}{(v_1, v_1)} \neq 0,$$

and the s-th superdiagonal of $H_{d,d-1}$ is indeed nonzero. □

In the preceding lemma we have assumed $s + 2 < d_{\min}(\mathbf{A})$, and after Definition 4.3.3 we have noted that the case $s + 2 > d_{\min}(\mathbf{A})$ is meaningless. It remains to clarify the situation for $s + 2 = d_{\min}(\mathbf{A})$.

If $s + 2 = d_{\min}(\mathbf{A})$, then for *any* inner product and *any* initial vector v, the condition (4.4.1) is trivially satisfied, i.e. $\mathbf{A}$ admits for any inner product an optimal recurrence of length *at most* $s + 2$. If $\mathbf{A}$ is, moreover, normal(s) with respect to some *given* inner product, then $\mathbf{A}$ admits for this inner product and any initial vector v of grade $d \geq s + 2$ an optimal $(s + 2)$-term recurrence. As shown in the proof of Lemma 4.4.2, the element $h_{1,d-1}$ of $H_{d,d-1}$ is nonzero in this case. Therefore $\mathbf{A}$ admits for the given inner product an optimal $(s + 2)$-term recurrence.

This completely characterises the special and in practice uninteresting case $s + 2 = d_{\min}(\mathbf{A})$. In the following summary of the *sufficient conditions* on $\mathbf{A}$ we will skip this case and formulate the result for $s + 2 < d_{\min}(\mathbf{A})$. This is made for convenience and for consistency with Theorem 4.5.4. In that result on the *necessary* conditions on $\mathbf{A}$ the case $s + 2 = d_{\min}(\mathbf{A})$ must be excluded.

Theorem 4.4.3

Let $\mathbf{A}$ be a linear operator on a finite-dimensional vector space $\mathcal{V}$ over $\mathbb{C}$ with a given inner product, and let $d_{\min}(\mathbf{A})$ denote the degree of the minimal polynomial of $\mathbf{A}$.

If $\mathbf{A}$ is normal(s) for some non-negative integer s, $s + 2 < d_{\min}(\mathbf{A})$, then $\mathbf{A}$ admits for the given inner product and any initial vector v an optimal recurrence of length at most $s + 2$, while for any v of grade at least $s + 2$, it admits an optimal $(s + 2)$-term recurrence.

Therefore, **A** being normal(s) represents a sufficient condition for **A** to admit for the given inner product an optimal $(s+2)$-term recurrence.

4.5 NECESSARY CONDITIONS

In this section we prove that the normal(s) condition is also necessary for **A** to admit an optimal $(s+2)$-term recurrence. We start our approach to this proof with two results (Theorem 4.5.1 and Lemma 4.5.2) that further characterise the normal(s) condition.

Theorem 4.5.1
*Let **A** be a linear operator on a finite-dimensional vector space $\mathcal{V}$ over $\mathbb{C}$ with a given inner product, let $d_{\min}(\mathbf{A})$ denote the degree of the minimal polynomial of **A**, and let s be a non-negative integer. Then the following two conditions are equivalent:*

*(1) **A** is normal(s).*
*(2) There exists an orthonormal basis of $\mathcal{V}$ consisting of eigenvectors of **A**, and **A** has exactly $d_{\min}(\mathbf{A})$ distinct eigenvalues.*
*Moreover, there exists a polynomial p_s of degree s with $p_s(\lambda_j) = \overline{\lambda}_j$ for each eigenvalue λ_j of **A**. The degree s is the smallest possible of all polynomials with this property, and the polynomial p_s is uniquely determined.*

Proof
We first show that item (1) implies item (2). Since **A** is normal(s), **A** is normal on $\mathcal{V}$. It is well known that in this case there exists an orthonormal basis of $\mathcal{V}$ consisting of eigenvectors of **A**; see, e.g. [216, Vol. 1, Theorem 4, p. 272]. From this it is clear that **A** has exactly $d_{\min}(\mathbf{A})$ distinct eigenvalues.

Let $\dim \mathcal{V} = N$ and let $y_1, \ldots, y_N$ be an orthonormal basis of $\mathcal{V}$ consisting of eigenvectors of **A** with corresponding eigenvalues $\lambda_1, \ldots, \lambda_N$. Then $\mathbf{A} y_j = \lambda_j y_j$ implies $q(\mathbf{A}) y_j = q(\lambda_j) y_j$ for any polynomial q. Let $\widehat{q}$ be the Newton–Lagrange interpolation polynomial that has value $\overline{\lambda}_j$ at each eigenvalue λ_j, $j = 1, \ldots, N$. Then $\widehat{q}$ is of degree at most $d_{\min}(\mathbf{A}) - 1$, since **A** has exactly $d_{\min}(\mathbf{A})$ distinct eigenvalues. In addition, for each pair y_i, y_j, $i, j = 1, \ldots, N$, of the basis vectors we get

$$(y_i, \widehat{q}(\mathbf{A}) y_j) = (y_i, \widehat{q}(\lambda_j) y_j) = (y_i, \overline{\lambda}_j y_j) = \lambda_j (y_i, y_j) = \lambda_j \delta_{ij} = \lambda_i \delta_{ij}$$
$$= \lambda_i (y_i, y_j) = (\lambda_i y_i, y_j) = (\mathbf{A} y_i, y_j) = (y_i, \mathbf{A}^\times y_j).$$

Since this holds for each pair y_i, y_j of the basis vectors of $\mathcal{V}$, we must have

$$(v, \widehat{q}(\mathbf{A}) w) = (v, \mathbf{A}^\times w) \quad \text{for all } v, w \in \mathcal{V},$$

and thus $\widehat{q}(\mathbf{A}) = \mathbf{A}^\times$. Since by assumption p_s is the polynomial of the smallest possible degree with $p_s(\mathbf{A}) = \mathbf{A}^\times$ (see Definition 4.4.1), and $\widehat{q}$ is of degree at most $d_{\min}(\mathbf{A}) - 1$, we see that $s \leq d_{\min}(\mathbf{A}) - 1$. If q_s is any polynomial of degree at most s with $q_s(\lambda_j) = \overline{\lambda}_j$ for each eigenvalue λ_j, then the polynomial $p_s - q_s$ is a polynomial of degree at most $d_{\min}(\mathbf{A}) - 1$ that has $d_{\min}(\mathbf{A})$ distinct roots. Therefore $p_s - q_s = 0$. This must also hold for $\widehat{q}$, giving $p_s = \widehat{q}$, and p_s is indeed uniquely determined.

We now show that item (2) implies item (1). Suppose that $\dim \mathcal{V} = N$ and that $y_1, \ldots, y_N$ is an orthonormal basis of $\mathcal{V}$ consisting of eigenvectors of $\mathbf{A}$ with corresponding eigenvalues $\lambda_1, \ldots, \lambda_N$. Let p_s be the (uniquely defined) polynomial of smallest possible degree with $p_s(\lambda_j) = \overline{\lambda}_j, j = 1, \ldots, N$. Then $p_s(\mathbf{A})y_j = p_s(\lambda_j)y_j, j = 1, \ldots, N$, and for each pair $y_i, y_j, i, j = 1, \ldots, N$, of the basis vectors we get

$$(y_i, p_s(\mathbf{A})y_j) = (y_i, p_s(\lambda_j)y_j) = (y_i, \overline{\lambda}_j y_j) = \lambda_j \delta_{ij} = \lambda_i \delta_{ij}$$
$$= \lambda_i (y_i, y_j) = (\lambda_i y_i, y_j) = (\mathbf{A}y_i, y_j) = (y_i, \mathbf{A}^\times y_j).$$

Since this holds for each pair y_i, y_j of the basis vectors of $\mathcal{V}$, we must have $(v, p_s(\mathbf{A})w) = (v, \mathbf{A}^\times w)$ for all $v, w \in \mathcal{V}$, and thus $p_s(\mathbf{A}) = \mathbf{A}^\times$. Since s is the smallest possible degree of a polynomial with this property, $\mathbf{A}$ is normal(s). □

If $\mathbf{A}$ is normal(s), then $\mathbf{A}^\times v = p_s(\mathbf{A})v \in \mathcal{K}_{s+1}(\mathbf{A}, v)$ for all vectors $v \in \mathcal{V}$. Our next lemma, which is a variant of [174, Lemma 4.3] (here extended to possibly singular $\mathbf{A}$), answers the question whether the converse of this statement is true as well.

Lemma 4.5.2
Let a non-negative integer s, $s + 2 \leq d_{\min}(\mathbf{A})$, be given. If

$$\mathbf{A}^\times v \in \mathcal{K}_{s+1}(\mathbf{A}, v) \quad \text{for all vectors } v \text{ of grade } d_{\min}(\mathbf{A}), \tag{4.5.1}$$

then $\mathbf{A}$ is normal(t) for some $t \leq s$.

Proof
First suppose that $\mathbf{A}$ is invertible. We will extend the result to the general case at the end of the proof.

Let $v \in \mathcal{V}$ be any vector of grade $d_{\min}(\mathbf{A})$ with respect to $\mathbf{A}$. Since $\mathbf{A}$ is invertible, $\mathbf{A}v$ is of grade $d_{\min}(\mathbf{A})$ with respect to $\mathbf{A}$. Consider any (fixed) scalar γ that is not an eigenvalue of $\mathbf{A}$. Then the operator $\mathbf{A} - \gamma \mathbf{I}$ is invertible and the vector $w_\gamma \equiv (\mathbf{A} - \gamma \mathbf{I})v$ is of grade $d_{\min}(\mathbf{A})$ with respect to $\mathbf{A}$. When (4.5.1) holds, there exist polynomials $p_\gamma, q,$ and r of degree at most s, which satisfy

$$\mathbf{A}^\times w_\gamma = p_\gamma(\mathbf{A})w_\gamma, \quad \mathbf{A}^\times (\mathbf{A}v) = q(\mathbf{A})(\mathbf{A}v), \quad \mathbf{A}^\times v = r(\mathbf{A})v. \tag{4.5.2}$$

Note that the polynomial p_γ depends on γ, but the polynomials q and r do not. Using (4.5.2) and $w_\gamma = (\mathbf{A} - \gamma \mathbf{I})v$ we obtain

$$\mathbf{A}^\times w_\gamma = p_\gamma(\mathbf{A})w_\gamma = p_\gamma(\mathbf{A})(\mathbf{A} - \gamma \mathbf{I})v = \mathbf{A}p_\gamma(\mathbf{A})v - \gamma p_\gamma(\mathbf{A})v,$$

and

$$\mathbf{A}^\times w_\gamma = \mathbf{A}^\times (\mathbf{A} - \gamma \mathbf{I})v = \mathbf{A}^\times (\mathbf{A}v) - \gamma \mathbf{A}^\times v = \mathbf{A}q(\mathbf{A})v - \gamma r(\mathbf{A})v.$$

Combining the last two identities gives

$$t_\gamma(\mathbf{A})v = 0, \quad \text{where} \quad t_\gamma(\lambda) \equiv \lambda \left(p_\gamma(\lambda) - q(\lambda)\right) - \gamma \left(p_\gamma(\lambda) - r(\lambda)\right).$$

The polynomial t_γ is of degree at most $s+1$. Since v is of grade $d_{\min}(\mathbf{A}) \geq s+2$ and $t_\gamma(\mathbf{A})v = 0$, we see that t_γ must be the zero polynomial, giving

$$\gamma(p_\gamma(\lambda) - r(\lambda)) = \lambda(p_\gamma(\lambda) - q(\lambda)).$$

Since the polynomial on the left-hand side is of degree at most s, the polynomial $p_\gamma - q$ is of degree at most $s - 1$. A straightforward algebraic manipulation then gives

$$\gamma(q(\lambda) - r(\lambda)) = (\lambda - \gamma)(p_\gamma(\lambda) - q(\lambda)). \tag{4.5.3}$$

The same construction can be made for every γ that is not an eigenvalue of $\mathbf{A}$. Each time we get an equation of the form (4.5.3). The polynomial $q - r$ on the left-hand side does not depend on the choice of γ. However, (4.5.3) shows that each $\gamma \neq 0$ that is not an eigenvalue of $\mathbf{A}$ is a root of $r - q$. Consequently, $r - q$ must be the zero polynomial.

From $q = r$ and the relations (4.5.2) we then get

$$\mathbf{A}^\times(\mathbf{A}v) = q(\mathbf{A})(\mathbf{A}v) = \mathbf{A}q(\mathbf{A})v = \mathbf{A}r(\mathbf{A})v = \mathbf{A}(\mathbf{A}^\times v),$$

where v is an arbitrary vector of grade $d_{\min}(\mathbf{A})$ with respect to $\mathbf{A}$. As we know from Lemma 4.2.11, there exists a basis of $\mathcal{V}$, say $v_1, \ldots, v_N$, consisting of vectors of grade $d_{\min}(\mathbf{A})$. Hence

$$\mathbf{A}^\times \mathbf{A} v_j = \mathbf{A}\mathbf{A}^\times v_j, \quad j = 1, \ldots, N,$$

so that the operators $\mathbf{A}^\times \mathbf{A}$ and $\mathbf{A}^\times \mathbf{A}$ are equal, $\mathbf{A}^\times \mathbf{A} = \mathbf{A}\mathbf{A}^\times$. In this way we proved that the (invertible) operator $\mathbf{A}$ commutes with its adjoint $\mathbf{A}^\times$, and therefore $\mathbf{A}$ is normal. This means that $\mathbf{A}^\times = p(\mathbf{A})$ for some polynomial p; see, e.g. [216, Vol. 1, p. 272, equation (64)]. From (4.5.1) we see that (if p is chosen with the minimal degree) the degree of the polynomial p is at most s, so that $\mathbf{A}$ is normal(t) for some $t \leq s$.

Now let $\mathbf{A}$ be an operator that is *not* invertible and that satisfies (4.5.1). Consider any scalar μ so that the operator $\mathbf{A} - \mu\mathbf{I}$ is invertible. We know from Lemma 4.2.6 that $d_{\min}(\mathbf{A}) = d_{\min}(\mathbf{A} - \mu\mathbf{I})$. Let v be a vector of grade $d_{\min}(\mathbf{A})$ with respect to $\mathbf{A} - \mu\mathbf{I}$. Then the proof of Lemma 4.2.6 shows that v is of grade $d_{\min}(\mathbf{A})$ with respect to $\mathbf{A}$. Hence $\mathbf{A}^\times v \in \mathcal{K}_{s+1}(\mathbf{A}, v)$ by (4.5.1), and we get

$$(\mathbf{A} - \mu\mathbf{I})^\times v = (\mathbf{A}^\times - \overline{\mu}\mathbf{I})v = \mathbf{A}^\times v - \overline{\mu}v \in \mathcal{K}_{s+1}(\mathbf{A}, v) = \mathcal{K}_{s+1}(\mathbf{A} - \mu\mathbf{I}, v).$$

Since $(\mathbf{A} - \mu\mathbf{I})^\times v \in \mathcal{K}_{s+1}(\mathbf{A} - \mu\mathbf{I}, v)$ for each vector v that is of grade $d_{\min}(\mathbf{A})$ with respect to $\mathbf{A} - \mu\mathbf{I}$, and $\mathbf{A} - \mu\mathbf{I}$ is invertible, the operator $\mathbf{A} - \mu\mathbf{I}$ is normal(t) for some $t \leq s$. This means that $(\mathbf{A} - \mu\mathbf{I})^\times = q_t(\mathbf{A} - \mu\mathbf{I})$ for some (uniquely defined) polynomial of the smallest possible degree $t \leq s$. Since $(\mathbf{A} - \mu\mathbf{I})^\times = \mathbf{A}^\times - \overline{\mu}\mathbf{I}$ we get $\mathbf{A}^\times = p_t(\mathbf{A})$ for the polynomial $p_t(\lambda) = q_t(\lambda - \mu) + \overline{\mu}$, which completes the proof. □

We now come to the proof of the main result of this section, namely that the normal(s) condition on $\mathbf{A}$ is necessary so that $\mathbf{A}$ admits an optimal $(s+2)$-term recurrence; see Theorem 4.5.4 below. The proof of this theorem is significantly more complicated than the proof of the sufficiency part in Theorem 4.4.3. In order

Necessary Conditions

to motivate our strategy of the proof, let us discuss the origin of the complications and the difficulties one faces when trying to prove the necessity part.

In the sufficiency part we know that $\mathbf{A}^\times = p_s(\mathbf{A})$ for some polynomial p_s of (the smallest possible) degree s. Using this relation in the formulas (4.3.4) of the Arnoldi recurrence we immediately see that $\mathbf{A}$ admits for the given inner product an optimal $(t+2)$-term recurrence for some $t \le s$. The additional argument in Lemma 4.4.2 then shows that there is indeed an initial vector for which the matrix $H_{d,d-1}$ is $(s+2)$-band Hessenberg, so that $t = s$.

In the necessity part we assume that $\mathbf{A}$ admits for the given inner product an optimal $(s+2)$-term recurrence. This means that for any initial vector the matrix $H_{d,d-1}$ is at most $(s+2)$-band Hessenberg, and that $H_{d,d-1}$ is $(s+2)$-band Hessenberg for at least one initial vector. We have to show that $\mathbf{A}$ is normal(s). In other words, from knowledge of the Arnoldi recurrence coefficients we must prove a result about the operator $\mathbf{A}$ itself. Let $d_{\min}(\mathbf{A}) > s+2$ and let v be any initial vector of grade $d \ge s+2$. Since $\mathbf{A}\mathcal{K}_d(\mathbf{A},v) \subseteq \mathcal{K}_d(\mathbf{A},v)$ there exist scalars $h_{1,d},\ldots,h_{d,d} \in \mathbb{C}$ such that

$$\mathbf{A}v_d = \sum_{j=1}^{d} h_{j,d}\, v_j.$$

Adding this additional equation to the Arnoldi relation $\mathbf{A}V_{d-1} = V_d H_{d,d-1}$ gives

$$\mathbf{A}V_d = V_d H_{d,d}, \quad \text{where } H_{d,d} \equiv \left[\; H_{d,d-1} \;\left|\; \begin{matrix} h_{1,d} \\ \vdots \\ h_{d,d} \end{matrix} \right. \right]. \tag{4.5.4}$$

In (4.5.4) the matrix $H_{d,d}$ is the matrix representation of the operator $\mathbf{A}$ on the cyclic subspace $\mathcal{K}_d(\mathbf{A},v)$ with respect to the orthogonal basis $v_1,\ldots,v_d$.

Note that we do *not* know whether the matrix $H_{d,d}$ is $(s+2)$-band Hessenberg. From the assumption that $\mathbf{A}$ admits for the given inner product an optimal $(s+2)$-term recurrence we only know the $(s+2)$-band Hessenberg structure for the matrix $H_{d,d-1}$, i.e. the submatrix consisting of the first $d-1$ columns of $H_{d,d}$. This means that we must prove a property of the operator $\mathbf{A}$ (namely that $\mathbf{A}$ is normal(s)), but we do not have complete information about the structure of its matrix representation. It is precisely this lack of information that leads to the difficulties in the proof of necessity.

The question of whether the square matrix $H_{d,d}$ is $(s+2)$-band Hessenberg leads to the problem of *reducibility* of $\mathbf{A}$ to $(s+2)$-band Hessenberg form. The relation between the different concepts of reducibility to $(s+2)$-band Hessenberg form and admissibility of an $(s+2)$-term recurrence is discussed in [426, Section 2]; see Fig. 2.2 in that paper for an illustration. The problem of reducibility is analysed in [423] and [175]. For reducibility of $\mathbf{A}$ to $(s+2)$-band Hessenberg form the proofs of necessity are easier (though still nontrivial), because the structure of the matrix representation of $\mathbf{A}$ is known completely.

Our proof of the necessary condition for $\mathbf{A}$ to admit an optimal $(s+2)$-term recurrence is a modified version of the proof given in [174, pp. 1330–1331]. We will use the *continuity* of the Arnoldi algorithm (4.3.3) to generate the 'missing information' about the last column of a certain matrix $H_{d,d}$. The continuity is apparent when one looks at the operations performed by the Arnoldi algorithm (additions, multiplications, square roots) and considers the fact that by construction no division by zero can occur. However, a technical complication is that a small change in the initial vector may change the grade of the vector with respect to $\mathbf{A}$, which will in turn change the number of vectors generated by the Arnoldi algorithm. In the original proof of Faber and Manteuffel [176] this problem is avoided by extending each sequence of Arnoldi vectors with zero vectors so that the resulting sequence consists of $d_{\min}(\mathbf{A})$ vectors. More precisely, if v is of grade d with respect to $\mathbf{A}$, then the Arnoldi algorithm (4.3.3) generates vectors $v_1, \ldots, v_d$ and one can set $v_j \equiv 0$ for $j = d+1, \ldots, d_{\min}(\mathbf{A})$. Using this extension, Faber and Manteuffel [176, Lemma 5] show that the sequence of vectors $v_1, v_2, \ldots, v_{d_{\min}(\mathbf{A})}$ generated by the Arnoldi algorithm (which is possibly extended by zero vectors) depends continuously on the initial vector. Here we will not require this extension, but we will use the following lemma, which easily follows from the continuity result of Faber and Manteuffel.

Lemma 4.5.3
Let $v = v(\gamma)$ be a vector that continuously depends on the variable $\gamma \in \mathbb{C}$, and suppose that there exists an integer k so that $d(\mathbf{A}, v(\gamma)) \geq k$ for all $\gamma \in \mathbb{C}$. If the Arnoldi algorithm (4.3.3) is applied to $\mathbf{A}$ and the initial vector $v(\gamma)$, then the kth vector $v_k(\gamma)$ is a continuous function of $\gamma \in \mathbb{C}$.

As an example consider eigenvectors y_1, y_2, y_3 of $\mathbf{A}$ corresponding to three distinct eigenvalues $\lambda_1, \lambda_2, \lambda_3$. Then $v(\gamma) \equiv y_1 + y_2 + \gamma y_3$ depends continuously on the parameter $\gamma \in \mathbb{C}$. For each fixed $\gamma \neq 0$, the minimal polynomial of $v(\gamma)$ is $(\lambda - \lambda_1)(\lambda - \lambda_2)(\lambda - \lambda_3)$, while the minimal polynomial of $v(0)$ is $(\lambda - \lambda_1)(\lambda - \lambda_2)$. Therefore $d(\mathbf{A}, v(\gamma)) \geq 2$ for all γ, and the (nonzero) vector $v_2(\gamma)$ generated by the Arnoldi algorithm with the initial vector $v(\gamma)$ is a continuous function of γ.

After these preliminary remarks we will now proceed to the proof that the normal(s) condition on $\mathbf{A}$ is indeed necessary for the existence of an optimal $(s+2)$-term recurrence.

Theorem 4.5.4
Let $\mathbf{A}$ be a linear operator on a finite-dimensional vector space $\mathcal{V}$ over $\mathbb{C}$ with a given inner product, let $d_{\min}(\mathbf{A})$ denote the degree of the minimal polynomial of $\mathbf{A}$, and let a non-negative integer s, $s+2 < d_{\min}(\mathbf{A})$, be given.
If $\mathbf{A}$ admits for the given inner product an optimal $(s+2)$-term recurrence, then $\mathbf{A}$ is normal(s).

Proof
Suppose that the linear operator $\mathbf{A}$ admits for the given inner product an optimal $(s+2)$-term recurrence. As in the proof given in [174, pp. 1330–1331] we proceed in two steps.

Step 1. Restriction of the operator **A** *to a cyclic subspace of dimension* $s + 2$.
Let v_1 be any given vector of grade $s + 2$ with respect to **A**. Denote by $\widehat{\mathbf{A}}$ the restriction of **A** to $\mathcal{K}_{s+2}(\mathbf{A}, v_1)$, i.e. the linear operator

$$\widehat{\mathbf{A}} : \mathcal{K}_{s+2}(\mathbf{A}, v_1) \to \mathcal{K}_{s+2}(\mathbf{A}, v_1), \qquad v \mapsto \mathbf{A}v \text{ for } v \in \mathcal{K}_{s+2}(\mathbf{A}, v_1).$$

Clearly, the degree of the minimal polynomial of $\widehat{\mathbf{A}}$ with respect to $\mathcal{K}_{s+2}(\mathbf{A}, v_1)$ is $d_{\min}(\widehat{\mathbf{A}}) = s + 2$. Consider on $\mathcal{K}_{s+2}(\mathbf{A}, v_1)$ the same inner product as on the whole space $\mathcal{V}$. We will now show that $\widehat{\mathbf{A}}^\times \widehat{y}_1 \in \mathcal{K}_{s+1}(\mathbf{A}, v_1)$ holds for any vector $\widehat{y}_1$ of grade $s + 2$ with respect to $\widehat{\mathbf{A}}$. Then Lemma 4.5.2 implies that the operator $\widehat{\mathbf{A}}$ is normal(t) for some $t \leq s$. In the second step of the proof we will extend the result for the operator $\widehat{\mathbf{A}}$ to the operator **A** which is defined on the whole space $\mathcal{V}$, and, finally, that $t = s$.

Let $\widehat{y}_1 \in \mathcal{K}_{s+2}(\mathbf{A}, v_1)$ be any vector of grade $s + 2$ with respect to $\widehat{\mathbf{A}}$. Clearly, $\widehat{y}_1$ is also of grade $s + 2$ with respect to **A** and $\mathcal{K}_{s+2}(\widehat{\mathbf{A}}, \widehat{y}_1) = \mathcal{K}_{s+2}(\mathbf{A}, v_1)$. Let $u_1 \in \mathcal{V}$ be any vector so that $\widehat{y}_1 + u_1$ is of grade $s + 3$ with respect to **A**. Such a vector u_1 can always be found,[1] since $d_{\min}(\mathbf{A}) > s + 2$. Then for each (fixed) nonzero $\gamma \in \mathbb{C}$ the vector

$$w_1(\gamma) = \widehat{y}_1 + \gamma u_1$$

is of grade $s + 3$ with respect to **A**, while $w_1(0) = \widehat{y}_1$ is of grade $s + 2$ with respect to **A**. We have $d(\mathbf{A}, w_1(\gamma)) \geq s + 2$ for each $\gamma \in \mathbb{C}$.

By Lemma 4.5.3 the application of the Arnoldi algorithm to **A** and $w_1(\gamma)$ leads to a nonzero vector $w_{s+2}(\gamma)$ that depends continuously on γ. By assumption, for each fixed $\gamma \neq 0$ the resulting matrix $H_{s+3,s+2}$ is an $(s + 2)$-band upper Hessenberg matrix. In particular, $h_{1,s+2} = 0$, which is equivalent to

$$(w_1(\gamma), \mathbf{A}w_{s+2}(\gamma)) = 0. \tag{4.5.5}$$

Since the vector $w_{s+2}(\gamma)$ depends continuously on the choice of γ, the inner product in (4.5.5) must be zero also for $\gamma = 0$, i.e.

$$0 = (w_1(0), \mathbf{A}w_{s+2}(0)) = (\widehat{y}_1, \mathbf{A}\widehat{y}_{s+2}) = (\widehat{y}_1, \widehat{\mathbf{A}}\,\widehat{y}_{s+2}), \tag{4.5.6}$$

where $\widehat{y}_{s+2}$ is the last basis vector generated by the Arnoldi algorithm with the initial vector $\widehat{y}_1$, which is of grade $s + 2$. (See Figure 4.3 for an illustration of this key idea of the proof.) Using $\widehat{\mathbf{A}}^\times : \mathcal{K}_{s+2}(\widehat{\mathbf{A}}, \widehat{y}_1) \to \mathcal{K}_{s+2}(\widehat{\mathbf{A}}, \widehat{y}_1)$, the adjoint operator of $\widehat{\mathbf{A}}$, we obtain from (4.5.6) the equation

$$0 = (\widehat{y}_1, \widehat{\mathbf{A}}\,\widehat{y}_{s+2}) = (\widehat{\mathbf{A}}^\times \widehat{y}_1, \widehat{y}_{s+2}).$$

Since $\widehat{y}_1, \widehat{y}_2, \ldots, \widehat{y}_{s+2}$ form an orthogonal basis of the cyclic subspace $\mathcal{K}_{s+2}(\widehat{\mathbf{A}}, \widehat{y}_1)$, and the vector $\widehat{\mathbf{A}}^\times \widehat{y}_1$ is by construction an element of this subspace, we must have

$$\widehat{\mathbf{A}}^\times \widehat{y}_1 \in \text{span}\{\widehat{y}_1, \ldots, \widehat{y}_{s+1}\} = \mathcal{K}_{s+1}(\widehat{\mathbf{A}}, \widehat{y}_1).$$

---

1. If $\widehat{y}_1$ is of the form (4.2.23), then there exists at least one index j with ξ_j strictly less than the number of vectors in the corresponding Jordan basis (since $\widehat{y}_1$ is not of 'full grade' with respect to **A**). Assume, for simplicity of notation and without loss of generality, that $j = 1$. Then the vector u_1 can be chosen as $(\mathbf{A} - \widetilde{\lambda}_1 \mathbf{I})^{\widetilde{d}_1 - \xi_1 - 1} \widetilde{v}_1$.

$$\begin{bmatrix} h_{1,1} & \cdots & \cdots & h_{1,s+1} & & 0 \\ 1 & \ddots & & \vdots & & h_{2,s+2} \\ & \ddots & \ddots & \vdots & & \vdots \\ & & 1 & h_{s+1,s+1} & & h_{s+1,s+2} \\ & & & 1 & & h_{s+2,s+2} \\ & & & & & 1 \end{bmatrix} \xrightarrow{\gamma=0} \begin{bmatrix} h_{1,1} & \cdots & \cdots & h_{1,s+1} & 0 \\ 1 & \ddots & & \vdots & * \\ & \ddots & \ddots & \vdots & \vdots \\ & & 1 & h_{s+1,s+1} & * \\ & & & 1 & * \end{bmatrix}$$

Figure 4.3 The key idea in the proof of Theorem 4.5.4. For each $\gamma \neq 0$ the Arnoldi algorithm applied to $\mathbf{A}$ and $w_1(\gamma)$ yields an $(s+2)$-band Hessenberg matrix $H_{s+3,s+2}$. In particular $h_{1,s+2} = 0$; see the matrix on the left-hand side. The vector $w_1(0)$ is of grade $s+2$ with respect to $\mathbf{A}$, and the Arnoldi algorithm applied to $\mathbf{A}$ and $w_1(0)$ yields a matrix $H_{s+2,s+1}$ given by the first $s+1$ columns of the matrix on the right-hand side. The continuity of the Arnoldi algorithm implies that $(w_1(0), \mathbf{A}w_{s+2}(0)) = 0$. Thus, if we extend $H_{s+2,s+1}$ by one column, the resulting $(s+2) \times (s+2)$ matrix, which represents the operator $\mathbf{A}$ on the cyclic subspace $\mathcal{K}_{s+2}(\mathbf{A}, w_1(0))$ with respect to the basis $w_1(0), \ldots, w_{s+2}(0)$, is (at most) $(s+2)$-band Hessenberg.

The last equation holds for *any* vector $\widehat{y}_1 \in \mathcal{K}_{s+2}(\mathbf{A}, v_1)$ of grade $s+2$ with respect to $\widehat{\mathbf{A}}$. Lemma 4.5.2 now implies that $\widehat{\mathbf{A}}$ is normal(t) for some $t \leq s$.

By Theorem 4.5.1, $\widehat{\mathbf{A}}$ has $d_{\min}(\widehat{\mathbf{A}}) = s+2$ distinct eigenvalues $\lambda_1, \ldots, \lambda_{s+2}$, with corresponding eigenvectors that are mutually orthogonal with respect to the inner product on $\mathcal{K}_{s+2}(\mathbf{A}, v_1)$ (which coincides with the inner product on $\mathcal{V}$). Moreover, there exists a uniquely determined polynomial p_t of (smallest possible) degree t such that $p_t(\lambda_j) = \overline{\lambda}_j, j = 1, \ldots, s+2$.

In summary, if we restrict $\mathbf{A}$ to any cyclic subspace of dimension $s+2$, we get an operator $\widehat{\mathbf{A}}$ that is normal(t) for some $t \leq s$. In particular, $\widehat{\mathbf{A}}$ has $s+2$ distinct eigenvalues and corresponding eigenvectors that are mutually orthogonal with respect to the given inner product. By definition, any eigenpair of $\widehat{\mathbf{A}}$ is an eigenpair of $\mathbf{A}$. To complete the proof, we will now use this information to show that $\mathbf{A}$ itself is normal(t), and finally that $t = s$.

Step 2. Extension to the whole space.
Consider a cyclic decomposition of $\mathcal{V}$ as in (4.2.14),

$$\mathcal{V} = \bigoplus_{j=1}^{\ell} \mathcal{K}_{c_j}(\mathbf{A}, w_j),$$

where $p_{w_1} = \varphi_\mathcal{V}$ and $p_{w_{j+1}}$ is a divisor of p_{w_j} for $j = 1, \ldots, \ell - 1$. Suppose that $\mathbf{A}$ has k distinct eigenvalues, $\widetilde{\lambda}_1, \ldots, \widetilde{\lambda}_k$, and consider the construction of the vector w_1 as in (4.2.12). For simplicity, let us slightly change the notation and write the previously defined w_1, c_1, p_{w_1} and $\mathcal{K}_{c_1}(\mathbf{A}, w_1)$ as

$$w_1 \equiv v_1 + \cdots + v_k,$$
$$c_1 \equiv d_1 + \cdots + d_k,$$
$$p_{w_1} \equiv p_{v_1} \cdots p_{v_k} = (\lambda - \widetilde{\lambda}_1)^{d_1} \cdots (\lambda - \widetilde{\lambda}_k)^{d_k},$$
$$\mathcal{K}_{c_1}(\mathbf{A}, w_1) \equiv \bigoplus_{j=1}^{k} \mathcal{K}_{d_j}(\mathbf{A}, v_j).$$

In particular, d_j is the size of the largest Jordan block of $\mathbf{A}$ corresponding to the eigenvalue $\widetilde{\lambda}_j$, $j = 1, \ldots, k$. We can assume without loss of generality that $d_1 \geq \cdots \geq d_k$. Our first goal is to show that $d_1 = 1$. If this holds, then $\mathcal{K}_{c_1}(\mathbf{A}, w_1)$ is a direct sum of one-dimensional cyclic subspaces. Since $p_{w_{j+1}}$ is a divisor of p_{w_j}, $j = 1, \ldots, \ell - 1$, this means that $\mathcal{V}$ completely decomposes into one-dimensional cyclic subspaces of $\mathbf{A}$, i.e. that $\mathbf{A}$ is diagonalisable.

We first suppose that $d_1 \geq s + 2$ and derive a contradiction. Let y_1 be any vector of grade $s + 2$ in the cyclic subspace $\mathcal{K}_{d_1}(\mathbf{A}, v_1)$. From Step 1 above (using Theorem 4.5.1) we know that the restriction $\widehat{\mathbf{A}}$ of $\mathbf{A}$ to the $(s + 2)$-dimensional cyclic subspace $\mathcal{K}_{s+2}(\mathbf{A}, y_1) \subseteq \mathcal{K}_{d_1}(\mathbf{A}, v_1)$ has $s + 2$ distinct eigenvalues. But then p_{y_1}, the minimal polynomial of y_1 which is of degree $s + 2$, has $s + 2$ distinct roots. At the same time, we know from Lemma 4.2.7 that p_{y_1} must be a divisor of the minimal polynomial of v_1. The latter is given by $p_{v_1} = (\lambda - \widetilde{\lambda}_1)^{d_1}$ and thus it has only one $(d_1$-fold) root. This gives a contradiction, so that indeed $d_1 < s + 2$.

Consequently, there exist a uniquely determined largest index m, $1 < m \leq k$, and a corresponding uniquely determined integer $\widetilde{d}_m$, $0 < \widetilde{d}_m \leq d_m$, so that

$$d_1 + \cdots + d_{m-1} + \widetilde{d}_m = s + 2.$$

Let $\widetilde{z}_m$ be any vector of grade $\widetilde{d}_m$ in the cyclic subspace $\mathcal{K}_{d_m}(\mathbf{A}, v_m)$. By construction, the minimal polynomial of this vector is given by $p_{\widetilde{z}_m} = (\lambda - \widetilde{\lambda}_m)^{\widetilde{d}_m}$. (If $\widetilde{d}_m = 0$, then $\widetilde{z}_m = 0$ and $p_{\widetilde{z}_m} = 1$.) Since the minimal polynomials of $v_1, \ldots, v_{m-1}, \widetilde{z}_m$ are relatively prime, the vector

$$\widetilde{w}_1 \equiv v_1 + \cdots + v_{m-1} + \widetilde{z}_m$$

is of grade $s + 2$ (cf. Lemma 4.2.9). We can now apply the previous Step 1, showing that the restriction $\widehat{\mathbf{A}}$ of $\mathbf{A}$ to the cyclic subspace $\mathcal{K}_{s+2}(\mathbf{A}, \widetilde{w}_1)$ has $s + 2$ distinct eigenvalues. To these eigenvalues correspond $s + 2$ eigenvectors that are mutually orthogonal with respect to the given inner product. This implies that $d_1 = \cdots = d_{m-1} = 1$.

Since in particular $d_1 = 1$, we have shown that $\mathbf{A}$ is diagonalisable, i.e. that $\mathbf{A}$ has a complete system of eigenvectors. Furthermore, we know that any $s + 2$ of these eigenvectors corresponding to distinct eigenvalues of $\mathbf{A}$ are mutually orthogonal. In the subspaces corresponding to a multiple eigenvalue we can find an orthogonal basis. Therefore $\mathbf{A}$ has a complete system of eigenvectors that are orthonormal with respect to the given inner product.

From Step 1 we know that for every subset of $s + 2$ distinct eigenvalues there exists a uniquely determined polynomial p_t of degree $t \leq s$ that satisfies $p_t(\lambda_j) = \overline{\lambda}_j$ for all eigenvalues λ_j in the subset. If we take any two subsets of $s + 2$ distinct eigenvalues having $s + 1$ eigenvalues in common, then the two corresponding polynomials (both of degree at most s) must be identical. Now take the first $s + 2$ distinct eigenvalues $\lambda_1, \ldots, \lambda_{s+2}$ and the corresponding uniquely defined polynomial p_t. Clearly, p_t remains the same for the subsets

$$\{\lambda_1, \ldots, \lambda_{s+2}\}, \quad \{\lambda_2, \ldots, \lambda_{s+3}\}, \quad \ldots, \quad \{\lambda_{k-s-1}, \ldots, \lambda_k\}.$$

Consequently, all the polynomials are identical, so that there indeed exists a uniquely determined polynomial p_t of degree $t \leq s$ with $p_t(\lambda_j) = \overline{\lambda}_j$ for all eigenvalues of $\mathbf{A}$. By Theorem 4.5.1, the operator $\mathbf{A}$ is normal(t) for some $t \leq s$.

If $t < s$, then by the sufficiency result in Theorem 4.4.3, **A** admits an optimal $(t+2)$-term recurrence. This would contradict our initial assumption that **A** admits an optimal $(s+2)$-term recurrence (recall from the discussion after Definition 4.3.3 that these conditions on **A** are mutually exclusive). Therefore we must have $t = s$, so that **A** indeed is normal(s). □

As mentioned above, the continuity of the Arnoldi algorithm (4.3.3) is an important ingredient in the proof of the necessity part given by Faber and Manteuffel [176]. In the paper [174] this ingredient is replaced by an argument that uses the continuity of the determinant function. The paper [174] contains a further proof of the necessity of the normal(s) condition, which is based on a more elementary construction using (Givens) rotation matrices. That proof requires the continuity of the rotations, and thus the continuity of the standard trigonometric functions sine and cosine.

In summary, all proofs of the necessity part known to us require the continuity of some transformation. A completely algebraic proof of the necessity of the normal(s) condition (as hoped for in [426]) has not yet been found. The situation appears to be comparable with the fundamental theorem of algebra, which cannot be proven without a continuity or, more generally, analytic ingredient; see, e.g. [152, p. 109].

4.6 MATRIX FORMULATION AND EQUIVALENT CHARACTERISATIONS

In this section we translate Theorems 4.4.3 and 4.5.4 into the language of matrices, and, following [426, Section 3], we characterise the resulting necessary and sufficient conditions.

In terms of matrices, the ingredients in the two theorems are a matrix $A \in \mathbb{C}^{N \times N}$ and a given inner product on $\mathbb{C}^N$. It is well known (see, e.g. [216, Chapter IX, Section 2, equations (4)–(8)]) that any given inner product on $\mathbb{C}^N$ is generated by some (uniquely defined) Hermitian positive definite (HPD) matrix $B \in \mathbb{C}^{N \times N}$ and thus can be written as

$$(x, y)_B \equiv y^* B x, \quad \text{for all } x, y \in \mathbb{C}^N.$$

When the matrix A has real entries we consider only *real* HPD matrices B, and real initial vectors v_1. In this case statements like 'for any initial vector v_1' should be read 'for any initial vector $v_1 \in \mathbb{R}^N$', the resulting polynomials have real coefficients, and the conjugate transpose, denoted by v^* for a vector v and M^* for a matrix M, coincides with the transpose.

If A is a given (real or complex) square matrix, and B is a given HPD matrix, the adjoint $A^\times$ of A with respect to the B-inner product, i.e. the *B-adjoint of A*, is given by

$$A^\times \equiv B^{-1} A^* B.$$

The essential condition in our context is that A is normal(s) with respect to the given inner product; cf. Definition 4.4.1. The following is the translation of this definition into the language of matrices.

Definition 4.6.1: Let A be a square matrix, and let B be an HPD matrix. If

$$A^\times = p_s(A), \qquad (4.6.1)$$

where p_s is a polynomial of the smallest possible degree s having this property, then A is called normal of degree s with respect to B, or, shortly, B-normal(s).

Based on this definition we can formulate Theorems 4.4.3 (sufficiency) and 4.5.4 (necessity) as follows.

Theorem 4.6.2
Let A be a square matrix with the degree of its minimal polynomial denoted by $d_{\min}(A)$. Let B be an HPD matrix, and let s be a non-negative integer, $s + 2 < d_{\min}(A)$. The matrix A admits for the given B an optimal $(s + 2)$-term recurrence if and only if A is B-normal(s).

As mentioned in Section 4.1, the first proof of this theorem was given by Faber and Manteuffel in 1984 [176]. They later generalised the result to methods based on definite bilinear forms that are not necessarily symmetric (or Hermitian) [177]. The proofs given in [177] are essentially the same as those in [176]. Related questions were investigated by Voevodin and Tyrtyshnikov in the early 1980s [645]; see also [643, 644]. Voevodin and Tyrtyshnikov considered the question of reducibility of A to a square banded Hessenberg matrix instead of the admissibility of an optimal short recurrence; see Section 4.5, following equation (4.5.4) for a brief discussion of the difference between these two properties. Moreover, the considerations of Voevodin and Tyrtyshnikov are restricted to nonderogatory matrices. Two additional proofs of the necessity part of Theorem 4.6.2 and thus of Theorem 4.5.4 are given in [174]. Our proof of Theorem 4.5.4 is similar to the first of these two proofs. The other proof in [174] is more constructive but it is valid only for $s + 3 < d_{\min}(A)$ instead of $s + 2 < d_{\min}(A)$. In other words, the case $s + 3 = d_{\min}(A)$ is not included in that other proof. This difference is discussed in [174, Section 7].

In practice A is given, and we ask whether there exists an HPD matrix B such that A is B-normal(s) for some $s \ll d_{\min}(A)$. To investigate this question we will now characterise the property that A is B-normal(s). We start with a general characterisation (cf. [426, Theorem 3.1]), which is a matrix analogue of Theorem 4.5.1 above.

Theorem 4.6.3
Let A be a square matrix, and let B be an HPD matrix. Then the following two assertions are equivalent:

(1) A is B-normal(s).
(2) (a) A is diagonalisable with the eigendecomposition $A = Y \Lambda Y^{-1}$ (without loss of generality we consider the eigenvalues and eigenvectors of A ordered so that equal eigenvalues form a single diagonal block in Λ),
    and

(b) using the eigenvector matrix Y of A, the matrix B^{-1} has the decomposition $B^{-1} = YDY^*$, where D is an HPD block diagonal matrix with block sizes corresponding to those of Λ,

and

(c) there exists a polynomial p_s of degree s such that $p_s(\Lambda) = \Lambda^*$, and s is the smallest degree of all polynomials with this property. The polynomial p_s is uniquely determined.

Proof
The statement can be proven by showing that items (2a)–(2c) in this theorem are equivalent with item (2) in Theorem 4.5.1.

Let W be the B-orthonormal eigenvector matrix of A given by (2) in Theorem 4.5.1. Then $W^*BW = I$, so that $B^{-1} = WW^*$. Now reorder the columns of W as requested in (2a) above and denote the result by $\widetilde{Y}$. Then any eigenvector matrix (with the same ordering of eigenvalues) of A can be written as $Y = \widetilde{Y}\widetilde{D}$, where $\widetilde{D}$ is a block diagonal matrix with the block sizes corresponding to those of Λ. Hence

$$B^{-1} = WW^* = \widetilde{Y}\widetilde{Y}^* = Y\widetilde{D}^{-1}\widetilde{D}^{-*}Y^* = YDY^*, \quad \text{where } D \equiv (\widetilde{D}^*\widetilde{D})^{-1},$$

which proves (2a) and (2b). Reversing the argument proves the other direction. The part (2c) translates from Theorem 4.5.1 without any change. □

Theorem 4.6.3 gives conditions on A and B such that A is B-normal(s). Now consider a *diagonalisable* matrix $A = Y\Lambda Y^{-1}$, where we use the block ordering of the eigenvalues of A on the diagonal of Λ as in condition (2a). Then we define the class of matrices satisfying condition (2b),

$$\mathcal{B} \equiv \{ (YDY^*)^{-1} : D \text{ is an HPD block diagonal matrix with the sizes of its blocks corresponding to the blocks of } \Lambda \}. \quad (4.6.2)$$

As shown in the proof of Theorem 4.5.1, a polynomial of minimal degree s satisfying condition (2c) is nothing but the (unique) interpolating polynomial q satisfying $q(\lambda_j) = \overline{\lambda}_j$ for all eigenvalues λ_j of A. Clearly, $p_s \equiv q$ is of degree $s \leq d_{\min}(A) - 1$, and s is uniquely determined by this construction. Note that for every $B \in \mathcal{B}$,

$$A^\times = B^{-1}A^*B = Y\Lambda^*Y^{-1}.$$

We summarise these consequences of Theorem 4.6.3 in the following corollary.

Corollary 4.6.4
Let A be a square matrix, and let B be an HPD matrix, such that A is B-normal(s). Then, using the notation of Theorem 4.6.3, $A = Y\Lambda Y^{-1}$, $s \leq d_{\min}(A) - 1$ is uniquely determined by the location of the eigenvalues of A, $B \in \mathcal{B}$ as defined in (4.6.2), and $A^\times = Y\Lambda^*Y^{-1}$.

We have seen that s is determined by the location of the eigenvalues of A. In practice we are interested only in s for which $s \ll d_{\min}(A)$. The following result, first obtained in a different, but mathematically equivalent formulation by Faber and Manteuffel [176, Lemma 3], characterises the two smallest possible values of s, namely $s = 0$ and $s = 1$.

Theorem 4.6.5
Let A be a square matrix.

(1) *There exists an HPD matrix B for which A is B-normal(0), if and only if $A = \alpha I$ for some nonzero $\alpha \in \mathbb{C}$.*
(2) *There exists an HPD matrix B for which A is B-normal(1), if and only if A is diagonalisable with $d_{\min}(A) \geq 2$ and A has collinear eigenvalues (i.e. all eigenvalues lie on a single straight line in the complex plane).*

Proof
If the matrix A is B-normal(0) for some given HPD matrix B, then from Theorem 4.6.3, A must be diagonalisable, and there exists a polynomial p_0 of degree zero, say $p_0(\lambda) \equiv \overline{\alpha}$ for some nonzero $\alpha \in \mathbb{C}$, that satisfies $p_0(\lambda_j) = \overline{\alpha} = \overline{\lambda}_j$ for all eigenvalues λ_j of A. Clearly, $A = \alpha I$. The other implication is a straightforward consequence of Theorem 4.6.3. Note that $A = \alpha I$ is B-normal(0) for *any* HPD matrix B.

Now suppose that A is B-normal(1) for some given HPD matrix B. From Theorem 4.6.3, A must be diagonalisable, and there exists a polynomial $p_1(\lambda) \equiv \alpha + \beta\lambda$, $\alpha, \beta \in \mathbb{C}$ with $\beta \neq 0$, such that $p_1(\lambda_j) = \overline{\lambda}_j$ for all eigenvalues $\lambda_1, \ldots, \lambda_m$ of A. Since one is the minimal degree of a polynomial with this property, A must have at least two distinct eigenvalues and $d_{\min}(A) \geq 2$. If A has exactly two distinct eigenvalues, then they are trivially collinear. Otherwise we determine the coefficient β using any two of the distinct eigenvalues of A, say λ_1 and λ_2,

$$\beta = \frac{\overline{\lambda}_2 - \overline{\lambda}_1}{\lambda_2 - \lambda_1}.$$

Clearly, $|\beta| = 1$ and we write for convenience $\beta = e^{\iota(2\varphi)}$, $\varphi \in [0, \pi)$. The coefficient β and therefore the angle φ are uniquely determined independently of the choice of the (distinct) eigenvalues above. We will now rotate the complex plane by the angle φ and show that after this rotation all rotated eigenvalues $e^{\iota\varphi}\lambda_j$, $j = 1, \ldots, m$, are located on a single line parallel to the real axis, which proves that λ_j, $j = 1, \ldots, m$, are located on the inversely rotated line. Indeed, using

$$\alpha + e^{2\iota\varphi}\lambda_j = \overline{\lambda}_j,$$

we easily get

$$e^{\iota\varphi}\lambda_j - e^{-\iota\varphi}\overline{\lambda}_j = -e^{-\iota\varphi}\alpha,$$

i.e. the imaginary part of $e^{\iota\varphi}\lambda_j$ is a constant independent of the index j. Consequently, all eigenvalues of A lie on a single straight line in the complex plane.

Conversely, suppose that the distinct eigenvalues $\lambda_1, \ldots, \lambda_m$, where $m \geq 2$, of the diagonalisable and nonsingular matrix A are collinear. Then there exist $\omega \in \mathbb{C}$ and $\varphi \in [0, \pi)$ such that $\lambda_j = \omega + \varrho_j e^{\iota\varphi}$ for some $\varrho_j \in \mathbb{R}, j = 1, \ldots, m$. An easy computation shows that the degree one polynomial

$$p_1(\lambda) \equiv (\overline{\omega} - e^{-2\iota\varphi}\omega) + e^{-2\iota\varphi}\lambda$$

satisfies $p_1(\lambda_j) = \overline{\lambda}_j$ for $j = 1, \ldots, m$. Since $m \geq 2$, the same property cannot hold for any polynomial of degree zero. Consequently, A is B-normal(1) for any HPD matrix B in the set $\mathcal{B}$. $\square$

We now return to the question of the existence of the optimal $(s + 2)$-term recurrences, and we consider the case $s = 1$ and $s + 2 = 3 < d_{\min}(A)$. Theorem 4.6.2 shows that A admits for a given HPD matrix B an optimal three-term recurrence if and only if A is B-normal(1). Theorem 4.6.5 then shows that there exists an HPD matrix B for which A is B-normal(1), and, consequently, A admits an optimal three-term recurrence, if and only if A is diagonalisable with collinear eigenvalues. Well-known classes of diagonalisable matrices with collinear eigenvalues are the Hermitian and skew-Hermitian matrices. Note that these matrices are unitarily diagonalisable, which results, with the choice $D = I$ in (4.6.2), in $B = I$. An optimal three-term recurrence algorithm for orthogonalising any Krylov sequence for a Hermitian matrix A is well known; see the Hermitian Lanczos algorithm (Algorithm 2.4.3) studied in Section 2.4.1.

Interesting examples of matrices A and $B \neq I$ for which A is B-normal(1) are given in the context of saddle point problems in [56, 186], with a generalisation presented in [421]. Here

$$A = \begin{bmatrix} A_1 & A_2^T \\ -A_2 & A_3 \end{bmatrix} \in \mathbb{R}^{(m+k) \times (m+k)}, \qquad (4.6.3)$$

where $A_1 \in \mathbb{R}^{m \times m}$ is symmetric positive definite, $A_2 \in \mathbb{R}^{k \times m}$ has full rank $k \leq m$, and $A_3 \in \mathbb{R}^{k \times k}$ is symmetric positive semidefinite (possibly zero). The matrix A can be written as

$$A = \begin{bmatrix} A_1 & 0 \\ 0 & A_3 \end{bmatrix} + \begin{bmatrix} 0 & A_2^T \\ -A_2 & 0 \end{bmatrix},$$

where the first and second matrix on the right-hand side are the symmetric and skew-symmetric parts of A, respectively. Since A_1 is positive definite and A_3 is positive semidefinite, the symmetric part of A has (real) non-negative eigenvalues. It follows that all eigenvalues of A have non-negative real parts.

An elementary computation shows that $A^T \neq p_1(A)$ for any polynomial p_1 of degree one. Thus, A is *not* I-normal(1). On the other hand, as shown in [421], if

$$2\|A_2\| < \lambda_{\min}(A_1) - \lambda_{\max}(A_3),$$

then the matrix

$$B(\gamma) \equiv \begin{bmatrix} A_1 - \gamma I_m & A_2^T \\ A_2 & \gamma I_k - A_3 \end{bmatrix}, \quad \text{where } \gamma \equiv \frac{\lambda_{\min}(A_1) + \lambda_{\max}(A_3)}{2}$$

is symmetric positive definite. This matrix satisfies $B(\gamma)A = A^T B(\gamma)$, so that the $B(\gamma)$-adjoint of A satisfies

$$A^\times = B(\gamma)^{-1} A^T B(\gamma) = A = p_1(A) \quad \text{for } p_1(\lambda) \equiv \lambda.$$

In other words, A is $B(\gamma)$-normal(1). Hence in this case A has collinear eigenvalues and there exists a conjugate gradient like descent method based on the $B(\gamma)$-inner product for solving linear systems with A. The paper [421] contains the complete derivation of this method as well as more general necessary and sufficient conditions for positive definiteness of $B(\gamma)$.

4.7 SHORT RECURRENCES AND THE NUMBER OF DISTINCT EIGENVALUES

For general B-normal(s) matrices A we will now study the relation between the integer s and the degree of the minimal polynomial $d_{\min}(A)$, i.e. the number of distinct eigenvalues. Throughout this subsection we consider a *diagonalisable* matrix A with minimal polynomial degree $d_{\min}(A) > 2$.

As shown above, such a matrix A is normal of degree s with respect to *any* HPD matrix $B \in \mathcal{B}$ (see (4.6.2)) where s is the smallest degree of a polynomial p_s for which $p_s(\Lambda) = \Lambda^*$; see condition (2c) in Theorem 4.6.3. Equivalently, s is determined as the smallest degree of a polynomial p_s such that the eigenvalues of A are roots of the *harmonic polynomial* $p_s(\lambda) - \bar{\lambda}$. In general, a harmonic polynomial is a function of the form $p(\lambda) + \overline{q(\lambda)}$, where p and q are polynomials, see, e.g. [560, Section 1.1.7].

We know from item (2) in Theorem 4.6.5, that $s = 1$ if and only if the eigenvalues of A are collinear. A closer look at the proof of this theorem reveals that for every given straight line in the complex plane there exists a harmonic polynomial $p_1(\lambda) - \bar{\lambda}$ that is zero on the entire line. Hence a harmonic polynomial can have infinitely many roots. On the other hand, the example $1 - \lambda + \bar{\lambda}$ shows that there exist non-constant harmonic polynomials that have no roots at all.

Let us now look at the case that the eigenvalues of A are *not* collinear. Then s must be larger than one. As shown by Khavinson and Świątek using techniques of complex dynamics, a harmonic polynomial $p_s(\lambda) - \bar{\lambda}$ with $s > 1$ may have at *most $3s - 2$* (distinct) roots [383, Theorem 1]; see [353] for an elementary proof of this result for $s = 2$. According to Geyer [231], this bound on the maximal number of roots is sharp for all $s > 1$. An example for the case $s = 3$ (taken from [423, Example 3.7]) is the following: The harmonic polynomial

$$p_3(\lambda) - \bar{\lambda} = -\frac{\lambda(\lambda^2 - 9)}{8} - \bar{\lambda}$$

has seven roots given by

$$0, \pm 1, \frac{\pm 5 \pm \iota\sqrt{7}}{2}.$$

The result of [383, Theorem 1] has the following fundamental consequence. If $d_{\min}(A) > 3s - 2$, where s is any integer larger than one, then there exists no polynomial p_t of degree $t \leq s$ that satisfies $p_t(\Lambda) = \Lambda^*$, since the eigenvalues of such an A cannot be roots of *any* harmonic polynomial $p_t(\lambda) - \bar{\lambda}$ with $t \leq s$.

By Theorem 4.6.3, a matrix A whose eigenvalues are not collinear and that satisfies $d_{\min}(A) > 3s - 2$ for some $s > 1$ cannot be normal of *any* degree $t \leq s$ with respect to *any* given B. An alternative way to state this consequence of [383, Theorem 1] is that if a matrix A is normal of degree $s > 1$ with respect to some given B, then $d_{\min}(A) \leq 3s - 2$. Using this fact together with Theorem 4.6.2 yields the following important result on the existence of optimal short recurrences.

Theorem 4.7.1
Let s be a given integer greater than one. If the eigenvalues of the diagonalisable matrix A are not collinear, and if the degree of the minimal polynomial of A satisfies $d_{\min}(A) > 3s - 2$, then there exists no HPD matrix B for which A admits an optimal recurrence of length $s + 2$ or less.

In practice $s \ll d_{\min}(A)$, and therefore we can formulate the following important consequence:

> Except for the diagonalisable matrices having collinear eigenvalues, which admit an optimal three-term recurrence, we know of no practically interesting matrix A (with reasonably large $d_{\min}(A)$) and no HPD matrix B, such that A admits for B an optimal $(s + 2)$-term recurrence with a small s.

Finally, we remark in order to avoid confusion, that even when the matrix A fails to be diagonalisable, it may still admit for *some* HPD matrix B and *some* initial vector v_1 an optimal $(s + 2)$-term recurrence with small s. For example, consider the transposed $N \times N$ Jordan block

$$A = \begin{bmatrix} \lambda & & & \\ 1 & \ddots & & \\ & \ddots & \ddots & \\ & & 1 & \lambda \end{bmatrix}, \quad \lambda \in \mathbb{C} \setminus \{0\}.$$

For $B = I$, and $v_1 = [1, 0, \ldots, 0]^T \in \mathbb{C}^N$, the matrix representation of the recurrence (4.3.3) is

… Other Types of Recurrences

$$v_1 = [1, 0, \ldots, 0]^T,$$

$$A \underbrace{\begin{bmatrix} 1 & & \\ & \ddots & \\ & & 1 \\ 0 & \cdots & 0 \end{bmatrix}}_{\equiv V_{N-1}} = \underbrace{\begin{bmatrix} 1 & & & \\ & \ddots & & \\ & & 1 & \\ & & & 1 \end{bmatrix}}_{\equiv V_N} \underbrace{\begin{bmatrix} \lambda & & & \\ 1 & \ddots & & \\ & \ddots & \lambda & \\ & & 1 & \end{bmatrix}}_{\equiv H_{N,N-1}},$$

$$V_N^T V_N = I, \quad N = \dim \mathcal{K}_N(A, v_1).$$

The matrix $H_{N,N-1}$ is 2-band Hessenberg, and hence A admits for $B = I$ and the particular vector $v_1 = [1, 0, \ldots, 0]^T$ an optimal 2-term recurrence. We stress, however, that the existence of this particular recurrence is only due to the special relationship between A, B and v_1. Since the $N \times N$ Jordan block A is not diagonalisable, it does not admit an optimal $(s+2)$-term recurrence with $s < N - 2$ for any HPD matrix B.

4.8 OTHER TYPES OF RECURRENCES

In the previous sections of this chapter, we have only considered recurrences of the form (4.3.3). An observing reader may ask whether there exist alternative and potentially more efficient ways to compute orthogonal Krylov subspace bases. In case the matrix A is B-normal(1) for some explicitly known HPD matrix B, there certainly is no practical need to search for a more efficient recurrence. In all other cases, the construction of alternative types of short recurrences is an interesting and practically relevant problem, that is addressed in this section.

We will first consider unitary matrices. Clearly, when U is unitary with $d_{\min}(U) \geq 3$, then the eigenvalues of U are *not* located on a single line in the complex plane, and hence there exists no HPD matrix B for which U is B-normal(1); see Theorem 4.6.5. Nevertheless, as we will demonstrate below, an orthogonal Krylov subspace basis for U and any initial vector v_1 can be computed via an efficient short recurrence. It should be stressed that this recurrence *is not* of the form (4.3.6) and therefore its existence is not in contradiction with the results derived in the previous sections.

4.8.1 The Isometric Arnoldi Algorithm

Let $U \in \mathbb{C}^{N \times N}$ be unitary, let $v \in \mathbb{C}^N$ be a vector of grade d with respect to U and of unit Euclidean norm. Let $v_1, \ldots, v_d$ be the basis of $\mathcal{K}_d(U, v)$ determined by the Arnoldi algorithm (4.3.3) with $B = I$ (i.e. the Euclidean inner product). Since $Uv_d \in \mathcal{K}_d(U, v)$, there exist scalars $h_{1,d}, \ldots, h_{d,d}$, such that

$$Uv_d = \sum_{m=1}^{d} h_{m,d} v_m.$$

Using this additional equation in (4.3.4) yields

$$v_1 = v,$$
$$UV_d = V_d H_{d,d}, \quad (4.8.1)$$
$$V_d^* V_d \text{ is } d \times d \text{ and diagonal,}$$

where $V_d \equiv [v_1, \ldots, v_d]$, and $H_{d,d}$ is a $d \times d$ unreduced upper Hessenberg matrix. In the following, we will slightly abuse the notation introduced in (4.3.3)–(4.3.4), and assume without loss of generality that the columns of V_d are scaled to have unit Euclidean norm, i.e. that $V_d^* V_d = I_d$.

Under this assumption,

$$H_{d,d}^* H_{d,d} = H_{d,d}^* V_d^* V_d H_{d,d} = V_d^* U^* U V_d = I_d,$$

so that $H_{d,d}$ is unitary. It is well known that there exist Givens rotations $G_1, \ldots, G_d$, such that $G_d^* G_{d-1}^* \cdots G_1^* H_{d,d} \equiv R_{d,d} \in \mathbb{C}^{d \times d}$ is upper triangular; see, e.g. [250, Section 5.2.4] on the Hessenberg QR factorisation via Givens rotations. Since the matrices $G_1, \ldots, G_d$ and $H_{d,d}$ are unitary, the upper triangular matrix $R_{d,d}$ is also unitary and hence diagonal (each unitary upper triangular matrix is a diagonal matrix).

Note that, due to the scaling applied in (4.8.1), $h_{n+1,n} = \|v_{n+1}\| / \|v_n\|$ for $n = 1, \ldots, d-1$, so that the subdiagonal elements of $H_{d,d}$ are positive. Hence there exist uniquely determined Givens rotations such that $R_{d,d} = I$, or, equivalently,

$$H_{d,d} = G_1 \cdots G_{d-1} G_d; \quad (4.8.2)$$

see [257, Section 1]. As stated in that paper, these rotation matrices G_n are of the form

$$G_n = \begin{bmatrix} I_{n-1} & & & \\ & -\gamma_n & \sigma_n & \\ & \sigma_n & \overline{\gamma}_n & \\ & & & I_{d-n-1} \end{bmatrix}, \quad n = 1, \ldots, d-1, \quad (4.8.3)$$

$$G_d = \begin{bmatrix} I_{d-1} & \\ & -\gamma_d \end{bmatrix}. \quad (4.8.4)$$

For $n = 1, \ldots, d-1$, the entries σ_n and γ_n of G_n satisfy $0 < \sigma_n \leq 1$ and $\gamma_n \in \mathbb{C}$, with $|\gamma_n| < 1$ and $\sigma_n^2 + |\gamma_n|^2 = 1$. Furthermore, $\gamma_d \in \mathbb{C}$ with $|\gamma_d| = 1$, and if we set $\gamma_0 = 1$, then

$$h_{m,n} = \begin{cases} -\overline{\gamma}_{m-1} \gamma_n \prod_{i=m}^{n-1} \sigma_i, & m < n+1, \\ \sigma_n, & m = n+1, \\ 0, & m > n+1, \end{cases} \quad (4.8.5)$$

where we set $\prod_{i=n}^{n-1} \sigma_i = 1$.

Other Types of Recurrences

We will next show that the orthonormal Krylov subspace basis vectors $v_1, \ldots, v_d$ satisfy a short recurrence relation. To this end, consider the nth columns on the left and right-hand side of (4.8.1) for any $n = 1, \ldots, d - 1$. Then the relation (4.8.5) yields

$$\sigma_n v_{n+1} = U v_n + \sum_{m=1}^{n} \left(\overline{\gamma}_{m-1} \gamma_n \prod_{i=m}^{n-1} \sigma_i \right) v_m$$

$$= U v_n + \gamma_n \widetilde{v}_n, \qquad (4.8.6)$$

where

$$\widetilde{v}_n \equiv \sum_{m=1}^{n} \left(\overline{\gamma}_{m-1} \prod_{i=m}^{n-1} \sigma_i \right) v_m, \quad n = 1, \ldots, d - 1. \qquad (4.8.7)$$

Note that (4.8.7) implies $\widetilde{v}_1 = v_1$. In particular, $\|\widetilde{v}_1\| = \|v_1\| = 1$, which we use in the following inductive argument to show that $\|\widetilde{v}_{n+1}\| = 1$ for $n = 1, \ldots, d - 1$. Since the vectors $v_1, \ldots, v_d$ form an orthonormal set,

$$\widetilde{v}_{n+1}^* \widetilde{v}_{n+1} = \sum_{m=1}^{n+1} \left(|\overline{\gamma}_{m-1}|^2 \prod_{i=m}^{n} \sigma_i^2 \right)$$

$$= |\overline{\gamma}_n|^2 + \sigma_n^2 \sum_{m=1}^{n} \left(|\overline{\gamma}_{m-1}|^2 \prod_{i=m}^{n-1} \sigma_i^2 \right)$$

$$= |\gamma_n|^2 + \sigma_n^2 \widetilde{v}_n^* \widetilde{v}_n$$

$$= 1.$$

To obtain the last equality we have used the induction hypothesis that $\|\widetilde{v}_n\| = 1$, as well as the relation $|\gamma_n|^2 + \sigma_n^2 = 1$, which holds for $n = 1, \ldots, d - 1$.

Furthermore, for $n = 1, \ldots, d - 1$, the vector $\widetilde{v}_{n+1}$ satisfies

$$\widetilde{v}_{n+1} = \sum_{m=1}^{n+1} \left(\overline{\gamma}_{m-1} \prod_{i=m}^{n} \sigma_i \right) v_m$$

$$= \overline{\gamma}_n v_{n+1} + \sum_{m=1}^{n} \left(\overline{\gamma}_{m-1} \prod_{i=m}^{n} \sigma_i \right) v_m$$

$$= \overline{\gamma}_n v_{n+1} + \sigma_n \sum_{m=1}^{n} \left(\overline{\gamma}_{m-1} \prod_{i=m}^{n-1} \sigma_i \right) v_m$$

$$= \overline{\gamma}_n v_{n+1} + \sigma_n \widetilde{v}_n. \qquad (4.8.8)$$

Next note that $\widetilde{v}_n \in \text{span}\{v_1, \ldots, v_n\}$, so that $\widetilde{v}_n$ is orthogonal to v_{n+1}. Using this in (4.8.6) shows that

$$0 = \widetilde{v}_n^* (\sigma_n v_{n+1}) = \widetilde{v}_n^* U v_n + \gamma_n \|\widetilde{v}_n\|^2,$$

from which we receive

$$\gamma_n = -\widetilde{v}_n^* U v_n, \quad n = 1, \ldots, d-1. \tag{4.8.9}$$

Combining (4.8.6)–(4.8.9) yields the isometric Arnoldi algorithm for computing the orthonormal basis $v_1, \ldots, v_d$ of $\mathcal{K}_d(U, v)$; see Algorithm 4.8.1.

The isometric Arnoldi algorithm was originally derived by Gragg [258]. Algebraically, it relies on an efficient LU factorisation of unitary upper Hessenberg matrices; see [540] for an early description of this idea. In spite of its algebraic derivation above, it has deep connections with Gauss quadrature and orthogonal polynomials on the unit circle [258]. In particular, it can be shown that the above statement of the isometric Arnoldi algorithm corresponds to Szegö's classical recursion [607, Theorem 11.4.2, p. 286] for generating orthogonal polynomials with respect to some measure on the unit circle. Watkins' survey paper [660, Section 14] also gives this recursion and further details; also see [661].

Similar relationships between a matrix algorithm and classical analysis are well known for the Hermitian Lanczos algorithm. As shown in detail in Chapter 3 in this book, this algorithm is closely related to Gauss quadrature and orthogonal polynomials on the real line. As Watkins points out, 'virtually everything that holds in the self-adjoint case has its counterpart in the unitary case' [660, p. 453].

Algorithm 4.8.1 Isometric Arnoldi algorithm

Input: unitary matrix $U \in \mathbb{C}^{N \times N}$, initial vector $v \in \mathbb{C}^N$. Let d be the grade of v with respect to U.
Output: orthonormal basis vectors $v_1, \ldots, v_d$ with span$\{v_1, \ldots, v_n\} = \mathcal{K}_n(U, v)$ for $n = 1, \ldots, d$.

Initialise: $v_1 = v/\|v\|$ and $\widetilde{v}_1 = v_1$.

For $n = 1, 2, \ldots$

$$\begin{aligned}
\gamma_n &= -\widetilde{v}_n^* U v_n, \\
\sigma_n &= (1 - |\gamma_n|^2)^{1/2}, \quad \text{(by construction } \sigma_n > 0\text{)} \\
v_{n+1} &= \sigma_n^{-1}(U v_n + \gamma_n \widetilde{v}_n), \\
\widetilde{v}_{n+1} &= \sigma_n \widetilde{v}_n + \overline{\gamma}_n v_{n+1}.
\end{aligned}$$

Obviously, linear systems of the form $Ux = b$ with a unitary matrix U do not have to be solved using the isometric Arnoldi algorithm. However, systems of the form $(U + \zeta I)x = b$, with unitary U and $\zeta \in \mathbb{C}$, represent a challenge in some important applications. For such systems, an efficient minimal residual Krylov subspace method based on the isometric Arnoldi algorithm has been derived by Jagels and Reichel [362, 363]. This method has been applied, e.g. in solving systems with the (shifted unitary) overlap operator in quantum chromodynamics (QCD); see, e.g. [15]. The main current use of the isometric Arnoldi algorithm seems to be in the context of unitary eigenproblems. Corresponding work can be found in [67, 89, 308].

Other Types of Recurrences

The isometric Arnoldi algorithm shows that orthonormal Krylov subspace bases for a unitary matrix U and any initial vector can be computed by a short recurrence, although U is in general not B-normal(s) with small s for any HPD matrix B. As mentioned above, this does not contradict the results obtained previously in this chapter, because the isometric Arnoldi recurrence is of a *different type* from the 'optimal' Arnoldi recurrence (4.3.3) studied previously. The difference will become more apparent in the following section, where we address the question of whether it is possible to extend the isometric Arnoldi type recurrence to more general classes of matrices.

4.8.2 $(s+2, t)$-term Recurrences

Consider the two coupled 2-term recurrences for v_{n+1} and $\tilde{v}_{n+1}$ in the isometric Arnoldi algorithm. If $\gamma_{n-1} \neq 0$, a simple rearrangement yields

$$\begin{aligned}
\sigma_n v_{n+1} &= U v_n + \gamma_n \tilde{v}_n \\
&= U v_n + \gamma_n (\sigma_{n-1}\tilde{v}_{n-1} + \overline{\gamma}_{n-1} v_n) \\
&= U v_n + \gamma_n \overline{\gamma}_{n-1} v_n + \frac{\gamma_n \sigma_{n-1}}{\gamma_{n-1}} (\sigma_{n-1} v_n - U v_{n-1}) \\
&= U v_n - \frac{\gamma_n \sigma_{n-1}}{\gamma_{n-1}} U v_{n-1} + \frac{\gamma_n}{\gamma_{n-1}} v_n,
\end{aligned} \qquad (4.8.10)$$

where in the last step we have used that $\sigma_{n-1}^2 + |\gamma_{n-1}|^2 = 1$.

In (4.8.10) we clearly see that isometric Arnoldi is different from the optimal Arnoldi recurrence (4.3.3): For isometric Arnoldi we either have to perform an additional multiplication with U, or we have to store the vector $U v_{n-1}$ instead of v_{n-1}. Note that the single recurrence (4.8.10) breaks down in the case $\gamma_{n-1} = 0$. An example for such a breakdown is given in [41, Section 4.3]. Such a breakdown is avoided when coupled 2-term recurrences, as in Algorithm 4.8.1, are used instead.

We will now discuss whether the isometric Arnoldi type recurrence can be generalised to other classes of matrices. We will present the main ideas only and refer for details to the work of Barth and Manteuffel [38, 40, 41, 42], on which most of our discussion is based.

Let A be a given square matrix, let B be a given HPD matrix, and let $t \geq 1$ be a (supposedly small) given integer. Suppose that, for some $n > t$, we have generated B-orthogonal vectors $v_1, \ldots, v_n$, so that the first j of them span $\mathcal{K}_j(A, v_1)$, for $j = 1, \ldots, n$. Then, for *any* given scalars $g_{n-t,n}, \ldots, g_{n-1,n}$, the next (non-normalised) Krylov basis vector v_{n+1}, giving span$\{v_1, \ldots, v_{n+1}\} = \mathcal{K}_{n+1}(A, v_1)$, can be generated as

$$v_{n+1} = A v_n - \sum_{m=1}^{n} h_{m,n} v_m - \sum_{k=n-t}^{n-1} g_{k,n} A v_k, \qquad (4.8.11)$$

$$h_{m,n} = \frac{\left(A v_n - \sum_{k=n-t}^{n-1} g_{k,n} A v_k, v_m\right)_B}{(v_m, v_m)_B}, \quad m = 1, \ldots, n. \qquad (4.8.12)$$

Note that (4.8.11)–(4.8.12) is just the Arnoldi recurrence (4.3.3), where we have added the sum $\sum_{k=n-t}^{n-1} g_{k,n} A v_k$. Of course, at this point we have saved nothing. The main idea now is to find conditions on A, so that a significant number of terms in the sum $\sum_{m=1}^{n} h_{m,n} v_m$ are equal to zero. More specifically, given an integer $s \geq 0$ (supposedly small), we search for conditions on A such that there exist scalars $g_{n-t,n}, \ldots, g_{n-1,n}$, resulting in $h_{m,n} = 0$ for $m = 1, \ldots, n-s-1$ or, equivalently,

$$\left(v_n - \sum_{k=n-t}^{n-1} g_{k,n} v_k, A^\times v_m \right)_B = 0, \quad \text{for } m = 1, \ldots, n-s-1. \quad (4.8.13)$$

If this holds, (4.8.11) reduces to

$$v_{n+1} = A v_n - \sum_{m=\max\{n-s,1\}}^{n} h_{m,n} v_m - \sum_{k=n-t}^{n-1} g_{k,n} A v_k. \quad (4.8.14)$$

We call a recurrence of the form (4.8.14) an $(s+2,t)$-*term recurrence*. For $t = 0$ the sum $\sum_{k=n-t}^{n-1} g_{k,n} A v_k$ drops out and (4.8.14) is equal to the optimal $(s+2)$-term recurrence (4.3.6). Hence the only cases of additional interest here are those with $t \geq 1$.

Note that the isometric Arnoldi recurrence (4.8.10) corresponds to (4.8.14) with $s = 0$ and $t = 1$. There we have $B = I$, and

$$U^\times = U^* = p_0(U)(q_1(U))^{-1} \quad \text{with} \quad p_0(\lambda) = 1 \quad \text{and} \quad q_1(\lambda) = \lambda.$$

In other words, the adjoint of U is a rational function in U, with the numerator and denominator polynomial of this rational function being of degrees $s = 0$ and $t = 1$, respectively.

More generally, suppose that for a given HPD matrix B the B-adjoint of a matrix A is of the form

$$A^\times = p_s(A)(q_t(A))^{-1}, \quad (4.8.15)$$

where p_s and q_t are polynomials of respective degrees s and t. Every Krylov basis vector v_m can be written as $v_m = \psi_m(A) v_1$, where ψ_m is a polynomial of (exact) degree m. For $m \geq t$, there exist polynomials φ_{m-t} and $\varphi_{t-1}^{(m)}$, of respective degrees $m - t$ and $t - 1$, such that

$$\psi_m = q_t \cdot \varphi_{m-t} + \varphi_{t-1}^{(m)}.$$

If (4.8.15) holds, we get

$$A^\times v_m = p_s(A) \varphi_{m-t}(A) v_1 + A^\times \varphi_{t-1}^{(m)}(A) v_1.$$

By construction, the first vector on the right-hand side is a linear combination of $v_1, \ldots, v_{m-t+s}$, while the second is a linear combination of $A^\times v_1, \ldots, A^\times v_t$.

A straightforward (but somewhat tedious) argument now shows that the conditions (4.8.13) reduce to

$$\left(v_n - \sum_{k=n-t}^{n-1} g_{k,n} v_k, A^\times v_m\right)_B = 0, \quad \text{for } m = 1,\ldots,t, \tag{4.8.16}$$

which can be written in form of the following linear system,

$$\begin{bmatrix} (v_{n-t}, A^\times v_1)_B & \cdots & (v_{n-1}, A^\times v_1)_B \\ \vdots & & \vdots \\ (v_{n-t}, A^\times v_t)_B & \cdots & (v_{n-1}, A^\times v_t)_B \end{bmatrix} \begin{bmatrix} g_{n-t,n} \\ \vdots \\ g_{n-1,n} \end{bmatrix} = \begin{bmatrix} (v_n, A^\times v_1)_B \\ \vdots \\ (v_n, A^\times v_t)_B \end{bmatrix};$$

cf. [41, equation (4.6)]. In summary, assuming that the condition (4.8.15) holds, one can derive a linear system of order t for computing t scalars $g_{n-t,n}, \ldots, g_{n-1,n}$, that yield an $(s+2,t)$-term recurrence of the form (4.8.14) for generating a B-orthogonal Krylov subspace basis for A.

For $t=1$ the above linear system is of order one, namely

$$(v_{n-1}, A^\times v_1)_B \, g_{n-1,n} = (v_n, A^\times v_1)_B.$$

Clearly, when $(v_{n-1}, A^\times v_1)_B = 0$, but $(v_n, A^\times v_1)_B \neq 0$, there exists no solution $g_{n-1,n}$ of the linear system, leading to a breakdown of the algorithm. In general, a breakdown occurs whenever the above linear system is inconsistent. However, similarly to isometric Arnoldi, the whole computation can be rearranged in the form of coupled recurrences instead of a single recurrence, which avoids the problem of breakdowns. A complete derivation is given in [41, Section 4.4].

We have only discussed the sufficiency of the condition (4.8.15). A partial solution to the question of necessary conditions is presented in [42]. The special case $B = I$ and $A^* = (q_t(A))^{-1}$, i.e. (4.8.15) with $B = I$ and $p_s(\lambda) = 1$, is treated in [691]. For such matrices the authors derive two different recurrences generalising isometric Arnoldi, and they prove sufficient and necessary conditions for these recurrences being short (of course, the crucial condition is that t is small). The proof of necessity appears to be closely related to the proof of necessity by Faber and Manteuffel [176]. In addition, the paper [691] studies relations between the condition $A^* = (q_t(A))^{-1}$ and the *displacement rank* of the matrix

$$K_n = [v, \ldots, A^{n-1}v]^* [v, \ldots, A^{n-1}v].$$

It is shown that if $A^* = (q_t(A))^{-1}$ for a polynomial of degree t, then the displacement rank of K_n is equal to $t+1$. In particular, when A is unitary, $t=1$ and thus the displacement rank of K_n is equal to two.

The main question of practical interest is whether there exist classes of matrices whose adjoint (for some given HPD matrix B) satisfies (4.8.15) with *small* integers $s \geq 0$ and $t \geq 1$. We have already considered the unitary matrices, for which we can use $B = I$, giving $s = 0$ and $t = 1$. Another example is the shifted unitary matrices of the form $A = U + \zeta I$ with a nonzero $\zeta \in \mathbb{C}$ and a unitary matrix U. Any such matrix satisfies $A^* = p_1(A)(q_1(A))^{-1}$, where $p_1(\lambda) = \bar{\zeta}\lambda + (1 - |\zeta|^2)$ and $q_1(\lambda) = \lambda - \zeta$, and hence $s = 1$ and $t = 1$. The resulting $(3, 1)$-term recurrence has been employed by Jagels and Reichel [362, 363] (this has already been noted in the previous section).

The condition (4.8.15) is studied in detail in [420]. In [420, Corollary 2.5] it is shown that for a given matrix A there exists an HPD matrix B such that $A^\times$ is a rational function in A if and only if A is diagonalisable. Moreover, it is shown that for A diagonalisable the HPD matrices B for which $A^\times$ is a rational function in A are completely characterised as in (2b) of our Theorem 4.6.3. If B is any such matrix, then $A^\times = r(A)$ holds for a rational function r if and only if $r(\Lambda) = \Lambda^*$ holds for the diagonal matrix Λ of the eigenvalues of A. Consequently, if $A^\times = r(A)$, then r is completely determined by the eigenvalues of A.

For a rational function $r = p/q$, where p and q are relatively prime polynomials (i.e. their only common divisor is the constant polynomial 1), its *McMillan degree* is defined as

$$\deg(r) \equiv \max\{\deg(p), \deg(q)\}.$$

To allow a rigorous characterisation of the condition (4.8.15), the *McMillan degree of a diagonalisable matrix A* is defined in [420, Definition 2.6] as the (uniquely determined) smallest McMillan degree of a rational function r, such that $r(\lambda) = \bar{\lambda}$ for all eigenvalues λ of A. To determinine this McMillan degree the results of [382, 523] on the zeros of *harmonic rational functions* $r(\lambda) - \bar{\lambda}$ can be applied, which leads to the following result; see [420, Theorem 3.6].

Theorem 4.8.1
Let A be a diagonalisable matrix with $n \geq 4$ distinct eigenvalues, and denote its McMillan degree by $d_r(A)$. Then the following hold:

(1) *If the eigenvalues are collinear or concyclic (i.e. lie on a single straight line or a single circle in the complex plane), then $d_r(A) = 1$.*
(2) *In all other cases, $d_r(A) \geq \lfloor n/5 + 1 \rfloor$.*

In other words, if the eigenvalues of a diagonalisable matrix A are neither collinear nor concyclic, then $d_r(A)$ is small if and only if A has only a small number of distinct eigenvalues. Related observations are made in [41, Theorem 3.1]. This means that $(s + 2, t)$-term recurrences with small $s > 0$ and $t \geq 1$ exist only for diagonalisable matrices having concyclic eigenvalues. Here the unitary and shifted unitary matrices represent the most important classes. Consequently, the practical gain of the $(s + 2, t)$-term recurrences in the context of solving linear algebraic systems is rather limited.

Finally, we mention that Barth and Manteuffel [38, 40, 41, 42] have also studied a more general class of matrices that satisfy $A^\times = p_s(A)(q_t(A))^{-1} + R$, where R is a low rank matrix. In related work, Beckermann and Reichel [53] have derived a

Other Types of Recurrences

short recurrence Arnoldi type algorithm for generating orthogonal Krylov subspace bases in case $A^* = A + R$, where R has low rank.

4.8.3 Generalised Krylov Subspaces

In this section we will consider whether it is possible to derive an efficient recurrence for computing an orthogonal basis of appropriate spaces $\mathcal{L}_m(A, v)$ containing $\mathcal{K}_n(A, v)$. For simplicity, we only consider orthogonality with respect to the Euclidean inner product, i.e. $B = I$. Most results easily generalise to a general B-inner product.

Following Elsner and Ikramov [161], we consider the sequence of vectors

$$\underbrace{v,}_{\text{1st layer}} \underbrace{Av, A^*v,}_{\text{2nd layer}} \underbrace{A^2v, AA^*v, A^*Av, (A^*)^2v,}_{\text{3rd layer}} \ldots \qquad (4.8.17)$$

We denote by L_j the set of the vectors in the jth layer, and we call the subspace

$$\mathcal{L}_n(A, v) \equiv \text{span}\{L_j : j = 1, \ldots, n\} \qquad (4.8.18)$$

the *n*th generalised Krylov subspace. Clearly,

$$\mathcal{K}_n(A, v) \subset \mathcal{L}_n(A, v), \quad n = 1, \ldots d,$$

where d is the grade of v with respect to A. We denote $l_n \equiv \dim \mathcal{L}_n(A, v)$, and we call

$$w_n \equiv l_n - l_{n-1}$$

the *width of the n*th *layer* (here we formally set $l_{-1} = 0$). Note that if A is Hermitian, then $\mathcal{L}_n(A, v) = \mathcal{K}_n(A, v)$ and $w_n = 1$ for $n = 1, \ldots, d$. If A is unitary, then $w_n \leq 2$, and if A is a general normal matrix, then $w_n \leq n$ for all n.

We are now going to state an algorithm for computing orthogonal bases $v_1, \ldots, v_{l_n}$ of $\mathcal{L}_n(A, v)$ for $n = 1, 2, \ldots$. This can in principle be achieved by applying the classical Gram–Schmidt orthogonalisation to the sequence (4.8.17), and by orthogonalising each vector with respect to all previously computed basis vectors. This procedure is greatly simplified by the following two observations.

First, the layer L_{n+1} in (4.8.17) is obtained from L_n by forming Ay and A^*y for each vector $y \in L_n$. Second, let $v_n \in \mathcal{L}_n(A, v) \setminus \mathcal{L}_{n-1}(A, v)$ be a basis vector corresponding to the nth layer. Then for all $y \in \mathcal{L}_j(A, v)$ with $1 \leq j \leq n-2$, $A^*y \in \mathcal{L}_{n-1}(A, v)$ and $Ay \in \mathcal{L}_{n-1}(A, v)$, which yields

$$(Av_n, y) = (v_n, A^*y) = 0,$$
$$(A^*v_n, y) = (v_n, Ay) = 0.$$

In other words, in the $(n+1)$th step the vectors Av_n and A^*v_n for each basis vector v_n corresponding to L_n are already orthogonal to all previous basis vectors corresponding to $L_1, \ldots, L_{n-2}$. Therefore the orthogonalisation only has to be performed with respect to the basis vectors corresponding to L_{n-1}, L_n, and the newly computed basis vectors corresponding to the current layer L_{n+1}. This yields the *generalised Lanczos algorithm* of Elsner and Ikramov [161, Algorithm 2] that is stated in Algorithm 4.8.2.

If the width w_n is zero at a certain step n (i.e. all vectors resulting from step (b) of Algorithm 4.8.2 are zero), then the algorithm terminates having computed an invariant subspace of both A and A^*. Note that the computed basis vectors depend on the ordering of the vectors in the sequence (4.8.17). We know of no analysis that shows which ordering should be preferred.

As mentioned above, when A is Hermitian, $\mathcal{L}_n(A, v) = \mathcal{K}_n(A, v)$ for $n = 1, \ldots, d$. In this case Algorithm 4.8.2 is mathematically equivalent with (4.3.3) for $B = I$, and it represents nothing but the Hermitian Lanczos algorithm (Algorithm 2.4.3 in this book). This is also why Elsner and Ikramov called their algorithm a 'generalised' Lanczos algorithm. While the Hermitian Lanczos algorithm is related to the computation of orthogonal polynomials on the real line, the generalised Lanczos algorithm, when applied to normal matrices, is related to the computation of orthogonal *polyanalytic polynomials*. A polyanalytic polynomial is a function of the form

$$\sum_{j=0}^{n} h_j(\lambda) \overline{\lambda}^j,$$

where each $h_j(\lambda)$ is a polynomial of degree j; see, e.g. [35]. More details on the relations between the generalised Lanczos algorithm and orthogonal polyanalytic

---

Algorithm 4.8.2 Generalised Lanczos algorithm

Input: square matrix $A \in \mathbb{F}^{N \times N}$, initial vector $v \in \mathbb{F}^N$.
Output: orthogonal basis vectors $v_1, \ldots, v_{l_n}$ of $\mathcal{L}_n(A, v)$, for $n = 1, 2, \ldots$.

$v_1 = v$

For $n = 1, 2, \ldots$

Let an orthogonal basis of $\mathcal{L}_n(A, v)$ be formed by the vectors $v_1, \ldots, v_{l_n}$. Then, for each $y \in \{v_{l_{n-1}+1}, \ldots, v_{l_n}\}$ in turn, perform the following steps:

(a) Compute $w = Ay$.
(b) Orthogonalise w with respect to $v_{l_{n-2}+1}, \ldots, v_{l_n}$ and the previously computed basis vectors corresponding to the $(n+1)$th layer.
(c) If the resulting vector of step (b) is nonzero, take it as the most recent basis vector corresponding to the $(n+1)$th layer.
(d) For the vector $w = A^* y$, repeat the steps (b) and (c) above.

polynomials can be found in [350]. Similarly to the isometric Arnoldi algorithm studied in the previous section, here again we can observe a deep connection between a matrix algorithm and a topic of classical analysis.

Let us continue with studying properties of the generalised Lanczos algorithm. For simplicity of notation we suppose that $d_{\min}(A) = N$. Then for any initial vector of grade N, the outcome of the Hermitian Lanczos algorithm is a relation of the form $AV_N = V_N H_{N,N}$, where $H_{N,N}$ is tridiagonal. In other words, the matrix representation $H_{N,N}$ of the linear operator $A : \mathbb{C}^N \to \mathbb{C}^N$ in the Lanczos basis is a tridiagonal matrix. Similarly, the generalised Lanczos algorithm applied to a nonsingular matrix A yields a matrix representation $G_{N,N}$ of A in the computed basis that is a *block tridiagonal* matrix with diagonal blocks of sizes w_n (recall that w_n denotes the width of the nth layer). Moreover, the numbers of nonzero elements in the ith row and the ith column of $G_{N,N}$ are both bounded by $w_{n-1} + w_n + w_{n+1}$, where n is the index of the layer to which the basis vector v_i corresponds. This bound results from the fact that orthogonalisation at each step has to be performed only with respect to the current and the two previous layers of vectors.

The generalised Lanczos algorithm can be applied to any matrix A, but it is particularly useful in the normal case. As shown in [161, Section 3], for each $N \times N$ normal matrix A and initial vector v, it holds $w_n < \sqrt{2N}$ for all n, so that the number of nonzero elements in each row and column of $G_{N,N}$ is bounded by $\sqrt{18N}$. Consequently, the total number of nonzero elements of $G_{N,N}$ is at most $\sqrt{18}N^{3/2}$. Because of this significant order reduction compared to the potentially N^2 nonzero elements of A, the matrix $G_{N,N}$ is called the *condensed form of A*. This form was used by Huhtanen [350] to develop a Krylov subspace type method for general normal matrices. The length of the recurrence for computing the orthogonal basis of $\mathcal{L}_n(A, v)$ is not fixed, but it is slowly growing with n. More precisely, in step n of Huhtanen's method, at most $\sqrt{8n}$ of the previous basis vectors have to be stored. Huhtanen and Larsen [351] use the same method for computing exclusion and inclusion regions for the eigenvalues of normal matrices.

When the normal matrix A satisfies some further condition in addition to the relation $A^*A - AA^* = 0$, then additional savings can arise in its condensed form. The obvious example is a Hermitian matrix A that satisfies $A - A^* = 0$, so that $w_n = 1$ for all n. Then the number of nonzero elements in each row and column of the condensed form is at most three, and we have the Hermitian Lanczos algorithm. A unitary matrix A satisfies $A^*A - I = AA^* - I = 0$, implying that $w_n \leq 2$ for all n, so that the number of nonzero elements in each row and column of its condensed form is at most six. In fact, it can be shown that in the unitary case the condensed form $G_{N,N}$ has at most four nonzero elements in each row and column, and hence that the bound $w_{n-1} + w_n + w_{n+1}$ for the number of nonzero elements is not sharp (note that the blocks are not necessarily dense due to the inner structure corresponding to the ordering in the individual layers).

The condensed form of unitary matrices is closely related to the isometric Arnoldi algorithm. These relations are extensively discussed in [660]. In general, the condensed form of a normal matrix A is typically very sparse when the eigenvalues of A lie on an algebraic curve of a low degree. This has been exploited by Huhtanen [349] as well as Faßbender and Ikramov [180], who used the condensed form to derive Krylov subspace type methods for normal matrices having their eigenvalues contained in a second degree curve in the complex plane.

4.9 REMARKS ON FUNCTIONAL ANALYTIC REPRESENTATIONS

In Section 4.2 we have studied the theory of (cyclic) invariant subspaces of a linear operator on a finite-dimensional complex vector space from a purely algebraic point of view. The main result is the decomposition of a vector space into cyclic subspaces (Theorem 4.2.8), which turns out to be equivalent to the Jordan canonical form of the given operator (Theorem 4.2.13). We will now complement the algebraic view of Section 4.2 by a brief discussion of the functional analytic approach to invariant subspaces that is based on the Cauchy integral formula. We will use matrix rather than operator notation. The material we present is well known and can be found in many books. In our treatment below we follow the presentation in [237, Chapter 8] and [239, Chapter 2].

Let $A \in \mathbb{C}^{N \times N}$ be given and let $s(A)$ denote any given subset of the eigenvalues of A. Suppose that we want to construct a projector onto the invariant subspace of A associated with the eigenvalues in the set $s(A)$. Let

$$A = S \begin{bmatrix} A_1 & 0 \\ 0 & A_2 \end{bmatrix} S^{-1}, \qquad (4.9.1)$$

where the spectrum of $A_1 \in \mathbb{C}^{k \times k}$ is equal to the given set $s(A)$, and the spectrum of the matrix $A_2 \in \mathbb{C}^{(N-k) \times (N-k)}$ does not intersect $s(A)$. In other words, the Jordan canonical form of the matrix A_1 contains all Jordan blocks corresponding to the eigenvalues in $s(A)$, while the Jordan canonical form of A_2 contains the remaining Jordan blocks of A. Hence k is equal to the sum of the sizes of all Jordan blocks corresponding to the eigenvalues in the set $s(A)$.

Then we can define the matrix

$$P_1 \equiv S \begin{bmatrix} I_k & 0 \\ 0 & 0 \end{bmatrix} S^{-1}. \qquad (4.9.2)$$

A simple computation shows that $P_1^2 = P_1$, i.e. P_1 represents a projection. Moreover, if we partition $S = [S_1, S_2]$, where $S_1 \in \mathbb{C}^{N \times k}$, then $\text{Range}(S_1)$ is the k-dimensional invariant subspace of A corresponding to the eigenvalues in the set $s(A)$. Equivalently, $\text{Range}(S_1)$ is the direct sum of the cyclic invariant subspaces of A corresponding to the Jordan blocks in the matrix A_1. Any vector $v \in \mathbb{C}^N$ can be decomposed as

$$v = P_1 v + (I - P_1) v,$$

where $I - P_1$ also represents a projection. We have

$$P_1 v \in \text{Range}(S_1) \quad \text{and} \quad (I - P_1) v \in \text{Range}(S_2),$$

This means that P_1 projects onto Range(S_1) along Range(S_2), i.e. orthogonally to Range(S_2)$^\perp$.

We will next represent the projector P_1 by an integral. To do this we need some notation. Let Γ be any simple, closed and positively oriented curve in the complex plane, and consider any function mapping from Γ to $\mathbb{C}^{N \times N}$ of the form $B(\lambda) = [b_{ij}(\lambda)]$, where each entry $b_{ij}(\lambda)$, $i, j = 1, \ldots, N$, is a continuous function of the complex variable $\lambda \in \Gamma$. Then we can define

$$\int_\Gamma B(\lambda)\, d\lambda \equiv [c_{ij}] \in \mathbb{C}^{N \times N}, \quad \text{where } c_{ij} \equiv \int_\Gamma b_{ij}(\lambda)\, d\lambda. \quad (4.9.3)$$

In other words, the integral over the contour Γ of the function $B(\lambda)$ is the complex $(N \times N)$-matrix whose entries are the integrals over Γ of the entries $b_{ij}(\lambda)$ of $B(\lambda)$. The techniques we will employ below are standard techniques from complex function theory that can be found in any book on this subject; see, e.g. [2, 475].

Suppose that the contour Γ contains the given subset of the eigenvalues $s(A)$ in its interior, and that all remaining eigenvalues are in the exterior of Γ. Then the *resolvent function* $(\lambda I_N - A)^{-1}$ is continuous on Γ. Substituting $(\lambda I_N - A)^{-1}$ for $B(\lambda)$ in (4.9.3), multiplying by $1/(2\pi \iota)$, and exploiting the decomposition (4.9.1) yields a matrix $C \in \mathbb{C}^{N \times N}$ of the form

$$\begin{aligned}C &\equiv \frac{1}{2\pi \iota} \int_\Gamma (\lambda I_N - A)^{-1}\, d\lambda \\ &= \frac{1}{2\pi \iota} S \begin{bmatrix} \int_\Gamma (\lambda I_k - A_1)^{-1}\, d\lambda & 0 \\ 0 & \int_\Gamma (\lambda I_{N-k} - A_2)^{-1}\, d\lambda \end{bmatrix} S^{-1}. \quad (4.9.4)\end{aligned}$$

We will show that C is equal to P_1 as defined in (4.9.2). By construction, the domain bounded by Γ contains no eigenvalues of the matrix A_2. It therefore can be shown that the resolvent $(\lambda I_{N-k} - A_2)^{-1}$ is analytic within this domain, and by Cauchy's integral theorem we get

$$\int_\Gamma (\lambda I_{N-k} - A_2)^{-1}\, d\lambda = 0 \in \mathbb{C}^{(N-k) \times (N-k)}.$$

To evaluate the other integral in (4.9.4) we note that for $|\lambda| > \|A_1\|$ we can expand the resolvent in a power series (see, e.g. [250, Section 2.3.4] for a proof),

$$\begin{aligned}(\lambda I_k - A_1)^{-1} &= \frac{1}{\lambda}(I_k - \lambda^{-1} A_1)^{-1} = \frac{1}{\lambda} \sum_{j=0}^\infty (\lambda^{-1} A_1)^j \\ &= \frac{1}{\lambda} I_k + \frac{1}{\lambda^2} A_1 + \frac{1}{\lambda^3} A_1^2 + \cdots \\ &\equiv \frac{1}{\lambda} I_k + F(A_1, \lambda). \quad (4.9.5)\end{aligned}$$

By construction, all eigenvalues of A_1, and hence all singularities of the resolvent function $(\lambda I - A_1)^{-1}$, are contained in the interior of Γ. Therefore we may replace, without changing the value of the integral, the contour Γ by any other closed and piecewise smooth contour that contains the interior of Γ. In particular, if $\widehat{\Gamma}$ is a circle centred at zero and with radius $\alpha > \|A_1\|$, then $|\lambda| > \|A_1\|$ for all $\lambda \in \widehat{\Gamma}$. Using the Cauchy formula (see also Chapter 3, Sections 3.3.3 and 3.4.2) and the expansion (4.9.5) leads to

$$\int_\Gamma (\lambda I_k - A_1)^{-1}\, d\lambda = \int_{\widehat{\Gamma}} \lambda^{-1} I_k\, d\lambda + \int_{\widehat{\Gamma}} F(A_1, \lambda)\, d\lambda$$
$$= (2\pi\iota) I_k + \int_{\widehat{\Gamma}} F(A_1, \lambda)\, d\lambda. \quad (4.9.6)$$

If $|\lambda| = \alpha > \|A_1\|$, we can bound the norm of the function $F(A_1, \lambda)$ defined in (4.9.5) as

$$\|F(A_1, \lambda)\| \le \frac{\|A_1\|}{\alpha^2} \frac{1}{1 - \|A_1\|/\alpha}.$$

Hence (4.9.6) yields

$$\left\| \frac{1}{2\pi\iota} \int_\Gamma (\lambda I_k - A_1)^{-1}\, d\lambda - I_k \right\| = \left\| \frac{1}{2\pi\iota} \int_{\widehat{\Gamma}} F(A_1, \lambda)\, d\lambda \right\|$$
$$\le \frac{1}{2\pi} \int_{\widehat{\Gamma}} \|F(A_1, \lambda)\, d\lambda\|$$
$$\le \frac{1}{2\pi} \frac{\|A_1\|}{\alpha^2} \frac{1}{1 - \|A_1\|/\alpha} 2\pi\alpha$$
$$= \frac{\|A_1\|}{\alpha} \frac{1}{1 - \|A_1\|/\alpha}.$$

This identity holds for any $\alpha > \|A_1\|$. For $\alpha \to \infty$ the right-hand side approaches zero. Therefore we must have

$$\frac{1}{2\pi\iota} \int_\Gamma (\lambda I_k - A_1)^{-1}\, d\lambda = I_k.$$

Using this in the definition of C in (4.9.4) shows that indeed $C = P_1$. We summarise the main results of this discussion in the following well-known theorem (see, e.g. [239, Proposition 2.4.3]).

Theorem 4.9.1
Let $A \in \mathbb{C}^{N \times N}$, and let $s(A)$ be a given subset of the eigenvalues of A. Let Γ be any simple, closed and positively oriented curve in the complex plane that contains the set $s(A)$ in its interior, and all other eigenvalues of A in its exterior. Then

$$P_1 = \frac{1}{2\pi\iota} \int_\Gamma (\lambda I_N - A)^{-1}\, d\lambda \in \mathbb{C}^{N \times N}$$

where P_1 is defined as in (4.9.2).

Suppose that A has k distinct eigenvalues, $\lambda_1, \ldots, \lambda_k$. Then we can find non-intersecting discs $\Delta_1, \ldots, \Delta_k$ with their respective centres at $\lambda_1, \ldots, \lambda_k$, so that each disc contains exactly one eigenvalue (namely the eigenvalue at the centre). Denote the boundary of the disc Δ_j by Γ_j, and define

$$\pi_j \equiv \frac{1}{2\pi \iota} \int_{\Gamma_j} (\lambda I_N - A)^{-1}\, d\lambda, \quad j = 1, \ldots, k.$$

Then π_j projects onto the A-invariant subspace generated by the Jordan basis vectors of A corresponding to the eigenvalue λ_j, $j = 1, \ldots, k$. One can show that

$$\pi_i \pi_j = \pi_j \pi_i = 0, \quad \text{if } i \neq j, \quad \text{and} \quad I_N = \sum_{j=1}^{k} \pi_j, \qquad (4.9.7)$$

In the functional analysis context these relations can be shown by applying certain *resolvent identities* to the integral representation of the projectors. Alternatively, a linear algebra based proof can easily be given using the matrix representation of the projectors as in (4.9.2).

In analogy to the decomposition (4.2.1), the relations (4.9.7) show that the whole space $\mathbb{C}^N$ is decomposed into the direct sum of A-invariant subspaces of the form $\text{Range}(\pi_j), j = 1, \ldots, k$. We remark that

$$\text{Range}(\pi_j) = \text{Kernel}\big((A - \lambda_j I_N)^{d_j}\big),$$

where d_j is the dimension of the largest cyclic subspace of A corresponding to the eigenvalue λ_j. This can be seen, e.g. by plugging in the Jordan canonical form of A; see Section 4.2.2, the second matrix in Figure 4.1. The dimension d_j is equal to the size of the largest Jordan block corresponding to the eigenvalue λ_j, or, equivalently, the power of the factor $\lambda - \lambda_j$ in the minimal polynomial of A with respect to $\mathbb{C}^N$.

The second equation in (4.9.7) can be written as

$$I_N = \frac{1}{2\pi \iota} \sum_{j=1}^{k} \int_{\Gamma_j} (\lambda I_N - A)^{-1}\, d\lambda.$$

This is a special case of the more general definition of the (primary) *matrix function* via the Cauchy integral formula. Assuming that a given function $f(\lambda)$ is analytic on and inside a simple, closed and positively oriented curve Γ that contains all discs Δ_j (and hence all distinct eigenvalues λ_j of A) in its interior, the matrix function $f(A)$ can be written as

$$f(A) = \frac{1}{2\pi \iota} \int_{\Gamma} f(\lambda)(\lambda I_N - A)^{-1}\, d\lambda = \frac{1}{2\pi \iota} \sum_{j=1}^{k} \int_{\Gamma_j} f(\lambda)(\lambda I_N - A)^{-1}\, d\lambda;$$

see [316, Definition 1.11] and also [237, Section 8.4]. In particular,

$$A = \frac{1}{2\pi \iota} \sum_{j=1}^{k} \int_{\Gamma_j} \lambda (\lambda I_N - A)^{-1} \, d\lambda.$$

Different representations of Hermitian matrices and operators using the Riemann–Stieltjes integral were given in Remark 3.5.1 in this book.

Both representations are equivalent for a Hermitian matrix A. Indeed, considering the spectral decomposition $A = Y\Lambda Y^*$ with $YY^* = Y^*Y = I_N$ and $\Lambda = \mathrm{diag}(\lambda_\ell)$, and assuming, for simplicity of notation, that all eigenvalues of A are distinct,

$$\pi_j = \frac{1}{2\pi \iota} \int_{\Gamma_j} (\lambda I_N - A)^{-1} \, d\lambda = Y \left(\frac{1}{2\pi \iota} \int_{\Gamma_j} (\lambda I_N - \Lambda)^{-1} \, d\lambda \right) Y^*$$

$$= Y \, \mathrm{diag}\left(\frac{1}{2\pi \iota} \int_{\Gamma_j} \frac{1}{\lambda - \lambda_\ell} \, d\lambda \right) Y^* = Y (e_j e_j^T) Y^* = y_j y_j^*, \quad j = 1, \ldots, N.$$

Analogously,

$$I_N = \frac{1}{2\pi \iota} \sum_{j=1}^{N} \int_{\Gamma_j} (\lambda I_N - A)^{-1} \, d\lambda = \sum_{j=1}^{N} y_j y_j^*,$$

$$A = \frac{1}{2\pi \iota} \sum_{j=1}^{N} \int_{\Gamma_j} \lambda (\lambda I_N - A)^{-1} \, d\lambda = \sum_{j=1}^{N} Y \, \mathrm{diag}\left(\frac{1}{2\pi \iota} \int_{\Gamma_j} \frac{\lambda}{\lambda - \lambda_\ell} \, d\lambda \right) Y^*$$

$$= \sum_{j=1}^{N} \lambda_j y_j y_j^*.$$

It is worth recalling that the integral representation of self-adjoint operators and operator or (Hermitian) matrix functions using the Riemann–Stieltjes integral (see Remark 3.5.1) is based on ideas that appeared many decades later than the Cauchy integral representation of functions; see the historical notes in Sections 3.3.5 and 3.4.3.

5

Cost of Computations Using Krylov Subspace Methods

> **cost.** the amount or equivalent paid or charged for something: price.
>
> —MERRIAM-WEBSTER, *Online Dictionary*

In the previous chapters we have focused on the mathematical background of Krylov subspace methods and on connections between these methods and classical mathematical topics, including projections, moments, quadratures, orthogonal polynomials, continued fractions, and invariant subspaces. So far, our description has mostly dealt with analytic properties that hold in exact arithmetic.

In this chapter we will consider computations using Krylov subspace methods. In practice the methods are used to obtain *acceptable approximate solutions of algebraic problems* (linear systems and eigenvalue problems) *at an acceptable cost*. The cost in this context is typically measured in terms of computer memory requirements and computation time. Our goals in this chapter are to present a more general concept of computational cost of Krylov subspace methods, and to study in detail the standard Krylov subspace methods CG and GMRES. The presentation therefore splits into two parts.

The first part starts with an introductory section entitled *Seeing the context is essential* (Section 5.1). This section sets the stage for the rest of the chapter, and it addresses some as yet unresolved questions that potentially form a research programme for years to come. We then discuss the main issues in the context of computational cost of Krylov subspace methods, namely the principal differences between direct and iterative computations (Section 5.2), the computational cost of algebraic iterations (Section 5.3), the difference between computational cost and the concepts of complexity theory (Section 5.4), and the general concept of convergence of iterative methods (Section 5.5).

The question of computational cost of practical computations brings in the *finite precision behaviour* of Krylov subspace methods. It will be explained why the aspects of convergence and numerical stability (effects of rounding errors) are inherently linked, and hence why in practical applications the exact and finite precision behaviour of Krylov subspace methods should not be treated as separate issues. Because of this link, evaluating the cost of computations using Krylov subspace methods represents a very involved problem. In addition, and possibly most important of all, for an algebraic problem that arises from mathematical modelling of complicated real-world phenomena, the decision of whether a computed approximate solution is acceptable cannot be addressed within the field of matrix computations alone. Therefore, as outlined in the introductory section of this chapter, evaluating cost of iterative computations requires a much wider context.

The second part of this chapter (Sections 5.6–5.10) will deal with CG and GMRES, both in exact and finite precision arithmetic. We will consider issues like nonlinearity in solving *linear* algebraic systems, the relationship between spectral information and convergence, maximal attainable accuracy, and delay of convergence. We will give mathematical theorems as well as numerical illustrations. We will also discuss some misunderstandings concerning the evaluation of convergence in the context of Krylov subspace methods. The issues raised for Krylov subspace methods can perhaps also contribute to the general discussion of concepts like complexity, computational cost, convergence, and stability in numerical analysis and scientific computing.

The chapter ends with remarks on some issues that are not covered in detail in this book and with an outlook to further work.

We started this book with quotes from Lanczos and Einstein about the inner nature of the problem. This chapter, which deals with computations using Krylov subspace methods, can be perhaps motivated by the following quote taken from Patrick J. Roache's book *Validation and Verification in Computational Science* [530, p. 387]:

> With the often noted tremendous increases in computer speed and memory, and with the less often acknowledged but equally powerful increases in algorithmic accuracy and efficiency, a natural question suggest itself. What are we doing with the new computer power? with the new GUI and other set-up advances? with the new algorithms? What *should* we do? ... *Get the right answer.*

5.1 SEEING THE CONTEXT IS ESSENTIAL

Consider the solution process of a real-world problem described in Section 1.1 of this book, which is shown schematically in Figures 1.1 and 1.2. In the first step of the solution process the mathematical model, given by integral and/or differential equations and describing some part of reality, has to omit less substantial relationships in order to be understandable and solvable. The second step typically consists of a discretisation which leads to a linear algebraic problem that can be treated on a computer. Here the inifinite-dimensional model is replaced by a finite-dimensional approximation. As an example, consider the step from the weak formulation of a boundary value problem in (2.5.21) to its discretised version in (2.5.22). Finally,

Seeing the Context is Essential

the linear algebraic problem is solved by some matrix computation algorithm. If we use Krylov subspace methods for this task, then this third step can be considered a (further) matching moments model reduction; see Chapter 3.

In this section we discuss some important issues in the solution process that are due to the second and third (i.e. the discretisation and computation) steps. We will use, for simplicity, the model problem (2.5.18)–(2.5.19) with zero Dirichlet boundary conditions, $g_D = 0$ in (2.5.19), and the discretisation using the Galerkin finite element method, as described in Section 2.5.2. Our observations are, however, relevant for other types of problems and other types of discretisations.

For convenience, we now briefly summarise key points from Section 2.5.2. In the Galerkin finite element method the solution u of the variational problem (2.5.21) is approximated by a linear combination of basis functions ϕ_j,

$$u \approx u_h = \sum_{j=1}^{N} \zeta_j \phi_j.$$

The coefficients ζ_j in this linear combination are the components of the exact solution x of a linear algebraic system $Ax = b$; see (2.5.24). In the large-scale mathematical modelling of real-world phenomena it is of vital importance that the system matrix A is sparse. The finite element method achieves this requirement since each basis function ϕ_j represents only a *local* approximation of the solution u. The global approximation property is restored by solving the linear algebraic system for the coefficients ζ_j and by forming the linear combination of all basis functions ϕ_j. Using a Krylov subspace method for (approximately) solving the linear algebraic system $Ax = b$ leads to an approximation x_n of x, and thus to an approximation $u_h^{(n)}$ of u_h. In the overall solution process we are therefore confronted with three types of errors:

- the *discretisation error* $u - u_h$, where u solves the variational problem and $u_h = \sum_{j=1}^{N} \zeta_j \phi_j$ is obtained from the exact solution $x = [\zeta_1, \ldots, \zeta_N]^T$ of $Ax = b$ (for simplicity we assume that these solutions exist and are unique, which is certainly true for the model problem (2.5.18)–(2.5.19) with zero Dirichlet boundary conditions);
- the *algebraic error* $x - x_n$ or $u_h - u_h^{(n)}$, where $x_n = [\zeta_1^{(n)}, \ldots, \zeta_N^{(n)}]^T$ is a computed approximation to x and $u_h^{(n)} = \sum_{j=1}^{N} \zeta_j^{(n)} \phi_j$ is the corresponding approximation to u_h;
- the *total error* $u - u_h^{(n)}$.

The three errors are related by the simple, yet fundamental equation

$$u - u_h^{(n)} = (u - u_h) + \left(u_h - u_h^{(n)}\right).$$

In words, the total error is the sum of the discretisation error and the algebraic error.

The total error must be evaluated in an appropriate function space corresponding to the mathematical modelling level. We will illustrate below that in the function space the algebraic error can have *strongly varying local components*. These can in fact dominate locally the total error, even when the globally measured algebraic error is significantly smaller than the globally measured discretisation error. This underlines a need for investigating the discretisation error in the regions of interest; see, e.g. [32, p. 417 and Chapter 6] and [36, Chapter 1]. Incorporating the algebraic error into the total error with considering the *locality and the interplay between the discretisation and the algebraic computation* represents a challenge that is important in particular in a mathematical justification of adaptive PDE solvers.

Remark 5.1.1
While in the given model problem the situation is simple and the energy norm is appropriate, in many practical cases the choice of an appropriate function space represents a crucial problem. The state-of-the-art theory of nonlinear partial differential equations focuses on solutions that are local in time and exist under the assumptions of sufficient smoothness. This typically gives little guidance for the (physically) meaningful evaluation of error. In order to get such guidance one must take into account the underlying (physical) principles. Models in continuum thermodynamics as well as thermodynamics of multi-component materials may serve as examples. Here the natural function spaces are determined in relation to the properties of the entropy and the rate of the entropy production; see [181, 182, 439].

The issues described above belong to the area of verification and validation; see, e.g. [31, 484, 531] (also see Section 1.1 in this book). Although the algebraic part of the error was occasionally mentioned in the past (see, e.g. [167, Sections 6.4–6.5 and Chapter 12]), it was rarely investigated, or its relevance was restricted to numerical stability issues and conditioning; see [580] and [75, Sections 9.5–9.7]. Some approaches that incorporate the algebraic part of the error and its comparison with the discretisation part into the error analysis will be referenced below.

The main purpose of the rest of this section is to examine whether the standard algebraic approaches can be easily incorporated into the framework of evaluation of the total error in functional spaces. A significant progress in understanding results of algebraic computations is related to the concept of the (algebraic) *backward error*. Instead of asking how close the computed result is to the exact one, the backward error analysis asks which problem is solved *exactly* by the computed result. In particular, let x_n be an approximation to the solution of the linear algebraic system $Ax = b$. The backward error analysis looks for the perturbed linear algebraic system with the perturbations δA and δb,

$$(A + \delta A) x_n = b + \delta b, \tag{5.1.1}$$

and answers the following question:

How close is the perturbed problem (5.1.1), which is solved *exactly* by x_n, to the original problem $Ax = b$, which is solved *approximately* by x_n?

As shown by Rigal and Gaches [526] (see also [315, Theorem 7.1]) the *normwise (relative) backward error* of x_n, defined by

$$\beta(x_n) \equiv \min \{ \beta \; : \; (A + \delta A) \, x_n = b + \delta b, \; \|\delta A\| \leq \beta \|A\|, \; \|\delta b\| \leq \beta \|b\| \, \}, \tag{5.1.2}$$

satisfies

$$\beta(x_n) = \frac{\|b - Ax_n\|}{\|b\| + \|A\| \, \|x_n\|} = \frac{\|\delta A_{\min}\|}{\|A\|} = \frac{\|\delta b_{\min}\|}{\|b\|}. \tag{5.1.3}$$

In other words, $\beta(x_n)$ is equal to the norm of the *smallest* relative perturbations $\delta A_{\min}$ in A and $\delta b_{\min}$ in b such that x_n exactly solves the perturbed system. Here $\|\cdot\|$ can be any vector and the corresponding induced matrix norm. The componentwise variant of the backward error can be found in [486]; see also [315, Chapter 7].

Although the concept of backward error arose from investigations of numerical instabilities (see the historical note in Section 5.8.2), it can be used irrespective of the source of the error (truncation and/or roundoff). The algebraic backward error ingeniously separates the properties of the method and even of the particular individual computations from the *conditioning* of the problem. Their combination allows us to estimate the size of the forward algebraic error $x - x_n$; see the essays [29, 669], [46, Section 3.2] and the monograph [315]. Arioli, Noulard, and Russo [13] used functional backward errors and extended the concept to function spaces; see also [8, 12] and [459, Section 4.3].

At first sight the inclusion of the algebraic backward error concept into estimates of the total error measured in the function space seems essentially to be a technical exercise. Because of the error of the model, the discretisation error and the uncertainties in the data, the system $Ax = b$ represents *a whole class* of admissible systems. Each system in this class corresponds (possibly in a stochastic sense) to the original real-world problem. One can therefore argue that as long as the algebraic backward error $\beta(x_n)$ in (5.1.2)–(5.1.3) is *small enough*, the computed algebraic solution x_n is, with respect to the subject of the mathematical modelling, as good as the solution x of $Ax = b$. The meaning of 'small enough' is sometimes intuitively interpreted as, say, an order of magnitude below the size of the discretisation error (all measured in the corresponding norms). It is worth pointing out that the balance between the discretisation and the algebraic errors is frequently evaluated globally (in norms).

However, the whole problem seems to be much more subtle. In particular, as mentioned above, in order to perform the computations efficiently we need tight *a posteriori* estimates of the *local* distribution of the total error which incorporate the algebraic error; see [368] for an example. In that respect the normwise backward error has a serious deficiency. The perturbed system $(A + \delta A)x_n = b + \delta b$ usually does not preserve the *inner structure* (including the physically meaningful relationships between the individual matrix and vector entries) of the algebraic system $Ax = b$, and therefore it also does not reflect the properties of the underlying mathematical model. We will now illustrate this difficulty and some of the other ideas outlined above using the following particular model problem.

Consider the two-dimensional Poisson equation

$$-\Delta u = 32(\eta_1 - \eta_1^2 + \eta_2 - \eta_2^2) \tag{5.1.4}$$

in the (open) unit square $(\eta_1, \eta_2) \in \Omega \equiv (0, 1) \times (0, 1)$ with zero Dirichlet boundary conditions, $u = 0$ on $\partial\Omega$. The problem has the nicely smooth exact solution

$$u(\eta_1, \eta_2) = 16\eta_1\eta_2(1 - \eta_1)(1 - \eta_2). \tag{5.1.5}$$

We discretise the problem using the Galerkin finite element method with linear basis functions ϕ_j on a regular triangular grid with the mesh size $h = 1/(m+1)$, where m is the number of inner nodes in each direction. We use the reverse orientation; see, e.g. [66, Figure 3.7]. For $j = 1, \ldots, m^2$ the basis function ϕ_j has a support of six triangle elements with the jth inner node as the central point. The approximate solution is given by

$$u_h = \sum_{j=1}^{N} \zeta_j \phi_j, \quad N = m^2;$$

cf. (2.5.23) and recall that we consider zero Dirichlet boundary conditions. The resulting linear algebraic system $Ax = b$ is of the form stated in (2.5.24)–(2.5.25). It is well known that a lexicographical ordering of the nodes leads to

$$A = [a(\phi_j, \phi_i)] = \text{tridiag}(-I, T, -I) \in \mathbb{R}^{N \times N}, \quad \text{with}$$
$$T = \text{tridiag}(-1, 4, -1) \in \mathbb{R}^{m \times m};$$

see, e.g. [167, Figure 15.11 in Section 15.1.4]. The matrix A is symmetric and positive definite, with its extreme eigenvalues given by

$$\lambda_{\min}(A) = 8 \sin^2\left(\frac{h\pi}{2}\right), \quad \lambda_{\max}(A) = 8 \sin^2\left(\frac{mh\pi}{2}\right);$$

see, e.g. [290, Chapter 4]. We assemble the right-hand side b using a two-dimensional Gaussian quadrature formula that is exact for polynomials of degree at most three.

For our numerical experiment we use $m = 50$ and thus A is of order 2500. Similar results can be obtained for any other choice of m. The extreme eigenvalues of A and the resulting condition number (with respect to the matrix 2-norm) are

$$\lambda_{\min}(A) = 7.5867 \times 10^{-3}, \quad \lambda_{\max}(A) = 7.9924, \quad \kappa(A) = 1.0535 \times 10^3.$$

Details on the relationship between the eigenvalues of the discretised operator and the original Laplace operator can be found, e.g. in [66]. We have computed the solution x of $Ax = b$ using the MATLAB backslash operator. Neglecting the algebraic error in this computation, the (closely approximated) squared energy norm of the discretisation error is

$$\|\nabla(u - u_h)\|^2 = a(u - u_h, u - u_h) = 5.8299 \times 10^{-3}.$$

The shape of the discretisation error is very similar to the shape of the solution; see (a) and (b) in Figure 5.1. As explained in the following remark, the discretisation error is, however, not as smooth as indicated by the plot (b) in Figure 5.1.

Remark 5.1.2
All figures shown in this section have been generated by the MATLAB `trisurf` command, which generates a triangular surface plot. The inputs of `trisurf` are the coordinates of the nodes in the given triangular mesh and the respective values of the plotted function at these nodes. In the plot the function values in the triangle interiors are interpolated linearly from the values at the nodes, and hence *the figures do not show the actual function values inside the triangles*. For the solution u the difference is not significant. In the case of the discretisation error $u - u_h$ (see (b) in Figure 5.1), the plot is, however, misleading. The discretisation error is not as smooth as suggested by the plot, but contains 'bubbles' inside the triangles, which can be (depending on the size of the error) significant. The same holds for the total errors shown in Figures 5.1 and 5.2.

Now we apply the CG method with $x_0 = 0$ to the discretised problem $Ax = b$ and stop the iteration when the normwise backward error drops below the level $\gamma\, h^\alpha$, i.e. when

$$\frac{\|b - Ax_n\|}{\|b\| + \|A\|\, \|x_n\|} < \gamma\, h^\alpha, \tag{5.1.6}$$

where $\gamma > 0, \alpha > 0$ are positive parameters and $\|\cdot\|$ denotes the 2-norm. Here we use the normwise relative backward error for stopping. The results reported in the table below show also the componentwise variant,

$$\max_i \frac{(|b - Ax_n|)_i}{(|A|\, |x_n| + |b|)_i}, \tag{5.1.7}$$

where $(y)_i$ denotes the ith entry of the vector y and $|\cdot|$ means that we take the corresponding matrix or vector with the absolute values of its entries; see [315, Theorem 7.3]. Using the componentwise variant for stopping the iteration does not lead to any significant change.

We want to examine quantitatively the intuitive expectation that the backward error being 'small enough' in relation to the size of the discretisation error ensures that the algebraic error does not noticeably affect the total error. For this purpose we have used $\alpha = 3$ and different values of γ. The leftmost column in the following table shows these values and some error measures computed at the step n when CG satisfied the stopping criterion (5.1.6):

γ	Total error $\|\nabla(u - u_h^{(n)})\|^2$	Algebraic error $\|x - x_n\|_A^2$	Algebraic error $\|x - x_n\|_2^2$	Backward error (5.1.7)
50.0	1.0195×10^{-2}	4.3656×10^{-3}	2.1960×10^{-2}	2.6831×10^{-2}
1.00	5.8444×10^{-3}	1.4503×10^{-5}	1.9486×10^{-4}	4.2274×10^{-4}
0.50	5.8304×10^{-3}	5.6043×10^{-7}	4.0382×10^{-6}	5.4886×10^{-5}
0.10	5.8299×10^{-3}	1.6639×10^{-8}	7.3513×10^{-8}	4.0418×10^{-6}
0.02	5.8299×10^{-3}	5.5286×10^{-10}	1.4415×10^{-10}	2.1091×10^{-6}

The five rows in this table correspond to 27, 35, 42, 50, and 56 CG steps, respectively.

For our choice of $m = 50$ the values $\alpha = 3$ and $\gamma = 50$ closely resemble the situation $\alpha = 2$ and $\gamma = 1$ that corresponds to the sizes of the inaccuracies in determining A and b being proportional to $h^2 = (51)^{-2}$. The table shows that with decreasing γ the A-norm of the algebraic error quickly drops very significantly below the energy norm of the discretisation error. Moreover, the Galerkin orthogonality property

$$\|\nabla(u - u_h^{(n)})\|^2 = \|\nabla(u - u_h)\|^2 + \|\nabla(u_h - u_h^{(n)})\|^2$$
$$= \|\nabla(u - u_h)\|^2 + \|x - x_n\|_A^2$$

(see (2.5.29) in Section 2.5.2) is satisfied up to a small inaccuracy caused by the computational errors in determining the solution of the discretised problem and the errors in the evaluation of the norms; this inaccuracy is proportional to machine precision. Therefore, except for $\gamma = 50$, the total error *measured in the energy norm* is dominated by the discretisation error, with the globally measured contribution of the algebraic error being significantly (even orders of magnitude) smaller.

The whole picture changes dramatically when one considers the local distribution of the error in the functional space. Figures 5.1–5.2 show the algebraic and total errors for our choice of parameters. For $\gamma = 50$ the discretisation error and the algebraic error measured in the energy norm are of similar size. In order to compare their local distributions, it is worth recalling that both u_h and $u_h^{(n)}$ are piecewise linear. Therefore their gradients as well as the gradient of the algebraic error $\nabla(u_h - u_h^{(n)})$ are piecewise constant. The norm $\|\nabla(u_h - u_h^{(n)})\| = \|x - x_n\|_A$ is nothing but the weighted sum of the size of the individual finite elements with the weights measuring how well the constant vector ∇u_h is approximated by the constant vector $\nabla u_h^{(n)}$.

In contrast to that, the gradient of the solution ∇u is not piecewise constant. When we evaluate the norm $\|\nabla(u - u_h)\|$ we integrate (over the individual elements) the error in the approximation of the gradient ∇u by the elementwise constant ∇u_h. Since the approximation of ∇u by ∇u_h is of a different character than the difference of the two elementwise constant vectors ∇u_h and $\nabla u_h^{(n)}$, there is no reason why the local distribution of the discretisation and the algebraic error should in general be similar.

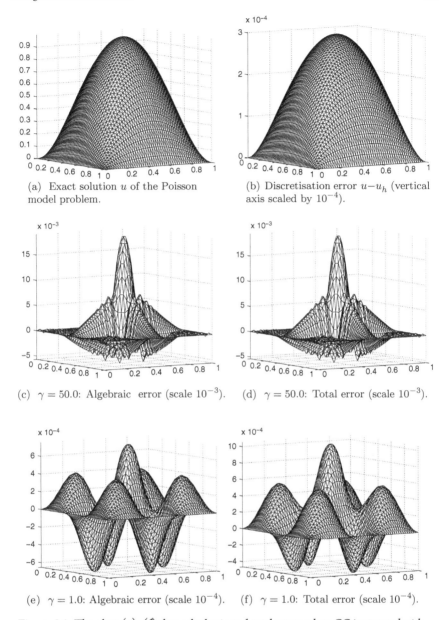

Figure 5.1 The plots (c)–(f) show algebraic and total errors when CG is stopped with $\alpha = 3$ and different values of γ in (5.1.6). Note that the vertical axes in these plots have different scales. Also note that while the algebraic error is piecewise linear, the total error is not. The plots of the total errors do not show the small 'bubbles' over individual elements; see Remark 5.1.2.

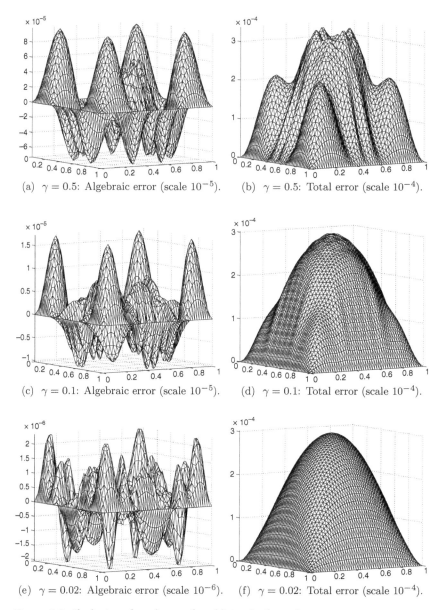

Figure 5.2 Algebraic and total errors for additional values of γ.

This is indeed demonstrated by our experiment. In spite of the comparable size of the values

$$\|\nabla(u - u_h)\|^2 \quad \text{and} \quad \left\|\nabla\left(u_h - u_h^{(n)}\right)\right\|^2 = \|x - x_n\|_A^2,$$

the shape of the total error for $\alpha = 3$ and $\gamma = 50$ is fully determined by its algebraic part. With decreasing γ the algebraic error gets smaller and it eventually becomes insignificant. But this happens only after $\|x - x_n\|_A^2$ drops seven orders of magnitude below the squared energy norm of the discretisation error $\|\nabla(u - u_h)\|^2$. We can also observe that using other (single number) measures of the algebraic error, such as the Euclidean norm or the componentwise backward error, would not make a difference in the observed phenomena.

Here we have used the simplest two-dimensional model problem with linear finite elements. The same reasoning can, however, be applied to a more general case. The particular choice of the model problem only helped to make the main point transparent. Because of the zero Dirichlet boundary conditions we have

$$\|\nabla(u - u_h)\|^2 = \|\nabla u\|^2 - \|\nabla u_h\|^2;$$

(see (2.5.30) in Section 2.5.2), where the first term on the right-hand side involves an integration of the 'bubbles' over the individual elements. Clearly, the norm cannot describe the differences in the shape of the errors.

It seems surprising that in our experiment the algebraic error exhibits such a strongly oscillating pattern. This is related to the approximation of the eigenvalues of A in the course of the CG computations. In brief, since the solution x is very smooth, it is dominated by components corresponding to the lower frequency eigenvectors; see, e.g. [66, Section 3]. The CG method decreases these dominating components, so that with increasing iteration numbers the dominance is shifted towards the more oscillatory eigenvectors. We will return to the oscillation pattern of the CG error in Section 5.9.4.

In conclusion, this simple example demonstrates:

1. For the given discretisation there is some iteration count n for which the contribution of the discretisation and the algebraic errors to the total error are in balance. It is, however, crucial to evaluate such balance in compliance with the goal of the whole mathematical modelling. Using standard norms (like the energy norm) can give misleading results.
2. The local distribution of the algebraic error can be very different from the local distribution of the discretisation error.
3. Increasing the number of the algebraic iterations after the total error becomes acceptable increases the computational cost without obtaining a better approximation of the original solution.
4. To make the mathematical modelling process efficient, the construction of stopping criteria for the algebraic solver must be linked with the original problem that determines the appropriate function space.

Roache writes in the context of verification and discretisation error [530, p. 292]: 'I noted the 'myth of the converged solution,' in the sense that different variables can converge at different rates, and that it would always be possible to derive some error measure that is exquisitely sensitive to discretization error, so that this measure is far from converged even when other, more benign measures are well converged.' In our experiment we see a clear demonstration of the same for the algebraic error, and, moreover, this simple model problem challenges the standard approaches to measuring the iteration error based on norms.

The stopping criteria in iterative algebraic solvers should be linked, in an optimal case, with fully computable and locally efficient (on individual elements) *a-posteriori* error bounds that allow us to keep an appropriate balance between the discretisation and the algebraic parts of the error; see, e.g. the discussion in the book by Bangerth and Rannacher [36], the papers [517, 448, 11, 10, 92, 93, 47, 368, 563, 297], the habilitation thesis [646], the PhD thesis [459], and the references given there. Although this goal seems highly ambitious and is certainly very difficult to achieve, the near future will certainly bring new exciting results in that direction.

Having in mind the broader context described in this section, we will investigate in the following sections the cost-affecting factors *within the algebraic computations*. This can be considered a contribution towards the desired further work.

5.2 DIRECT AND ITERATIVE ALGEBRAIC COMPUTATIONS

Householder's classical textbook *Principles of Numerical Analysis* contains the following definition [345, p. 65]: 'A direct method for solving an equation or system of equations is any finite sequence of arithmetic operations that will result in an exact solution.' According to this definition, Krylov subspace methods that have the finite termination property are direct methods. And indeed, Householder discussed the CG method of Hestenes and Stiefel in the section 'Direct Methods' of his book.

There are numerous examples of similar treatments of CG and other Krylov subspace methods even in the very recent literature. This point of view, though formally correct and backed up by leading authorities such as Householder, can lead to a great deal of confusion when applying Krylov subspace methods for solving practical problems and, related to that, when trying to compare the cost of direct and iterative methods for solving linear algebraic systems.

To avoid confusion we purposely do not distinguish between direct and iterative *methods*, but between direct and iterative *computations*. In the following we will explain what is meant by these terms, and we will discuss the practical implications of this distinction.

In direct computations the input data are typically transformed in a predetermined and mostly data independent finite number of subtasks (such as elimination or factorisation steps), which need to be performed in order to get the desired result. In this way we can obtain the factors in matrix factorisations such as LU, Cholesky factorisation, or QR factorisation, and we can also compute the solution of $Ax = b$. In exact arithmetic, the exact solution is attained within a finite number of arithmetic operations. The essential point is that *intermediate steps do not produce a meaningful intermediate approximation* to the solution of the whole problem.

The situation is fully transparent in computations with full matrices. In the case of sparse matrices the individual subtasks can depend on the particular data, which also holds for the cost of solving the whole problem. Sparsity of the original data typically evolves during their transformations by creating new nonzero elements (matrix fill-in). This can considerably affect the cost of the individual subtasks. Nevertheless, for a given class of problems such as solving $Ax = b$ with A symmetric positive definite, the number of subtasks (elimination or factorisation steps) to be performed remains especially independent of the particular data (matrix A and the right-hand side b).

In iterative computations, a sequence of *approximate solutions* to the problem is constructed with the aim of stopping the computation whenever the approximation error, i.e. the difference between the computed approximation and the desired solution, is small enough. As pointed out in Section 5.1, the approximation error should be measured in an appropriate context-dependent way. This may reach far beyond the algebraic problem itself.

This principal difference between direct and iterative computations was clearly understood by leading researchers in the dawn of modern scientific computing; see, e.g. the paper [196, pp. 149–150], which was published before the appearance of the first Krylov subspace methods. As mentioned above, the finite termination property of these methods has led to some confusion about the nature of direct and iterative computations in the context of solving linear algebraic problems. Further historical details are given in Section 5.2.1. It is worth noticing that no such misunderstanding arose in the context of eigenvalue computations. This is apparently related to the fact that such computations are *in principle* iterative due to the Abel–Galois Theorem from 1824 (Abel) resp. 1830 (Galois); see, e.g. [688, Section 2.6.5]. For nice illustrations we refer to [213, 214].

Many practitioners who solve systems of linear algebraic equations resulting from discretisations of application problems prefer direct computations (i.e. computation using direct methods) over any form of iterative computations. For an early statement supporting this view we refer to the conclusions in the remarkable publication [163] from 1959. Direct computations are robust and easy to perform using licensed or even freely distributed software packages. In contrast to that, iterative computations must always be tuned to the problem, including the investigation of the proper preconditioner and stopping criteria. Such tuning can be costly (in particular because highly qualified human work is required) and time consuming. These facts are indisputable. There are, however, practical problems where direct methods with their predetermined numbers of necessary computational subtasks are simply *too expensive*. The borderline is not strict, and this motivates comparisons of direct and iterative approaches in the literature. Such comparisons occasionally also use, among valid points on robustness and simplicity of application which are in favour of the direct computations, arguments which are not fully relevant.

In particular, consistently with the prevailing approach to a-posteriori error estimates in the literature on numerical solution of PDEs, direct methods are often considered to be highly accurate. In other words, it is assumed that they *always* produce computed results with a negligible algebraic error. This is admitted in [580], where the editors mention that a part of the error analysis in solving elliptic PDEs is missing and that there is a need to investigate the accuracy of the related algebraic computations. As a consequence of considering direct computations highly accurate, an iterative computation is often compared with a direct one on the basis of the same (presumably high) accuracy level. This, however, misses the point that while in direct computations the accuracy of the output is *predetermined* by the input data and the method (including its particular implementation), in iterative

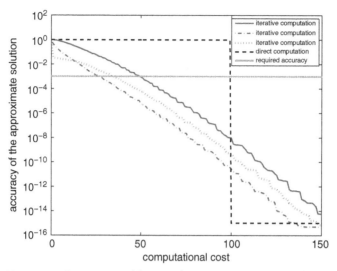

Figure 5.3 Comparison of direct and iterative computations when solving linear algebraic systems. The computational cost depends in iterative computations on the required accuracy and on the right-hand side b, which is illustrated by three convergence curves obtained for the same matrix and three different right-hand sides.

computations the accuracy of the output is *under control*, because the computation can be stopped whenever the properly evaluated accuracy is achieved.

Related to that, the cost of directly solving $Ax = b$ is essentially independent of the particular right-hand side b, and thus the cost essentially accounts for the *worst-case b*. This is not true for iterative computations. Since the right-hand side b represents some real-world conditions (e.g. discretised boundary conditions and outer forces in solving boundary value problems), it very rarely represents the worst-case data. Tuning the iterative computation to the particular data or class of data, in order to do 'justice to the inner nature of the problem'[1] can then lead to a significant advantage over the direct computation. The situation is illustrated schematically in Figure 5.3. Further discussions of this point are postponed to Section 5.4. A short but truly enlightening discussion of related issues is given in the section on robustness of codes in the monograph [530, pp. 360–361].

Another argument found in the literature states that the main advantage of iterative over direct computations is that the memory (mostly) and computational requirements are moderate. In iterative computations the matrix A need not be explicitly available, since usually only a function evaluating the multiplication with A is required. While this can indeed be a decisive advantage over direct computations, it is not what is typically meant by the given argument. The origin of the viewpoint can be found as early as in the 1950s, when the computer memory size was so restrictive that the low memory requirements of iterative algorithms used at

---

[1]. See the quote from the letter of Lanczos to Einstein from March 9, 1947, stated at the beginning of this book, which serves as its general motto.

that time made them competitive with direct computations, even for small problems; see, e.g. the comments of Hestenes [312, p. 171] and the essay of van der Vorst [635, p. 33]. It should, however, be pointed out that the memory and computational requirements of iterative computations are not automatically lower in comparison with direct computations. In iterative computations using long recurrences, such as in the GMRES method, the computational and memory requirements can become excessive, and the computations can fail unless a sufficiently accurate approximate solution is reached quickly enough.

Evaluating cost of iterative computations combines several different issues, each linked with specific approaches and tools:

- Computational cost of individual iterations, which can be constant per iteration, or iteration dependent.
- *A-priori* upper bounds for the computational cost that depend on characteristics of the problem to be solved. This gives basic information on whether the problem is efficiently solvable using the given approach, and it helps to identify a suitable combination of the iterative method and preconditioner. It should consider possible effects of rounding errors.
- *A-posteriori* evaluation of the computational error, its incorporation into the total error (see Section 5.1) and construction of the appropriate stopping criteria. This *must* include rounding error analysis. Otherwise, the a-posteriori error estimates may not be justified in real computations. Whenever possible, reliability of the computed results should not be based only on standard verifications on simple model problems.

After the following historical note we will discuss general aspects of these three issues. The rest of the book will then present analyses of computations using the CG and GMRES methods.

5.2.1 Historical Note: Are the Lanczos Method and the CG Method Direct or Iterative Methods?

We now complement the historical note in Section 2.5.7 by a discussion of the early understanding (before the work of Reid from 1971 [521]) of direct and iterative computations on the examples of the Lanczos method and the CG method.

As indicated in Section 2.5.7, the early researchers on Krylov subspace methods, in particular Lanczos, Hestenes, and Stiefel, thought of their methods as being *iterative*. Let us first consider Lanczos' point of view. Before he joined the Institute of Numerical Analysis (INA) in early 1949 he had spent three years as a Senior Engineer at Boeing Airplane Company in Seattle. He thus knew about industrial problems and the need for efficient computations. His INA project 11.1/1–49–AE2 (authorised on February 15, 1949) on solving linear algebraic equations was entitled

'*Approximate solution* of sets of arbitrary simultaneous algebraic equations' (our emphasis). The original project description given in the INA quarterly report [473] is as follows:

> *Objective:* To develop a practical and economical method for solving simultaneous algebraic equations by matrix iteration.
>
> *Background:* Such sets of equations arise repeatedly in all branches of mathematics.
>
> *Comments:* Related to Project 11.1/1–49–AE1 [Lanczos' second project 'Determination of characteristic values of matrices'] in that the same technique of iterating a trial vector to obtain a set of linear equations having a recurrent matrix is used. An approximation is introduced by truncating the iteration of the trial vector and thus obtaining a recurrent system of lower order than the one employed in Project 11.1/1–49–AE1.

One can see the basic ideas of the Krylov subspaces ('iterating a trial vector'), of computing an approximate solution only ('by truncating the iteration'), and even of the projected system ('a recurrent system of lower order'). It is hardly possible to give a more concise description of a research topic that was to keep generations of numerical analysts busy.

In his 1952 paper [406] on solving linear algebraic systems, Lanczos wrote that one can 'pursue our procedure as a sequence of successive approximations which may be terminated at any point where the residual has dropped down below a preassigned limit' (p. 38), and that 'we may expect fast convergence ... if the components of [the initial vector] b_0 in the direction of certain principal axes are small' (p. 40). He thus clearly advocated the idea of iterative methods; cf. also our quote from his paper in Section 2.6 above. His ideas were widely communicated in academia and industry. As an example one can consider Horvay's 100-page report of 1953 [343] that was published by the United States Atomic Energy Commission and (according to the Introduction) based on the authors' lectures at the General Electric Company, Brooklyn Polytechnic Institute, MIT, Brown University, and several other occasions. The report contains a complete presentation of Lanczos' methods for solving eigenvalue problems and linear algebraic systems as well as many numerical examples. Horvay stated the possibility of early termination when the residual norm is small enough (p. 69) and he included a numerical example using the 4×4 discrete 1D Laplacian $A = \text{tridiag}(-1, 2, -1)$ that showed the explicitly computed residual vectors r_1, r_2, r_3 and $r_4 = 0$ (Table II, p. 67).

In their 1952 paper [313], Hestenes and Stiefel treated the iterative nature of their own method in even greater detail than Lanczos did in [406]. Theorems 6:1 and 6:3 respectively show that the CG approximation minimises the energy norm of the error and that the Euclidean norm of the error decreases at every step. Immediately after these results one can read (see p. 417 of their paper):

> This result establishes the CG method as a method of successive approximations and justifies the procedure of stopping the algorithm before the final step is reached.

This point is illustrated by numerical examples in the paper's final section. For example, a table on p. 434 shows a comparison of the Euclidean error norms of CG and of Gaussian elimination in a computation with a 4 × 4 example matrix. Hestenes and Stiefel also referred to a numerical example from the manuscript of Hochstrasser's dissertation (supervised by Stiefel and Rutishauser and published 1954 [328]): 'The largest system that has been solved by the cg-method is a linear, symmetric system of 106 difference equations. The computation was done on the Zuse relay-computer at the Institute for Applied Mathematics in Zurich. The estimate obtained in the 90-th step was of sufficient accuracy to be acceptable.' In this case they did not mention, however, how the 'sufficient accuracy' was measured.

It is important to note that they warned against using the residual as a measure of the 'goodness' of the computed approximation [313, p. 410 and Theorem 18:3, p. 432]. In particular, they wrote:[2]

> However, this measure is not a reliable one because, as will be seen in Section 18, it is possible to construct cases in which the squared residual $\|r_n\|^2$ increases at each step (except for the last) while the length of the error vector $x - x_n$ decreases monotonically.

A numerical example illustrating this quote is shown in Section 5.9.3; see the CG residual norms shown in Figures 5.20 and 5.21. In addition Lanczos warned, in his almost forgotten paper on the application of Chebyshev polynomials in solving linear algebraic systems, that a small residual norm does not necessarily mean a small error, when he wrote that 'the mere smallness of the residual vector cannot guarantee the closeness of the solution' [407, p. 132]. It seems that these warnings were not seriously taken into account. Even Hochstrasser used the Euclidean residual norm to measure the convergence of CG in his experiments, and he stopped the iteration when this norm had dropped by a factor of 10^{-4} [328, p. 41]. Similarly, in a 1953 survey paper (containing 131 references), Forsythe wrote about the CG method that 'in practice the residual r_p may become so small for some $p < n$ that it is unnecessary even to complete the n steps theoretically required to compute $A^{-1}b$' [192, p. 316]. Unfortunately, the residual norm still appears to be the most popular measure of convergence in CG computations.

In spite of the remarks of Lanczos, Hestenes, and Stiefel on the iterative nature of their methods, the latter were considered *direct* by many researchers in the 1950s and 1960s. This was due to the methods' finite termination property, which in theory guarantees that they exactly solve a linear algebraic system with a finite number of operations. We have already seen this view of CG in the quotes from Householder's books given in Sections 2.5.7 and 5.2. Another example of this type can be found in the 1963 book of Faddeev and Faddeeva (Russian original of 1960), where CG is described as an 'exact' (i.e. direct) method that 'breaks down as soon as a certain residual r_k turns out to be zero, i.e. as soon as the approximation coincides with the exact solution of the system' [178, p. 393]. No other termination

---

2. We have adapted the notation in this quote to ours.

criterion or information on how to measure the error during the iteration is given. In fact, we find on p. 396: 'Computational control is achieved by computing the scalar products (s_i, As_i) or (r_i, r_j), which should equal zero.' (Here the vector s_i denotes the ith direction vector of CG.)

Since roundoff destroys the finite termination property in practical computations, the methods of Lanczos, Hestenes, and Stiefel were treated with suspicion in particular by mathematicians (cf. also the quotes given in Section 2.5.7). Another aspect certainly added to the fact that the true iterative nature of the methods was not fully realised early on. Large and sparse linear systems as those we routinely solve today did not appear when the papers of Lanczos, Hestenes, and Stiefel were published. It was the beginning of the computer area, with just six 'large scale digital computers' in existence in July 1947 [122].

To judge what 'large scale' meant at that time, consider that in the summer of 1949 the computation of eigenvalues and eigenvectors of an 8×8 symmetric matrix using a method of Hestenes and Karush took no less than 11 hours on the IBM CPC (Card Programmed Calculator) [314, p. 19]. The first fully operational stored-program electronic computer in the USA, called SEAC (Standards Eastern Automatic Computer), started its operation in May 1950. SEAC was constructed over a period of almost 15 months by the Machine Development Laboratory of the US National Bureau of Standards in Washington DC, and it preceded the delivery of the first (self-storage) commercial computer by one year [385]. Initially SEAC had 3072 bytes of internal memory. Loading a 16×16 matrix and a Monte Carlo-type program for inverting it into memory took 21 minutes, printing out the result took 11 minutes in 1951 [613]. The INA researchers in Los Angeles started having hands-on computer experience in August 1950, when the SWAC (Standards Western Automatic Computer) became operational. But due to budget constraints this machine, developed on the basis of the original 1945 design by Turing, only reached an acceptable level of reliability in 1953 [18]. This is probably why the only 'large scale' example in the paper of Hestenes and Stiefel was computed in Zürich and not in Los Angeles, where the paper was written.

The widespread considerations about Krylov subspace methods until the 1971 paper of Reid become apparent from the following two quotes. In 1951, Rosser, from 1949–1950 head of the INA, summarised discussions at his institute on the CG method [535, p. 64]: 'If at the end of n [$=$ size of A] steps, the r_n is not sufficiently small (either from round off or from mistakes), there is nothing whatever to bring the process to a termination. One can simply let the procedure continue until r_m is sufficiently small, even if this may in some cases require a value of m greater than n. . . . It thus appears that in actual computation, our schemes are iterative schemes which will not ordinarily terminate in any finite number of steps, but which may be expected to give a very close approximation in n steps.' Almost the same viewpoint was expressed by Ginsburg in 1971 [672, pp. 61–62]: 'Since roundoff errors prevent termination after n steps as stated by the theory, and consequently the CG-method has to be treated as if it were an infinite iterative process, proper termination is not trivial.'

For completeness it should be added that between the original works of Lanczos, Hestenes, and Stiefel published between 1950 and 1952, and the 1971 paper of

Reid there were some exceptional, though not properly recognised works, that emphasised the iterative nature of CG and related methods. One example is the presentation in Vorobyev's moment-oriented functional-theoretic approach from 1958 [651]; see Section 3.7 for more details. Another remarkable example is (despite some points bearing the sign of the time of its writing) the comprehensive exposition of Engeli, Ginsburg, Rutishauser, and Stiefel on iterative solution of self-adjoint boundary value problems from 1959 [163]. Rutishauser's description of the theory of gradient methods represents a truly fundamental piece of work, which should be placed next to the founding papers of Hestenes, Stiefel, and Lanczos. In addition, Ginsburg's contribution contains what may be the first published convergence plots of CG and other iterative methods, with different convergence characteristics (Euclidean norm of the residual, A-norm, and Euclidean norm of the error) plotted against iteration steps and computing time. One should also mention the 1958 paper of Stiefel [581], which, although written independently, develops several ideas presented already in Lanczos' paper of 1953 [407].

5.3 COMPUTATIONAL COST OF INDIVIDUAL ITERATIONS

In matrix computations the price is typically identified with the memory requirements and the execution time of the algorithm that accomplishes the task at hand. Here we put the memory requirements aside (some of their aspects were briefly discussed in Section 5.2) and focus on the execution time of the *individual iterations*, with some comments relevant to direct computations. Among the many factors that determine the timing, the following are particularly important:

(a) How well does the computation match the given computer resources?
(b) How many computer arithmetic operations are required?

Remark 5.3.1
Nowadays, many tasks within computers and their parts are performed concurrently. This leads to a significant level of parallelism in performing computations, ranging from the execution of individual arithmetic operations within an individual processor to large-scale numerical simulations. The state-of-the-art analysis of the limitations of parallel processor architectures and views on possible future developments can be found, e.g. in [460]. Parallel computer architectures and algorithms have been thoroughly studied since the 1960s; an overview of computer architectures and the related principles of building up efficient algorithms can be found, e.g. in the monograph [352]. Algorithmic developments and using parallel computer architectures for solving various real-world problems are documented, e.g. in the proceedings [611, 556, 69]. For a remarkable survey of the early algorithmic developments within numerical linear algebra we refer to [305]. An interesting method for performance evaluation of parallel computers and algorithms was developed by Hockney [333, Section 1.3]. It took into account the speed of computer arithmetic as well as the overhead needed for the transfers of data and communication in general. Using measurements performed on real machines together with identification of parameters of models of parallel computations, it allowed accurate estimation of the performance of particular methods and

their implementation on the given computer architectures [329, 593, 330, 332, 331]. Since algorithmic issues related to parallel computer architectures are out of the scope of this book, we will not consider them further. For additional overviews we refer to [132, 141, 142], and for an example of the state-of-the-art developments, to [620] and the references given there.

Factor (a) above depends to a large extent (in both the sequential and the parallel context) on the effectiveness of using the computer's memory in relation to the performed arithmetic operations, and, in particular, on the effectiveness of exploiting the memory cache system. The key issue is the probability that, with requesting some data from the main memory, the subsequent memory requests will concern some data within a sufficiently local neighbourhood (measured by the distance in the address space). The locality of the processed data is not a static feature of the input data, but a dynamic feature that is determined by the input data and the given algorithm. In spite of these complications, there is no principal problem in evaluating (a). It is only a matter of technicalities, of the level of detail of the particular implementation and computer architecture we want to work with, and last but not least, of the time which we wish to invest.

Factor (a) is of particular importance for evaluating the computational cost of direct computations applied to sparse matrices. Modern algorithms in this area are based on elaborate factorisation techniques that only perform arithmetic with nonzero entries and aim at preserving sparsity during the computation. This typically requires a lot of non-numeric processing and sophisticated data handling. Since incomplete factorisations are often used as preconditioners, factor (a) in practice plays an important role also for iterative methods. Moreover, the performance of Krylov subspace methods depends crucially on the speed of computing the required matrix–vector products. This in turn depends on the speed of data communication, both within the memory system of a single processor, or between multiple processors in a parallel system.

Because data communication often represents a bottleneck in large-scale applications, there has been an increasing interest in 'communication-avoiding' formulations of algorithms in dense as well as sparse linear algebra. In direct computations with dense or some particularly structured matrices the task is substantially different from the case of *general sparse matrices*, where the dependence of the number of arithmetic operations to be performed on input data (intermediate results) represents an additional challenging complication. In iterative computations, the communication-avoiding approach is based on working with *sufficiently large blocks of iterates*. This couples

- the efficiency of performing arithmetic operations within the sufficiently large blocks with
- the (nontrivial) assumption that the convergence of the iterative computation and stopping criteria can be handled using blocks of iterates as the smallest possible units instead of individual iterations. Replacing individual iterations by blocks can in general affect the efficiency due to performing extra arithmetic operations which may in fact not be needed.

Any such approach should therefore be compared with the standard approaches that focus on individual iterations and use, as the key ingredient, powerful (often problem-inspired) preconditioners.

A detailed study of communication-avoiding implementations of Krylov subspace methods can be found in Hoemmen's PhD thesis of 2010 [334]; see the related conference paper [462] for a brief summary. A numerical stability analysis and a theoretical as well as a comprehensive practical comparison with the stability and convergence properties of standard implementations, which would include solving real-world mathematical modelling based problems, is yet to be done. Therefore the question of whether or not the communication-avoiding approach can be competitive in solving practical problems is, in our opinion, open.

Factor (b), i.e. the number of computer arithmetic operations, is in the sequential context approximated by the number of *floating point operations (flops)*; see, e.g. [250, Section 1.2.4], or [315, Section 1.1]. When measured in flops only, the computational cost of iteratively solving $Ax = b$ is therefore equal to the sum of the flops required for the individual steps until the computation is stopped. Unless the computation fails, the iteration is stopped when the algorithm has produced an approximation with the error *within the user-specified tolerance*. This tolerance is determined by the choice of the measure for the method's performance on the one hand, and the accuracy level that should be reached on the other. As explained on the model problem in Section 5.1, both the convergence measure and the accuracy level should be considered in relation to the problem (or the class of problem) that is to be solved.

The flops for each iteration step can be determined from the description of an iterative algorithm. Many publications contain tables with the number of arithmetic operations for a single step of different iterative methods. This resembles a view that does not take into account the difference between direct and iterative computations; see Section 5.2. It is important to distinguish between computations based on short recurrences (with fixed computational cost and memory requirements per iterative step) on the one hand, and long recurrences on the other. Apart from that, the tables mentioned above offer very little information about the cost of practical computations. They ignore the problem-dependent cost of the matrix–vector products as well as the cost of preconditioning, which can be dominating. We feel that presenting tables of arithmetic operations for a single step of various iterative methods may create a misleading impression that iterative computations can simply be performed by assembling individual iteration steps without thorough investigation of the particular methods and their tuning to the given problems. This is dangerous especially for students and nonspecialists who have little experience with solving difficult practical problems.

Our viewpoint emphasises *efficiency* in solving challenging practical problems. If the problem is easy to solve, then the matter of cost is trivial and efficiency is not an issue. It becomes essential in cases where the cost of computations is so high that it limits the whole mathematical modelling of the investigated phenomena. Then the cost must be kept under control, and this must be done without compromising the desired accuracy of computed results.

5.4 PARTICULAR COMPUTATIONS AND COMPLEXITY

Linear algebraic systems arise in a huge variety of applications and have vastly different properties. Often these properties are inherited from the underlying application through the mathematical model and its discretisation. The matrix A may be symmetric or unsymmetric, dense or sparse, have special structural features, like symmetry with respect to some inner product or bilinear form, or it may be well- or ill-conditioned. If an ill-conditioned matrix A results from the discretisation of a differential operator, then small singular values and their singular vectors correspond in many cases to low frequencies (in the sense that the singular vectors resemble the standard low frequency Fourier trigonometric basis functions); see, e.g. [66] or [290, Chapter 4]. If, on the other hand, the matrix A arises from the discretisation of an integral operator with a sufficiently smooth kernel, then large singular values and their singular vectors correspond to low frequencies, while the small singular values and their singular vectors to high frequencies; see, e.g. [295, Section 2.5], [296, 325]. In the latter case the original infinite-dimensional problem is often ill-posed.

Naturally, it turns out that an iterative method can be highly efficient for one linear algebraic system, and hopelessly inefficient for another. Using three standard Krylov subspace methods for general nonsymmetric matrices (GMRES [544], CGS [576], and CGN [313]), Nachtigal, Reddy, and Trefethen [469] gave a convincing numerical illustration of this fact by a set of eight well chosen (though artificial) examples of linear algebraic systems. The difference in the behaviour of the three iterative methods when applied to these systems is illustrative. It ranges from 'all methods good' over 'method X wins' and 'method X loses' (for each of the three methods), to 'all methods bad'. This means that in the experiments of Nachtigal, Reddy, and Trefethen the different methods required significantly different numbers of steps until they reached the specified stopping tolerance. The conclusion is that the three algorithms are 'genuinely distinct in their behavior' [469, p. 793]. These experiments, and many other similar ones that have been reported in the literature, support the following point of view:

> When we think of the computational cost of iteratively solving a linear algebraic system $Ax = b$, we always consider a *particular* method (in an appropriate implementation) with a *particular* given data A, b, and x_0.

This viewpoint represents a significant conceptual difference to the classical task of *complexity theory* (of computer science), which is the examination of certain problem classes and the determination of *lower bounds* on the cost of *any* algorithm for the solution of *every* problem in the class. Developments in complexity theory and its relationship with computational mathematics are described, for example, in [46, Section 2.4], [65], and the many references given there. A typical problem class studied in the numerical linear algebra context is 'solve a linear algebraic system with a general $(N \times N)$-matrix'. It can be shown that this class has the same *complexity*, usually denoted by N^ω, as the class 'multiply two general

$(N \times N)$-matrices'; see, e.g. [90, Chapter 16]. The elusive best possible constant ω is called the *exponent of matrix multiplication*. An algorithm devised by Coppersmith and Winograd in 1990 [119] shows that ω is at most 2.376. On the other hand, ω must be at least 2, since all N^2 entries of the general $(N \times N)$-matrix must be a part of the computation. In the early 2000s, group-theoretic methods devised by Cohn, Uhmans, et al. [117, 118] have revived the area of fast matrix multiplication; see [532] for a brief summary. A numerical stability analysis of the corresponding algorithms has been performed in [131]. The paper [336] gives a survey of complexity theory and the related stability analysis for matrix multiplication and similar problem classes.

Note that here we again see a concept developed for direct computations. By 'solve' in the above problem description one means to compute (in exact arithmetic) the exact solution. In iterative computations the exact solution may be not attainable (even theoretically), and if it is, as in Krylov subspace methods, there is typically no reason whatsoever to compute it. As argued in Sections 5.1–5.3, an iterative computation should stop when the error drops below the acceptable accuracy. Therefore there is hardly any meaningful lower bound on the cost of iterative computations (as in the classical complexity theory). The cost is strongly data dependent.

More relevant for our context is the *complexity theory of numerical analysis*. According to Smale, this theory studies *upper bounds* for 'the number of arithmetic operations required to pass from the input to the output of a numerical problem' [575, p. 523]. Here the subject is the examination of basic algorithms rather than problem classes. (In theoretical computer science, this is sometimes called an algorithm's *run-time analysis*.) We give a brief illustration considering factorisation methods for solving $Ax = b$ with a general *dense* $(N \times N)$-matrix A. It is well known that the LU factorisation with partial pivoting costs $2N^3/3 + \mathcal{O}(N^2)$ flops, while the QR factorisation costs $4N^3/3 + \mathcal{O}(N^2)$ flops. When A is symmetric and positive definite, the Cholesky factorisation can be applied, which costs $N^3/3 + \mathcal{O}(N^2)$ flops. After each such factorisation, a linear algebraic system with an upper triangular matrix needs to be solved, which costs an additional $\mathcal{O}(N^2)$ flops. The computed solution $\widetilde{x}$ is available only after the last computational step is performed. Unlike in iterative methods, there are no intermediate approximations. Consequently, a direct solve of $Ax = b$ with a general dense matrix A and any of the mentioned methods costs $\mathcal{O}(N^3)$ flops. This upper bound on the number of operations is the *complexity* of these methods in Smale's definition; see [575, p. 534].

Related concepts have been studied in the context of iterative methods. For example, Chou [112] measured the 'goodness' of an algorithm by the number of required steps n so that the error $x - x_n$ is in a given norm smaller than a given tolerance ϵ. An iterative algorithm is called (strongly) optimal if it achieves the minimal number of steps to compute an ϵ-approximation x_n of the solution *for every matrix in a given class and every possible right-hand side b*. It is shown in [112] (see [629, Sections 19.4–19.5] for summary of related results) that if the error is measured by the Euclidean residual norm, i.e. $\|b - Ax_n\| = \|x - x_n\|_{A*A}$, then the algorithms based on Krylov subspaces are (in some sense) almost optimal when compared with algorithms based on subspaces spanned by any vectors of the form $b, Az_1, \ldots, Az_{n-1}$ (for $z_0 = b$ and $z_j = Az_{j-1}, j = 1, \ldots, n-2$, this gives the Krylov subspace $\mathcal{K}_n(A, b)$). The conclusion drawn in [112] is that 'Krylov information has almost minimal complexity'.

These approaches are of some theoretical interest, but, as Smale points out, 'where iterative methods are needed, the complexity theory is not fully developed' [575, p. 533]. Our discussion here is intentionally *not* directly related to this underdeveloped topic, which seems to be, for the reasons presented above (see also Sections 5.1–5.3), too involved. Rather, as pointed out above, we focus on the *cost of particular computations*. We emphasise:

> Our concept of the computational cost is not attributed to a method or its algorithmic realisation, but to the *application* of the method in the given algorithmic realisation to a given particular data.

In computational sciences and engineering the success or failure of an approach often depends on whether hidden structural dependencies within the given problem can be uncovered and exploited. If this is possible, the particular problem that reflects a real-world situation can be efficiently solvable even when the generalised abstract (worst-case) mathematical problem of the same class is not. This point fully applies, in our opinion, also to problems in numerical linear algebra, where the iterative computations provide the most convincing examples; we again refer to Lanczos' letter to Einstein.

5.5 CLOSER LOOK AT THE CONCEPT OF CONVERGENCE

In the text above we have pointed out that any evaluation of the computational cost of iterative computations must consider their *convergence behaviour*. On some examples we will now examine different understanding, descriptions, and approaches to the analysis of convergence.

5.5.1 Linear Stationary Iterative Methods

Let us start with a brief look at the classical linear stationary iterative methods of the first order described, e.g. in the works of Young and Varga [683, 680, 639, 684, 682]. These methods are based on the idea of *splitting* of the matrix (operator) A into an easily invertible part and the rest. Suppose that $A = K - L$ is such a splitting, where K is invertible. Then $Ax = b$ is equivalent to

$$x = K^{-1}Lx + K^{-1}b,$$

which leads to the *stationary iterative method*

$$x_n = K^{-1}Lx_{n-1} + K^{-1}b, \quad n = 1, 2, \ldots$$

There are numerous variations and extensions of this general scheme. Many of them are based on so-called relaxation techniques, with the *Successive Overrelaxation*

Method (SOR), introduced simultaneously in 1950 by Frankel [199] and Young [683], forming an important example.

All these methods have in common that the nth error satisfies

$$x - x_n = M^n(x - x_0), \quad n = 1, 2, \ldots,$$

where M is the *iteration matrix* ($M = K^{-1}L$ in the simplest case). Since the nth error is given by the product of the nth power of M and the initial error, the behaviour of these iterative methods after some 'transient phase' is essentially *linear*. Apart from trivial cases, even in exact arithmetic these methods do not terminate with the exact solution of the given system. Therefore it is fully justified to analyse their convergence behaviour using asymptotics and to study limits for $n \to \infty$.

Much of the theory in this context is concerned with determining the corresponding linear *rate of convergence*; see, e.g. the description in [291, Section 3.3.2]. Note that for consistent vector and matrix norms $\|\cdot\|$,

$$\frac{\|x - x_n\|}{\|x - x_0\|} \leq \|M^n\|.$$

The quantity

$$\sigma_n \equiv \left(\frac{\|x - x_n\|}{\|x - x_0\|}\right)^{1/n} \leq \|M^n\|^{1/n} \qquad (5.5.1)$$

is called the *average reduction factor* per iteration of the successive error norms. If $\|M^n\| < 1$, one can define the *average rate of convergence* for n iterations with the matrix M by

$$R(M^n) \equiv -\log\left(\|M^n\|^{1/n}\right) = \frac{-\log(\|M^n\|)}{n},$$

so that $\sigma_n \leq e^{-R(M^n)}$. Then the *asymptotic rate of convergence* is given by

$$\lim_{n \to \infty} R(M^n) \equiv -\log(\rho(M)),$$

where $\rho(M)$ is the spectral radius of the iteration matrix M. Varga's book [639] gives a thorough treatment of the relevant theory of linear stationary and semi-iterative methods, in particular the Chebyshev semi-iterative method, which will be described in the following subsection. Our exposition in the following greatly benefits from the contribution of Rutishauser in [163, Chapter II, pp. 24-49]. Nice descriptions can also be found in [684, 292, 22] and [291, Chapter 7].[3]

---

3. In his lectures on iterative methods at Stanford University, Gene Golub used to develop the CG method by first explaining the role of linear iterations, then considering the Chebyshev acceleration, and finally stating CG using a formally similar setting but imposing a crucial CG global minimisation property. We thank Lek Heng Lim for providing us with a record of Golub's lectures.

5.5.2 Richardson Iteration and Chebyshev Semi-iteration

Since in later sections we will describe the relationship between the Chebyshev semi-iterative method and the CG method, we here consider $Ax = b$ with A symmetric positive definite. It needs to be pointed out, however, that linear stationary methods and the Chebyshev semi-iterative method are by no means restricted to this class; see, e.g. the second (revised and expanded) edition of Varga's book [640, Chapter 5] and Young's book [684, Chapter 11].

Suppose that, using the previous notation, we split $A = K - L$ with $K \equiv \omega I$, $L \equiv \omega I - A$, and ω being a nonzero real parameter. This yields the stationary iterative method

$$x_n = M x_{n-1} + \omega^{-1} b, \quad n = 1, 2, \ldots, \quad M \equiv I - \omega^{-1} A. \tag{5.5.2}$$

Denote by $0 < \lambda_1 < \lambda_N$ the extremal eigenvalues of A. Then a simple computation shows that the parameter $\omega = (\lambda_1 + \lambda_N)/2$ minimises the spectral radius of the iteration matrix M. The iteration with this choice of ω is called the *stationary Richardson iteration*. Its iteration matrix satisfies

$$\rho(M) = \frac{\lambda_N - \lambda_1}{\lambda_N + \lambda_1} = \frac{\kappa(A) - 1}{\kappa(A) + 1}, \quad \kappa(A) = \frac{\lambda_N}{\lambda_1}.$$

If A is ill-conditioned, i.e. $\lambda_N \gg \lambda_1$, then $\rho(M)$ is very close to 1 and the convergence of this iteration can be exceedingly slow.

The nth error of the stationary Richardson iteration can be written as

$$x - x_n = M^n (x - x_0) = (I - \omega^{-1} A)^n (x - x_0) \equiv \varphi_n^R(A)(x - x_0),$$

where the *iteration polynomial* $\varphi_n^R(\lambda) \equiv (1 - \omega^{-1}\lambda)^n$ is of degree n and satisfies $\varphi_n^R(0) = 1$. The only (n-fold) root of this polynomial is at the midpoint of the interval $[\lambda_1, \lambda_N]$ that contains all eigenvalues of A. Already Richardson in his paper of 1911 [524] observed that faster convergence may be obtained by considering a polynomial of the form

$$\varphi_n(\lambda) = (1 - \omega_1^{-1}\lambda) \cdots (1 - \omega_n^{-1}\lambda),$$

where the values $\omega_1, \ldots, \omega_n$ can be distinct. However, he did not provide a mathematical procedure for choosing these values.[4]

In the early 1950s, several researchers discovered independently that using the n roots of the nth shifted Chebyshev polynomial of the first kind, defined by

$$\chi_n(\lambda) = \cos\left(n \arccos\left(\frac{2\lambda - \lambda_N - \lambda_1}{\lambda_N - \lambda_1}\right)\right) \quad \text{for} \quad \lambda \in [\lambda_1, \lambda_N], \tag{5.5.3}$$

---

[4]. In fact, Richardson proposed to determine a set of values by drawing the graph of the polynomial for some arbitrary values and then 'altering them or adding new ones until the maxima and minima of the curve were all sufficiently small', and he suggested distribution of the values 'fairly uniformly' [524, p. 321].

represent a good choice for the values $\omega_1, \ldots, \omega_n$ (historical details are given in Section 5.5.3). When the n roots of $\chi_n(\lambda)$ are used as parameters $\omega_1, \ldots, \omega_n$, the iteration polynomial is given by the scaled and shifted Chebyshev polynomial $\chi_n(\lambda)/\chi_n(0)$. The resulting iterative method, which can be implemented using the well-known three-term recurrence relation of the Chebyshev polynomials, is called the *Chebyshev semi-iterative method*. It is sometimes interpreted as an acceleration of the Richardson iteration by means of Chebyshev polynomials.

The justification for using Chebyshev polynomials comes from the fact that the scaled and shifted Chebyshev polynomial $\chi_n(\lambda)/\chi_n(0)$ represents the unique solution of the minimisation problem

$$\min_{\substack{\varphi(0)=1 \\ \deg(\varphi) \leq n}} \max_{\lambda \in [\lambda_1, \lambda_N]} |\varphi(\lambda)|.$$

In other words, among all nth degree polynomials that are normalised at zero, the scaled and shifted Chebyshev polynomial has the minimal maximum norm on the smallest interval containing the eigenvalues of A (recall that $0 < \lambda_1 < \lambda_N$). The original proof of the minimising property of the Chebyshev polynomials is due to Vladimir Markov [446], but the standard reference[5] for the proof in the numerical analysis literature is the 1950 paper of Flanders and Shortley [189]; see [684, Theorem 3.1, pp. 303–304]. For a thorough coverage of the extremal properties of Chebyshev polynomials we refer to [528, Chapter 2] (see in particular Theorem 2.1 on p. 56 and Theorem 2.20 with the subsequent inequality (2.37) on pp. 92–93).

Let us now look at the error of the Chebyshev semi-iterative method. We recall that the shifted Chebyshev polynomial defined in (5.5.3) satisfies

$$|\chi_n(\lambda)| \leq 1, \quad \lambda \in [\lambda_1, \lambda_N], \quad \chi_n(\lambda_N) = 1,$$

and

$$\chi_n(0) = \frac{1}{2}\left[\left(\frac{\sqrt{\kappa(A)} - 1}{\sqrt{\kappa(A)} + 1}\right)^n + \left(\frac{\sqrt{\kappa(A)} + 1}{\sqrt{\kappa(A)} - 1}\right)^n\right].$$

This equation follows easily from the well-known alternative representation (see, e.g. [528, Section 1.1])

$$\chi_n(\gamma) = \frac{1}{2}\left((\gamma + (\gamma^2 - 1)^{\frac{1}{2}})^n + (\gamma + (\gamma^2 - 1)^{\frac{1}{2}})^{-n}\right),$$

where we have used the change of variable $\gamma \equiv \frac{2\lambda - \lambda_N - \lambda_1}{\lambda_N - \lambda_1}$ for ease of notation. For later comparisons with the CG method we now use the A-norm, but we note that in

---

5. In 1960, Forsythe and Wasow [194, p. 227] exclaimed: 'Markoff's result is continually being rediscovered by enterprising numerical analysts!'

the context of semi-iterative methods the error can be evaluated in any consistent norm. To obtain a convergence bound we start with

$$\|x - x_n\|_A = |\chi_n(0)|^{-1} \|A^{1/2} \chi_n(A)(x - x_0)\|$$
$$\leq |\chi_n(0)|^{-1} \|\chi_n(A)\| \|x - x_0\|_A,$$

where $\|\cdot\|$ denotes the 2-norm. An application of the above relations for the Chebyshev polynomial gives

$$\frac{\|x - x_n\|_A}{\|x - x_0\|_A} \leq |\chi_n(0)|^{-1} \|\chi_n(A)\|$$

$$= |\chi_n(0)|^{-1} \max_{1 \leq j \leq N} |\chi_n(\lambda_j)| \qquad (5.5.4)$$

$$\leq |\chi_n(0)|^{-1} \max_{\lambda \in [\lambda_1, \lambda_N]} |\chi_n(\lambda)| = |\chi_n(0)|^{-1} \qquad (5.5.5)$$

$$= 2 \left[\left(\frac{\sqrt{\kappa(A)} - 1}{\sqrt{\kappa(A)} + 1} \right)^n + \left(\frac{\sqrt{\kappa(A)} + 1}{\sqrt{\kappa(A)} - 1} \right)^n \right]^{-1} \qquad (5.5.6)$$

$$\leq 2 \left(\frac{\sqrt{\kappa(A)} - 1}{\sqrt{\kappa(A)} + 1} \right)^n. \qquad (5.5.7)$$

This bound can already be found in Rutishauser's wonderful exposition of theory of gradient methods published in 1959 [163, formulas (II.23) and (II.32)]. Let us summarise the previous results.

Theorem 5.5.1
If the Chebyshev semi-iterative method with initial approximation x_0 is applied to $Ax = b$, where A is symmetric positive definite, then the nth approximation x_n satisfies

$$\frac{\|x - x_n\|_A}{\|x - x_0\|_A} \leq 2 \left(\frac{\sqrt{\kappa(A)} - 1}{\sqrt{\kappa(A)} + 1} \right)^n. \qquad (5.5.8)$$

We see that the Chebyshev semi-iterative method potentially has a significantly smaller asymptotic convergence factor than the stationary Richardson iteration. The factor in the case of Richardson iteration contains the condition number $\kappa(A)$, while for the Chebyshev semi-iterative method it is the square root of this number.

5.5.3 Historical Note: Semi-iterative and Krylov Subspace Methods

Flanders and Shortley in 1950 [189] were probably the first who used Chebyshev polynomials for the convergence acceleration of iterative methods; see [684, Supplementary discussion on pp. 385–386]. Since they were mainly interested in eigenvalues, they applied their techniques to a homogeneous system with a singular matrix. Shortly after their work, numerous publications on the use of Chebyshev polynomials in the context of iterative methods followed. A 1950 paper of Gavurin written in Russian was translated into English in 1952 [229]. Young made an important contribution to this line of research in 1954 [681] and in several further papers in the mid 1950s. In [681, Section 2] he essentially derived what we have shown in Theorem 5.5.1 above (without explicitly stating the handy formula on the right-hand side of (5.5.8)). Independently of Young, Lanczos applied Chebyshev polynomials for the 'preliminary purification' of the initial error in his iterative method published in 1952 [406, Section 5, in particular pp. 42 and 44]. According to a footnote on p. 42, the idea of using Chebyshev polynomials for the solution of linear systems had been suggested to him by Kincaid in 1947.

Another study of Lanczos, dealing with the application of Chebyshev polynomials in the solution of large-scale linear algebraic systems, appeared in 1953 [407]. As mentioned several times above, this is a fascinating piece of work, which describes polynomial preconditioning, warns against using the residual as a measure of convergence, and discusses the problems of noise amplification and emphasises the importance of scaling the system matrix. As noted in the commentary by Saad [410, pp. 3-527 to 3-528], it has been rarely referenced in later works. In addition to these researchers, von Neumann derived a three-term recurrence method for Hermitian matrices based on Chebyshev polynomials including its convergence analysis. His handwritten (undated) notes were published posthumously in the appendix of the 1958 report [63] and again in the 1959 paper [64].

It is interesting to note that polynomial preconditioning was described even earlier by Lamberto Cesari in his papers of 1937 [99, 100] on the general theory of linear stationary iterative methods based on splittings. His papers were published in Italian and they were referenced in several influential later works including the books by Householder [346] and Faddeev and Faddeeva [178].[6]

A simple rearrangement shows that the iterates of the stationary Richardson iteration (5.5.2) can be written as $x_n = x_{n-1} + \omega^{-1} r_{n-1}$, where $r_{n-1} = b - A x_{n-1}$. Analogously, the nth approximation in the Chebyshev semi-iterative method can be written as the $(n-1)$th approximation plus a more sophisticated linear combination of the previous residual vectors. Rutishauser [163] and others used the term 'gradient method' in this context, since the residual r_{n-1} corresponding to the approximation x_{n-1} is the negative gradient of the quadratic form

$$\mathcal{F}(z) = \frac{1}{2} z^* A z - z^* b \text{ evaluated at } x_{n-1};$$

see (2.5.33) in the derivation of CG in Section 2.5.2. The term 'semi-iterative method' was coined by Varga in 1957 [638].

Early on, the SOR method was considered independently of the Chebyshev semi-iterative method. For example, Young's two papers of 1954 ([680] on SOR and [681] on Chebyshev semi-iteration) referenced each other, but they

---

6. We thank Michele Benzi for sharing his knowledge on this topic and for providing copies of the original works by Cesari.

developed the two approaches parallel to each other and with few connections. As Varga in [638], Golub compared the two types of methods in his PhD thesis of 1959 [241]. Varga's visit to the University of Illinois at the invitation of Golub's thesis advisor Taub led to the two seminal papers of 1961 [251, 252], which showed that the Chebyshev semi-iterative method and Young's SOR method 'are from a certain point of view remarkably similar' [251, p. 148]. For further details we refer to the original papers as well as the remarks of Varga and the commentary of Greenbaum in Golub's Selected Works [243, pp. 35–45]. That book also contains reprints of [251, 252].

It is a historical coincidence that the Chebyshev semi-iterative method, the SOR method, and the first Krylov subspace methods were independently developed at almost the same time. In particular, the Flanders and Shortley paper [189], Young's thesis [683], and Frankel's paper [199] on SOR, and Lanczos' paper on eigenvalue computations [405] were all published in 1950. The 1950s and 1960s then saw a stormy development of stationary and semi-iterative methods, which is well documented by Varga's book of 1962 [639]. By this time, due to unfortunate misunderstandings and somewhat surprisingly, Krylov subspace methods had not yet entered the main stream of numerical mathematics and scientific computing. As explained in the historical notes in Sections 2.5.7 and 5.2.1, this only happened after the 1971 paper of Reid.

It is interesting to read that in 1958 even Stiefel considered the approach using Chebyshev polynomials (he mentioned Flanders and Shortley and explicitly cited Young [681]) 'most successful' if A is well conditioned and a good lower bound on the smallest eigenvalue is known [581, p. 11]. On the contrary, the conclusions in the seminal paper of Engeli, Ginsburger, Rutishauser, and Stiefel [163] are entirely in favour of the CG method. In his contribution to this work, Rutishauser discussed in detail polynomial preconditioning (with references to [189] and [406]) under the name 'smoothing', which preprocesses the original problem in order to filter out the influence of the large eigenvalues; see [163, pp. 35–38]. He then suggested eliminating the effect of the smallest eigenvalues by a further preprocessing using Chebyshev polynomials and thereby considered 'deflation techniques' many decades before they appeared in the main stream of computational mathematics; see the end of Section 5.6.4 for some references on such techniques.

The fact that the development of stationary and semi-iterative methods for some time overshadowed Krylov subspace methods nicely demonstrates that mathematics, as almost all human activities, is subject to temporary fashions. In the late 1950s and early 1960s, one rarely finds balanced coverage of both types of iterative computations in a single publication. For example, the Hestenes and Stiefel paper is not even cited in [639]. (An outstanding exception is [163].) The fascinating wiggles in the historical development of algebraic iterative computations are also related to the development of numerical analysis of partial differential equations. Here the method of finite differences dominated the field for several decades. One can observe a change in the trend in favour of the finite element method that occurred in the late 1970s and early 1980s, and hence at about the same time that Krylov subspace methods became more popular.

5.5.4 Krylov Subspace Methods: Nonlinear Methods for Linear Problems

The following general remarks about the nature of Krylov subspace methods are useful to keep in mind when reading the rest of the book, which will focus on the specific methods CG and GMRES.

As described in Chapter 2, in a Krylov subspace method we typically have

$$x_n \in x_0 + \mathcal{K}_n(A, r_0), \quad n = 1, 2, \ldots$$

The approximation x_n is determined by an orthogonality condition of the form

$$r_n = b - Ax_n \perp \mathcal{C}_n,$$

where $\mathcal{C}_n$ is a given n-dimensional subspace. Because of this projection, well-defined Krylov subspace methods have the *finite termination property*. Hence their (exact precision) behaviour at each individual step anticipates, in some sense, the whole computation until the very end when the exact solution is to be found. It is immediately clear that the analysis of Krylov subspace methods requires mathematical tools that are very different from the ones used for the linear stationary or semi-iterative methods described above. In Krylov subspace methods no limit for $n \to \infty$ can be formed. Therefore it is in general meaningless to talk about an 'asymptotic convergence rate' when a Krylov subspace method is applied to a *single* given system $Ax = b$. (Situations where an asymptotic approach is useful will be described at the end of Section 5.6.2 and in Section 5.7.6.)

In addition, the usual goal when applying a Krylov subspace method is to find a sufficiently accurate approximate solution in a number of steps that is *significantly smaller* than the system dimension. This is true even for methods based on short recurrences, such as the CG method. Some interesting remarks supporting this point are given by Hackbusch in [291, Section 2.4.2, in particular p. 270]. As a consequence, we need to be interested in the convergence from the very beginning of the iteration. This 'transient behaviour' is typically disregarded in the analysis of stationary and semi-iterative methods. As we will see later in this chapter, the early stage of convergence of Krylov subspace methods can depend significantly on the right-hand side b and the initial approximation x_0; see also Figure 5.3 in Section 5.2. Hence no analysis based on the matrix *only* can be sufficient for achieving a complete understanding of the behaviour of the Krylov subspace methods.

As described in Chapter 3, Krylov subspace methods are inherently linked with moments and hence they are *highly nonlinear*. The behaviour of an observed convergence characteristic (e.g. the residual norm) can be far from smooth, even when this characteristic is minimised by the respective method. For example, the Euclidean residual norm in the GMRES method can completely stagnate for many subsequent steps at *any stage* of the iteration; see Section 5.7.4.[7] The A-norm of the error in

---

[7]. In light of this proven fact, it seems surprising that the occurrence of stagnation of GMRES (and other Krylov subspace methods) is still treated as a remarkable special case even in very recent commercial software packages; see, e.g. the `flag=3` indicator in the implementations of all Krylov subspace methods in MATLAB R2011a.

CG computations can exhibit a staircase-like behaviour, where repeatedly very slow convergence for a few steps is followed by a sharp drop of the error norm; see Section 5.9.1 below.

In the context of computations using Krylov subspace methods we are often confronted with the following, seemingly paradoxical situation:

1. In many instances a nonlinear problem like a system of nonlinear algebraic equations or a nonlinear optimisation problem is solved by forming a sequence of *linearisations.* The outer iteration deals with the *local* approximation of the nonlinearity. The inner linear computational core (in our context the computation using a Krylov subspace method) determines the parameters of the outer iteration step. Convergence is evaluated *locally* based on the decrease (at the given individual step) of the functional that determines the quality of the approximate solution.
2. Krylov subspace methods used for solving the linear algebraic systems are by their nature *nonlinear.* This immediately follows from their formulation as projections onto Krylov subspaces, or from their relationship to the problem of moments; see Chapters 2 or 3 respectively.

We strongly believe, contrary to the spirit of some descriptions found in the literature (see, e.g. the influential exposition in [435, Section 10.8]), that Krylov subspace methods for solving linear algebraic problems can not be considered simplifications of nonlinear methods or merely modifications of methods like the stationary Richardson iteration, Chebyshev semi-iteration, or the classical method of steepest descent. Because of some kind of inertia within mathematical disciplines and insufficient communication between them, the principal difference between stationary and semi-iterative iterative methods on the one hand, and Krylov subspace methods (used as iterative methods) on the other, is still not always accepted. We hope that the previous chapters and the following analysis of CG and GMRES present reasonable arguments for overcoming this mathematically unjustified situation.

5.6 CG IN EXACT ARITHMETIC

In this section we describe properties of the CG method that hold in *exact arithmetic*. For convenience, let us first recall some basic facts and results on CG that were stated earlier in the book.

Let $Ax = b$ be a given linear algebraic system with an HPD matrix A. The approximate solution $x_n \in x_0 + \mathcal{K}_n(A, r_0)$ generated by the CG method with initial approximation x_0 is characterised by the orthogonality property

$$x - x_n \perp_A \mathcal{K}_n(A, r_0),$$

and the equivalent optimality property is

$$\|x - x_n\|_A = \min_{z \in x_0 + \mathcal{K}_n(A, r_0)} \|x - z\|_A;$$

see case (1) in Theorem 2.3.1. The method terminates with $x_d = x$, where d is the grade of the initial residual $r_0 = b - Ax_0$ with respect to A.

When the Hermitian Lanczos algorithm (Algorithm 2.4.3) is applied to A and r_0, we get at each step $n < d$ a relation of the form

$$AV_n = V_n T_n + \delta_{n+1} v_{n+1} e_n^T,$$

where the columns of V_n form an orthonormal basis of $\mathcal{K}_n(A, r_0)$ and T_n is an $n \times n$ Jacobi matrix; see (2.4.5) and the following discussion. Using this relation we can write the nth approximate solution as (see (2.5.5))

$$x_n = x_0 + V_n T_n^{-1} (\|r_0\| e_1).$$

In Section 2.5.1 we have derived the CG algorithm of Hestenes and Stiefel (Algorithm 2.5.1) from this formula. The CG algorithm computes an orthogonal basis of the Krylov subspace $\mathcal{K}_n(A, r_0)$ given by the residual vectors $r_0, \ldots, r_{n-1}$ and an A-orthogonal basis of the same subspace given by the direction vectors $p_0, \ldots, p_{n-1}$. These vectors satisfy a pair of coupled two-term recurrences.

Since T_n contains coefficients of the three-term recurrence for the Lanczos vectors, one can expect deep links between the CG method and orthogonal polynomials, continued fractions and Gauss quadrature. These links were studied in Chapter 3, in particular in Section 3.5. The Lanczos vectors $v_1, \ldots, v_{n+1}$ are given as

$$v_{j+1} = \varphi_j(A) v_1, \quad j = 0, 1, \ldots, n, \tag{5.6.1}$$

where $\varphi_0(\lambda) \equiv 1$ and each $\varphi_j(\lambda)$ is a polynomial of exact degree j; see (2.4.1). The Lanczos vectors are mutually orthogonal with respect to the Euclidean inner product; see (2.4.2). Moreover, $\|v_j\| = 1$ for $j = 1, \ldots, n+1$, and hence it is easy to see that the polynomials

$$\varphi_0(\lambda), \varphi_1(\lambda), \ldots, \varphi_n(\lambda)$$

form the sequence of orthonormal polynomials (3.3.1) with respect to the inner product (3.2.8) determined by the eigendecomposition of A and the initial vector v_1; see Section 3.5. Their *monic* counterparts

$$\psi_0(\lambda), \psi_1(\lambda), \ldots, \psi_n(\lambda)$$

(see (3.2.7)) can then be written in the form

$$\psi_0(\lambda) \equiv 1, \quad \psi_j(\lambda) = \left(\lambda - \theta_1^{(j)}\right) \cdots \left(\lambda - \theta_j^{(j)}\right), \quad j = 1, \ldots, n;$$

see Theorem 3.4.1. Here $\theta_1^{(j)}, \ldots, \theta_j^{(j)}$ are the j eigenvalues of the Jacobi matrix T_j. These eigenvalues are distinct (see Theorem 3.4.1) and they are called the *Ritz values* of A determined at the nth iteration step. In Section 5.7.1 we will give a general characterisation of Ritz values based on the orthogonality condition.

As shown in (2.5.7), the CG residual vectors are proportional to the Lanczos vectors. In particular, by construction we have

$$v_{j+1} = (-1)^j \frac{r_j}{\|r_j\|}, \quad j = 0, 1, \ldots, n; \qquad (5.6.2)$$

see Algorithms 2.4.3 and 2.5.1. The CG residual vectors satisfy $r_j \in r_0 + A\mathcal{K}_j(A, r_0)$ and therefore

$$r_j = \varphi_j^{CG}(A) r_0, \quad j = 0, 1, \ldots, n,$$

where $\varphi_j^{CG}(\lambda)$ is the normalised counterpart of the Lanczos polynomial $\varphi_j(\lambda)$ (and of the monic $\psi_j(\lambda)$) such that $\varphi_j^{CG}(0) = 1$, i.e. $\varphi_0^{CG}(\lambda) \equiv 1$ and

$$\varphi_j^{CG}(\lambda) = \frac{\left(\lambda - \theta_1^{(j)}\right) \cdots \left(\lambda - \theta_j^{(j)}\right)}{(-1)^j \theta_1^{(j)} \cdots \theta_j^{(j)}}, \quad j = 1, \ldots, n. \qquad (5.6.3)$$

5.6.1 Expressions for the CG Errors and Their Norms

We will now derive several useful and mostly well-known expressions for the errors in the CG method and their norms. The recurrence formula for the nth approximation is given by $x_n = x_{n-1} + \alpha_{n-1} p_{n-1}$. From this we easily get by induction that

$$x - x_n = x - x_k - \sum_{j=k+1}^{n} \alpha_{j-1} p_{j-1} \quad \text{for each } k = 0, 1, \ldots, n. \qquad (5.6.4)$$

(Note that for $k = n$ the sum on the right-hand side is empty, hence equal to zero, and the equality is trivial.) For $n = d$ we have $x_d = x$, and this gives

$$x - x_k = \sum_{j=k+1}^{d} \alpha_{j-1} p_{j-1}, \quad k = 0, 1, \ldots, d. \qquad (5.6.5)$$

Let us rewrite the two recurrences for the direction and the residual vectors in the CG algorithm as

$$p_{j-1} = r_{j-1} + \omega_{j-1} p_{j-2} \quad \text{and} \quad A p_{j-1} = \alpha_{j-1}^{-1}(r_{j-1} - r_j)$$

(recall that $\alpha_j > 0$ for each $j = 0, 1, \ldots, d$). Using these recurrences and the vectors' *local* orthogonality properties we obtain

$$(p_{j-1}, A p_{j-1}) = (r_{j-1} + \omega_{j-1} p_{j-2}, A p_{j-1}) = (r_{j-1}, A p_{j-1})$$
$$= \left(r_{j-1}, \alpha_{j-1}^{-1}(r_{j-1} - r_j)\right)$$
$$= \alpha_{j-1}^{-1} \|r_{j-1}\|^2 \qquad (5.6.6)$$

(which is equivalent to the first equation in (2.5.43)), so that

$$\|x - x_k\|_A^2 = (x - x_k, A(x - x_k)) = \sum_{j=k+1}^{d} \alpha_{j-1}^2 (p_{j-1}, Ap_{j-1}) \qquad (5.6.7)$$

$$= \sum_{j=k+1}^{d} \alpha_{j-1} \|r_{j-1}\|^2, \quad k = 0, 1, \ldots, n. \qquad (5.6.8)$$

We thus have shown the following result.

Theorem 5.6.1
Using the previous notation, the approximations in the CG method satisfy

$$\|x - x_k\|_A^2 = \sum_{j=k+1}^{n} \alpha_{j-1} \|r_{j-1}\|^2 + \|x - x_n\|_A^2 \quad \text{for } 0 \le k \le n \le d, \qquad (5.6.9)$$

where d is the grade of the initial residual with respect to A, i.e. the step at which CG terminates with giving, in exact arithmetic, the exact solution. In particular, for $k = n - 1$,

$$\|x - x_{n-1}\|_A^2 - \|x - x_n\|_A^2 = \alpha_{n-1} \|r_{n-1}\|^2, \qquad (5.6.10)$$

which quantifies the strict decrease of the A-norm of the error in each step of the CG method.

The relation expressed in (5.6.9) has already been shown by Hestenes and Stiefel in [313, Theorem 6:1, equation (6:2)]. It has been derived or applied by many other authors; see, e.g. [8, 25, 135, 245, 598]. Our derivation of (5.6.9) above relied on orthogonality properties that hold only in exact arithmetic. In particular, we have used the local orthogonality in the derivation of (5.6.6), and the *global* A-orthogonality among the direction vectors in the derivation of (5.6.8).

As shown in [598], one can prove (5.6.9) using only *local* orthogonality relations among consecutive residuals and direction vectors. Indeed, using (5.6.4) for $k = n - 1$ as well as (5.6.6) and $(r_{j-1}, p_{j-1}) = \|r_{j-1}\|^2$ (see (2.5.37)) we get

$$\|x - x_n\|_A^2 = \|(x - x_{n-1}) - (\alpha_{n-1} p_{n-1})\|_A^2$$
$$= \|x - x_{n-1}\|_A^2 - 2\Re((A(x - x_{n-1}), \alpha_{n-1} p_{n-1})) + \alpha_{n-1}^2 (p_{n-1}, Ap_{n-1})$$
$$= \|x - x_{n-1}\|_A^2 - 2\Re(\alpha_{n-1}(r_{n-1}, p_{n-1})) + \alpha_{n-1} \|r_{n-1}\|^2$$
$$= \|x - x_{n-1}\|_A^2 - \alpha_{n-1} \|r_{n-1}\|^2,$$

which is nothing but (5.6.10). The recursive application of this equation now gives (5.6.9). Note that previously we have obtained (5.6.10) as a special case of (5.6.9). However, since (5.6.10) can be derived using only local orthogonality properties, it can be shown that (5.6.9) also holds (up to a small insignificant inaccuracy) in finite precision computations with the CG method; see [598].

We point out that (5.6.9) can be helpful for estimating the error in the CG method. It expresses the difference $\|x - x_k\|^2 - \|x - x_n\|^2$ as a sum of $n - k$ products of the quantities α_j and $\|r_j\|^2$, which are readily available during CG computations; see [598] and [457, Section 5.3].

In Section 3.5 we have analysed the model reduction properties of CG, which are related to projecting the possibly large system $Ax = b$ using Krylov subspaces to the potentially small systems $T_n t_n = \|r_0\| e_1, n = 1, 2, \ldots$ As in Section 3.5 we now assume, for simplicity of notation, that A has N distinct eigenvalues,

$$\lambda_1 < \lambda_2 < \cdots < \lambda_{N-1} < \lambda_N,$$

and that $d = N$, i.e. that the initial residual r_0 has nonzero components in the direction of all N associated mutually orthogonal eigenvectors. Equivalently, the distribution function $\omega(\lambda)$ determined by A and r_0 has N points of increase with N corresponding positive weights $\omega_1, \ldots, \omega_N$; see (3.1.6) and (3.5.2).

The quantitative equivalence of the CG method and the Gauss–Christoffel quadrature was, in Section 3.5, summarised in the relations (3.5.8)–(3.5.11). Considering $k = 0$ in (5.6.9) we obtain, after dividing both sides by $\|r_0\|^2$,

$$\frac{\|x - x_0\|_A^2}{\|r_0\|^2} = \frac{1}{\|r_0\|^2} \sum_{j=1}^n \alpha_{j-1} \|r_{j-1}\|^2 + \frac{\|x - x_n\|_A^2}{\|r_0\|^2}.$$

Combining this with (3.5.8)–(3.5.9) gives an extension of (3.5.11),

$$\frac{1}{\|r_0\|^2} \sum_{j=1}^n \alpha_{j-1} \|r_{j-1}\|^2 = \sum_{j=1}^n \omega_j^{(n)} \left\{\theta_j^{(n)}\right\}^{-1} = (e_1, T_n^{-1} e_1) = -\mathcal{F}_n(0),$$

(5.6.11)

and, taking $n = N$,

$$\frac{1}{\|r_0\|^2} \sum_{j=1}^N \alpha_{j-1} \|r_{j-1}\|^2 = \sum_{j=1}^N \omega_j \lambda_j^{-1} = (e_1, T_N^{-1} e_1) = -\mathcal{F}_N(0) \quad (5.6.12)$$

In (5.6.11), $\theta_1^{(n)}, \ldots, \theta_n^{(n)}$ are the n eigenvalues of T_n and $\omega_1^{(n)}, \ldots, \omega_n^{(n)}$ are the squared first components of the corresponding normalised eigenvectors. These $2n$ numbers are respectively the n roots of the nth orthogonal polynomial $\varphi_n(\lambda)$ and the corresponding n weights of the n-node Gauss–Christoffel quadrature associated with the distribution function $\omega(\lambda)$; see Theorem 3.4.1. As in (3.3.18), $\mathcal{F}_n(0)$ denotes the nth convergent of the continued fraction $\mathcal{F}_N(0)$. Subtracting (5.6.11) from (5.6.12) and using (5.6.8) now gives the following result.

Corollary 5.6.2
Using the previous notation, and in particular the assumption $d = N$, we get for $n = 1, 2, \ldots, N$,

$$\|x - x_n\|_A^2 = \sum_{j=n+1}^N \alpha_{j-1} \|r_{j-1}\|^2$$

$$= \|r_0\|^2 \left((e_1, T_N^{-1} e_1) - (e_1, T_n^{-1} e_1)\right) \quad (5.6.13)$$

$$= \|r_0\|^2 \left(\sum_{j=1}^N \omega_j \lambda_j^{-1} - \sum_{j=1}^n \omega_j^{(n)} \left\{\theta_j^{(n)}\right\}^{-1}\right). \quad (5.6.14)$$

Equation (5.6.13) was used in the 1970s by Golub and his co-workers [125, 126]. The expression in (5.6.14) indicates the complicated relation between the convergence of CG and the convergence of the Ritz values towards the eigenvalues of A. Another way of stating this relationship is given in the following result; see [457, Theorem 3.1].

Theorem 5.6.3
Using the previous notation, the residual and error vectors of the CG method satisfy, for $n = 1, \ldots, N$,

$$\|r_n\|^2 = \|r_0\|^2 \sum_{j=1}^{N} \prod_{\ell=1}^{n} \left(1 - \frac{\lambda_j}{\theta_\ell^{(n)}}\right)^2 \omega_j,$$

$$\|x - x_n\|^2 = \|r_0\|^2 \sum_{j=1}^{N} \prod_{\ell=1}^{n} \left(\frac{1}{\lambda_j} - \frac{1}{\theta_\ell^{(n)}}\right)^2 \omega_j,$$

$$\|x - x_n\|_A^2 = \|r_0\|^2 \sum_{j=1}^{N} \prod_{\ell=1}^{n} \left(\frac{1}{\lambda_j^{1/2}} - \frac{\lambda_j^{1/2}}{\theta_\ell^{(n)}}\right)^2 \omega_j.$$

Proof
As shown in the introductory part of Section 5.6, the nth error in the CG method is of the form $x - x_n = \varphi_n^{CG}(A)(x - x_0)$. All relations now follow by straightforward computations using $r_n = A(x - x_n)$ for $n = 0, 1, \ldots, N$ and the form of the CG polynomial $\varphi_n^{CG}(\lambda)$ given in (5.6.3). □

The assumption $d = N$ (made above for simplicity of notation) is now no longer needed, and hence we again consider a general initial residual r_0 of grade d with respect to A. Our next goal is to show that not only the A-norms, but also the Euclidean norms of the errors in the CG method, decrease strictly monotonically. This property was shown by Hestenes and Stiefel in their original paper of 1952 [313]. It did not get the same attention as other properties of the CG method and it was rediscovered several times in the following years. For the proof we need to summarise further relations satisfied by the residual and direction vectors.

Lemma 5.6.4
Consider a linear algebraic system $Ax = b$, where A is HPD, and an initial approximation x_0 so that $r_0 = b - Ax_0$ is of grade d with respect to A. Then the residual and direction vectors in the CG method satisfy:

(1) $p_n = \|r_n\|^2 \sum_{j=0}^{n} \|r_j\|^{-2} r_j$, for $n = 0, 1, \ldots, d-1$.
(2) $\|p_n\|^2 = \|r_n\|^4 \sum_{j=0}^{n} \|r_j\|^{-2}$, for $n = 0, 1, \ldots, d-1$.
(3) $(p_n, r_\ell) = 0$ for $n < \ell$.
(4) $(p_n, r_\ell) = \|r_n\|^2$ for $n \geq \ell$.
(5) $(p_n, p_\ell) = \|r_\ell\|^2 \|p_n\|^2 \|r_n\|^{-2}$ for $n \leq \ell$.

Proof
(1) follows inductively from the recurrence relation satisfied by the direction vectors,

$$p_n = r_n + \frac{\|r_n\|^2}{\|r_{n-1}\|^2} p_{n-1} = \|r_n\|^2 \left(\frac{1}{\|r_n\|^2} r_n + \frac{1}{\|r_{n-1}\|^2} r_{n-1} + \frac{1}{\|r_{n-2}\|^2} p_{n-2} \right)$$

$$= \cdots$$

$$= \|r_n\|^2 \left(\frac{1}{\|r_n\|^2} r_n + \frac{1}{\|r_{n-1}\|^2} r_{n-1} + \frac{1}{\|r_{n-2}\|^2} r_{n-2} + \cdots + \frac{1}{\|r_0\|^2} r_0 \right),$$

where in the last step we have used that $p_0 = r_0$.

(2)–(4) follow directly from (1) and the mutual orthogonality of the residual vectors. Finally, (5) can be shown exploiting all previous relations,

$$(p_n, p_\ell) = \left(p_n, \|r_\ell\|^2 \sum_{j=0}^{\ell} \|r_j\|^{-2} r_j \right) = \|r_\ell\|^2 \sum_{j=0}^{\ell} \|r_j\|^{-2} (p_n, r_j)$$

$$= \|r_\ell\|^2 \sum_{j=0}^{n} \|r_j\|^{-2} (p_n, r_j) = \|r_\ell\|^2 \|r_n\|^2 \sum_{j=0}^{n} \|r_j\|^{-2}$$

$$= \|r_\ell\|^2 \|r_n\|^2 \|p_n\|^2 \|r_n\|^{-4},$$

which completes the proof. □

The relations (1)–(5) in the previous lemma were all known to Hestenes and Stiefel and can be found in their paper; see respectively their equations (5:2), (5:3c), (5:4a) and (5:6a). Using (5) and some of the previous results we now relate the A-norms and the Euclidean norms of the errors in the CG method.

Theorem 5.6.5
Using the notation of Lemma 5.6.4, the errors in the CG method satisfy

$$\frac{\|x - x_{n-1}\|^2 - \|x - x_n\|^2}{\|p_{n-1}\|^2} = \frac{\|x - x_{n-1}\|_A^2 + \|x - x_n\|_A^2}{\|p_{n-1}\|_A^2}, \tag{5.6.15}$$

for $n = 1, \ldots, d - 1$. In particular, the Euclidean norm of the error in the CG method decreases strictly monotonically.

Proof
For $n = 1, \ldots, d - 1$, the difference of two consecutive (squared) Euclidean error norms in the CG method can be written as

$$\|x - x_{n-1}\|^2 - \|x - x_n\|^2 = \|x - x_n + (x_n - x_{n-1})\|^2 - \|x - x_n\|^2$$

$$= \|x_n - x_{n-1}\|^2 + 2\Re\left((x - x_n, x_n - x_{n-1})\right)$$

$$= \alpha_{n-1}^2 \|p_{n-1}\|^2 + 2\Re\left(\left(\sum_{j=n+1}^{d} \alpha_{j-1} p_{j-1}, \alpha_{n-1} p_{n-1}\right)\right)$$

$$= \alpha_{n-1}^2 \|p_{n-1}\|^2 + 2\Re\left(\alpha_{n-1} \sum_{j=n+1}^{d} \alpha_{j-1} (p_{j-1}, p_{n-1})\right)$$

$$= \frac{\alpha_{n-1} \|p_{n-1}\|^2}{\|r_{n-1}\|^2}\left(\alpha_{n-1}\|r_{n-1}\|^2 + 2\left(\sum_{j=n+1}^{d} \alpha_{j-1}\|r_{j-1}\|^2\right)\right)$$

$$= \frac{\|p_{n-1}\|^2}{\|p_{n-1}\|_A^2}\left(\|x - x_{n-1}\|_A^2 - \|x - x_n\|_A^2 + 2\|x - x_n\|_A^2\right).$$

The assertion now follows by a simple rearrangement. In this derivation we have used (5.6.8) with $k = n$, (5.6.10), (5) in Lemma 5.6.4, and the formulas $x_n - x_{n-1} = \alpha_{n-1} p_{n-1}$ and $\alpha_{n-1} = \|r_{n-1}\|^2 / \|p_{n-1}\|_A^2$ from the CG algorithm. □

This result was shown by Hestenes and Stiefel in [313, Theorem 6:3]. After stating their Theorem 6:3 they stressed that CG is 'a method of successive approximations' [313, p. 417], i.e. an *iterative method*; see the historical note in Section 5.2.1.

5.6.2 Eigenvalue-based Convergence Results for CG

We will now derive and discuss eigenvalue-based convergence results for the CG method. As shown in the introductory part of Section 5.6, the nth CG residual vector is of the form $r_n = \varphi_n^{CG}(A) r_0$ for the CG polynomial $\varphi_n^{CG}(\lambda)$ of exact degree n that is normalised at zero. Since $r_n = A(x - x_n)$, the nth error is given by

$$x - x_n = \varphi_n^{CG}(A)(x - x_0).$$

The polynomial $\varphi_n^{CG}(\lambda)$ is defined uniquely by the CG minimisation problem

$$\|x - x_n\|_A = \|\varphi_n^{CG}(A)(x - x_0)\|_A = \min_{\substack{\varphi(0)=1 \\ \deg(\varphi) \leq n}} \|\varphi(A)(x - x_0)\|_A. \quad (5.6.16)$$

Let $A = Y \mathrm{diag}(\lambda_j) Y^*$ be an eigendecomposition of the HPD matrix A with the positive eigenvalues $\lambda_1 \leq \lambda_2 \leq \cdots \leq \lambda_{N-1} \leq \lambda_N$ and $Y^* Y = I$. Let $d_{\min}(A)$ denote the degree of the minimal polynomial of A (i.e. the number of distinct eigenvalues), let d be the grade of the initial residual $r_0 = b - A x_0$ with respect to A, and let $A^{1/2} = Y \mathrm{diag}(\lambda_j^{1/2}) Y^*$. Then, for $n = 0, 1, \ldots, d$,

$$\|x - x_n\|_A = \left\|\varphi_n^{CG}(A)(x - x_0)\right\|_A = \left\|A^{1/2}\varphi_n^{CG}(A)(x - x_0)\right\|$$

$$= \min_{\substack{\varphi(0)=1 \\ \deg(\varphi)\le n}} \|A^{1/2}\varphi(A)(x - x_0)\|$$

$$= \min_{\substack{\varphi(0)=1 \\ \deg(\varphi)\le n}} \|\varphi(A)A^{1/2}(x - x_0)\| \tag{5.6.17}$$

$$\le \|x - x_0\|_A \min_{\substack{\varphi(0)=1 \\ \deg(\varphi)\le n}} \|\varphi(A)\| \tag{5.6.18}$$

$$= \|x - x_0\|_A \min_{\substack{\varphi(0)=1 \\ \deg(\varphi)\le n}} \max_{1\le j\le N} |\varphi(\lambda_j)|.$$

The following theorem gives some important properties of this bound.

Theorem 5.6.6
Using the previous notation, the nth relative A-norm of the error in the CG method satisfies

$$\frac{\|x - x_n\|_A}{\|x - x_0\|_A} \le \min_{\substack{\varphi(0)=1 \\ \deg(\varphi)\le n}} \max_{1\le j\le N} |\varphi(\lambda_j)|, \quad n = 0, 1, \ldots, d. \tag{5.6.19}$$

This bound is sharp in the sense that for every $n = 0, 1, \ldots, d_{\min}(A)$ there exists an initial approximation x_0 so that (5.6.19) is an equality. Moreover, for every $n = 1, \ldots, d_{\min}(A) - 1$, there exist $n + 1$ distinct eigenvalues $\widehat{\lambda}_1, \ldots, \widehat{\lambda}_{n+1}$ of A, such that

$$\min_{\substack{\varphi(0)=1 \\ \deg(\varphi)\le n}} \max_{1\le j\le N} |\varphi(\widehat{\lambda}_j)| = \left(\sum_{k=1}^{n+1} \prod_{\substack{j=1 \\ j\ne k}}^{n+1} \frac{|\widehat{\lambda}_j|}{|\widehat{\lambda}_j - \widehat{\lambda}_k|}\right)^{-1}. \tag{5.6.20}$$

Proof
The bound (5.6.19) was shown above. For the other assertions we refer to the original work of Greenbaum [268]. □

The fact that the bound (5.6.19) is sharp means that its right-hand side, evaluated for $n = 0, 1, \ldots, d_{\min}(A)$, gives an 'envelope' for all possible CG convergence curves (with respect to the relative A-norm of the error) that can be attained for the given matrix A. In other words, the right-hand side of (5.6.19) completely describes the *worst-case behaviour* of the CG method. However, the convergence behaviour for a specific initial approximation can be very different from the worst case. This will be illustrated by a numerical example in the following section.

The equality expressed in (5.6.20) is well known in approximation theory and follows from standard results in that area; see, e.g. [136, Theorem 2.4 and Corollary 2.5] (see also [428] for a discussion). Note that the right-hand side can only be computed when *all* eigenvalues of A are known. Therefore (5.6.20) is of limited interest in practical applications.

One can estimate the min–max expression in (5.6.20) by replacing the discrete set of eigenvalues of A by the (continuous) interval $[\lambda_1, \lambda_N]$, and by using the scaled and shifted Chebyshev polynomials (5.5.3). Similarly to the derivation of the convergence bound for the Chebyshev semi-iterative method in (5.5.4)–(5.5.7) this leads to

$$\frac{\|x - x_n\|_A}{\|x - x_0\|_A} \leq \min_{\substack{\varphi(0)=1 \\ \deg(\varphi) \leq n}} \max_{1 \leq j \leq N} |\varphi(\lambda_j)| \tag{5.6.21}$$

$$\leq \min_{\substack{\varphi(0)=1 \\ \deg(\varphi) \leq n}} \max_{\lambda \in [\lambda_1, \lambda_N]} |\varphi(\lambda)| \tag{5.6.22}$$

$$= |\chi_n(0)|^{-1} \max_{\lambda \in [\lambda_1, \lambda_N]} |\chi_n(\lambda)| = |\chi_n(0)|^{-1}$$

$$\leq 2 \left(\frac{\sqrt{\kappa(A)} - 1}{\sqrt{\kappa(A)} + 1} \right)^n,$$

where $\kappa(A) = \lambda_N / \lambda_1$.

Corollary 5.6.7
Using the previous notation, the A-norm of the nth error in the CG method is bounded by

$$\frac{\|x - x_n\|_A}{\|x - x_0\|_A} \leq 2 \left(\frac{\sqrt{\kappa(A)} - 1}{\sqrt{\kappa(A)} + 1} \right)^n, \quad n = 0, 1, \ldots, d. \tag{5.6.23}$$

The bound (5.6.23) is possibly the most widely known convergence result for the CG method. It can be found in most books on iterative methods; see, e.g. [272, Theorem 3.11, pp. 51–52] for a statement and the complete proof. Without giving the final formula on the right-hand side of (5.6.23), the bound is essentially present in the papers by Meinardus from 1963 [449, Satz 2, p. 18] and Kaniel from 1966 [374, Theorem 4.3, p. 373]. Both papers bound the min–max polynomial approximation problem on the discrete set of the eigenvalues of A in terms of the scaled and shifted Chebyshev polynomial on the interval $[\lambda_1, \lambda_N]$, i.e. they consider the step from (5.6.21) to (5.6.22) in our derivation.

To our knowledge, the first occurrence of (5.6.23) in a direct relationship with CG was in a 1967 paper of Daniel [127, Theorem 1.2.2] (his other paper of 1967 [128] contains the bound as well, but it was written after [127]). An often cited reference for the proof is Luenberger's book of 1973 [437, p. 187]; see, e.g. [250, Theorem 10.2.6] for such a citation. Pointing to Daniel's papers [127, 128] as the original sources, Luenberger had described the formula already in his 1969 book [435, Section 10.8, pp. 296 and 311]. It is worth mentioning that there he referred to the 'orthogonalisation of moments' as the underlying principle of the CG method, and that he pointed to Vorobyev's book [651] for details.

A comparison with Theorem 5.5.1 shows that the CG bound (5.6.23) is identical to the convergence bound for the Chebyshev semi-iterative method that was already known to Rutishauser, and without expressing it precisely in the form (5.6.23) also to Lanczos, von Neumann, Young, and others in the 1950s; see the historical note in Section 5.5.3. It was clear that due to the minimisation property of the CG method, the bound for the relative A-norm of the error of the Chebyshev semi-iterative method must hold also for CG. In retrospect, it seems that techniques for stationary and semi-iterative methods, which were in high fashion in the 1960s, were at that time directly transferred to the analysis of CG. Although mathematically correct, this led to some misunderstandings of the true nature of the CG method. In particular, since the bound for CG is the same as the one for the Chebyshev semi-iterative method, and since the right-hand side of (5.6.23) decreases linearly, it is tempting *but mathematically questionable* to call (in analogy with (5.5.1)) the quantity

$$\frac{\sqrt{\kappa(A)} - 1}{\sqrt{\kappa(A)} + 1}$$

the 'average reduction factor' for n iterations of the CG method, and to identify the behaviour of CG with the corresponding 'rate of convergence'. Nevertheless, this practice became standard in numerous publications. An early example can be found in Luenberger's book of 1969 [435, p. 296]. It affected many later expositions of the CG method and therefore probably contributed to the unfortunate and widespread identification of the convergence of CG with the bound (5.6.23) in the literature until the present day.

As outlined in Section 5.5.4, any 'linear asymptotic' approach to the convergence analysis of Krylov subspace methods applied to a *single* linear algebraic system $Ax = b$ is problematic (see the end of this section for remarks on rigorous and well justified asymptotic approaches). We will now add some CG specific details that go beyond the general discussion in Section 5.5.4.

1. The CG iteration polynomials $\varphi_n^{CG}(\lambda)$ that give the minimum in (5.6.16) are equal to the (normalised) Lanczos polynomials $\varphi_n(\lambda)/\varphi_n(0)$, for $n = 1, \ldots, d$, which are orthogonal with respect to the Riemann–Stieltjes inner product corresponding to the distribution function $\omega(\lambda)$ that is determined by the matrix A and the initial residual r_0; see the introductory part of Section 5.6. The construction of the CG iteration polynomials is based on information about *all* eigenvalues of A and *all* projections of r_0 on the corresponding invariant subspaces; see Theorem 5.6.3.
2. The derivation of the upper bound (5.6.23) contains two steps where the nature of the minimisation problem can be substantially changed:
    (i) In the step from (5.6.17) to (5.6.18) we separate the CG iteration polynomial from the initial error. In terms of the distribution function $\omega(\lambda)$ this means that we separate its points of increase from the associated weights.
    (ii) In the step from (5.6.21) to (5.6.22) we take out of consideration any information about the distribution of the eigenvalues $\lambda_2, \ldots, \lambda_{N-1}$ within the interval $[\lambda_1, \lambda_N]$. This means that we no longer consider a distribution function having finitely many points of increase that occur at the eigenvalues of A.

3. The resulting minimisation problem (5.6.22) is solved by the scaled and shifted Chebyshev polynomials $\chi_n(\lambda)/\chi_n(0)$. They are orthogonal with respect to continuous and discrete inner products determined by the extremal eigenvalues λ_1 and λ_N and further information that is *completely unrelated* to the specific data A, b, x_0. Details on the orthogonality properties of Chebyshev polynomials can be found, e.g. in [528, Section 1.5, pp. 30–31, and Exercises 1.5, pp. 49–50] and [124, Theorem 4.5.20, pp. 461–462]. Instructive comments particularly on the difference between minimisation over a discrete set and over an interval had already been given by Lanczos in 1952 [406, Section 5, p. 46]. For related nice illustrations we refer also to [402]; see, e.g. Figure 1.1 on p. 7.

As a consequence, there is no reason whatsoever why the bound (5.6.23) that is relevant for the Chebyshev semi-iteration should be, in general, descriptive for the convergence of the CG method (a nice point on this is made by Hackbusch [291, Section 9.4.3]). As a matter of fact, (5.6.23) often does not even describe the worst-case convergence behaviour of CG. This will be illustrated numerically in the following section.

While (5.6.23) is not descriptive in general, it at least shows that CG converges quickly when $\kappa(A)$ is small. This motivates the term *preconditioning*, that refers to techniques by which the given system $Ax = b$ is modified in order to obtain an equivalent system whose matrix has a smaller condition number than A. Note, however, that (5.6.23) does *not* imply that CG converges slowly when $\kappa(A)$ is large. Moreover, a smaller condition number does *not* necessarily mean that CG converges faster.

One of the earliest appearances of the term 'preconditioning' is in a 1953 paper of Forsythe. He described confusions in this context that apparently are still not fully resolved [192, p. 318]: 'The belief is widespread that the condition of a system (1) [$Ax = b$] has a decisive influence on the convergence of an iterative solution and on the accuracy of a direct solution; this cannot always be true. Even when it is true for an iterative process, it may be possible actually to take advantage of the poor condition of (1) in converting the slow process into an accelerated method which converges rapidly. There is a great need for clarification of the group of ideas associated with 'condition.' With the concept of 'ill-conditioned' systems $Ax = b$ goes the idea of 'preconditioning' them.'

Remark 5.6.8
Based on Meinardus' paper [449], Li characterised in 2008 [417] all initial approximations x_0 so that for a given step $n < d_{\min}(A)$ the relative A-norm of the error in the CG method satisfies

$$\frac{\|x - x_n\|_A}{\|x - x_0\|_A} = 2\left[\left(\frac{\sqrt{\kappa(A)} - 1}{\sqrt{\kappa(A)} + 1}\right)^n + \left(\frac{\sqrt{\kappa(A)} + 1}{\sqrt{\kappa(A)} - 1}\right)^n\right]^{-1}$$

(cf. (5.5.6) for the expression on the right-hand side and Theorem 5.6.6 for the solution of a different, although somewhat related, attainability problem). His analysis shows that this exact equality occurs only for very special distributions of eigenvalues of A and corresponding entries of the initial residual, which can be explicitly constructed using the roots of the shifted Chebyshev polynomials. Moreover, if the exact equality occurs at step n, then CG terminates with the exact solution in the next step, i.e. $r_{n+1} = 0$. As pointed out by Li, this result had already been shown by Kaniel in 1966 [374, Theorem 4.4]. It implies that there exists no single system $Ax = b$ and initial x_0 such that the equality holds for every step $n = 1, \ldots, d$.

It is often observed in computations with the CG method that the speed of convergence (on average) increases during the iteration; see the following section for a numerical example and Sections 5.6.5 and 5.9.1 for clarification of possible irregularities in the CG convergence behaviour. This means that the reduction of the error per step is (on average) increasing, or, equivalently, that the sequence

$$\frac{\|x - x_n\|_A}{\|x - x_{n-1}\|_A}, \quad n = 1, 2, \ldots, \qquad (5.6.24)$$

is (on average) decreasing. When this occurs the CG method is said to converge *superlinearly*. We have written '$n = 1, 2, \ldots$' in (5.6.24), and this indicates the mathematical difficulties involved in describing and analysing 'superlinear convergence behaviour'. For a given system $Ax = b$ the sequence of ratios in (5.6.24) ends (in exact arithmetic) after finitely many steps. Therefore the term 'superlinear convergence' cannot be considered in any sense that assumes an infinite process.

An instructive way to address the issue of superlinear convergence is to consider model problems arising from discretised differential equations. By refining the mesh one obtains a sequence of linear algebraic problems that may get arbitrarily large, and thus can be the subject of asymptotic analysis. For an example of such an approach using the finite difference discretisation of the one-dimensional Poisson problem we refer to the papers by Naiman and his co-workers [470, 471]. A more general framework for asymptotic convergence analysis of Krylov subspace methods has been established and applied (mostly to HPD matrices and the CG method) by Beckermann and Kuijlaars [50, 51, 52]. A highly readable survey of their techniques and related work is given in [402]. Similarly to Naiman et al. they do not consider a single matrix A, but a sequence of matrices A_N of size $N \times N$, where N tends to infinity. They require that the matrices A_N have some 'asymptotic eigenvalue distribution'; for details see [50, Assumption 2.1, p. 6] or [51, Conditions (i)–(iii), p. 304]. Denoting the spectrum of A_N by $\Lambda(A_N)$, they study (for $n < N$) the value

$$\min_{\substack{\varphi(0)=1 \\ \deg(\varphi) \leq n}} \max_{\lambda \in \Lambda(A_N)} |\varphi(\lambda)|.$$

In the case of an HPD matrix A_N, this value gives an attainable worst-case bound for the nth relative A_N-norm of the CG error; see Theorem 5.6.6 above. Beckermann and Kuijlaars are interested in results when both n and N tend to infinity in such a way that n/N tends to a fixed number in the interval $(0, 1)$. We will give an example in the following section.

5.6.3 Illustrations of the Convergence Behaviour of CG

In this section we present three examples that illustrate the convergence behaviour of the CG method and some of the theoretical results presented above. The first two examples have previously appeared in a similar form in [427, Section 3.1.1] (also see the earlier investigations in [52, 470, 471]), while the third example is adapted from [51, Fig. 1 and Corollary 3.2].

EXAMPLE 1

Worst-case CG and the bound (5.6.23). As shown in Theorem 5.6.6, the *worst-case behaviour* of CG for a given matrix A depends completely on its eigenvalues, and it is described by the right-hand side of the bound (5.6.19); see also (5.6.20). We will illustrate this bound and compare it with (5.6.23) using the two sets of eigenvalues given by

$$\lambda_j^{(1)} = j^2/400, \quad j = 1, 2, \ldots, 20, \tag{5.6.25}$$

$$\lambda_j^{(2)} = \log(j)/\log(20), \quad j = 2, 3, \ldots, 20, \quad \lambda_1^{(2)} = 1/400. \tag{5.6.26}$$

Both sets of eigenvalues are contained in the interval $[1/400, 1]$. The first set has more eigenvalues to the left end of this interval (intuitively, a 'cluster' at zero), and the second set has more eigenvalues to the right end (intuitively, a 'cluster' at one). Each HPD matrix A having the eigenvalues (5.6.25) or (5.6.26) has the condition number $\kappa(A) = 400$.

Figure 5.4 shows the worst-case convergence of CG, i.e. the right-hand side of the bound (5.6.19) for the two sets of eigenvalues. We have used the routine cheby0

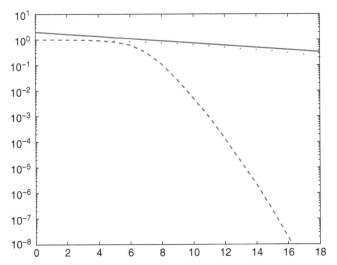

Figure 5.4 The worst-case CG convergence for the two sets of eigenvalues given in (5.6.25) and (5.6.26) are shown by the dots and the dashed curve, respectively. The solid curve shows the upper bound (5.6.23) based on the condition number of A.

of the MATLAB semi-definite programming package SDPT3 [619][8] to compute the worst-case curves. Apparently, the worst-case convergence for the set (5.6.26), which is shown by the dashed curve, is much faster than for the set (5.6.25), which is shown by dots. This is due to the fact that the set (5.6.25) has many more eigenvalues close to zero, where the CG iteration polynomial is normalised by construction. The bound (5.6.23) represents an upper bound on the worst-case behaviour of CG. The solid line in Figure 5.4 shows this bound for $\kappa(A) = 400$. It is close to the worst-case convergence for the eigenvalues (5.6.25), but it is not descriptive for the worst-case convergence for the eigenvalues (5.6.26).

EXAMPLE 2

Influence of the initial error. In this experiment we show that the initial error can have a significant influence on the convergence of the CG method. We consider the matrix

$$A = \text{tridiag}(-1, 2, -1)$$

that arises from the discretisation of the one-dimensional Poisson equation using central finite differences (or standard linear finite elements) on a uniform mesh. We choose a mesh consisting of 120 interior points, so that A is of size 120×120. We construct two different right-hand sides, $b^{(1)}$ and $b^{(2)}$, so that the corresponding solutions $x^{(j)}$ of $Ax^{(j)} = b^{(j)}$, $j = 1, 2$, exhibit a significantly different level of smoothness.[9] To describe how the right-hand sides are constructed, let $A = Y\Lambda Y^*$ be an eigendecomposition of A with $Y^* Y = I$. The vector $b^{(1)}$ is given as a linear combination of all eigenvectors, where each coefficient in this combination is equal to one, i.e. $b^{(1)} = Y[1, \ldots, 1]^T$. The solution corresponding to this right-hand side is

$$x^{(1)} = A^{-1}b^{(1)} = Y\Lambda^{-1}[1, \ldots, 1]^T.$$

This is a very smooth vector, whose components are shown by the dashed line in the right part of Figure 5.5. The smoothness comes from the fact that the contribution of the highly oscillating eigenvectors corresponding to large eigenvalues of A (see, e.g. [66]) is scaled out by the multiplication with Λ^{-1}. The second right-hand side is given by $b^{(2)} = Y\Lambda[1, \ldots, 1]^T$, so that the corresponding solution vector is

$$x^{(2)} = Y[1, \ldots, 1]^T.$$

As shown by the solid line in Figure 5.5, this vector is highly non-smooth. This is caused by the fact that in the linear combination of eigenvectors that forms $x^{(2)}$ the highly oscillating vectors are not scaled out.

---

8. In all experiments with SDPT3 reported in this chapter we have used version 4.0 of February 2009.

9. Since we deal with discrete vectors, the level of smoothness is more an intuitive feeling than a rigorous mathematical concept. It would be possible to find analogues of the vectors in the continuous Poisson problem, and thereby make the notion of smoothness precise, but this is not required for the purpose of this example.

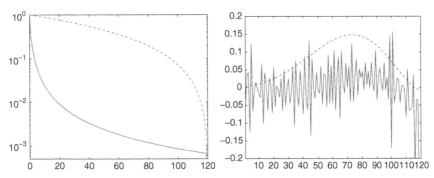

Figure 5.5 Convergence of CG for the one-dimensional Poisson model problem and two different right-hand sides (left) corresponding to the two solutions (right).

We choose $x_0 = 0$ in the CG method, so that the initial error is equal to the solution. The relative A-norms of the errors when CG is applied to the two linear algebraic systems $Ax^{(j)} = b^{(j)}, j = 1, 2$, are shown in the left part of Figure 5.5. The CG computations were performed using double reorthogonalisation of the residual vectors, which simulates exact arithmetic; see [280]. The qualitative difference in the convergence behaviour of CG is striking. While the speed of convergence increases during the iteration for the smooth initial error (dashed line), it decreases for the non-smooth initial error (solid line).

This particular example of the one-dimensional Poisson problem has been analysed in several publications and from different points of view; see, e.g. [52, 429, 470, 471]. Using results of [470], it was shown in [429, Section 4], that the errors of CG applied to $Ax^{(1)} = b^{(1)}$ (with $x_0 = 0$) satisfy

$$\frac{\|x^{(1)} - x_n\|_A}{\|x^{(1)} - x_{n-1}\|_A} = \left(\frac{N-n}{N-n+3}\right)^{1/2}, \quad n = 1, \ldots, N.$$

The right-hand side decreases strictly monotonically with the iteration step n, which gives an analytic proof that CG converges superlinearly in this case.

For a discussion of the difference between generic and specific right-hand sides (or initial errors) from the asymptotic point of view we refer to [50, 51]. A different point related to the oscillating pattern of the eigenvectors of the discrete 2D Laplacian and measuring the convergence of the CG method was mentioned in Section 5.1. It will be discussed in more detail in Section 5.9.4.

EXAMPLE 3

An example of Beckermann and Kuijlaars. Following the notation of Beckermann and Kuijlaars in [50, 51], let

$$E_n(\Lambda(A_N)) \equiv \min_{\substack{\varphi(0)=1 \\ \deg(\varphi) \leq n}} \max_{\lambda \in \Lambda(A_N)} |\varphi(\lambda)|,$$

where $\Lambda(A_N)$ is the spectrum of a given $N \times N$ matrix A_N. As mentioned at the end of Section 5.6.2, Beckermann and Kuijlaars studied this quantity for N and n

tending to infinity, while n/N approaches a fixed number in the interval $(0, 1)$. Stating their general results would require us to introduce notation and mathematical concepts that go beyond the scope of this book. We strongly recommend their original papers to interested readers and here restrict ourselves to an illustration of their approach using one of their illustrative examples.

As shown in [50, Corollary 3.2], for the uniformly distributed set of eigenvalues $\{1, 2, \ldots, N\}$ and for every fixed $t \in (0, 1)$,

$$\lim_{\substack{n,N \to \infty \\ n/N \to t}} \frac{1}{n} \log\bigl(E_n(\{1, 2, \ldots, N\})\bigr) = -\frac{(1 + t) \log(1 + t) + (1 - t) \log(1 - t)}{2t}.$$

Similar to Beckermann and Kuijlaars [50, Section 3] we can now forget about asymptotics for a moment, choose a 'large' N and consider the approximation

$$E_n(\{1, 2, \ldots, N\}) \approx \exp\left(-n \frac{(1 + t) \log(1 + t) + (1 - t) \log(1 - t)}{2t}\right)$$
$$= \exp\left(-\frac{(N + n) \log(1 + n/N) + (N - n) \log(1 - n/N)}{2}\right),$$

where we have replaced t by n/N in the last expression.

Figure 5.6 shows numerical results obtained for $N = 100$. The dashed line represents the above estimate of $E_n(\{1, 2, \ldots, 100\})$ for $n = 0, 1, 2 \ldots$ The worst-case convergence curve of CG for the eigenvalues $1, 2, \ldots, 100$ (again computed

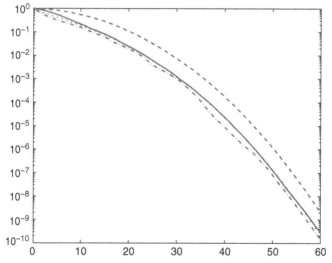

Figure 5.6 The estimate of Beckermann and Kuijlaars is shown by the dashed line. The worst-case behaviour of CG (solid) for the eigenvalues $\{1, 2, \ldots, 100\}$, and the actual CG convergence for $Ax = b$ with $A = \text{diag}(1, 2, \ldots, N)$, $b = [1, \ldots, 1]^T$ and $x_0 = 0$ (dots), are almost the same. The dashed-dotted line shows the CG convergence for $Ax = b$, where b is a random vector.

using the MATLAB package SDPT3) is shown by the solid line. The actual CG convergence curve with $x_0 = 0$ for $Ax = b$, where $A = \text{diag}(1, 2, \ldots, 100)$ and $b = [1, \ldots, 1]^T$, is shown by the dots; the dashed-dotted line shows the CG convergence curve for A and a 'random' b generated by randn in MATLAB.

Remarkably, the asymptotic expression of Beckermann and Kuijlaars is a good estimate of the worst-case curve of CG, even for this rather small choice of N. Moreover, we observe that the convergence for the right-hand side vector consisting of all ones is almost identical to the worst-case convergence for the given matrix A.

5.6.4 Outlying Eigenvalues and Superlinear Convergence

In the literature, a pronounced superlinear convergence behaviour of the CG method (and of Krylov subspace methods in general) is often related to spectra with 'outlying eigenvalues', i.e. spectra where a few (largest or smallest) eigenvalues are well separated from all other eigenvalues. A typical argument in this context is that once such an eigenvalue, say λ_N, is well approximated by a root of the CG iteration polynomial, the method further on behaves as though this eigenvalue had not been present to begin with, and therefore the iteration speed must increase. Let us look at such arguments more closely. We will restrict ourselves to large outlying eigenvalues, and we briefly comment on the case of small outlying eigenvalues at the end of this section.

The key ingredient in analyses that are found in the literature is the observation that for any polynomial $q_\ell(\lambda)$ of degree $\ell \leq n$ and with $q_\ell(0) = 1$ one can write

$$\min_{\substack{\varphi(0)=1 \\ \deg(\varphi) \leq n}} \max_{1 \leq j \leq N} |\varphi(\lambda_j)| \leq \min_{\substack{\varphi(0)=1 \\ \deg(\varphi) \leq n-\ell}} \max_{1 \leq j \leq N} |q_\ell(\lambda_j)\varphi(\lambda_j)|. \qquad (5.6.27)$$

The polynomial that solves the minimisation problem on the right-hand side is composed of the fixed part $q_\ell(\lambda)$ and a part from the minimisation over a set of polynomials of lower degree. We therefore refer to this choice as a *composite polynomial*.

In particular, the polynomial

$$q_\ell(\lambda) \equiv \prod_{j=N-\ell+1}^{N} \left(1 - \frac{\lambda}{\lambda_j}\right), \qquad (5.6.28)$$

satisfies $q_\ell(0) = 1$, $|q_\ell(\lambda_j)| < 1$ for $j = 1, \ldots, N - \ell$, and $q_\ell(\lambda_j) = 0$ for $j = N - \ell + 1, \ldots, N$. For this polynomial we get, analogously to (5.6.21), the following bound on the worst-case convergence of CG,

$$\min_{\substack{\varphi(0)=1 \\ \deg(\varphi) \leq n-\ell}} \max_{1 \leq j \leq N} |q_\ell(\lambda_j)\varphi(\lambda_j)| \leq \max_{1 \leq j \leq N-\ell} |q_\ell(\lambda_j)| \min_{\substack{\varphi(0)=1 \\ \deg(\varphi) \leq n-\ell}} \max_{1 \leq j \leq N-\ell} |\varphi(\lambda_j)|$$

$$\leq 2 \left(\frac{\sqrt{\kappa_\ell(A)} - 1}{\sqrt{\kappa_\ell(A)} + 1}\right)^{n-\ell}, \qquad (5.6.29)$$

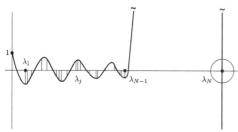

Figure 5.7 Illustration of a composite polynomial that has a root at the large outlying eigenvalue λ_N. The vertical bars indicate the values of the composite polynomial at the individual eigenvalues. The reason why λ_N is encircled will become apparent in Section 5.9.1; see Figure 5.18.

where $\kappa_\ell(A) \equiv \lambda_{N-\ell}/\lambda_1$. This quantity is sometimes called the 'effective condition number', which, as described above, supposedly governs the CG convergence speed after ℓ large outlying eigenvalues have been well approximated by the CG iteration polynomial. An illustration for $\ell = 1$ of the composite polynomial $q_1(\lambda)\varphi(\lambda)$, where $q_1(\lambda) = 1 - \lambda_N^{-1}\lambda$ and $\varphi(\lambda)$ satisfies $\varphi(0) = 1$ and is small on the eigenvalues $\lambda_1, \ldots, \lambda_{N-1}$, is shown in Figure 5.7.

Worst-case convergence bounds of the form (5.6.29) have been used by Axelsson [19], Jennings [364], Jennings and Malik [365], and many others to qualitatively and quantitatively explain superlinear convergence in CG computations in the case of large outlying eigenvalues; see also [27, 28], [24, p. 24–28], and the surveys in [483, 594]. Jennings had already warned in 1977, that such convergence bounds 'do not make any allowance for the effect of rounding error, which may slow down the convergence rate' [364, p. 72]. In fact, as we will show in Section 5.9.1, the validity of this approach in the context of finite precision computations is indeed questionable.

Before further discussing bounds based on composite polynomials we will consider an alternative approach that quantitatively relates different CG computations. The following result is motivated by an idea originally presented by van der Sluis and van der Vorst [633].

Theorem 5.6.9
Let A be an HPD matrix with eigenvalues $0 < \lambda_1 \leq \cdots \leq \lambda_N$ and an eigendecomposition of the form $A = Y \mathrm{diag}(\lambda_j) Y^*$, where $Y^* Y = I$ and $Y = [y_1, \ldots, y_N]$. Let x_n be the nth CG approximation for $Ax = b$ and the initial approximation x_0. Let the initial error be given by

$$x - x_0 = \sum_{j=1}^{N} \zeta_j y_j.$$

On the other hand, let $\tilde{x}_n$ be the nth CG approximation for $Ax = b$ and the initial approximation $\tilde{x}_0$, where the initial error is given by

$$x - \tilde{x}_0 = \sum_{j=1}^{N-\ell} \zeta_j y_j,$$

i.e. $x - \tilde{x}_0$ is the projection of $x - x_0$ onto the $N - \ell$-dimensional subspace span$\{y_1, \ldots, y_{N-\ell}\}$. Then for any iteration step $n \geq \ell$,

$$\|x - \tilde{x}_n\|_A \leq \|x - x_n\|_A \leq \|x - \tilde{x}_{n-\ell}\|_A, \qquad (5.6.30)$$

where the left inequality holds also for $n = 0, 1, \ldots, \ell - 1$.

Proof
First note that for any vector $v = Y[\eta_1, \ldots, \eta_N]^T$ and any polynomial $\varphi(\lambda)$ we have

$$\|\varphi(A)v\|_A^2 = \|A^{1/2}\varphi(A)v\|^2 = \left\|\mathrm{diag}(\lambda_j^{1/2})\varphi(\mathrm{diag}(\lambda_j))[\eta_1, \ldots, \eta_N]^T\right\|^2$$

$$= \sum_{j=1}^{N} \lambda_j |\varphi(\lambda_j)\,\eta_j|^2.$$

We will use this formula repeatedly in this proof.

Let $\tilde{\varphi}_n^{CG}(\lambda)$ and $\varphi_n^{CG}(\lambda)$ denote the nth CG iteration polynomials for the system $Ax = b$ and the initial approximations $\tilde{x}_0$ and x_0, respectively. Then, for any $n = 0, 1, 2, \ldots$,

$$\|x - \tilde{x}_n\|_A^2 = \left\|\tilde{\varphi}_n^{CG}(A)(x - \tilde{x}_0)\right\|_A^2 \leq \left\|\varphi_n^{CG}(A)(x - \tilde{x}_0)\right\|_A^2$$

$$= \sum_{j=1}^{N-\ell} \lambda_j \left|\varphi_n^{CG}(\lambda_j)\,\zeta_j\right|^2 \leq \sum_{j=1}^{N} \lambda_j \left|\varphi_n^{CG}(\lambda_j)\,\zeta_j\right|^2$$

$$= \left\|\varphi_n^{CG}(A)(x - x_0)\right\|_A^2 = \|x - x_n\|_A^2,$$

which proves the left inequality in (5.6.30). Note that in the first inequality in this derivation we have used the optimality of the CG polynomial $\tilde{\varphi}_n^{CG}(\lambda)$ with respect to the initial error $x - \tilde{x}_0$.

To prove the right inequality we consider a polynomial $q_\ell(\lambda)$ as in (5.6.28), so that for every $n \geq \ell$,

$$\|x - x_n\|_A^2 = \sum_{j=1}^{N} \lambda_j \left|\varphi_n^{CG}(\lambda_j)\,\zeta_j\right|^2 \leq \sum_{j=1}^{N} \lambda_j \left|q_\ell(\lambda_j)\tilde{\varphi}_{n-\ell}^{CG}(\lambda_j)\,\zeta_j\right|^2$$

$$= \sum_{j=1}^{N-\ell} \lambda_j \left|q_\ell(\lambda_j)\tilde{\varphi}_{n-\ell}^{CG}(\lambda_j)\,\zeta_j\right|^2 \leq \sum_{j=1}^{N-\ell} \lambda_j \left|\tilde{\varphi}_{n-\ell}^{CG}(\lambda_j)\,\zeta_j\right|^2$$

$$= \left\|\tilde{\varphi}_{n-\ell}^{CG}(A)(x - \tilde{x}_0)\right\|_A^2 = \|x - \tilde{x}_{n-\ell}\|_A^2,$$

and the proof is finished. $\square$

This theorem shows that the CG computation for $Ax = b$ with x_0 is from the $(\ell + 1)$th iteration at least as fast as the computation for $Ax = b$ with $\tilde{x}_0$ from the start. Here $\tilde{x}_0$ is such that the influence of the ℓ largest eigenvalues of A has been

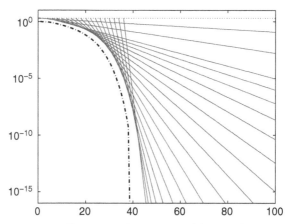

Figure 5.8 Illustration of a sequence of the bounds (5.6.29) with $\ell = 0, 2, 4, \ldots$, i.e. an increasing number of deflated eigenvalues at the upper end of the spectrum.

removed (deflated). We can divide the right inequality in (5.6.30) by $\|x - x_0\|_A$, use that $\|x - x_0\|_A \geq \|x - \widetilde{x}_0\|_A$, and obtain (for $n \geq \ell$) the same bound as in (5.6.29) above,

$$\frac{\|x - x_n\|_A}{\|x - x_0\|_A} \leq \frac{\|x - \widetilde{x}_{n-\ell}\|_A}{\|x - x_0\|_A} \leq \frac{\|x - \widetilde{x}_{n-\ell}\|_A}{\|x - \widetilde{x}_0\|_A} \leq 2 \left(\frac{\sqrt{\kappa_\ell(A)} - 1}{\sqrt{\kappa_\ell(A)} + 1} \right)^{n-\ell}.$$

This bound is illustrated in Figure 5.8. The dashed-dotted line shows a CG convergence curve computed using double reorthogonalisation of the residual vectors, which simulates exact arithmetic; cf. [280]. The lines originating at the points $[\ell, 2] = [0, 2], [2, 2], [4, 2], \ldots$ show the corresponding convergence bounds for $\ell = 0, 2, 4, \ldots$ (note that the value of the bound always starts at 2). These bounds form a close envelope for the actual (exact arithmetic) convergence curve.

Following the previous arguments, one may argue that the convergence rate of CG should increase whenever roots of $\varphi_n^{CG}(\lambda)$ approximate closely some large outlying eigenvalues of A. The roots of the CG iteration polynomial $\varphi_n^{CG}(\lambda)$ are equal to the eigenvalues $\theta_1^{(n)}, \ldots, \theta_n^{(n)}$ of the Jacobi matrix T_n, i.e. the Ritz values of A determined at the nth CG iteration. Hence the occurrence of superlinear convergence should be related to the convergence of Ritz values to the eigenvalues of A. This has been thoroughly investigated by van der Sluis and van der Vorst [633]. They considered approximation of an arbitrary eigenvalue of A, with a particular emphasis on the lower end of the spectrum. In the notation of Theorem 5.6.9, they compared the iterates $x_{\ell+m}$, $m = 1, 2, \ldots$, of the CG computations for A, b and x_0 with the CG computations for A, b, and the initial error constructed by projecting $x - x_\ell$ onto the subspace span$\{y_{\ell+1}, \ldots, y_N\}$. Their results show that in exact arithmetic there is indeed a connection between the convergence of Ritz values *at the lower end of the spectrum* and superlinear convergence behaviour of CG (we will comment on the upper end below).

The following result (cf. [457, Theorem 3.3]) addresses the link between convergence of CG and convergence of the Ritz values in a different way; cf. also Theorem 5.6.3.

Theorem 5.6.10
Consider a linear algebraic system $Ax = b$, where A is an HPD matrix with eigenvalues $\lambda_1 \leq \cdots \leq \lambda_N$. Let x_0 be such that r_0 is of grade d with respect to A. Then for each $n = 1, \ldots, d$ there exists $\xi_n \in [\lambda_1, \lambda_N]$, such that the nth error of CG satisfies

$$\|x - x_n\|_A^2 = \frac{\|r_0\|^2}{\xi_n^{2n+1}} \sum_{i=1}^{N} \omega_i \prod_{j=1}^{n} \left(\lambda_i - \theta_j^{(n)}\right)^2, \qquad (5.6.31)$$

where $\theta_1^{(n)}, \ldots, \theta_n^{(n)}$ are the nth Ritz values and $\omega_1, \ldots, \omega_N$ are the weights of the distribution function $\omega(\lambda)$ determined by A and r_0 (see (3.5.2)).

Proof
It was shown in Theorem 3.5.2 that

$$\frac{\|x - x_n\|_A^2}{\|r_0\|^2} = E_\omega^n(\lambda^{-1}),$$

where the right-hand side is the approximation error of the n-node Gauss–Christoffel quadrature for $f(\lambda) = \lambda^{-1}$ with the distribution function $\omega(\lambda)$. The assertion now immediately follows from the formula (3.2.21) for $f(\lambda) = \lambda^{-1}$, with the interval of integration containing in its interior $[\lambda_1, \lambda_N]$, and the fact that the polynomial $\psi_n(\lambda)$ in that formula is given by $\psi_n(\lambda) = (\lambda - \theta_1^{(n)}) \cdots (\lambda - \theta_n^{(n)})$. □

The right-hand side of (5.6.31) shows that once a Ritz value $\theta_j^{(n)}$ is (very) close to an eigenvalue λ_i of A, the component of the initial residual in the direction of the corresponding eigenvector (represented by ω_i) is eliminated. In exact arithmetic this indeed can explain the superlinear convergence behaviour of CG for spectra with outlying eigenvalues.

However, van der Sluis and van der Vorst, who focused on the lower end of the spectrum, stated with full clarity in 1986 [633, p. 559]:

> Our theory, of course, applies to the upper end as well. But in practice, due to the presence of rounding errors, the situation there is much more cumbersome ... convergence of the Ritz value ... and consequently very rapid loss of orthogonality ... spoils any convergence pattern that we might observe in exact computation. Therefore, convergence of Ritz values at the upper end of the spectrum will rarely lead to impressive increases of the rate of convergence of the CG process.

It is worth pointing out that similar considerations can already be found in Lanczos' paper of 1952 [406, pp. 39 and 46]. In 1959, Rutishauser thoroughly examined the possibilities of preconditioning by polynomials that filter out the smallest eigenvalues (see also Section 5.5.3) and, concerning the upper end of the spectrum, concluded [163, p. 36]: 'In numerical computations however, exact smoothing, i.e. the total elimination of the contribution of all eigenvalues above a given limit Λ, can practically never be achieved.' One should also read the experimental part of [163] written by Ginsburg.

The effects of rounding errors indeed drastically limit the practical applicability of bounds like (5.6.29), originally presented in [19, 364, 365]. In short, unless the matrix A has a very special distribution of eigenvalues such that the effects of rounding errors are minimal, or unless the number of iterations is very small, these bounds are *practically useless*. We come back to this in Section 5.9.1, where we explain what happens in finite precision arithmetic.

Let us mention that for outlying eigenvalues on the *lower end* of the spectrum it is observed that the results developed in [19, 364, 365] are relevant even in finite precision arithmetic computations. However, as already noticed by Jennings [364, Section 7, in particular formula (7.9)], in that case they offer a limited insight. To see this point, consider a polynomial $q_\ell(\lambda)$ that has roots at the ℓ smallest eigenvalues. For this polynomial we have no guarantee that it is small on the rest of the spectrum, and in fact we can expect that $|q_\ell(\lambda_j)| \gg 1$ for $j = \ell + 1, \ldots, N$.

Finally, we remark that the concept of a polynomial with roots at (or close to) the smallest eigenvalues appears frequently in the context of the so-called 'deflated' Krylov subspace methods. Here such polynomials are constructed implicitly, e.g. by adding computed eigenvector approximations corresponding to the smallest eigenvalues to the computed Krylov subspaces. The goal to accelerate the speed of convergence has been successfully reached in many practical applications. The numerical observations and intuitive arguments are, however, rarely accompanied by a quantitative analysis. In particular, it is often unclear what features of the problem really need to be removed (or deflated) to guarantee that the convergence rate of a Krylov subspace method increases.

Deflation techniques have been studied intensively since the mid 1980s. One of the first to apply deflation to the CG method was Nicolaides [482]. Numerous papers on the same subject followed; see, e.g. [166, 546, 653]. As pointed out by Gutknecht [288], some of the early papers using the idea of deflation, in particular [143, 440], were initially almost unnoticed and are not frequently quoted in the later literature. Here one should also recall the work of Engeli, Ginsburg, Rutishauser, and Stiefel from 1959 [163], and Lanczos from 1952 [406, 407]; see Section 5.5.3. For further details and additional references on deflation techniques applied to Krylov subspace methods we refer to, e.g. [288, 33, 450, 218].

5.6.5 Clustered Eigenvalues and the Sensitivity of the Gauss–Christoffel Quadrature

We have seen above that the distribution of the eigenvalues of A has a strong influence on the convergence behaviour of CG. In particular, the eigenvalues completely determine the worst-case behaviour; see Theorem 5.6.6 and the bound (5.6.20). A rather intriguing situation occurs when the eigenvalues appear in several tight and well separated clusters. In some publications it has been suggested that for the (approximate) solution of the minimisation problem in (5.6.20) and, consequently, for the convergence of CG computations, it does not matter much whether A has single (well separated) eigenvalues or tight (well separated) clusters of eigenvalues. For examples we refer to [354, Section 9], [379, p. 18 and Exercise 2.8.5], and [395, p. 261].

In order to examine the situation we will present a numerical example. Let A be an HPD diagonal matrix with the N real and positive diagonal entries $\lambda_1 < \lambda_2 < \cdots < \lambda_{N-1} < \lambda_N$ (being the eigenvalues of A), where

$$\lambda_j = \lambda_1 + \frac{j-1}{N-1}(\lambda_N - \lambda_1)\gamma^{N-j}, \quad j = 2, 3, \ldots, N-1, \qquad (5.6.32)$$

for given $\lambda_N > \lambda_1 > 0$ and $\gamma > 0$. Matrices with spectra of the form (5.6.32) appeared in [594, Section 2], and they have been used for experiments with the CG and Lanczos methods in a number of publications; see, e.g. [272, 280, 453, 457, 598]. With $\gamma = 1$ the eigenvalues are uniformly distributed in the interval $[\lambda_1, \lambda_N]$, while $\gamma < 1$ yields a spectrum where most of the eigenvalues accumulate close to the left end of the spectrum. Nice examples of preconditioned matrices having spectra with analogous distribution of eigenvalues can be found in the two closely related papers [27] and [28].

For this experiment we choose

$$N = 34, \quad \lambda_1 = 0.1, \quad \lambda_N = 100, \quad \gamma = 0.65.$$

Then 24 of the 34 eigenvalues are inside the interval $[0.1, 1.04]$, 33 of the 34 eigenvalues are inside the interval $[0.1, 63.1]$, and λ_N is an outlying eigenvalue on the rightmost part of the spectrum. The resulting diagonal matrix A has the condition number $\kappa(A) = 1000$. To set up a linear algebraic system $Ax = b$ we consider a randomly generated right-hand side b with normally distributed entries ranging between 0 and 1.

Remark 5.6.11
We emphasise that in experiments illustrating the convergence rate of iterative methods applied to practical problems, random right-hand sides or initial residuals should be avoided as much as possible. In most applications a problem-related right-hand side is given which is anything but random. In the example here we are not interested in measuring the performance of the CG method. The right-hand side is chosen to be random in order to demonstrate that the illustrated phenomena are *not* linked to a specific right-hand side. Some additional comments on possible consequences of choosing a random initial approximation x_0 are given in Section 5.8.3.

In addition to the system $Ax = b$ we consider two systems, $By = b_1$ and $Cz = b_2$, where B and C have tight clusters of eigenvalues, each cluster corresponding to an eigenvalue of A. The $2N \times 2N$ diagonal matrix B is obtained by 'splitting' each of the diagonal entries of A into two diagonal entries of B. More precisely, $B = \mathrm{diag}(\mu_1, \ldots, \mu_{2N})$, where

$$\mu_{2j-1} = \lambda_j - \Delta, \quad \mu_{2j} = \lambda_j + \Delta, \quad j = 1, 2, \ldots, N,$$

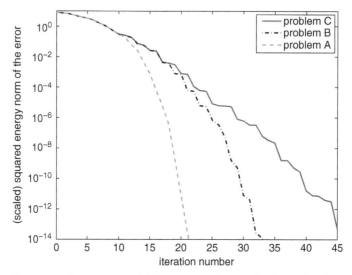

Figure 5.9 Convergence of CG in exact arithmetic for three related linear algebraic systems, where the matrices B and C have tight clusters of eigenvalues with two respectively five eigenvalues corresponding to each individual eigenvalue of the matrix A.

and $\Delta = 10^{-8}$ (supposing that $\Delta \ll \lambda_1$, the exact value of Δ is unimportant). The right-hand side b_1 is obtained from b by splitting each individual entry of b into two parts of the same absolute value, one corresponding to μ_{2j-1} and one to μ_{2j}, and then normalising so that $\|b_1\| = \|b\|$. The system $Cz = b_2$ is generated analogously, with each of the diagonal entries of A split into *five* close diagonal entries of C.

Figure 5.9 shows the convergence of the CG method. The dashed curve shows the squared energy-norm of the error for the system with the matrix A scaled by the squared Euclidean residual norm, i.e. $\|x - x_n\|_A^2 / \|r_0\|^2$. The analogous quantities for the systems with B and C are shown by the dashed-dotted and solid curves, respectively. In each case the initial approximation x_0 is the zero vector. The differences in the behaviour seen in Figure 5.9 do *not* result from rounding errors; the CG computations were performed with double reorthogonalisation of the residual vectors. The following can be observed in this figure:

1. The behaviour of CG for the three linear algebraic systems is quantitatively and qualitatively very different, although the three system matrices have essentially the same condition number and even very close eigenvalues. We have not plotted the bound (5.6.23) based on the condition number, but it is clear that this bound does not describe well any of the three CG convergence curves.
2. The convergence curve for the matrix C shows a staircase-like behaviour (to a lesser extent this also holds for the matrix B), which nicely demonstrates the nonlinear behaviour of CG; see Section 5.5.4 for general comments. Given the behaviour at a single step, or in a few consecutive

steps, nothing can in general be said about the behaviour in the subsequent iterations. Moreover, the experiment demonstrates that nonlinear behaviour of Krylov subspace methods is not related to 'undesirable' properties of the matrix (like indefiniteness, ill conditioning, or non-normality). Apparently, such behaviour can occur even for well conditioned, diagonal, and symmetric positive definite matrices. In order to avoid any confusion, we again emphasise that here we deal with *mathematical* phenomena that are not affected by rounding errors.

3. Most importantly for the point we are going to make in this section, the experiment shows that the arguments (made at the beginning of the section) about the behaviour of CG in the presence of eigenvalue clusters is *in general not valid*. More precisely, it is not true that CG applied to a matrix with t tight eigenvalue clusters behaves similarly as though it were applied to a matrix with t distinct eigenvalues representing the individual clusters. In particular, with t clusters of eigenvalues, t iterations of the CG method *may not* produce a reasonably accurate approximation to the exact solution.

If we replace individual eigenvalues (here the eigenvalues of A) by very tight clusters (the eigenvalues of B or C), or, vice versa, if tight clusters (the eigenvalues of B or C) are replaced by single representatives (the eigenvalues of A), then the *convergence behaviour of the CG method can change dramatically*. This fact has been clearly presented by Greenbaum [269]. It means that a 'bird's eye view' of the spectrum is not sufficient to determine the actual behaviour of the method.

Finite precision computations with CG in the presence of large outlying eigenvalues will be analysed in Section 5.9.1.

We will now reformulate the previous experiment in terms of the Gauss–Christoffel quadrature and discuss the sensitivity of the latter method. We know that the CG method can be interpreted as a matrix formulation of the Gauss–Christoffel quadrature, and that the quadrature error for the integrated function $f(\lambda) = \lambda^{-1}$ is equal to the squared A-norm of the CG error divided by the squared Euclidean norm of the initial residual; see Section 3.5 and in particular (3.5.9).

In the previous experiment we have replaced single well separated eigenvalues of the matrix A by tight well separated eigenvalue clusters of the matrices B and C, and we have correspondingly modified the right-hand sides of the linear algebraic systems. In terms of the distribution function determined by the system matrix and the initial residual, these changes mean that we have replaced the original function $\omega_A(\lambda)$ determined by A and b by closely related (in the sense described above) functions $\omega_B(\lambda)$ determined by B and b_1, and $\omega_C(\lambda)$ determined by C and b_2. We have seen that the exact arithmetic convergence behaviour of CG for the three systems was significantly different.

For a given (fixed) number of quadrature nodes, say n, the quadrature approximations

$$I^n_{\omega_A}(\lambda^{-1}), \quad I^n_{\omega_B}(\lambda^{-1}), \quad I^n_{\omega_C}(\lambda^{-1})$$

to the integral of the same function $f(\lambda) = \lambda^{-1}$ but with different distribution functions $\omega_A(\lambda), \omega_B(\lambda), \omega_C(\lambda)$ can differ substantially,

$$I_{\omega_A}(\lambda^{-1}) \equiv \int \lambda^{-1} d\omega(\lambda) = I_{\omega_A}^n(\lambda^{-1}) + \frac{\|x - x_n^A\|_A^2}{\|r_0\|^2},$$

$$I_{\omega_B}(\lambda^{-1}) \equiv \int \lambda^{-1} d\omega_B(\lambda) = I_{\omega_B}^n(\lambda^{-1}) + \frac{\|x - x_n^B\|_B^2}{\|r_0\|^2},$$

$$I_{\omega_C}(\lambda^{-1}) \equiv \int \lambda^{-1} d\omega_C(\lambda) = I_{\omega_C}^n(\lambda^{-1}) + \frac{\|x - x_n^C\|_B^2}{\|r_0\|^2}.$$

For the close distribution functions (in the sense described in the experiment), the values of the integrals to be approximated are very close. Indeed, here

$$I_{\omega_A}(\lambda^{-1}) \approx 7.7335,$$
$$|I_{\omega_A}(\lambda^{-1}) - I_{\omega_B}(\lambda^{-1})| \approx 7.4607 \times 10^{-14},$$
$$|I_{\omega_A}(\lambda^{-1}) - I_{\omega_C}(\lambda^{-1})| \approx 1.4122 \times 10^{-13}.$$

However, as we see in Figure 5.10, the differences between the quadrature approximations, i.e. the values

$$\left|I_{\omega_A}^n(\lambda^{-1}) - I_{\omega_B}^n(\lambda^{-1})\right| \quad \text{and} \quad \left|I_{\omega_A}^n(\lambda^{-1}) - I_{\omega_C}^n(\lambda^{-1})\right|$$

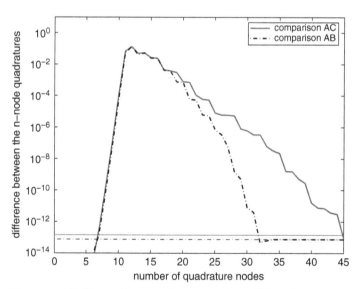

Figure 5.10 Difference between the Gauss–Christoffel quadrature errors for the integrated function $f(\lambda) = \lambda^{-1}$ and for the three distribution functions determined by the linear algebraic systems given above, where the matrices B and C have tight clusters of eigenvalues with two respectively five eigenvalues corresponding to each individual eigenvalue of the matrix A.

are for the number of quadrature nodes between 8 and 30, respectively 8 and 44 significantly larger (even many orders of magnitude) than the difference between the corresponding integrals (the latter are plotted by the horizontal lines in Figure 5.10).

As demonstrated by this example, the Gauss–Christoffel quadrature can be highly sensitive to small changes of the distribution function that change its support (i.e. the number of points of increase). Precisely this point was made in [487]. The proof of the main result in that paper is based on what we have stated as Theorem 3.2.7 in this book.

Numerical examples in [487] demonstrate that the difference between Gauss–Christoffel quadrature approximations (using the same number of quadrature nodes) of integrals with close distribution functions $\omega(\lambda)$ and $\widetilde{\omega}(\lambda)$ can be many orders of magnitude larger than the difference between the integrals being approximated, i.e.

$$|I_\omega^n(f) - I_{\widetilde{\omega}}^n(f)| \gg |I_\omega(f) - I_{\widetilde{\omega}}(f)|.$$

This can happen for analytic integrands and discontinuous, continuous, and even analytic distribution functions. For further details and connections to different problems such as *computation of the Gauss–Christoffel quadrature* we refer to [487].

The sensitivity of the Gauss–Christoffel quadrature described in [487] seems counterintuitive: the last thing one would like to see is that a small change of the approximated integral causes changes of its quadrature approximation that are larger by orders of magnitude. A quadrature method with this property must be given a special attention in any application where the inaccuracies in determining the problem to be integrated can enlarge the *support of the corresponding distribution function*. As the CG method, the Gauss–Christoffel quadrature is a perfectly tuned tool matching the maximal number of the underlying moments. As other perfectly tuned tools, it should be used with proper care.

Many problems remain open. In particular, it would be of great theoretical and practical interest to characterise quantitatively classes of problems for which the described sensitivity occurs.

5.7 GMRES IN EXACT ARITHMETIC

Assuming exact arithmetic we study in this section the convergence behaviour of the GMRES method [544]. More precisely, we study the mathematical properties of the projection process

$$x_n \in x_0 + \mathcal{K}_n(A, r_0) \quad \text{and} \quad r_n \perp A\mathcal{K}_n(A, r_0), \quad n = 1, \ldots, d, \qquad (5.7.1)$$

where $A \in \mathbb{F}^{N \times N}$ is a general nonsingular matrix, $b \in \mathbb{F}^N$ and the initial residual $r_0 = b - Ax_0$ is of grade d with respect to A. Results in this section apply to any Krylov subspace method that is based on a projection process of the form (5.7.1).

As shown in case (3) of Theorem 2.3.1, the projection process (5.7.1) is for a nonsingular matrix A well defined at each step $n = 1, \ldots, d$, and the equivalent optimality property is

$$\|r_n\| = \min_{z \in x_0 + \mathcal{K}_n(A, r_0)} \|b - Az\|. \tag{5.7.2}$$

It is easy to see that the nth residual vector is of the form $r_n = \varphi_n^{MR}(A) r_0$, where $\varphi_n^{MR}(\lambda)$ is a (uniquely determined) polynomial of degree at most n with $\varphi_n^{MR}(0) = 1$. Here the acronym 'MR' stands for 'minimal residual'. Using polynomials, equation (5.7.2) can equivalently be written as

$$\|r_n\| = \|\varphi_n^{MR}(A) r_0\| = \min_{\substack{\varphi(0) = 1 \\ \deg(\varphi) \le n}} \|\varphi(A) r_0\|. \tag{5.7.3}$$

At first sight the minimisation problem (5.7.3) looks very similar to the CG minimisation problem (5.6.16). There are, however, some significant differences between the two problems. Some of these differences are indicated by our discussion of the moment matching properties of the Hermitian Lanczos and the Arnoldi algorithm in Chapter 3. Recall that the Arnoldi algorithm is used within GMRES to generate orthonormal Krylov subspace bases; see the implementation of GMRES in Algorithm 2.5.2. As shown in Section 3.7.3, the reduced model generated by the Arnoldi algorithm matches only half as many moments as the reduced model generated by the Hermitian Lanczos algorithm that is used within CG. Similarly to Hermitian Lanczos, the Arnoldi algorithm generates orthogonal polynomials with respect to a discrete inner product; see (2.4.2). However, as mentioned in Section 3.7.4, the corresponding generalised Gauss–Christoffel quadrature approximations (in the complex plane) do not provide analogous quantitative relationships as in the case of the CG method and the Gauss–Christoffel quadrature on the real line. In particular, they do not provide tools based on distribution functions with the corresponding quadrature nodes and weights that would allow us to analyse GMRES in a similar fashion to CG.

5.7.1 Ritz Values and Harmonic Ritz Values

We start our investigations with a closer look at the GMRES iteration polynomials $\varphi_n^{MR}(\lambda)$. Polynomials that are mathematically characterised by (5.7.3) have been studied in several papers and under different names. Freund [201] and Fischer [183] (whose focus was on real symmetric matrices) called them *kernel polynomials*. This term had already been used by Stiefel in 1958 [581]. Manteuffel and Otto [441] used the term *residual polynomials*, Paige, Parlett, and van der Vorst [494] used *MR polynomials* (they also considered real symmetric matrices), and Goossens and Roose [254] used *GMRES residual polynomials*.

The proof of the following theorem, which we skip for brevity, is in some way present in most of these works. Our formulation is adapted from Freund's paper [201, Section 5], which contains a complete derivation for the case of a general matrix A.

Theorem 5.7.1
Consider a linear algebraic system $Ax = b$, where $A \in \mathbb{F}^{N \times N}$ is nonsingular and $b \in \mathbb{F}^N$, and let the initial residual $r_0 = b - Ax_0$ be of grade d with respect to A. For any $n \leq d$, consider the nth GMRES approximation x_n to the solution x as in Section 2.5.5,

$$x_n = x_0 + V_n t_n, \quad \text{where} \quad t_n = (\underline{H}_{n,n}^* \underline{H}_{n,n})^{-1} \underline{H}_{n,n}^* (\|r_0\| e_1);$$

see (2.5.54). Then the corresponding nth GMRES residual polynomial is given by

$$\varphi_n^{MR}(\lambda) = \frac{\det(\lambda H_{n,n}^* - \underline{H}_{n,n}^* \underline{H}_{n,n})}{(-1)^n \det(\underline{H}_{n,n}^* \underline{H}_{n,n})}, \tag{5.7.4}$$

and hence the zeros of $\varphi_n^{MR}(\lambda)$ are the eigenvalues of the generalised eigenvalue problem

$$(\underline{H}_{n,n}^* \underline{H}_{n,n}) z = \lambda H_{n,n}^* z. \tag{5.7.5}$$

To comment on this result we first recall from Section 2.5.5 that the $(n+1) \times n$ unreduced upper Hessenberg matrix $\underline{H}_{n,n}$ has full rank n, so that $\underline{H}_{n,n}^* \underline{H}_{n,n}$ is nonsingular. If additionally the matrix $H_{n,n} = V_n^* A V_n$ is nonsingular, then the generalised eigenvalue problem (5.7.5) is equivalent to

$$H_{n,n}^{-*} (\underline{H}_{n,n}^* \underline{H}_{n,n}) z = \lambda z.$$

In this case the GMRES residual polynomial $\varphi_n^{MR}(\lambda)$ is of exact degree n and its n roots are the eigenvalues of the invertible $n \times n$ matrix $H_{n,n}^{-*}(\underline{H}_{n,n}^* \underline{H}_{n,n})$.

However, for a general (non-HPD) matrix A we can encounter a *singular* matrix $H_{n,n}$. A simple example is given by

$$A = \begin{bmatrix} 0 & 1 \\ 1 & 0 \end{bmatrix}, \quad V_1 = [v_1] = \begin{bmatrix} 1 \\ 0 \end{bmatrix}, \quad \text{so that} \quad V_1^* A V_1 = H_{1,1} = [0].$$

Using $H_{1,1} = [0]$ in (5.7.4) shows that the corresponding GMRES iteration polynomial is $\varphi_1^{MR}(\lambda) \equiv 1$. In general, the determinant expression in the numerator on the right-hand side of (5.7.4) shows that a singular matrix $H_{n,n}$ implies that the degree of $\varphi_n^{MR}(\lambda)$ is strictly less than n.

As shown by Brown [85, Theorem 3.1], the occurrence of a singular matrix $H_{n,n}$ is necessary and sufficient for an *exact stagnation* of GMRES at step n with $x_{n-1} = x_n$ and hence $r_{n-1} = r_n$. An example of exact stagnation of GMRES had previously been presented by Saad and Schultz in [544, p. 865]. We will show in Section 5.7.4 below that there exist linear algebraic systems for which exact stagnation of GMRES can occur for any number of subsequent steps and at any point during the iteration.

Numerous works contain analyses of the exact stagnation of GMRES (see, e.g. [428, 454, 455, 568, 571, 686, 687]), and, starting with Brown's paper [85], some authors have related the exact stagnation of GMRES to the breakdown of the Krylov subspace method FOM; see Section 2.5.6. More generally, Brown showed that a nonsmooth behaviour of the Euclidean residual norm in the FOM method (which he referred to as 'Arnoldi's method') is related to a near stagnation of GMRES [85, Theorem 5.1]. An analogous 'peak-plateau behaviour' was later shown for other pairs of methods including CG/MINRES and BiCG/QMR; see [121] or [272, Section 5.4 and Exercise 5.1]. Expressions for the FOM and GMRES residual norms involving the Hessenberg matrices that occur in the Arnoldi algorithm were derived in [456]. For a comparison of the residual norms of GMRES, FOM, BiCG, and QMR we refer to [327].

We will now study whether the roots of the GMRES iteration polynomial can be interpreted as certain Ritz values. Previously in this book, the Ritz values of an HPD matrix A were defined as the eigenvalues of the Jacobi matrix T_n that results from the application of the Hermitian Lanczos algorithm to A and the initial vector $v_1 = r_0/\|r_0\|$. These values are the roots of the nth CG iteration polynomial $\varphi_n^{CG}(\lambda)$; see (5.6.3). Let $\theta_j^{(n)}$ be such a Ritz value and $z_j^{(n)} \neq 0$ be a corresponding Ritz vector,

$$T_n z_j^{(n)} = \theta_j^{(n)} z_j^{(n)}.$$

Using $T_n = V_n^* A V_n$ and $V_n^* V_n = I_n$, this equation can be written as

$$V_n^* \left(A y_j^{(n)} - \theta_j^{(n)} y_j^{(n)} \right) = 0, \quad \text{where} \quad y_j^{(n)} \equiv V_n z_j^{(n)}.$$

The pair $(\theta_j^{(n)}, y_j^{(n)})$ is thus characterised by the orthogonality condition

$$A y_j^{(n)} - \theta_j^{(n)} y_j^{(n)} \perp \mathcal{K}_n(A, r_0), \quad \text{where} \quad y_j^{(n)} \in \mathcal{K}_n(A, r_0),$$

which motivates the following generalisation.

Definition 5.7.2: The (real or complex) number θ is called a *Ritz value* of the matrix $A \in \mathbb{C}^{N \times N}$ with respect to a given subspace $\mathcal{W} \subseteq \mathbb{C}^N$ and with a corresponding *Ritz vector* $y \neq 0$, if

$$A y - \theta y \perp \mathcal{W}, \quad \text{where} \quad y \in \mathcal{W}.$$

In the terminology of this definition, the previously studied Ritz values of an HPD matrix A determined at the nth iteration step of the CG or the Lanczos method are called the Ritz values of A with respect to $\mathcal{K}_n(A, r_0)$.

To study the roots of the GMRES iteration polynomials, i.e. the eigenvalues of (5.7.5), we consider the Arnoldi relation $AV_n = V_{n+1}\underline{H}_{n,n}$; see (2.5.53). Using the orthonormality of the columns of V_{n+1} we see that

$$\underline{H}_{n,n}^* \underline{H}_{n,n} = \underline{H}_{n,n}^* V_{n+1}^* V_{n+1} \underline{H}_{n,n} = (AV_n)^* AV_n.$$

In addition, we know that $H_{n,n}^* = V_n^* A^* V_n$. Therefore, if $(\mu_j^{(n)}, z_j^{(n)})$ is an eigenpair of the generalised eigenvalue problem (5.7.5), then $\mu_j^{(n)} \neq 0$ and the equation (5.7.5) for this eigenpair can be written as

$$\begin{aligned}
0 &= (AV_n)^* \left(AV_n z_j^{(n)} - \mu_j^{(n)} V_n z_j^{(n)}\right) \\
&= -\mu_j^{(n)}(AV_n)^* \left(A^{-1} y_j^{(n)} - (\mu_j^{(n)})^{-1} y_j^{(n)}\right),
\end{aligned}$$

where $y_j^{(n)} \equiv AV_n z_j^{(n)} \in A\mathcal{K}_n(A, r_0)$. Equivalently, the pair $(\mu_j^{(n)}, y_j^{(n)})$ is characterised by

$$A^{-1} y_j^{(n)} - \left(\mu_j^{(n)}\right)^{-1} y_j^{(n)} \perp A\mathcal{K}_n(A, r_0), \quad \text{where} \quad y_j^{(n)} \in A\mathcal{K}_n(A, r_0). \tag{5.7.6}$$

Using Definition 5.7.2, this can be summarised as follows.

Corollary 5.7.3
The roots of the nth GMRES iteration polynomial $\varphi_n^{MR}(\lambda)$ are given by the reciprocals of the Ritz values of A^{-1} with respect to the subspace $A\mathcal{K}_n(A, r_0)$.

If A is Hermitian (or real symmetric), then the sum of the Ritz values of A with respect to $\mathcal{K}_n(A, r_0)$ can be interpreted as a weighted *arithmetic* mean of the eigenvalues of A; see Proposition 3.4.8 in Section 3.4.2. Similarly, the reciprocals of the Ritz values of A^{-1} correspond to a weighted *harmonic* mean[10] of the eigenvalues of A. This motivated the authors of [494] to call the latter values *harmonic Ritz values*. The paper [494] focuses on the real symmetric case, and a generalisation to the non-Hermitian case was presented in [254]. Besides the GMRES context, harmonic Ritz values play an important role in iterative methods for eigenvalue problems, and in particular in the Jacobi–Davidson method; see [572] and the references given there for details.

In summary, even for an HPD matrix A, the relation between the roots of the iteration polynomial $\varphi_n^{MR}(\lambda)$ and the eigenvalues of A is much more complicated than for CG and its iteration polynomial $\varphi_n^{CG}(\lambda)$. This gives another indication of the significant differences between CG and GMRES that we have mentioned above.

5.7.2 Convergence Descriptions Based on Spectral Information

In this section we will briefly review how spectral information is used for describing the convergence behaviour of GMRES. Our presentation benefits from the survey papers [427, 595]. A very nice description of the material related to the rest of Section 5.7 can be found in the survey paper by Eiermann and Ernst [154].

As above, we consider a linear algebraic system $Ax = b$ with a general nonsingular matrix A, and an initial residual $r_0 = b - Ax_0$ of grade d with respect to A. Let

---

[10]. The arithmetic and harmonic means of n nonzero real numbers $\mu_1, \ldots, \mu_n$ are respectively defined as $n^{-1} \sum_{j=1}^n \mu_j$ and $n(\sum_{j=1}^n \mu_j^{-1})^{-1}$.

$A = YJY^{-1}$, $J = \text{diag}(J_1, \ldots, J_k)$, be a Jordan decomposition of A. Then (5.7.3) leads to the GMRES convergence bound

$$\|r_n\| = \min_{\substack{\varphi(0)=1 \\ \deg(\varphi) \leq n}} \|\varphi(A) r_0\| = \min_{\substack{\varphi(0)=1 \\ \deg(\varphi) \leq n}} \|Y\varphi(J) Y^{-1} r_0\| \tag{5.7.7}$$

$$\leq \|Y\| \, \|Y^{-1} r_0\| \min_{\substack{\varphi(0)=1 \\ \deg(\varphi) \leq n}} \max_{1 \leq i \leq k} \|\varphi(J_i)\| \tag{5.7.8}$$

$$\leq \kappa(Y) \|r_0\| \min_{\substack{\varphi(0)=1 \\ \deg(\varphi) \leq n}} \max_{1 \leq i \leq k} \|\varphi(J_i)\|, \tag{5.7.9}$$

for $n = 1, \ldots, d$, where $\kappa(Y) \equiv \|Y\| \, \|Y^{-1}\|$ is the 2-norm condition number of Y. This bound was presented (for diagonalisable A) by Saad and Schultz in [544, Proposition 4] and even earlier by Elman [157] in the context of the generalised conjugate residual (GCR) method that is mathematically equivalent to GMRES (see Section 5.10 for some comments on methods that are mathematically equivalent to GMRES). We formally state some properties of this bound.

Theorem 5.7.4
Using the previous notation, the relative Euclidean residual norm in the GMRES method satisfies

$$\frac{\|r_n\|}{\|r_0\|} \leq \kappa(Y) \min_{\substack{\varphi(0)=1 \\ \deg(\varphi) \leq n}} \max_{1 \leq i \leq k} \|\varphi(J_i)\|. \tag{5.7.10}$$

If A is normal, i.e. unitarily diagonalisable, with eigenvalues $\lambda_1, \ldots, \lambda_N$, then

$$\frac{\|r_n\|}{\|r_0\|} \leq \min_{\substack{\varphi(0)=1 \\ \deg(\varphi) \leq n}} \max_{1 \leq i \leq N} |\varphi(\lambda_i)|, \tag{5.7.11}$$

and in this case the bound is sharp in the sense that for each iteration step $n = 1, \ldots, d-1$ there exists an initial approximation $x_0^{(n)}$, so that equality holds in (5.7.11).

Proof
The first bound was shown in (5.7.9), the second follows trivially in the normal case. For the sharpness result we refer to the two independent proofs given in [277, 372]. □

We will now describe the implications of this result for different classes of matrices. We start with the case of an HPD matrix A. For such a matrix the right-hand side of (5.7.11) coincides with the worst-case bound for the CG method given in (5.6.19)–(5.6.20). Hence in the HPD case the worst-case convergence behaviour of GMRES can be analysed using the techniques presented in Section 5.6.2. In addition, the worst-case convergence of GMRES and CG, measured respectively by the relative Euclidean residual norm and the relative A-norm of the error, is identical. It is important to note that, given the iteration step n, the worst-case behaviour of CG and GMRES is attained, in general, for different initial residuals. Examples using the discretised 1D reaction-diffusion equation and the resulting symmetric positive definite tridiagonal Toeplitz matrices are given in [429].

When A is Hermitian *indefinite* it has positive and negative real eigenvalues, say

$$\lambda_1 \leq \cdots \leq \lambda_s < 0 < \lambda_{s+1} \leq \cdots \leq \lambda_N.$$

It is a well-known result of approximation theory (see, e.g. [136, Chapter 3, Theorem 2.4 and Corollary 2.5]) that the value on the right-hand side of (5.7.11) is equal to the right-hand side of (5.6.20) for $n+1$ distinct eigenvalues of A. To obtain an upper bound one can determine two intervals (instead of one in the HPD case) containing all eigenvalues and excluding the origin,

$$I^- \cup I^+ \equiv [\lambda_1, \lambda_s] \cup [\lambda_{s+1}, \lambda_N],$$

which leads to the bound

$$\min_{\substack{\varphi(0)=1 \\ \deg(\varphi) \leq n}} \max_{1 \leq i \leq N} |\varphi(\lambda_i)| \leq \min_{\substack{\varphi(0)=1 \\ \deg(\varphi) \leq n}} \max_{z \in I^- \cup I^+} |\varphi(z)|.$$

When the two intervals I^- and I^+ have the same length, $\lambda_N - \lambda_{s+1} = \lambda_s - \lambda_1$, it can be shown that

$$\min_{\substack{\varphi(0)=1 \\ \deg(\varphi) \leq n}} \max_{z \in I^- \cup I^+} |\varphi(z)| \leq 2 \left(\frac{\sqrt{|\lambda_1 \lambda_N|} - \sqrt{|\lambda_s \lambda_{s+1}|}}{\sqrt{|\lambda_1 \lambda_N|} + \sqrt{|\lambda_s \lambda_{s+1}|}} \right)^{[n/2]},$$

where $[n/2]$ denotes the integer part of $n/2$; see, e.g. [537, 659] or [272, pp. 53–54], [160, Section 6.2.4]. For an illustration of this bound consider the situation $|\lambda_1| = \lambda_N = 1$ and $|\lambda_s| = \lambda_{s+1}$. Then the 2-norm condition number of A is $\kappa(A) = \lambda_{s+1}^{-1}$ and the term on the right-hand side reduces to

$$2 \left(\frac{\kappa(A) - 1}{\kappa(A) + 1} \right)^{[n/2]}.$$

This expression corresponds to the value of the CG bound (5.6.23) at iteration step $[n/2]$ for an HPD matrix having all its eigenvalues in the interval $[\lambda_{s+1}^2, 1]$, and thus the condition number λ_{s+1}^{-2}. Hence the convergence bound for an indefinite matrix with the condition number κ needs in this case twice as many steps to decrease to a prescribed value as the bound for a definite matrix with the condition number κ^2. Although neither of the two bounds is sharp, this comparison indicates that solving Hermitian indefinite problems represents a significant challenge; see [537] or the survey on the numerical solution of saddle point problems given in [55]. In the general case when the two intervals I^- and I^+ have different lengths, the explicit solution of the min–max approximation problem on $I^- \cup I^+$ becomes quite complicated, and no simple and explicit bound on its value is known. For more information on this problem we refer to Fischer's book [183, Chapter 3], and for an application in the context of the Stokes problem to [160, Chapter 6].

If A is a general normal matrix (or is close to normal in the sense that $\kappa(Y) \approx 1$), then the right-hand side of (5.7.10) depends completely (or mostly) on the eigenvalues of A. In such cases the eigenvalues can be used to obtain a good estimate of the worst-case convergence behaviour of GMRES. Intuitively, the value of the right-hand side of (5.7.11) should be smaller when the eigenvalues $\lambda_1, \ldots, \lambda_N$ are farther away from the origin (where the GMRES iteration polynomial is normalised). Finding a close estimate of the right-hand side of (5.7.11), however, represents in a general case of complex eigenvalues a more difficult problem than in the real case; see, e.g. [543, Section 6.11] and the description in [184].

Finally, if A is far from normal in the sense that $\kappa(Y)$ is large, then the bound (5.7.10) may fail to provide any reasonable information about the actual behaviour of GMRES. The reason is that in the derivation from (5.7.7) to (5.7.9), we have separated the polynomial minimisation problem from all other information about the (generalised) eigenvectors and the initial residual r_0, and additionally we have separated r_0 from the inverse transformation matrix Y^{-1}. For a closer look at this problem with the bound (5.7.10) we assume for simplicity that A is diagonalisable with an eigendecomposition $A = Y\Lambda Y^{-1}$, $\Lambda = \text{diag}(\lambda_1, \ldots, \lambda_N)$. Let us further assume that the eigenvectors of A are normalised, $\|y_j\| = 1$ for $j = 1, \ldots, N$. If we denote $Y^{-1} r_0 = [\xi_1, \ldots, \xi_N]^T$, then

$$r_n = \varphi_n^{MR}(A) r_0 = Y \varphi_n^{MR}(\Lambda) Y^{-1} r_0 = \sum_{j=1}^{N} (\varphi_n^{MR}(\lambda_j) \xi_j) y_j.$$

Here the nth GMRES residual is written as a linear combination of the normalised eigenvectors of A, and the values of the iteration polynomial at the eigenvalues of A represent *multiplicative corrections to the individual components* of the vector $Y^{-1} r_0$. If Y is ill-conditioned, then some of these components can be significantly larger (in modulus) than $\|r_0\|$. On the other hand, since the GMRES residual norms are monotonically decreasing, we have $\|r_n\| \leq \|r_0\|$ for all n. This means that in case of an ill-conditioned matrix Y the linear combination forming the vector r_n in general contains a *significant cancellation*. Due to the two separation steps in its derivation (see (5.7.7)–(5.7.9)), the minimisation problem in the bound (5.7.10), which is based on the eigenvalues only, cannot reflect such cancellations. Consequently, this minimisation problem can lack any relationship with the problem that is actually minimised by GMRES, even in the worst case. We illustrate this by a numerical example adapted from [427, Example 3.3].

Example 5.7.5
We consider $N \times N$ tridiagonal Toeplitz matrices of the form

$$A_\alpha = \text{tridiag}(-1, \alpha, -1) \quad \text{and} \quad B_\alpha = \text{tridiag}(-\alpha, \alpha, -1/\alpha),$$

where $\alpha \geq 2$ is a real parameter. Both A_α and B_α have the same eigenvalues given by

$$\alpha - 2\cos(j\pi/(N+1)), \quad j = 1, \ldots, N. \qquad (5.7.12)$$

Both matrices are diagonalisable, with the corresponding normalised eigenvectors given by

$$y_j^{(A)} = \left(\frac{2}{N+1}\right)^{\frac{1}{2}} \left[\sin\frac{j\pi}{N+1}, \ldots, \sin\frac{Nj\pi}{N+1}\right]^T, \quad j = 1, \ldots, N,$$

$$y_j^{(B)} = D y_j^{(A)} / \|D y_j^{(A)}\|, \quad j = 1, \ldots, N, \quad D \equiv \mathrm{diag}(\alpha, \alpha^2, \ldots, \alpha^N).$$

These well-known formulas and an analysis of the GMRES convergence behaviour for tridiagonal Toeplitz matrices can be found in [424]; see also [418, 419]. The matrix A_α is symmetric (and hence normal) and for $\alpha \geq 2$ it is positive definite. Its eigenvectors $y_1^{(A)}, \ldots, y_N^{(A)}$ form an orthonormal basis of $\mathbb{R}^N$. The matrix B_α is not normal and its eigenvector matrix $Y_\alpha^{(B)} = [y_1^{(B)}, \ldots, y_N^{(B)}]$ has a condition number of order α^{N-1}. Increasing α means that the departure from normality of the matrix B_α increases (in the sense that its eigenvector matrix $Y_\alpha^{(B)}$ becomes more ill-conditioned).

We apply GMRES with $x_0 = 0$ to the two linear algebraic systems

$$A_\alpha x = [1, \ldots, 1]^T \quad \text{and} \quad B_\alpha x = [1, \ldots, 1]^T$$

using $N = 40$ and the two values $\alpha = 3, 5$. The relative GMRES residual norms resulting from these computations are shown in the left part of Figure 5.11. The solid lines show the convergence for the two systems with A_α. Here faster convergence corresponds to larger α. The dots show the worst-case GMRES residual norms at every iteration step for the normal matrices A_α, i.e. the right-hand side of (5.7.11) for the eigenvalues (5.7.12) and the two values of α. These curves have been computed with the MATLAB package SDPT3 [619]. The GMRES convergence for the two systems with B_α is shown by the dashed lines.

We observe that the GMRES convergence rate for the systems with the normal matrices A_α increases when α is increased and hence the spectrum moves away from the origin. The GMRES convergence curves in these cases are very close to the worst case,

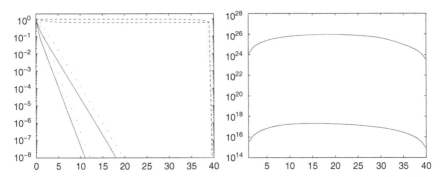

Figure 5.11 Left: Relative GMRES residual norms for the normal matrices A_α (solid) and the non-normal matrices B_α (dashed) for $\alpha = 3, 5$ and $r_0 = [1, \ldots, 1]^T$. For A_α the faster convergence corresponds to larger α. For B_α the choice of α does not really make a difference for the behaviour of GMRES. The worst-case GMRES residual norm at every iteration step n is for A_α shown by dots. Right: Absolute values of the components of the vectors $(Y_\alpha^{(B)})^{-1}[1, \ldots, 1]^T$ for $\alpha = 3$ and $\alpha = 5$.

which is determined by the spectrum only. Thus, the spectrum indeed gives relevant information about the actual behaviour of the GMRES method.

The situation is completely different for the non-normal matrices B_α. We first note that the bound (5.7.10) is quantitatively useless in these cases because the factor $\kappa(Y_\alpha^{(B)})$ is huge for the values of N and α we consider (the value of α^{N-1} for $N=40$ is about 10^{18} for $\alpha=3$ and 10^{27} for $\alpha=5$). But even qualitatively the bound (5.7.10) does not provide reasonable information. The actual convergence of GMRES for the two matrices B_α is almost identical, although the spectrum of B_α coincides with the one of A_α, and hence from the eigenvalues alone, one would expect an increasing rate of convergence when α increases.

On the right part of Figure 5.11 we show the absolute values of the components of the vector $(Y_\alpha^{(B)})^{-1}[1,\ldots,1]^T$. For $\alpha=3$ these components range around the value 10^{16}, for $\alpha=5$ they range around 10^{25}. In both cases the linear combination

$$r_n = \sum_{j=1}^N (\varphi_n^{MR}(\lambda_j))(Y_\alpha^{(B)})^{-1}[1,\ldots,1]^T)\, y_j^{(B)}$$

contains a huge cancellation since $\|r_n\| \leq \|r_0\| = \sqrt{N} \approx 6.325$. As described above, in such cases the minimisation problem in the bound (5.7.10) has hardly anything in common with the minimisation problem that is actually solved by GMRES.

This example illustrates the difficulties with the bound (5.7.10) mentioned above. It shows that this bound can be *quantitatively and qualitatively* useless when the matrix eigenvectors are ill-conditioned. In fact, as Trefethen stressed in [623, p. 384], 'if the matrix is far from normal, ... there may be no good scientific reason for attempting to analyze the problem in terms of eigenvalues and eigenvectors.' Using spectral information in such cases amounts to a significant distortion of the geometry of the finite-dimensional space $\mathbb{F}^N$. Using a distorted geometry, the obtained bounds may have very little relevance for the original quantities like the Euclidean residual norms.

5.7.3 More General Approaches

Some approaches for understanding at least the worst-case convergence of GMRES for non-normal matrices are based on the following upper bound:

$$\underbrace{\max_{r_0 \neq 0} \min_{\substack{\varphi(0)=1 \\ \deg(\varphi) \leq n}} \frac{\|\varphi(A)r_0\|}{\|r_0\|}}_{\text{worst-case GMRES at step } n} \leq \min_{\substack{\varphi(0)=1 \\ \deg(\varphi) \leq n}} \max_{r_0 \neq 0} \frac{\|\varphi(A)r_0\|}{\|r_0\|} = \underbrace{\min_{\substack{\varphi(0)=1 \\ \deg(\varphi) \leq n}} \|\varphi(A)\|}_{\text{ideal GMRES}}. \quad (5.7.13)$$

The *ideal GMRES approximation problem* on the right-hand side above was introduced by Greenbaum and Trefethen in order to 'disentangle [the] matrix essence of the process from the distracting effects of the initial vector' [282, p. 362]. It is known that (5.7.13) is an equality when A is normal [277, 372], and equality was shown for some non-normal matrices as well [612]. But there also exist examples of non-normal matrices so that the right-hand side of (5.7.13) is larger (even arbitrarily larger) than the left-hand side [173, 618]. In general it is an open question as to whether the ideal GMRES approximation is equal or close to the worst-case GMRES for practically relevant and reasonably large classes of non-normal matrices.

Several researchers have suggested choosing sets $\Omega \subset \mathbb{C}$ or $\widehat{\Omega} \subset \mathbb{C}$ that are somehow associated with A and to bound the ideal GMRES approximation problem from above or below by scalar approximation problems on these sets, i.e.

$$c \min_{\substack{\varphi(0)=1 \\ \deg(\varphi) \leq n}} \max_{\lambda \in \Omega} |\varphi(\lambda)| \leq \min_{\substack{\varphi(0)=1 \\ \deg(\varphi) \leq n}} \|\varphi(A)\| \leq \widehat{c} \min_{\substack{\varphi(0)=1 \\ \deg(\varphi) \leq n}} \max_{\lambda \in \widehat{\Omega}} |\varphi(\lambda)|.$$

Here c and $\widehat{c}$ should be some moderate real positive factors that may depend on A and possibly on n. Because of the normalisation condition $\varphi(0) = 1$, the chosen sets Ω and $\widehat{\Omega}$ must exclude the origin. We will now briefly survey several choices for non-normal matrices A.

Trefethen [621] suggested choosing $\widehat{\Omega}$ as the ϵ-*pseudospectrum* of A. For a real parameter $\epsilon > 0$ this set is defined by

$$\Lambda_\epsilon(A) \equiv \{\lambda \in \mathbb{C} : \|(\lambda I_N - A)^{-1}\| \geq \epsilon^{-1}\},$$

where for an eigenvalue λ of A we set $\|(\lambda I_N - A)^{-1}\| = +\infty$. Alternatively, the set $\Lambda_\epsilon(A)$ consists of the eigenvalues of all matrices $A + E$, where E is an arbitrary perturbation with $\|E\| \leq \epsilon$. The set $\Lambda_0(A)$ is equal to the spectrum of A. The history, properties, and numerous applications of ϵ-pseudospectra are described in the monograph of Trefethen and Embree [626]. For $\epsilon > 0$ the ϵ-pseudospectrum of A contains all eigenvalues of A in its interior. Denoting by $\partial \Lambda_\epsilon(A)$ the boundary of $\Lambda_\epsilon(A)$, the matrix function $\varphi(A)$ can be written as

$$\varphi(A) = \frac{1}{2\pi\iota} \int_{\partial \Lambda_\epsilon(A)} \varphi(\lambda)(\lambda I_N - A)^{-1} \, d\lambda;$$

see, e.g. the end of Section 4.9 in this book. Using this formula one easily obtains the upper bound

$$\min_{\substack{\varphi(0)=1 \\ \deg(\varphi) \leq n}} \|\varphi(A)\| \leq \frac{L}{2\pi \epsilon} \min_{\substack{\varphi(0)=1 \\ \deg(\varphi) \leq n}} \max_{\lambda \in \Lambda_\epsilon(A)} |\varphi(\lambda)|, \qquad (5.7.14)$$

where L denotes the arc length of $\partial \Lambda_\epsilon(A)$. Further details and the complete derivation are given, e.g. in [469]. In order to make the right-hand side of (5.7.14) reasonably small one must choose ϵ large enough to make the constant $L/2\pi\epsilon$ small, but small enough to make the set $\Lambda_\epsilon(A)$ not too large. An example given in [281, pp. 98–99] shows that this can be a challenging problem. As demonstrated for example in [623, 626], the ϵ-pseudospectrum is very useful in a wide range of applications beyond the context of iterative methods.

Another approach is based on the *field of values*[11] of A, which is defined by

$$F(A) \equiv \{v^* A v \;:\; \|v\| = 1, \; v \in \mathbb{C}^N\};$$

see, e.g. [342, Chapter 1]. If we denote by

$$\nu(F(A)) \equiv \min_{\lambda \in F(A)} |\lambda|$$

the distance of $F(A)$ from the origin in the complex plane, then

$$\min_{\substack{\varphi(0)=1 \\ \deg(\varphi)\leq n}} \|\varphi(A)\| \leq \left(1 - \nu(F(A))\nu(F(A^{-1}))\right)^{n/2}; \qquad (5.7.15)$$

see, e.g. [154]. If the Hermitian part of A, i.e. the matrix $M = (A + A^*)/2$, is positive definite, then a special case of (5.7.15) is

$$\min_{\substack{\varphi(0)=1 \\ \deg(\varphi)\leq n}} \|\varphi(A)\| \leq \left(1 - \frac{\lambda_{\min}(M)}{\lambda_{\max}(A^* A)}\right)^{n/2}.$$

This result was first shown by Elman for the GCR method in [157]; see also [156]. The field of values is always convex and it contains the convex hull of the eigenvalues of A. Because of the requirement $0 \notin F(A)$, the bound (5.7.15) is useless for indefinite matrices. Interesting applications of the field of values to the GMRES convergence analysis for linear algebraic systems arising from the Galerkin FEM discretisation of elliptic PDEs were given by Klawonn and Starke in [387, 578]. Eiermann presented a field of values based convergence analysis in a more abstract setting in [153].

A generalisation of the field of values of A is the *polynomial numerical hull*, introduced by Nevanlinna [477, Section 2.10] and defined as

$$\mathcal{H}_n(A) = \{\lambda \in \mathbb{C} \;:\; \|\varphi(A)\| \geq |\varphi(\lambda)| \text{ for all polynomials } \varphi \text{ of degree } \leq n\}.$$

It can be shown that $\mathcal{H}_1(A) = F(A)$. The set $\mathcal{H}_n(A)$ provides the *lower* bound

$$\min_{\substack{\varphi(0)=1 \\ \deg(\varphi)\leq n}} \max_{\lambda \in \mathcal{H}_n(A)} |\varphi(\lambda)| \leq \min_{\substack{\varphi(0)=1 \\ \deg(\varphi)\leq n}} \|\varphi(A)\|. \qquad (5.7.16)$$

The determination of the polynomial numerical hull represents a highly nontrivial problem. Greenbaum and her co-workers [91, 172, 273, 274, 275, 276] have obtained theoretical results about $\mathcal{H}_n(A)$ for Jordan blocks, banded triangular Toeplitz matrices, and block diagonal matrices with triangular Toeplitz blocks;

---

11. The concept of the field of values was introduced under the German name 'Wertvorrat' by Toeplitz in 1918 [617]. In the same paper Toeplitz introduced normal matrices (or bilinear forms).

see also [612]. A larger applicability of the bound (5.7.16) depends on whether there exists a sufficiently large class of matrices for which these sets can be efficiently described.

The approach of analysing the worst-case GMRES via bounding the ideal GMRES approximation problem is certainly useful to obtain a-priori convergence estimates in terms of some properties of A, and possibly to analyse the effectiveness of preconditioning techniques. However, none of the bounds stated above really *characterises* (in terms of properties of A) the convergence behaviour of GMRES in the non-normal case. In the following section we will describe a different approach to investigating the link between spectral information and GMRES, which will lead to some surprising insights into the behaviour of the method.

5.7.4 Any Nonincreasing Convergence Curve is Possible with Any Eigenvalues

As previously, we consider a nonsingular matrix $A \in \mathbb{F}^{N \times N}$ and an initial residual r_0 of grade d with respect to A. Suppose that for each $n = 1, \ldots, d$ the first n columns of the matrix $W_d = [w_1, \ldots, w_d]$ form an orthonormal basis of the subspace $A\mathcal{K}_n(A, r_0)$. Then the nth GMRES residual vector $r_n \in r_0 + A\mathcal{K}_n(A, r_0)$ can be written as

$$r_n = r_0 + W_n s_n, \quad n = 1, \ldots, d,$$

for some (uniquely determined) vector $s_n = [\xi_1, \ldots, \xi_n]^T$. Since A is nonsingular we have $\mathcal{K}_d(A, r_0) = A\mathcal{K}_d(A, r_0)$; see item (2) in Lemma 2.2.2. Hence $r_0 \in A\mathcal{K}_d(A, r_0)$, and we can expand r_0 in the orthonormal basis $w_1, \ldots, w_d$,

$$r_0 = \sum_{j=1}^{d} (r_0, w_j) w_j. \tag{5.7.17}$$

This leads to the following lemma.

Lemma 5.7.6
Using the notation above, for $n = 0, 1, \ldots, d$ the nth GMRES residual vector satisfies

$$r_n = r_0 - \sum_{j=1}^{n} (r_0, w_j) w_j = r_0 - \sum_{j=1}^{n} (r_{j-1}, w_j) w_j = \sum_{j=n+1}^{d} (r_0, w_j) w_j. \tag{5.7.18}$$

Hence, in particular, the vector r_{n-1} is orthogonally decomposed as

$$r_{n-1} = (r_{n-1}, w_n) w_n + r_n = (r_0, w_n) w_n + r_n, \tag{5.7.19}$$

so that

$$\|r_{n-1}\|^2 - \|r_n\|^2 = |(r_0, w_n)|^2, \quad n = 1, \ldots, d. \tag{5.7.20}$$

Proof
We have already shown (5.7.18) for $n = 0$. For each $n = 1, \ldots, d$ we have $r_n \perp A\mathcal{K}_n(A, r_0)$, which is equivalent to

$$0 = W_n^* r_n = W_n^*(r_0 + W_n s_n) = W_n^* r_0 + s_n,$$

and thus $s_n = -[(r_0, w_1), \ldots, (r_0, w_n)]^T$. Consequently,

$$r_n = r_0 + W_n s_n = \sum_{j=1}^{d}(r_0, w_j) w_j - \sum_{j=1}^{n}(r_0, w_j) w_j = \sum_{j=n+1}^{d}(r_0, w_j) w_j,$$

for $n = 1, \ldots, d$. By construction,

$$(r_0, w_j) = (r_{j-1} - W_{j-1} s_{j-1}, w_j) = (r_{j-1}, w_j), \quad j = 1, \ldots, n.$$

This proves (5.7.18) for $n = 1, \ldots, d$. The relations (5.7.19) and (5.7.20) follow immediately. $\square$

From (5.7.19) we see that GMRES exactly stagnates at step n if and only if r_0 (or, equivalently, r_{n-1}) is orthogonal to the vector w_n. A *complete stagnation* of GMRES from step 1 to step n happens if and only if r_0 is orthogonal to $w_1, \ldots, w_n$ and hence to the whole subspace $A\mathcal{K}_n(A, r_0)$.

For simplicity of notation we assume in the following that $d = N$. The relation (5.7.20) shows that the coefficients of r_0 in the linear combination (5.7.17) satisfy

$$|(r_0, w_n)| = \left(\|r_{n-1}\|^2 - \|r_n\|^2\right)^{1/2}, \quad n = 1, \ldots, N. \tag{5.7.21}$$

This key observation will be used several times below.

Following the approach in [278, Section 2] we will now construct a linear algebraic system $Ax = b$ for which GMRES with $x_0 = 0$ (which is chosen for simplicity and without loss of generality) attains a prescribed (nonincreasing) convergence curve. Moreover, we will construct the matrix A so that it has a prescribed set of N nonzero complex numbers (not necessarily distinct) as its eigenvalues.

To start the construction suppose we are given *any* N real numbers

$$f_0 \geq f_1 \geq \cdots \geq f_{N-1} > 0.$$

We define $f_N \equiv 0$ and

$$g_n \equiv \left(f_{n-1}^2 - f_n^2\right)^{1/2}, \quad n = 1, \ldots, N.$$

In addition, we consider *any* given N nonzero complex numbers $\lambda_1, \ldots, \lambda_N$ (not necessarily distinct). These will be the eigenvalues of the matrix A we are going to construct. We define the numbers $\alpha_0, \alpha_1, \ldots, \alpha_{N-1} \in \mathbb{C}$ by

$$(\lambda - \lambda_1) \cdots (\lambda - \lambda_N) = \lambda^N - \sum_{j=0}^{N-1} \alpha_j \lambda^j. \tag{5.7.22}$$

In other words, $\alpha_0, \alpha_1, \ldots, \alpha_{N-1}$ will turn out to be the coefficients of the characteristic polynomial of A. Since the numbers $\lambda_1, \ldots, \lambda_N$ are nonzero we have $\alpha_0 \neq 0$.

Let $w_1, \ldots, w_N$ be *any* given orthonormal basis of $\mathbb{C}^N$ and let $W_N = [w_1, \ldots, w_N]$ be the corresponding unitary matrix. Let $b \in \mathbb{C}^N$ be any vector such that

$$W_N^* b = [g_1, \ldots, g_N]^T, \tag{5.7.23}$$

then

$$\|b\| = \|W_N^* b\| = \left(\sum_{j=1}^N g_j^2\right)^{1/2} = f_0.$$

Alternatively, we may first choose a vector b with $\|b\| = f_0$ and then determine any unitary matrix W_N with the property (5.7.23). We will now construct a matrix A having the eigenvalues $\lambda_1, \ldots, \lambda_N$, so that for each $n = 1, \ldots, N$ the first n columns of the matrix W_N form a basis of $A\mathcal{K}_n(A, b)$. Then the residual norms of GMRES applied to $Ax = b$ with $x_0 = 0$ must by construction satisfy (5.7.21), and, finally, $\|r_n\| = f_n, n = 1, \ldots, N$.

The idea of the construction is similar to the approach in Section 3.7, where we defined a linear operator using its images with respect to a given basis of a vector space. Since $w_N^* b = g_N = f_{N-1} > 0$, the vectors $b, w_1, \ldots, w_{N-1}$ are linearly independent and form a basis of $\mathbb{C}^N$. Using this basis we define the linear operator $\mathbf{A}$ on $\mathbb{C}^N$ by the following N equations:

$$\mathbf{A} b \equiv w_1,$$
$$\mathbf{A} w_1 \equiv w_2,$$
$$\vdots$$
$$\mathbf{A} w_{N-2} \equiv w_{N-1},$$
$$\mathbf{A} w_{N-1} \equiv \alpha_0 b + \alpha_1 w_1 + \cdots + \alpha_{N-1} w_{N-1}.$$

The matrix representation of $\mathbf{A}$ with respect to the basis $b, w_1, \ldots, w_{N-1}$ is given by

$$C \equiv \begin{bmatrix} 0 & \cdots & 0 & \alpha_0 \\ 1 & \ddots & \vdots & \vdots \\ & \ddots & 0 & \alpha_{N-2} \\ & & 1 & \alpha_{N-1} \end{bmatrix}. \tag{5.7.24}$$

This matrix is the companion matrix of the polynomial (5.7.22), and hence it has the eigenvalues $\lambda_1, \ldots, \lambda_N$. Using the matrix $B = [b, w_1, \ldots, w_{N-1}]$ for a change of bases we now define the matrix

$$A \equiv BCB^{-1},$$

which means that A is the matrix representation of the linear operator $\mathbf{A}$ with respect to the standard basis of $\mathbb{C}^N$. The matrices A and C are similar and therefore they have the same eigenvalues. Moreover, a straightforward computation shows

$$Ab = w_1,$$
$$Aw_1 = A^2 b = w_2,$$
$$\vdots$$
$$Aw_{N-2} = A^{N-1} b = w_{N-1},$$

so that for each $n = 1, \ldots, N$ the vectors $w_1, \ldots, w_n$ indeed form an orthonormal basis of $A\mathcal{K}_n(A, b)$. In summary, we have shown the following result which first appeared in [278, Theorem 2.1].

Theorem 5.7.7
For any N positive numbers $f_0 \geq f_1 \geq \cdots \geq f_{N-1} > 0$ and any N nonzero complex numbers $\lambda_1, \ldots, \lambda_N$, not necessarily distinct, there exists a matrix $A \in \mathbb{C}^{N \times N}$ with eigenvalues $\lambda_1, \ldots, \lambda_N$ and a vector $b \in \mathbb{C}^N$ with $\|b\| = f_0$, so that GMRES applied to $Ax = b$ with $x_0 = 0$ has the residual norms $\|r_n\| = f_n$, $n = 0, 1, \ldots, N - 1$.

The main point can be formulated as follows:

> Any nonincreasing convergence curve is possible for GMRES for a matrix having any prescribed set of eigenvalues.

Note that the prescribed convergence curve for A is attained by GMRES for the *corresponding* right-hand side vector b. For a different right-hand side $\tilde{b}$ (and the same matrix A) the GMRES convergence curve can be different from the prescribed one. As an extreme example suppose that we have constructed A and b so that GMRES completely stagnates from iterations 1 to $N - 1$, i.e. $f_0 = f_1 = \cdots = f_{N-1}$. If $\tilde{b}$ is any eigenvector of A, then GMRES applied to $Ax = \tilde{b}$ with $x_0 = 0$ converges to the solution in one step. We point out that there exist analogous results with a matrix having nonzero singular values; see [154, Section 6.3, Lemma 6.9].

The next result, which first appeared in [14, Theorem 2.1], extends Theorem 5.7.7 by giving a *complete parameterisation* of all matrices A with prescribed eigenvalues and right-hand sides b so that GMRES attains prescribed residual norms.

Theorem 5.7.8
Consider N given positive numbers $f_0 \geq f_1 \geq \cdots \geq f_{N-1} > 0$ and N nonzero complex numbers $\lambda_1, \ldots, \lambda_N$, not necessarily distinct. Let $A \in \mathbb{C}^{N \times N}$ and $b \in \mathbb{C}^N$. Then the following three assertions are equivalent:

  (1) The eigenvalues of A are $\lambda_1, \ldots, \lambda_N$ and GMRES applied to $Ax = b$ with $x_0 = 0$ yields the residual norms $\|r_n\| = f_n$ for $n = 0, 1, \ldots, N - 1$.

(2) $A = W_N R_N C R_N^{-1} W_N^*$ and $b = W_N h$, where
- W_N is a unitary matrix,
- C is the companion matrix of the polynomial

$$q(\lambda) \equiv (\lambda - \lambda_1) \cdots (\lambda - \lambda_N) \equiv \lambda^N - \sum_{j=0}^{N-1} \alpha_j \lambda^j$$

(this matrix is stated in (5.7.24) above),
- $h = [g_1, \ldots, g_N]^T$, where $g_n \equiv (f_{n-1}^2 - f_n^2)^{1/2}, n = 1, \ldots, N$, and we set $f_N \equiv 0$,
- R_N is nonsingular and upper triangular such that $R_N s = h$, where

$$s = [\xi_1, \ldots, \xi_N]^T \quad \text{and} \quad p(\lambda) \equiv \left(1 - \frac{\lambda}{\lambda_1}\right) \cdots \left(1 - \frac{\lambda}{\lambda_N}\right) \equiv 1 - \sum_{j=1}^N \xi_j \lambda^j,$$

i.e. $\xi_n = -\alpha_n/\alpha_0$, $n = 1, \ldots, N-1$, and $\xi_N = 1/\alpha_0$.

(3) $A = W_N Y C Y^{-1} W_N^*$ and $b = W_N h$, where W_N, C and h are defined as in (2),

$$Y \equiv R_N C^{-1} = \begin{bmatrix} g_1 & & \\ \vdots & R & \\ g_{N-1} & & \\ g_N & 0 & \end{bmatrix},$$

where $g_1, \ldots, g_N$ are defined as in (2), and R is an $(N-1) \times (N-1)$ nonsingular upper triangular matrix.

Proof
We first prove that (1) implies (2). Since $\|r_{N-1}\| = f_{N-1} > 0$, the Krylov subspace $\mathcal{K}_N(A, b)$ has dimension N, and since A is nonsingular we have $A\mathcal{K}_N(A, b) = \mathcal{K}_N(A, b)$. Hence the matrix $B \equiv [Ab, \ldots, A^N b]$ has full rank N. We consider a QR factorisation of this matrix,

$$B = \widetilde{W}_N \widetilde{R}_N,$$

where $\widetilde{W}_N$ is unitary and $\widetilde{R}_N$ is nonsingular and upper triangular. Since GMRES applied to $Ax = b$ with $x_0 = 0$ produces the residual norms $\|r_n\| = f_n$ for $n = 1, \ldots, N$, and the first n columns of $\widetilde{W}_N$ form an orthonormal basis of $A\mathcal{K}_n(A, b)$, $n = 1, \ldots, N$, our observation made in (5.7.21) above implies that

$$\widetilde{W}_N^* b = [\eta_1, \ldots, \eta_N]^T, \quad |\eta_n| = g_n, \quad n = 1, \ldots, N.$$

Hence there exists a diagonal matrix $\Gamma = \text{diag}(\gamma_1, \ldots, \gamma_N)$ with $|\gamma_n| = 1$ for $n = 1, \ldots, N$, so that $\Gamma^* \widetilde{W}_N^* b = h$. We set $W_N \equiv \widetilde{W}_N \Gamma$, which is a unitary matrix.
Next note that $AB = BC$, or, equivalently,

$$A = BCB^{-1} = \widetilde{W}_N \widetilde{R}_N C \widetilde{R}_N^{-1} \widetilde{W}_N^* = W_N R_N C R_N^{-1} W_N^*,$$

where $R_N \equiv \Gamma^* \widetilde{R}_N$ is nonsingular and upper triangular. Since the polynomial $p(\lambda)$ is a multiple of the characteristic polynomial of A (which is given by $q(\lambda)$), we have $p(A) = 0$ and hence $p(A)b = 0$, which can be written as

$$b = \sum_{j=1}^{N} \xi_j A^j b = Bs.$$

Using this equation we obtain

$$W_N R_N s = (\widetilde{W}_N \Gamma)(\Gamma^* \widetilde{R}_N) s = \widetilde{W}_N \widetilde{R}_N s = Bs = b = W_N h,$$

so that indeed $R_N s = h$.

We will now show that (2) implies (1). By construction,

$$W_N^* b = h = [g_1, \ldots, g_N]^T,$$

where $g_n = (f_{n-1}^2 - f_n^2)^{1/2}$, $n = 1, \ldots, N$. If for each $n = 1, \ldots, N$ the first n columns of W_N form an orthonormal basis of $A\mathcal{K}_n(A, b)$, then, using the observation (5.7.21), the statement (1) follows by construction. The trick is to use

$$C^{-1} = \begin{bmatrix} -\alpha_1/\alpha_0 & & 1 & & \\ \vdots & & 0 & \ddots & \\ -\alpha_{N-1}/\alpha_0 & & & \ddots & 1 \\ 1/\alpha_0 & & & & 0 \end{bmatrix},$$

i.e. the first column of C^{-1} is given by the vector s. From $Cs = e_1$ and the structure of C we then easily get $C^n s = e_n$ for $n = 1, \ldots, N-1$. Therefore,

$$A^n b = \left(W_N R_N C R_N^{-1} W_N^*\right)^n b = W_N R_N C^n R_N^{-1} W_N^* b = W_N R_N C^n s$$
$$= W_N R_N e_n \in \text{span}\{w_1, \ldots, w_n\}, \quad n = 1, \ldots, N-1,$$

because R_N is upper triangular. Clearly, the vectors $w_1, \ldots, w_n$ form an orthonormal basis of $A\mathcal{K}_n(A, r_0)$, and the second part of the proof is finished.

In order to prove the equivalence of (2) and (3) it suffices to write

$$A = W_N R_N C R_N^{-1} W_N^* = W_N \left(R_N C^{-1}\right) C \left(R_N C^{-1}\right)^{-1} W_N^* \equiv W_N Y C Y^{-1} W_N,$$

and to realise that given a nonsingular upper triangular matrix R of order $(N-1)$, we can uniquely define the nonsingular upper triangular matrix R_N of order N such that R forms its left principal submatrix and $R_N s = h$ (the last column of R_N is uniquely determined by this equation). Then reading the previous sequence of equations from right to left finishes the proof. □

Note that there is an important difference between (2) and (3) in the previous theorem. In (2) the matrix R_N needs to satisfy the intriguing condition $R_N s = h$, which links the spectral information and the prescribed convergence curve. In (3), on the other hand, the spectral information, the prescribed convergence curve, and the free parameters contained in the matrix R are entirely separated. This feature was applied, e.g. in [150], where the authors showed that any GMRES convergence curve is possible for any prescribed set of Ritz values (in all iterations). Moreover, they derived a parameterisation of the class of matrices and initial vectors, so that the Arnoldi algorithm generates a prescribed sequence of Ritz values in all iterations.

The previous results prove what has been mentioned in Section 5.7.2 and illustrated in particular by Example 5.7.5: eigenvalues *alone* do not determine the convergence behaviour of the GMRES. The same convergence curve can be attained for matrices having very different eigenvalues, and, vice versa, matrices with the same eigenvalues can be linked with very different GMRES convergence curves. As pointed out also by Eiermann and Ernst [154, Sections 6.3–6.4], this fact may seem in conflict with the common wisdom linking the distribution of eigenvalues with the convergence of iterative methods. Therefore some remarks are in place:

1. In many practical applications one can observe a correlation between the eigenvalues of a (non-normal) matrix A and the convergence rate of GMRES. Considering the results in this section, linking eigenvalues and the convergence of GMRES for a non-normal matrix A must always be based on convincing arguments. They can consist, e.g. of demonstrating a special relationship between the initial residual (or the right-hand side if $x_0 = 0$) and the eigenvectors of A. Such a relationship may compensate for the influence of ill-conditioned eigenvectors, so that the approximation problem actually solved by GMRES is indeed (close to) an approximation problem on the eigenvalues of A. For an example we refer to [149].

2. The results in this section show that any general worst-case analysis of GMRES must include more information than the eigenvalues alone. What other information should be used? In our opinion, there is no single answer. In some cases ϵ-pseudospectra, the field of values or the polynomial numerical hull can be used to find close estimates of the norm of $p(A)$ and hence of the *ideal GMRES approximation*. As mentioned after equation (5.7.13), however, there exist examples in which there is an arbitrarily large gap between the worst-case GMRES and ideal GMRES. Then GMRES *for any initial residual* performs better than any analysis based only on polynomials in the matrix A can explain. It is still unclear whether this holds beyond academic examples. Nevertheless, it suggests possible serious limitations of the ideal GMRES analysis for understanding the practical GMRES behaviour.

5.7.5 Convection–diffusion Example

We will now illustrate convergence behaviour of GMRES using a discretised convection–diffusion model problem of the form

$$-\nu \Delta u + w \cdot \nabla u = 0 \text{ in } \Omega = (0,1) \times (0,1), \quad u = g \text{ on } \partial\Omega, \quad (5.7.25)$$

where $\nu > 0$ is a scalar diffusion coefficient, and w is a vector velocity field called the 'wind'. We are interested in the *convection dominated case* where $\nu \ll \|w\|$. In this case either very fine meshes (at least in some regions of the domain depending on the wind and the boundary conditions) or special discretisation techniques must be used to avoid non-physical oscillations of the numerical solution. Details can be found, e.g. in the books [465, 534] and the survey paper [602]. One of these techniques is the popular streamline-diffusion finite element method (SDFEM), which is also called streamline upwind Petrov–Galerkin (SUPG) method [84, 348]. In this method the stabilisation consists of an additional diffusion term of the form $\langle \delta w w^T \nabla u, \nabla v \rangle$ that is added to the bilinear form in the weak formulation of (5.7.25); $\langle \cdot, \cdot \rangle$ denotes the $\mathcal{L}^2(\Omega)$ inner product. Here $\delta > 0$ is the *stabilisation parameter*, and the diffusity tensor $\delta w w^T$ means that the stabilisation only acts in the direction of the flow (hence the name 'streamline-diffusion' method). The SUPG discretisation of (5.7.25) leads to a discrete operator of the form

$$\nu A_d + A_c + \delta A_s, \qquad (5.7.26)$$

where $A_d = \langle \nabla \phi_j, \nabla \phi_i \rangle$, $A_c = \langle w \cdot \nabla \phi_j, \phi_i \rangle$, and $A_s = \langle w \cdot \nabla \phi_j, w \cdot \nabla \phi_i \rangle$ represent the diffusion, convection, and stabilisation terms, respectively. The functions ϕ_j are the finite element basis functions.

Several authors have considered the model problem (5.7.25) with a *constant vertical wind* $w = [0, 1]^T$, and the SUPG method with bilinear finite elements on a regular grid with m inner nodes in each direction; see, e.g. [158, 159, 168, 187, 425] and [160, Chapter 3]. Using a vertical ordering of the unknowns (i.e. parallel to the direction of the wind) and a discrete sine transformation, it can be shown that the discrete operator (5.7.26) of size $m^2 \times m^2$ is in this case *orthogonally similar* to a block-diagonal matrix A_δ consisting of m nonsymmetric tridiagonal Toeplitz blocks $T_1, \ldots, T_m$,

$$A_\delta = \mathrm{diag}(T_1, \ldots, T_m) \in \mathbb{R}^{N \times N}, \quad N \equiv m^2,$$
$$T_j = \mathrm{tridiag}(\gamma_j, \lambda_j, \mu_j) \in \mathbb{R}^{m \times m}, \quad j = 1, \ldots, m.$$

The entries $\gamma_j, \lambda_j, \mu_j, j = 1, \ldots, m$, are known explicitly in terms of the diffusion coefficient ν, the mesh size $h = 1/(m+1)$ and the stabilisation parameter δ; see, e.g. [425, Section 2.1].

A standard choice of the stabilisation parameter (see [425, Section 2] and the references given) is

$$\delta = \delta_* \equiv \frac{h}{2}\left(1 - \frac{1}{P_h}\right),$$

where $P_h \equiv h\|w\|/(2\nu) = h/(2\nu)$ is the *mesh Péclet number*, which relates the ratio of the size of the wind and the diffusion parameter to the size of the grid elements. With this choice and with $P_h \gg 1$, which is realistic in the convection-dominated case, one can show that

$$\frac{|\lambda_j|}{|\gamma_j|} \approx 1 \gg \frac{|\mu_j|}{|\gamma_j|}, \quad j = 1, \ldots, m;$$

see [425, Lemma 3.2 and Figures 3.1–3.2]. Hence for large P_h each diagonal block T_j looks like a slightly perturbed (transposed) Jordan block, which means that the matrix A_{δ_*} is highly non-normal in the sense that it has very ill-conditioned eigenvectors.

Since the right-hand side in the convection–diffusion equation in (5.7.25) is zero, the right-hand side of the SUPG discretised linear algebraic system depends only on the (Dirichlet) boundary condition in (5.7.25). It had been observed numerically that GMRES applied to the discretised system initially converges slowly, which led Ernst to conjecture in [168, p. 1094] that the length of this initial phase 'is goverened by the time it takes for boundary information to pass from the inflow boundary across the domain following the streamlines of the velocity field.' This conjecture was confirmed analytically in [425]. In particular, boundary values were given explicitly in [425, Example 2.2] that lead to a slow initial phase of GMRES convergence lasting exactly k steps for each $k = 0, 1, \ldots, m - 1$.

A numerical illustration is shown in Figure 5.12 (see [425, Fig. 1.1] for a similar experiment). For the diffusion coefficient $\nu = 0.005$ in (5.7.25) and a SUPG discretisation with $m = 20$ inner nodes in each direction, the figure shows the relative residual norms of GMRES (with $x_0 = 0$) applied to $A_{\delta_*} x = b_{\delta_*}^{(j)}, j = 1, \ldots, 20$, where the right-hand sides $b_{\delta_*}^{(j)}$ result from 20 different boundary conditions; see [425, Example 2.2].

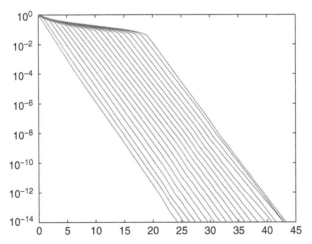

Figure 5.12 GMRES convergence curves (with $x_0 = 0$) for the SUPG discretised convection–diffusion model problem with the same discretised operator A_{δ_*} of order $m^2 \times m^2$ (here with $m = 20$ and $\nu = 0.005$ in (5.7.25)), but m different boundary conditions that lead to m different right-hand sides $b_{\delta_*}^{(j)}, j = 1, \ldots, m$.

It is clearly impossible to explain the convergence behaviour of GMRES in the initial phase using only the given matrix A_{δ_*}. In order to obtain relevant information about the convergence of GMRES in the initial phase one must include the right-hand side into the analysis. This has been done in [425], and hence the paper represents an example where the influence of the initial residual (here equal to the right-hand side) on the convergence of GMRES has been explained at least to some extent.

In many applications such as the example in this section, the right-hand vector of the linear algebraic system represents boundary conditions or outer forces. We believe that a systematic investigation of their influence on the convergence of algebraic iterative methods used for solving discretised problems would help both in solving practical problems and in obtaining a better understanding of the behaviour of methods like GMRES.

5.7.6 Asymptotic Estimates for the GMRES Convergence

In Section 5.6 we have described 'asymptotic' convergence results for the CG method; see in particular the example of the work of Beckermann und Kuijlars given in Section 5.6.3. Similar ideas have also been applied in the convergence analysis of GMRES. We will give a very brief summary of the main ideas. More information and many references can be found in the highly readable survey papers by Driscoll, Toh, and Trefethen [145] and Kuijlaars [402]. Interesting applications together with a review of the related theory can be found in the paper by Gragg and Reichel [261].

One typically starts with bounding the actual GMRES convergence by using the ideal GMRES approximation (see (5.7.13)),

$$\frac{\|r_n\|}{\|r_0\|} \leq \min_{\substack{\varphi(0)=1 \\ \deg(\varphi) \leq n}} \|\varphi(A)\|.$$

In the next step the ideal GMRES is approximated using some compact and continuous set Ω, which is related with A and which always contains the eigenvalues of A,

$$\min_{\substack{\varphi(0)=1 \\ \deg(\varphi) \leq n}} \|\varphi(A)\| \approx \min_{\substack{\varphi(0)=1 \\ \deg(\varphi) \leq n}} \max_{z \in \Omega} |\varphi(z)|.$$

Possible choices of Ω include an ϵ-pseudospectrum, the field of values or the polynomial numerical hull of A; see Section 5.7.2. Because of the normalisation $\varphi(0) = 1$, the set Ω must exclude the origin. Furthermore, Ω should not completely surround the origin (i.e. it should not separate 0 and ∞ in the extended complex plane). If this is satisfied, then

$$\rho(\Omega) \equiv \lim_{n \to \infty} \left(\min_{\substack{\varphi(0)=1 \\ \deg(\varphi) \leq n}} \max_{z \in \Omega} |\varphi(z)| \right)^{1/n} < 1.$$

The number $\rho(\Omega)$ is called the *asymptotic convergence factor* of the set Ω. It is possible to determine (or at least estimate) $\rho(\Omega)$ using methods from potential theory[12] or, if Ω is simply connected, using conformal mappings. In the first case

$$\rho(\Omega) = \exp(-g_\Omega(0)),$$

where g_Ω is the Green's function of the set Ω, and in the second case

$$\rho(\Omega) = |\Psi(0)|^{-1},$$

where Ψ is a conformal map from the exterior of Ω to the exterior of the unit disc that satisfies $\Psi(\infty) = \infty$. In either case $\rho(\Omega)$ corresponds to some measure of the distance of the set Ω to the origin. A larger distance of Ω to the origin means a smaller $\rho(\Omega)$ and vice versa. For a short description of these techniques we refer to [145], a more extensive discussion of the potential theory ideas is given in [402]. Finally, the asymptotic convergence factor is used in the convergence analysis of GMRES, where it is hoped that

$$\frac{\|r_n\|}{\|r_0\|} \approx (\rho(\Omega))^n \qquad (5.7.27)$$

represents a reasonable estimate.

Driscoll, Toh and Trefethen [145] identified six essential approximation steps that lead from the left-hand side of (5.7.27) to the right-hand side. Their analysis shows where information is lost and how serious this loss of information is in some examples. They justify the asymptotic approach to the convergence analysis of GMRES to some extent, and clearly point out the risks and downsides.

We have already discussed some of the six approximation steps identified in [145] previously in this section. In particular, we have mentioned that the actual convergence of GMRES for a given system $Ax = b$, and even the worst-case GMRES for A, can be arbitrarily faster than described by the ideal GMRES for A. This is one of the reasons why the asymptotic approach cannot be expected to give sharp bounds on (worst-case) GMRES in general. In addition, there are two points we consider essential for understanding the convergence of GMRES, which are not emphasised in [145], but which need to be kept in mind when using the asymptotic formula (5.7.27).

---

12. It is interesting to note that Stieltjes' work already foreshadowed some of the related ideas. As described in [631, Section 3] (see also the additional material in [630]), Stieltjes considered in 1885 an expression in n variables that attains its minimum when these variables are equal to the zeros of a certain Jacobi polynomial, and he interpreted this expression as the electrostatic energy for a system with $n + 2$ charges, of which n are fixed and two are given. As explained in [631, Section 3], the electrostatic energy interpretation of zeros of orthogonal polynomials leads via a 1918 paper of Schur to the potential theory concepts of the transfinite diameter and the logarithmic capacity, which were introduced by Fekete in 1923.

First, as argued many times in this book, the behaviour of Krylov subspace methods in general and hence GMRES in particular is *highly nonlinear*. As shown in Section 5.7.4, nonlinearity in the behaviour of GMRES can theoretically occur (for a non-normal matrix A) at any stage during the iteration, since any convergence curve is possible for GMRES. It is impossible to capture a nonlinear behaviour with a linear estimate of the form (5.7.27).

Second, in computations using long recurrences (e.g. the GMRES method), one is mostly interested in the convergence behaviour in the *initial phase*. The reason is practical—one can simply not afford many iterations which require storing all computed basis vectors. Even if the matrix A is very sparse (it may not be assembled at all), the basis vectors are in general dense, and they are of length N. If N is very large, then using more than a few such vectors is computationally unfeasible. As illustrated in Section 5.7.5, the behaviour of GMRES in the initial phase can be highly intriguing, and it may depend on the initial residual. An asymptotic analysis, however, disregards the influence of the initial residual, and it typically can not provide information about the behaviour in the early stages of the computation.

5.8 ROUNDING ERRORS AND BACKWARD STABILITY

In numerical computations we usually *do not compute exactly*. Although rounding errors are small when we add two positive real numbers of about the same value or when we multiply two real values, their effects can be in general computations disastrous. In spite of this fact, and with some exceptions, like in matrix computations (numerical linear algebra), the effects of rounding errors are in computational mathematics often not considered an important part of the error analysis. Even in direct methods for numerical solving of algebraic problems, effects of rounding errors are sometimes considered 'small', although there are simple examples proving the opposite, e.g. [22, Example A.6, pp. 608–609].

In *iterative computations*, rounding errors were for a long time considered unimportant as a matter of principle, and iterative methods were by their nature considered 'self-correcting'. Some sources attribute the term 'self-correcting' to Gauss, and refer to his letter to Gerling written on December 26, 1823 [220]; see [191] for an English translation and historical remarks as well as [636, Section 1.1] for an illustrative description of Gauss' ideas. In fact, this letter contains the relaxation method of Gauss, which is most likely the first iterative method for solving linear algebraic systems. While Gauss in this letter did not use the term 'self-correcting' or any similar formulation, he performed his computations only approximately at each iteration on purpose in order to save unnecessary arithmetic (of course, he was using only mental arithmetic). Hence he was clearly aware of the 'self-correcting' property of his method.

This property was also noticed by researchers in the early days of numerical analysis, when relaxation methods were the only known iterative methods. For instance, Hotelling wrote in 1943 that iterative methods have 'the pleasing characteristic that mistakes do not necessarily spoil the whole calculation, but tend to be corrected at later stages' [344, p. 5]. He added: 'This of course does not mean that there is no penalty for mistakes. They have an obvious tendency to prolong the number of repetitions required, and if repeated at late stages may actually prevent realisation of a substantially correct result.' In 1948 Fox, Huskey, and Wilkinson [196] distinguished between direct methods, and 'indirect methods' that construct 'successive approximations' of the unknowns. They stated that 'mistakes in the indirect process can *merely* delay the convergence to an accurate solution, while mistakes in direct methods may completely invalidate the solution obtained' [196, p. 150]. Similarly, Fox argued in 1948 that in 'all methods of iteration, mistakes are not serious. If some residual is incorrect, the mistake will be found in the final check and further relaxation will be necessary' [195, p. 258]. In current terminology, the self-correcting property holds (to some extent) for linear stationary methods of the first order such as the stationary Richardson iteration; see (5.5.2). Nevertheless, because of the effect of rounding errors, the terminology 'self-correcting methods' should be used with care. This fact has already been pointed out by Householder [346, Introduction to Chapter 4, p. 91]. His formulation can, however, be interpreted incorrectly if a part of it is taken out of context.

It must be stressed that the concept of 'self-correction' cannot be extended to Krylov subspace methods, which would most probably be classified as direct methods by the authors of [196]. This was clearly explained by Lanczos in his seminal work published in 1952 [406, pp. 39 and 46], where he pointed out loss of orthogonality due to rounding errors and discussed reorthogonalisation as a costly remedy. Instead of reorthogonalisation he advocated use of his algorithms 'in blocks', which in the terminology used nowadays means performing restarts whenever 'the rounding errors have done too much damage to the orthogonality'. He, however, also mentioned the price: 'Then, however, we lose in convergence and the number of iterations has to be extended.'

Insightful results describing the effects of rounding errors to the CG method can also be found in the founding paper by Hestenes and Stiefel [313]. Section 8 of that paper is entitled 'Propagation of Rounding-Off Errors in the cg-Method', and contains concepts that reappeared in the literature several decades later. In particular, Hestenes and Stiefel recognised that the numerical instability arose not due to *simple accumulation* of elementary rounding errors, but due to *propagation* of the elementary roundoff from the given iteration step in subsequent steps. They studied this propagation and came to the point that 'there exists always an initial residual r_0 such that ... the algorithm is stable with respect to the propagation of rounding-off errors.' This was explained rigorously due to the work of Paige and Scott published almost 30 years later; see Section 5.9 in this book, [492, 558] for original works, and [457, Section 4.2] for a summary and the explanation of the full context. In [313, Sections 8.2–8.4, 9 and 19] Hestenes and Stiefel investigated several possible refinements and modifications of the CG method, and they explained why such modifications can lead also to worsening of the numerical behaviour. Additionally they mentioned on p. 422, relation (8.25), the possibility of reorthogonalisation at the obvious price of correcting all direction vectors. It is worth noticing that they, without saying, avoided the simultaneous reorthogonalisation of direction vectors and residuals. Everyone who has tried can confirm that they took the right approach. Interestingly, Stiefel began his paper from 1958 [581] as follows:

> In recent years many computational procedures have been developed in order to handle problems of linear algebra, in particular for solving linear equations and computing eigenvalues. The principal need for this research in a relatively elementary field of mathematics has arisen in the use of high-speed computational equipment, where numerical stability and the self-correcting features of a numerical method are very important.

Then Stiefel advocated iterative computations and emphasised the role of the theory of orthogonal polynomials. His Introduction ends with the sentence, 'Hence in working with polynomials we introduce into pure mathematics the practical properties of our computational equipment.' All this sounds very modern even now, except for the disturbing phrase 'self-correcting features' used in connection with 'numerical stability'. Possibly Stiefel meant the self-correcting features as adaptivity of the algorithms based on orthogonal polynomials (that would correspond to the content of the paper).

As mentioned in Section 5.1, in the numerical PDE literature rounding errors are often not considered a part of the evaluated error; see, e.g. the introduction to the collection [580] devoted to the state-of-the-art error-controlled adaptive finite element methods in solid mechanics. In this context one should point out the lasting effort of Babuška and his collaborators for incorporating the algebraic errors in *direct computations* into the practice of finite element methods; see, e.g. [29, 30]. When the conditioning of the algebraic problem is kept under control, e.g. by some form of adaptive preconditioning as in [23], then intuitively the negative effects of rounding errors are under control (which might not always be the case, as shown below). A comprehensive description of the issue from the numerical PDE perspective can be found in the classical monograph by Brenner and Scott [75, Chapter 9, in particular Sections 9.5–9.7]. The authors show that if an appropriate scaling can be used and the mesh is non-degenerate, the condition number of the linear algebraic system resulting from the finite element discretisation need not grow unacceptably as the mesh is refined. However, as shown in Sections 5.1–5.6 above, the convergence properties of CG are *not* described by the condition number of the algebraic system to be solved. The same is true for the size and the local distribution of the algebraic error (due to truncation of the iteration process or rounding errors, or both). The application to the CG method presented in [75, Section 9.7] is relevant for small condition numbers. But this can not always be achieved. To the other extreme, rounding errors are sometimes associated (without clear indicators) with unpleasant behaviour of complex computer codes that are poorly understood.

Understanding of rounding errors is important in all computations. Computing an approximate solution without providing information about the possible size of the computational error is risky. As a consequence, users of software that relies upon methods and algorithms with incomplete error analysis and indicators may, without warning, experience some failures or, even worse, may in practice use results that may not be sufficiently accurate or may even be incorrect.

Obviously, a complete analysis of all sources of errors is often out of our hands. What we mean is that one should always be aware of limitations and possible uncertainty of the computed results, so that appropriate measures can be taken to verify their accuracy and validate their relevance. In iterative computations, analytic considerations about convergence and stopping criteria in solving real-world problems should take into account that rounding errors can *delay convergence* and *limit the attainable accuracy*.

In this section we will describe some general issues related to effects of rounding errors in solving linear algebraic systems. We start with a brief recollection of the approach used in direct computations, discuss the history of the backward error concept, and then point out similarities and differences to the approach in iterative computations.

5.8.1 Rounding Errors in Direct Computations

One of the most important and widely studied direct methods is the LU factorisation (Gaussian elimination) with row interchanges (column pivoting); see, e.g. [133, Sections 2.3 and 2.4], [250, Section 3.4], [315, Chapter 9], [625, Lectures 20–22], [662, Sections 1.8 and 2.7] for modern presentations of this method. The LU factorisation is a canonical example for demonstrating the ideas of rounding error analysis in the context of direct methods. In exact arithmetic, the method computes a factorisation of the form

$$PA = LU, \qquad (5.8.1)$$

where P is a permutation matrix, L is a lower triangular, and U is an upper triangular matrix. If A is nonsingular, the uniquely determined solution of $Ax = b$ is then given by

$$x = U^{-1}(L^{-1}Pb),$$

which can be computed by solving the two triangular systems with L and U. Typically only the solution with U has to be performed; that with L is implicitly done while transforming A to upper triangular form.

The strategy with row interchanges ensures that all entries l_{ij} of L satisfy $|l_{ij}| \leq 1$. However, it does not guarantee a similar property for the entries of U. A major role in the numerical stability analysis of the algorithm is played by the *element growth factor γ*, defined as the ratio of the largest and the smallest element of U (both in absolute value). It is well known (see, e.g. [315, Section 9.4]) that there exist matrices for which γ grows exponentially with the matrix size N. In fact, γ can become as large as 2^{N-1}. If γ is very large, then the numerical solution $\widetilde{x}$ of $Ax = b$ computed in finite precision arithmetic is likely to be spoiled by rounding errors. On the other hand, if γ is small or moderate, then the computed solution is in practice often very accurate.

Following Wilkinson [668, 671], the results can be stated in the following quantitative form. In finite precision arithmetic characterised by machine precision ε, the computed solution $\widetilde{x}$ satisfies

$$(A + \delta A)\widetilde{x} = b, \qquad (5.8.2)$$

where

$$\|\delta A\| \leq \gamma \varepsilon \nu_1(N)\|A\|, \qquad \|b - A\widetilde{x}\| \leq \gamma \varepsilon \nu_2(N) \|\widetilde{x}\|, \qquad (5.8.3)$$

where $\nu_1(N)$ and $\nu_2(N)$ stand for (positive) low-degree polynomial expressions in N. Here we use, for consistency with the rest of the text, the Euclidean vector norm and the matrix spectral norm. The original papers as well as most of the related literature use the ∞-norms; for a detailed review see [315, Chapter 9]. The terms $\nu_1(N)$ and $\nu_2(N)$ reflect technicalities of the worst-case rounding error analysis, but do not contain other useful information.

A fundamental message is encoded in (5.8.2)–(5.8.3): The rounding errors of the numerical floating point calculations appear as (possibly tiny) perturbations δA of the original data given by the matrix A, so that the method *in exact arithmetic but applied to the perturbed data* would produce the actually computed solution $\widetilde{x}$. This is a typical result of a *backward error analysis*; see also (5.1.1)–(5.1.3). It tells us which problem is solved *exactly* by the approximation we have *computed*.

Computing the approximate solution $\widetilde{x}$ requires the whole factorisation (5.8.1) and subsequent solution of the resulting triangular systems; see Section 5.2. The element growth factor γ can be tested during the computation. If the computation is completed, then an *a-posteriori* argument combined with perturbation analysis bringing into account conditioning of the problem shows to what extent the user-specified accuracy of the computed solution is guaranteed; see, e.g. [315, Chapters 7 and 9, in particular Section 9.7]. In the *sparse* LU factorisation one must also take into account, in addition to the matrix fill-in, the size of the intermediate results. Hence in general sparse direct solvers, numerical stability issues affect the cost of particular computations. This is also true, e.g. in solving discrete ill-posed problems arising from the Fredholm integral equation of the first kind with smooth kernels, with image or signal reconstruction as typical application areas; see [164, 294, 295].

Let us now consider a second example of a direct computation, namely the QR factorisation of matrix $B \in \mathbb{R}^{N \times M}$ with full column rank $M \leq N$, which is often used to solve least squares problems of the form

$$\min_{x \in \mathbb{R}^M} \|r\| \quad \text{subject to} \quad Bx = b - r. \qquad (5.8.4)$$

This factorisation is given by

$$B = QR, \quad \text{where } Q^T Q = QQ^T = I \text{ and } R \text{ is upper triangular.}$$

There exist two major variants of computing this factorisation.

In the *Householder QR factorisation algorithm* (and, analogously, in the QR factorisation algorithm using Givens rotations), one starts with $B_1 \equiv B$ and applies a sequence of elementary Householder matrices $P_k \equiv I - 2(w_k w_k^T)$ with $\|w_k\| = 1$ in order to obtain matrices

$$B_{k+1} = P_k B_k, \quad k = 1, \ldots, M,$$

so that, at the final step M,

$$\begin{bmatrix} R \\ 0 \end{bmatrix} = P_M \ldots P_2 P_1 B, \quad \text{where } R \text{ is upper triangular;}$$

see [315, Chapter 18], [346, Sections 5.4 and 5.5], and [670, Chapter 4, Sections 46–47]. The orthogonal matrix Q is given as a product of elementary Householder matrices,

$$Q \equiv (P_M \ldots P_2 P_1)^T = P_1 P_2 \ldots P_M.$$

Since elementary Householder matrices computed in finite precision arithmetic are very close to their exact precision counterparts, the near-orthogonality of the properly computed individual P_k, $k = 1, \ldots, M$, is inherited also by their product Q, which is, however, rarely formed explicitly. The computed upper triangular factor R satisfies

$$B + \delta B = \widehat{Q} \begin{bmatrix} R \\ 0 \end{bmatrix}, \tag{5.8.5}$$

where $\widehat{Q}$ is an *exactly* orthogonal matrix, and

$$\|\delta B\| \le \frac{\varepsilon\, v_3(M, N)}{1 - \varepsilon\, v_3(M, N)} \|B\|, \tag{5.8.6}$$

$$\|Q - \widehat{Q}\| \le \frac{\varepsilon\, v_4(M, N)}{1 - \varepsilon\, v_4(M, N)}. \tag{5.8.7}$$

Analogously to (5.8.3), here and in the rest of this section the $v_j(M, N)$ stand for (positive) low-degree polynomial expressions in M, N.

In the *Gram–Schmidt QR factorisation algorithm* the first column q_1 of Q is computed by normalising the first column of B. The subsequent columns q_j, $j = 2, \ldots, M$, result from orthogonalising the j-th column of B, denoted by b_j, against the previously computed $q_1, \ldots, q_{j-1}$, followed by normalisation. In the *classical Gram–Schmidt (CGS)* variant, the orthogonalisation coefficients are computed independently of each other, i.e.

$$w_j = b_j - (b_j, q_1) q_1 - \cdots - (b_j, q_{j-1}) q_{j-1}$$
$$= \left(I - q_1 q_1^T - \cdots - q_{j-1} q_{j-1}^T \right) b_j,$$
$$q_j = w_j / \|w_j\|, \quad j = 2, \ldots, M,$$

while in the *modified Gram–Schmidt (MGS)* variant, the orthogonalisation is performed recursively, i.e.

$$w_j = \left(I - q_{j-1} q_{j-1}^T \right) \cdots \left(I - q_1 q_1^T \right) b_j,$$
$$q_j = w_j / \|w_j\|, \quad j = 2, \ldots, M;$$

see, e.g. [60, Section 2.4.2] or [315, Chapter 19]. Different orders of orthogonalisation can be applied under different circumstances, but this feature is not studied here.

In exact arithmetic, the CGS and MGS variants are equivalent. However, their results may differ significantly in finite precision computations. A basic rounding error analysis of the MGS QR factorisation algorithm was given in [59], and it was further extended in [62]. It was shown that the computed upper triangular factor R satisfies

$$B + \delta B = \widehat{Q} R, \tag{5.8.8}$$

where $\widehat{Q} \in \mathbb{R}^{N \times M}$ has *exactly* orthonormal columns, and

$$\|\delta B\| \leq \frac{\varepsilon \, v_5(M, N)}{1 - \varepsilon \, v_5(M, N)} \|B\|, \tag{5.8.9}$$

$$\|Q - \widehat{Q}\| \leq \frac{\varepsilon \, v_6(M, N) \kappa(B)}{1 - \varepsilon \, v_6(M, N) \kappa(B)}, \tag{5.8.10}$$

$$\|I - Q^T Q\| \leq \frac{\varepsilon \, v_7(M, N) \kappa(B)}{1 - \varepsilon \, v_7(M, N) \kappa(B)}. \tag{5.8.11}$$

Although the orthogonality of the computed Q is in the MGS algorithm not preserved on the same level as in the Householder QR factorisation, the accuracy of the MGS upper triangular factor R is comparable to the one from the Householder QR. This somewhat surprising fact follows from the observation that the MGS QR factorisation of B is *numerically equivalent* to the Householder QR factorisation applied to B augmented with a square matrix of zero elements on top; see [62, Section 2] for details.

A similar relationship is not known for the CGS QR factorisation. Nevertheless, in a paper from 2005 [234] it was shown that the loss of orthogonality in the CGS QR factorisation is bounded by

$$\|I - Q^T Q\| \leq \frac{\varepsilon v_8(M, N) \kappa^2(B)}{1 - \varepsilon v_8(M, N) \kappa^2(B)}, \tag{5.8.12}$$

which solved a problem that had been open for almost forty years.

Analogously to Gaussian elimination, the rounding error analysis of the QR factorisation brings, besides the derived bounds, a fundamental insight that is more important than the bounds themselves. With respect to the computed upper triangular factor R, the MGS QR factorisation is proven to provide as good results as the Householder variant. The fundamental relationship between the MGS QR of the matrix B and the Householder QR of the matrix B augmented with a square zero matrix on the top also reveals the detailed mechanics of the loss of orthogonality in MGS QR; see [62, Section 4, in particular Theorem 4.1]. As we will see in Section 5.10, these results can be applied in the rounding error analysis of the GMRES method.

5.8.2 Historical Note: Wilkinson and the Backward Error Concept

With the first 'automatic digital computers' in the mid 1940s, researchers performing numerical computation lost close contact with intermediate results. Unlike the previously used desk calculators, computers were supplied with a code and input data in order to produce a result without any human interaction. The uncertainty caused by the separation of human and computer work grew into a fear of numerical instabilities when Hotelling wrote in 1943 that the error in a computed matrix inverse might grow exponentially with the problem size [344]; see also [505, p. 23]. This triggered a search for numerically stable algorithms for solving systems of linear algebraic equations, and it showed the need for a mathematically rigorous rounding error analysis. Von Neumann and Goldstine responded to this situation with their seminal paper of 1947 [649], in which they analysed the accuracy of the inverse of a symmetric positive definite matrix A computed via an elimination algorithm. They showed that the computed inverse X satisfies

$$\|AX - I\| \leq \mathcal{O}\left(N^2\right) \varepsilon \kappa(A). \qquad (5.8.13)$$

The analysis in [649] can be considered the first modern numerical stability analysis. A comprehensive analysis of the content and historical context of this paper was given by Grcar in [266]. Together with Turing's comparably fundamental work of 1948 [627], which addressed the matrix elimination algorithm as the LDU factorisation, worked with an analogy of the condition number of a matrix, and indicated the idea of preconditioning, it inspired many further developments. The history of Gaussian elimination is described, in addition to [266], in Grcar's papers [265, 267].

While the papers of von Neumann, Goldstine, and Turing contained the general idea of the backward error, they did not develop it further. A possible explanation, given in retrospect by Wilkinson in his historical review written in 1971 [669], is that the paper [649] is devoted primarily to matrix inverses. Though one can (now!) easily formulate a backward error result for each individual column of A^{-1}, it seems much more difficult to formulate and prove a backward error for the whole inverse, because the perturbation matrices change with each individual column.

In [669, p. 557] Wilkinson wrote that if von Neumann and Goldstine had treated a single linear system $Ax = b$ rather than the whole inverse A^{-1}, then 'I'm sure that backward error analysis would have got off to an earlier start.' Another reason why the paper of von Neumann and Goldstine did not lead to a 'quick-start' of the area of numerical stability analysis in general, and backward error analysis in particular, is its length and significant level of difficulty, characterised by Wilkinson as 'not exactly bedside reading' [669, p. 551]. According to him, the paper led to a 'mood of resignation' of his contemporaries, and he summed up a typical reaction of the time as follows: 'If it takes von Neumann 80 pages of closely reasoned argument to obtain an error bound for an algorithm which is one of the simplest in linear algebra, then we can scarcely expect to deal with any appreciable number of algorithms in the same rigorous fashion and must, in general, rely on empirical assessments of stability.' In the late 1950s, even leaders of the field like Newman and Todd [478,

p. 466] found that 'it is tedious and difficult to make error estimates in a calculation of significant size. The number of genuine general error estimates which have been made is very small indeed, even in the case of matrix calculations, which are in many respects the basic and simplest calculations. Our feeling however is that it is not economically possible to carry out rigorous error estimates for all computational problems.'

Considering the mood of the time, Wilkinson's work that led to (5.8.2)–(5.8.3) and other backward error results can be considered as epochal, or 'as one of the miracles of numerical computation' [45, p. 1]. Today, the backward error analysis has been established as a rigorous analytical tool of fundamental importance. The significance of Wilkinson's approach is not in formulas and theorems, but in the methodology he developed. As Wilkinson pointed out in 1971 [669, p. 567],

> ... the bound itself is usually the least important part ... The main object of such an analysis is to expose the potential instabilities, if any, of an algorithm, so that, hopefully, from the insight thus obtained one might be led to improved algorithms.

In the case of Gaussian elimination numerical stability analysis reveals the importance and technical details of the element growth factor, which affects practical implementations and leads to built-in indicators for adaptation of computation (in sparse direct solvers implementing a variant of the LU factorisation) or indicators of possible failures of the method. For a thorough description with many further details on the historical development of the error analysis of Gaussian elimination we refer to [315, Section 9.13].

In the following section we will explain different uses of the backward error concept in iterative computations. Besides analysing the effects of rounding errors, it can also be used for evaluating accuracy of the computed approximation in a *truncated* iteration. As indicated in Section 5.1, its relationship to error evaluation in the associated functional space (in the numerical solution of PDEs) is rather intriguing.

In the rest of the book we will focus on the *normwise backward error*. Many additional issues that we omit here are discussed, e.g. in the paper by Bunch [88] and the book by Higham [315, Chapter 7, in particular Section 7.6]. The history of rounding error analysis of iterative methods is summarised, with a detailed investigation of stationary iterative methods, in [315, Chapter 17]; see also [269, 280, 317, 318, 457].

5.8.3 The Backward Error Concept in Iterative Computations

In iterative computations one must specify the context in which the backward error concept is to be applied. We will distinguish two principally different situations.

1. After performing n iterations we wish to evaluate the accuracy of the computed approximation x_n. Unlike direct computations, where the computed approximation $\widetilde{x}$ differs from the exact solution x of $Ax = b$ because of roundoff, in iterative computations x_n is in general different

from x even in exact arithmetic. An evaluation of accuracy in iterative computations therefore has to take truncation *and* rounding errors into account; see Sections 5.2–5.4.

2. After performing n iterations we wish to evaluate how much the computed approximations $x_1, \ldots, x_n$ are affected by rounding errors. Here we are not interested in the truncation error, but we ask one (or both) of the following questions:

    (a) What is the limit that restricts the achievable accuracy of the computed approximations? Obviously, such a limit, called the *maximal attainable accuracy*, always exists, and it is related to the machine precision unit. In view of our arguments presented in Section 5.4, the question can be more precisely phrased as: what is the maximal attainable accuracy for the given method, its implementation and its application to the given data?

    (b) Provided that the maximal attainable accuracy has not yet been reached, how are the computed approximations $x_1, \ldots, x_n$ affected by rounding errors? Here it is in general very difficult to answer how much, for a *fixed* iteration $j = 1, \ldots, n$, the computed approximation x_j differs from its exact precision counterpart. This question can be answered, to some extent, for stationary iterative methods; see [315, Section 17]. Because of their strong nonlinearity, however, an analogous forward error analysis for Krylov subspace methods can hardly lead to informative bounds. Therefore it is better to ask: how much is the convergence *delayed* in steps 1 through n because of rounding errors? As we will see in Section 5.9.1, this question can indeed be answered in important cases such as CG computations.

In the rest of this section we will comment briefly on the backward error at the given fixed iteration step n. The questions on delay of convergence and maximal attainable accuracy due to rounding errors will be addressed in the following sections.

Let x_n be the computed approximation to the solution x of the linear algebraic system $Ax = b$. Then one can easily see that x_n solves the system with the perturbed right-hand side

$$Ax_n = b - r_n, \quad r_n = b - Ax_n. \tag{5.8.14}$$

Therefore the algebraic inaccuracy measured by the residual norm can be interpreted in terms of solving accurately the system with the same matrix and the modified right-hand side $b + \delta b$, with the size of the modification $\|\delta b\| = \|r_n\|$. A straightforward calculation gives a bound for the forward error,

$$\frac{\|x - x_n\|}{\|x\|} \leq \|A\| \, \|A^{-1}\| \frac{\|r_n\|}{\|b\|} = \kappa(A) \frac{\|r_n\|}{\|b\|}. \tag{5.8.15}$$

The relative size of the perturbation of the right-hand side in (5.8.14) is $\|r_n\|/\|b\|$. With $x_0 = 0$ this is the widely used relative residual norm $\|r_n\|/\|r_0\|$.

But when $x_0 \neq 0$, the relative residual norm lacks this backward error interpretation, and for $\|r_0\| \gg \|b\|$ it represents, in general, a rather insecure measure of convergence. In fact, a nonzero x_0 that contains no useful information about the solution x (e.g. a random x_0) might lead to $\|r_0\| \gg \|b\|$ with r_0 completely 'biased' towards x_0. Such a choice can create an illusion of fast convergence to a high relative accuracy measured by the relative residual norm. This holds in particular for the initial phase of the GMRES iteration.

For a numerical illustration we consider a linear algebraic system $Ax = b$ with A being the matrix FS1836 from MatrixMarket.[13] This is a non-normal and ill-conditioned ($\kappa(A) \approx 10^{11}$) matrix of order 183. In order to show that the results are not related to a specific right-hand side, we use a right-hand side b with its entries generated randomly (using the normal distribution). We approximate the exact solution x of $Ax = b$ using the MATLAB backslash operator, which is sufficiently accurate for evaluating the norm of the error in this experiment. We plot the convergence characteristics of GMRES applied to $Ax = b$ in Figure 5.13. The solid line shows the relative 2-norm of the residual ($\|r_n\|/\|r_0\|, n = 0, 1, 2, \ldots$) for a random initial approximation x_0. Choosing a random x_0 leads to

$$\|r_0\| \approx 10^9 \gg \|b\| \approx 10^1.$$

We observe that the residual norm decreases rapidly in the initial phase of the iteration. On the other hand, the corresponding relative 2-norm of the error ($\|x - x_n\|/\|x - x_0\|, n = 0, 1, 2, \ldots$), which is plotted by the dashed line in Figure 5.13, stays essentially constant in the initial phase. Thus, the GMRES method initially eliminates the artificial bias in the vector r_0 that is due to the nonzero initial approximation, while the approximation x_n does not get any closer to the exact solution. Further examples and discussions can be found in [500]; see in particular Figures 7.9 and 7.10 in that paper.

In a personal communication Hegedüs suggested that a simple way around this difficulty is to rescale the initial approximation. Given a nonzero preliminary initial approximation x_p, determine the scaling parameter $\zeta_{\min}$ as

$$\zeta_{\min} = \frac{b^* A x_p}{\|A x_p\|^2}. \tag{5.8.16}$$

Then, by setting $x_0 = \zeta_{\min} x_p$, we ensure that

$$\|r_0\| = \|b - \zeta_{\min} A x_p\| = \min_\zeta \|b - \zeta A x_p\| \leq \|b\|.$$

The extra cost for implementing this little trick is negligible; it should be used whenever a nonzero x_0 is considered. In the previous example, the application of the Hegedüs trick indeed scales down the irrelevant bias in the initial approximation. Using (5.8.16), the relative residual norm no longer creates an illusion of fast

---

13. http://math.nist.gov/MatrixMarket/

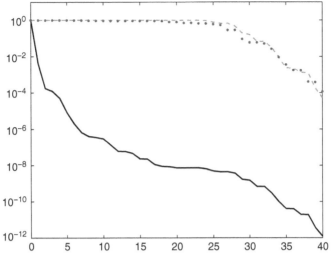

Figure 5.13 Convergence characteristics of GMRES applied to a linear algebraic system with the matrix FS1836 from MatrixMarket. The relative error norm (dashed line) shows that the approximation x_n does not approach the solution x in the initial phase of the iteration. A random initial approximation, however, can create an illusion of fast convergence when measured by the relative residual norm (solid line). This illusion can be avoided when using the Hegedüs trick (dots).

convergence; see the dots in Figure 5.13. In spite of the existence of this simple remedy, we recommend using $x_0 = 0$ if no problem-related initial approximation is available. In our example this would lead to a residual norm curve that is almost identical to the dots in Figure 5.13.

Allowing perturbations in both A and b will lead to the backward error, where the computed approximation x_n solves the system

$$(A + \delta A_{\min})x_n = b + \delta b_{\min}, \qquad (5.8.17)$$

and

$$\delta A_{\min} = \frac{\|A\| \, \|x_n\|}{\|b\| + \|A\| \, \|x_n\|} \frac{r_n x_n^*}{\|x_n\|^2}, \qquad (5.8.18)$$

$$\delta b_{\min} = -\frac{\|b\|}{\|b\| + \|A\| \, \|x_n\|} r_n, \qquad (5.8.19)$$

represent the *minimal normwise relative perturbations* in the sense given by (5.1.2); see, e.g. [315, Theorem 7.1]. Their relative size is given by $\|\delta A\|/\|A\| = \|\delta b\|/\|b\| = \beta(x_n)$; see (5.1.3). An elementary calculation then gives a bound for the forward error,

$$\frac{\|x - x_n\|}{\|x\|} \leq \left(\frac{\|A^{-1}\| \, \|b\|}{\|x\|} + \kappa(A) \right) \frac{\beta(x_n)}{1 - \kappa(A)\beta(x_n)}$$

$$\leq 2\kappa(A) \frac{\beta(x_n)}{1 - \kappa(A)\beta(x_n)}; \qquad (5.8.20)$$

see, e.g. [315, Theorem 7.2].

While in (5.8.14) the backward error is considered only in the right-hand side, here it is considered in both the matrix and the right-hand side; see (5.8.17)–(5.8.19). Its relative size $\|\delta A\|/\|A\|$ and $\|\delta b\|/\|b\|$ may not, in general be equal. For the weighted approach see, e.g. [498, 497] and [315, Chapter 7]). A comparison of the forward bounds (5.8.15) and (5.8.20) requires some thought; see, e.g. [315, Section 17.5]. In any case they represent worst-case bounds involving the condition number $\kappa(A)$ of A, and therefore they can in practice significantly overestimate the actual size of the forward error.

5.9 ROUNDING ERRORS IN THE CG METHOD

In spite of the comments by Lanczos, Hestenes, and Stiefel in 1952 (see the introductory part of Section 5.8), a thorough analysis of rounding errors in the Lanczos and CG methods did not appear until the PhD thesis of Paige in 1971 [489]. His thesis and the subsequent papers [490, 491, 492] enabled almost all further development; see the description including the historical context in the survey [457], in the PhD Thesis [689], and in the monograph [453]. We do not intend to repeat the same material here, but rather concentrate on several main ideas with consequences on evaluating the cost of CG computations.

5.9.1 Delay of Convergence

The question about delay of convergence asks whether the finite precision computation requires more iterations than its exact precision counterpart in order to reach the prescribed accuracy. A numerical illustration is shown in Figure 5.14. The dashed line shows the relative A-norm of the error $\|x - x_n\|_A / \|x - x_0\|_A$ in a computation with CG using double reorthogonalisation [280]. We use the same matrix A as in Figure 5.9 and a different (randomly generated) right-hand side b; the initial approximation is $x_0 = 0$. The solid line in Figure 5.14 shows the corresponding error norms for this problem and the standard implementation of the CG method by Hestenes and Stiefel [313]; see Algorithm 2.5.1 in this book. We can see, for example, that in exact arithmetic (simulated by the dashed line) the relative A-norm of the error reaches the level of 10^{-8} in 22 steps, while in finite precision arithmetic rounding errors cause a delay of 15 additional steps.

Before explaining how rounding errors cause this delay of convergence, we first return to the behaviour of CG computations in *exact arithmetic* in the presence of tight clusters of eigenvalues, which we have already discussed in Section 5.6.5. We will use a modification of the experiment presented in that section to explain how the delay of convergence occurs. Consider the linear algebraic system $Ax = b$,

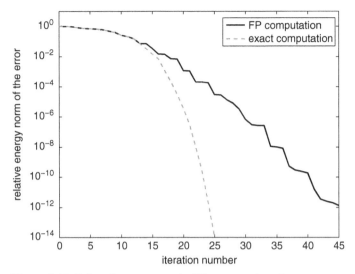

Figure 5.14 Delay of convergence in CG computations due to rounding errors. The finite precision CG computations (solid line; here illustrated by the relative energy norm) differ from the exact CG (dashed line) quantitatively and qualitatively (see the staircase-like behaviour of the solid line).

where A is a diagonal matrix based on the parameters $N = 25, \lambda_1 = 0.1, \lambda_N = 100$ and $\gamma = 0.65$ in (5.6.32), and b is a random vector. In addition, consider two systems

$$B^{(N)} y^{(N)} = b_1^{(N)} \quad \text{and} \quad C^{(N)} z^{(N)} = b_2^{(N)}.$$

Unlike in Section 5.6.5, here $B^{(N)}$ is obtained from A by 'splitting' the largest diagonal entry into two diagonal entries of $B^{(N)}$, i.e.

$$B^{(N)} = \text{diag}(\lambda_1, \ldots, \lambda_{N-1}, \lambda_N - \Delta, \lambda_N), \quad \Delta = 10^{-14}.$$

The matrix $C^{(N)}$ is obtained in an analogous way by 'splitting' λ_N into five close diagonal entries,

$$C^{(N)} = \text{diag}(\lambda_1, \ldots, \lambda_{N-1}, \lambda_N - 4\Delta, \lambda_N - 3\Delta, \lambda_N - 2\Delta, \lambda_N - \Delta, \lambda_N).$$

The right-hand side vectors $b_1^{(N)}, b_2^{(N)}$ are obtained by splitting the original entry of b corresponding to λ_N in the way described in Section 5.6.5.

Figure 5.15 shows the convergence of the CG method measured by the relative A-norm of the error. As in Figure 5.9, the differences in the behaviour do *not* result from rounding errors. We see that at the 13th iteration step the dashed-dotted line corresponding to the system with the matrix $B^{(N)}$ (at earlier steps it coincides with the solid line) separates from the dashed line corresponding to the system with the matrix A. At this iteration the CG computation fully recognises that the matrix

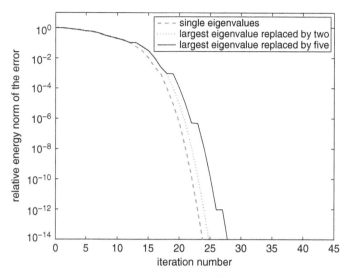

Figure 5.15 Delay of convergence in *exact* CG computations due to the presence of a single, well separated and tight cluster of eigenvalues. The dashed, dashed-dotted and solid lines show the exact CG behaviour with the single largest eigenvalue, and with this eigenvalue split into clusters of two and five eigenvalues, respectively.

$B^{(N)}$ has two eigenvalues $\lambda_N - \Delta$ and λ_N close to each other, and, starting from this iteration, each of them is closely approximated by the corresponding Ritz value (the eigenvalue of the computed Jacobi matrix; see Section 5.6). The sum of the weights $\omega_{12}^{(13)} + \omega_{13}^{(13)}$ in the tight cluster of the two Ritz values $\theta_{12}^{(13)}$ and $\theta_{13}^{(13)}$ must approximate the weight ω_N of the largest eigenvalue λ_N; see the discussion of questions **C1**–**C3** in [457, Section 4.4] and the original work of Wülling [678, 679], which resolved the conjectures posed in the IMA Report [596] from 1992. Thus, it can be seen from the formula (5.6.14) for the A-norm of the error, that replacing the single Ritz value $\theta_{12}^{(12)}$ with the weight $\omega_{12}^{(12)} \approx \omega_N$ at the step 12 by the two Ritz values $\theta_{12}^{(13)}$ and $\theta_{13}^{(13)}$ with the corresponding weights at the step 13, can not have a significant effect on the A-norm of the error. In our experiment the rest of the sum in (5.6.14) does not change much, i.e.

$$\sum_{j=1}^{11} \omega_j^{(12)} \left\{\theta_j^{(12)}\right\}^{-1} \approx \sum_{j=1}^{11} \omega_j^{(13)} \left\{\theta_j^{(13)}\right\}^{-1},$$

which represents a typical behaviour (there can be, however, an observable moderate decrease of the size of the error). In summary, at the given iteration step the CG computation nearly stagnates, i.e.

$$\|x - x_{12}\|_A \approx \|x - x_{13}\|_A.$$

This can also be explained by the fact that because at step 13 the two Ritz values $\theta_{12}^{(13)}$ and $\theta_{13}^{(13)}$ are placed close to the largest eigenvalue λ_N, only 11 Ritz values (and hence the same number as at step 12) remain to approximate the rest of the spectrum. Therefore, one can hardly expect a large drop of the size of the error from the 12th to the 13th step. This near stagnation means that convergence is delayed by one step. Because there is no other extra eigenvalue of $B^{(N)}$ in comparison to A, the fast convergence resumes at the next iteration step 14, and the dashed-dotted line copies the dashed line with the delay of one iteration step.

For the system with the matrix $C^{(N)}$ a similar behaviour can be observed, but here it occurs repeatedly, since we have 'split' the largest eigenvalue of A into five close eigenvalues. Further delays occur at iterations 18, 22, and 26. More generally, if each eigenvalue of A is split into a tight cluster, as in Section 5.6.5, then the same phenomenon takes place for all clusters (provided that we perform enough iteration steps). As an example, Figure 5.9[14] illustrates such superpositions. We again emphasise that here rounding errors *do not* play a role.

We will now explain how all this is related to effects of *rounding errors* in CG computations. For clarity of the exposition we will only consider cases with single eigenvalues and thus avoid the subtleties in the numerical approximation of multiple eigenvalues. In finite precision arithmetic CG computations, tight clusters of Ritz values corresponding to single eigenvalues are formed as a consequence of propagation of rounding errors throughout the iterations. The delay of convergence is then determined by the *rank-deficiency* of the computed Krylov subspaces [269, 280]. More precisely, the CG method in *exact arithmetic* determines an orthogonal basis of the Krylov subspace $\mathcal{K}_n(A, r_0)$ given by the residual vectors. Because of rounding errors, orthogonality and even linear independence of the computed residual vectors is (quickly) lost, so that these computed residual vectors can span a subspace of smaller dimension. Given a fixed accuracy level, this loss of dimension (or rank-deficiency) determines how many iterative steps the solid curve in Figure 5.14 lags behind the dashed curve. In *exact* CG computations with the matrices $A, B^{(N)}$, and $C^{(N)}$ (as well as A, B, and C in Figure 5.9), the delay is determined by the total number of the Ritz values in the clusters, decreased by the number of clusters (here a single well-separated Ritz value is considered a cluster with only one member, and therefore it does not count for determining the delay). In finite precision computations the situation is analogous.

As shown by Paige in his seminal paper [492], rank-deficiency and hence delay can occur only when multiple copies of the same (simple) eigenvalue of A are formed using the computed Krylov subspace. Paige's work, which started with his ingenious PhD thesis of 1971 [489], was the basis for Greenbaum's backward-like theory for the CG method [269]. This theory formally proves what is described above and illustrated in Figures 5.14 and 5.15:

> The finite precision CG computation behaves like the exact CG computation for a matrix having its eigenvalues replaced by tight clusters.

---

14. When comparing that figure with Figure 5.15, recall that it shows the squared A-norm of the error scaled by the squared Euclidean norm of the initial residual.

These clusters can contain different numbers of eigenvalues depending on the position of the original eigenvalue in the spectrum as well as on the given iteration step. For a summary of the backward-like analysis of Greenbaum and of the subsequent results published by other authors, including the relationship to the sensitivity of the Gauss–Christoffel quadrature (see Section 5.6.5), we refer to [457, Section 4.3]. The augumented problem in [493] might offer another insight and this may eventually lead to further quantifications that can extend the results published in [483].

A further illustration is given in Figure 5.16, which shows results of an experiment analogous to the one shown in Figure 5.15. The dashed line in the left part of Figure 5.16 shows results of the *exact* CG computation (more precisely, as above, of the reorthogonalised computations where the effect of roundoff is negligible) for a matrix A based on the parameters

$$N = 25, \quad \lambda_1 = 0.1, \quad \lambda_N = 100, \quad \gamma = 0.65$$

in (5.6.32). The solid line shows the exact CG, but now with *all* eigenvalues of A replaced by tight clusters (here $\Delta = 10^{-14}$). In the right part of Figure 5.16 we compare the exact CG (dashed line; the same as in the left part) with the *finite precision* CG computation for the matrix A. We observe that the solid lines in the two parts of the figure are clearly very similar.

In the left part of Figure 5.17 we show again the exact CG curve from Figure 5.16 (dashed line) and the curve generated by the points

$$\left(\text{rank}(K_n(A,b)); \; \|x - x_n\|_A / \|x - x_0\|_A\right), \quad n = 1, 2, \ldots,$$

from the finite precision CG computations. This means that we have shifted the points

$$(n; \; \|x - x_n\|_A / \|x - x_0\|_A), \quad n = 1, 2, \ldots,$$

which form the solid curve in the right part of Figure 5.16, by the rank-deficiency of the computed Krylov subspace and hence by the number of delayed iteration

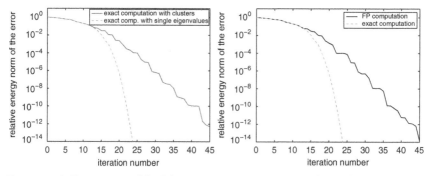

Figure 5.16 Comparison of the delay in exact CG computations due to the presence of clusters of eigenvalues (left) with the delay of convergence in finite precision CG computations due to roundoff (right).

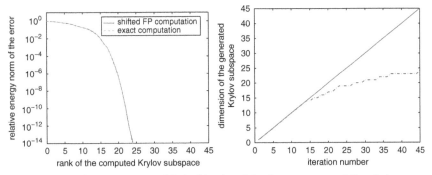

Figure 5.17 Left: The exact CG (dashed line) and the finite precision CG with the points of the convergence curve shifted horizontally by the number of delayed iteration steps (solid line). Right: The delay of convergence at a given iteration step is equal to the rank deficiency of the computed Krylov subspace, which here can be seen by the vertical difference between the solid and dashed-dotted lines.

steps. Here $\text{rank}(K_n(A,b))$ is given by the number of singular values of the matrix $[r_0, r_1, \ldots, r_{n-1}]$ (consisting of the computed residual vectors) that are greater or equal to 0.1. Thus, $\text{rank}(K_n(A,b))$ counts the number of residual vectors that are 'strongly linearly independent'. Using a value different from 0.1 would not change the main point, but the match between the two curves would not be as perfect as shown in Figure 5.17. The dashed-dotted line in the right part of this figure shows $\text{rank}(K_n(A,b))$, and the difference $n - \text{rank}(K_n(A,b))$, which can be seen as the vertical difference between the solid and dashed-dotted lines, gives the delay of convergence.

As mentioned above (see the introductory part of Section 5.8), Lanczos, Hestenes, and Stiefel had already suggested reorthogonalisation as a (costly) remedy for eliminating the deterioration of convergence due to roundoff. One of the main results of Paige's PhD thesis [489] says that the *orthogonality can be lost only in the direction of the converged Ritz vectors*; see [492] and [457, Section 4.2, Theorem 4.3]. This was used by Parlett, Scott, Simon, and others to develop efficient reorthogonalisation strategies; see [457, Section 4.5]. Here we will recall another remarkable consequence of the fact that the loss of orthogonality is so nicely structured. *Assume* that in the steps 1 though $N - 1$ no Ritz value has converged to an eigenvalue of A. Then there is also no converged Ritz vector and, remarkably, there is no significant loss of orthogonality among the CG residuals (or, with normalisation, among the Lanczos vectors). Scott proved in 1979 (in the context of the Lanczos method) that for any given symmetric matrix there is indeed a starting vector such that no Ritz value converges (in exact arithmetic) until the very last step; for details we refer to Theorem 3.4.12, to the paper [558] and to Section 4.2 of the survey [457]. Using such a starting vector in finite precision computations gives an experimental justification for the conjecture of Hestenes and Stiefel mentioned at the beginning of Section 5.8, that there always exists an initial residual for which the CG computation is numerically stable.

These results underline that effects of rounding errors in CG computations depend not only on the matrix A, but they can depend also on the right-hand side b

and on the initial residual $r_0 = b - Ax_0$. It is therefore clear that evaluation of the cost of practical computations using the CG method must consider the particular data, and that it can not be restricted to using information about the matrix A only (let alone using just 'single number information' such as the condition number $\kappa(A)$). This raises a question about validity of the bounds that use (in addition to the condition number) some additional *partial* information about eigenvalue distribution. This question is addressed next.

5.9.2 Delay of Convergence can Invalidate Composite Convergence Bounds

Delay of convergence has important consequences for the practical relevance of some CG convergence bounds studied in Section 5.6.4. As we will see in this section, improperly interpreted bounds used out of context can lead to false conclusions.

The CG convergence analysis based on composite polynomials (see (5.6.27)–(5.6.29)) led to the bound

$$\frac{\|x - x_n\|_A}{\|x - x_0\|_A} \leq 2 \left(\frac{\sqrt{\kappa_\ell(A)} - 1}{\sqrt{\kappa_\ell(A)} + 1} \right)^{n-\ell}, \quad n = \ell, \ell+1, \ell+2, \ldots, \quad (5.9.1)$$

where $\kappa_\ell(A) = \lambda_{N-\ell}/\lambda_1$. An easy manipulation of (5.9.1) gives the following widely known theorem; see, e.g. [19, relation (3.6) and Section 4], [364, relation (5.9)], and the recent paper [577, Theorem 2.5].

Theorem 5.9.1
Considering the previous notation, after

$$n = \ell + \left\lceil \frac{1}{2} \ln\left(\frac{2}{\delta}\right) \sqrt{\frac{\lambda_{N-\ell}}{\lambda_1}} \right\rceil$$

iteration steps, where $\lceil \cdot \rceil$ denotes rounding up to the closest integer, the exact CG will produce the approximate solution x_n satisfying

$$\frac{\|x - x_n\|_A}{\|x - x_0\|_A} \leq \delta.$$

An analogous consideration leads the authors of [443, Remark 2.1] to the statement that 'a few "bad eigenvalues" will have almost no effect on the asymptotic convergence of the method', and they refer to [28, 27].

All this assumes exact arithmetic. We will now explain why the statement of Theorem 5.9.1 is *essentially useless for finite precision CG computations* (apart from rare and practically uninteresting cases, where the CG convergence is essentially linear and unaffected by rounding errors). In our explanation we will use, for simplicity, the composite polynomial $q_1(\lambda)\chi_{n-1}(\lambda)$, where $q_1(\lambda)$ is monic and has the only root λ_N and $\chi_{n-1}(\lambda)$ is the (scaled) Chebyshev polynomial of degree $n-1$ shifted to the interval $[\lambda_1, \lambda_{N-1}]$; cf. Section 5.6.4 and Figure 5.7. Using the spectral decomposition $A = Y\text{diag}(\lambda_j)Y^*$ and $x - x_0 = Y[\zeta_1, \ldots, \zeta_N]^T$, we can write

$$\|x - x_n\|_A^2 = \sum_{j=1}^{N} \lambda_j \left|\varphi_n^{CG}(\lambda_j)\, \zeta_j\right|^2 \qquad (5.9.2)$$

$$\leq \sum_{j=1}^{N} \lambda_j \left|q_1(\lambda_j)\chi_{n-1}(\lambda_j)\, \zeta_j\right|^2. \qquad (5.9.3)$$

The equality (5.9.2) appeared earlier in the proof of Theorem 5.6.9, and the upper bound (5.9.3) follows trivially from the optimality of the CG iteration polynomial. This bound is the starting point for investigations of the CG convergence rate based on the composite polynomial and it can eventually lead to (5.9.1); see (5.6.27)–(5.6.29). Now recall from Section 5.9.1, that the finite precision CG behaviour for a given matrix can be viewed as the exact CG behaviour for a matrix where each of the original eigenvalues λ_j is replaced by a tight cluster of eigenvalues, say, $\widetilde{\lambda}_{n_j+1}, \ldots, \widetilde{\lambda}_{n_j+m_j}$, where m_j is the number of the eigenvalues in the jth cluster. Thus, when we take into account rounding errors and consider a finite precision CG computation, the expression (5.9.3) contains values $|q_1(\widetilde{\lambda}_\ell)\chi_{n-1}(\widetilde{\lambda}_\ell)|$ for the m_N eigenvalues $\widetilde{\lambda}_\ell$ that are located in the cluster around λ_N. However, these values $|q_1(\widetilde{\lambda}_\ell)\chi_{n-1}(\widetilde{\lambda}_\ell)|$ will be *very* large (even for small n) because by construction the polynomial $q_1(\lambda)\chi_{n-1}(\lambda)$ has a huge gradient at the point λ_N. This is visualised in Figure 5.18. It shows that the effect of rounding errors will make *any* CG convergence bound based on the composite polynomial meaningless in practical computations.

We point out that also the bound (5.9.1), which is based on the composite polynomial approach, but without explicit reference to the values of this polynomial close to λ_N, is meaningless in finite precision arithmetic (although here the reasons are different). This is demonstrated in Figure 5.19 which shows, similarly

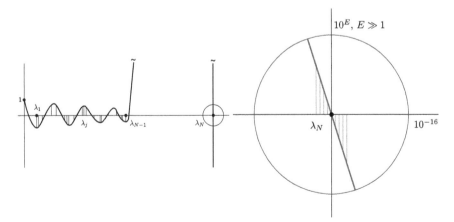

Figure 5.18 Left: The composite polynomial with one root at the centre of the very tight cluster of large outlying eigenvalues. Right: Magnification of the cluster around λ_N. If the values of the composite polynomial at the eigenvalues in this cluster are taken into account, then, due to the huge gradient of the polynomial at λ_N, they will make the upper bound (5.9.3) meaningless.

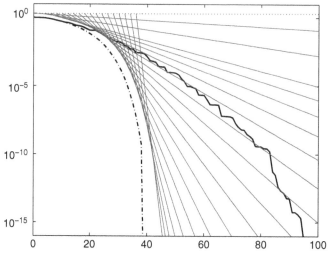

Figure 5.19 Sequence of the bounds (5.6.29) as in Figure 5.8 with an increasing number of deflated eigenvalues at the upper end of the spectrum compared with the finite precision CG computations. None of the straight lines representing the bounds offers an insight into the finite precision CG convergence plotted by the bold solid line.

to Figure 5.8, the sequence of bounds (5.9.1) with $\ell = 0, 2, 4, \ldots$ (straight solid lines) and compares them with the actual finite precision CG convergence curve (bold solid line). We see that *none* of the straight lines representing the sequence of bounds offers any relevant information about the actual finite precision CG convergence rate.

These problems cannot be resolved by placing more roots of the composite polynomial near λ_N, unless the number of such roots is equal to the number of eigenvalues in the cluster. This number is, however, not known a priori and in addition it grows with the iteration step n; see [269, 457].

The content of this section also demonstrates the place of numerical experiments in *theoretical* computational mathematics; see, e.g. Figure 5.19. As mentioned in the discussion following (5.6.29), Jennings, Axelsson, and others confronted their formulas with experiments, and Jennings had already warned in 1977 that the developed convergence bounds based on composite polynomials might not be applicable to finite precision computations [364]. Theorem 5.9.1 and Figures 5.18–5.19, show that his warning should be respected.

5.9.3 Maximal Attainable Accuracy

Examining the maximal attainable accuracy of an iterative computation parallels examining the effects of rounding errors in direct methods. However, in the case of iterative computations there is no *a-priori* known finite sequence of operations that leads to a 'final' approximate solution. When performed for a sufficiently large number of steps, an iterative computation typically reaches a point after which a continuation does not lead to a more accurate computed approximation. In the subsequent steps the size of the error stagnates, and sometimes it even starts to diverge. Estimating the maximal attainable accuracy of iterative computations therefore requires, in general, different techniques from numerical stability analysis of direct methods; see, e.g. the pioneering works of Woźniakowski [676, 677], Bollen [71, 72, 70], Higham and Knight [317, 318]. Surveys and further references can be found in [272, Section 7.3], [315, Chapter 17], and [457, Section 5.4].

The maximal attainable accuracy of an iterative computation can strongly depend on the algorithmic realisation of the given method. Mathematically equivalent algorithms can behave, in general, differently in finite precision arithmetic. The same is true also in direct computations where, however, the unnecessary loss of accuracy can be attributed to more or less obvious implementation mistakes. In iterative computations, the sources of errors are much less obvious. An example is given in [61], where four different iterative algorithms for solving the system of the normal equations $A^T A x = A^T b$ are studied. When solving such systems, the explicit formation of the matrix $A^T A$ is avoided (apart from some very special cases), and the algorithms perform two matrix–vector multiplications per step; the first one of the form Ap, and the second of the form $A^T q$. Numerical experiments in [61, Section 6] demonstrate that the attainable accuracy using the four mathematically equivalent algorithms can differ by up to a factor of $\kappa(A)$. In particular, with $\kappa(A) = 10^8$, for one algorithm mathematically equivalent to CG applied to the system of normal equations (CGLS1) the final Euclidean norm of the error is 10^{-9}, while for another one (CGLS2) this norm never drops below 10^{-1}. Other studies can be found, e.g. in [573, 271, 289, 574, 609].

In practice it is essential to know whether the maximal attainable accuracy is below or above the user-specified tolerance. In the second case, the given problem is *numerically unsolvable* by the given algorithmic implementation of the given method.

We will present several illustrations using the linear systems described in [289], which were constructed in order to demonstrate the possible differences between the three-term and two-term recurrences used in implementations of the CG method; the details of their construction are not important here. Figure 5.20 shows the relative residual norm $\|r_n\|/\|r_0\| = \|b - Ax_n\|/\|b\|$ (here $x_0 = 0$) for the exact CG (bold solid line) and for three finite precision computations using the CG implementations of Hestenes and Stiefel (Algorithm 2.5.1 in this book; solid line), Rutishauser (dotted line), and Hageman and Young (dashed line); see the details and references in Section 2.5.1 and in [289]. As above, the exact CG (bold solid line) is for our purpose sufficiently accurately simulated using the CG implementation with double reorthogonalisation; for details see [280] and [289, Section 6].

The figures illustrate that the CG residual norm is far from monotonic. This gives an example of the warning Hestenes and Stiefel issued in their 1952 paper [313]; see the quote from their paper given in the historical note in Section 5.2.1. In addition to its possibly strong non-monotonic behaviour, the residual norm in the CG method has (unlike the energy norm of the error) no physical interpretation.

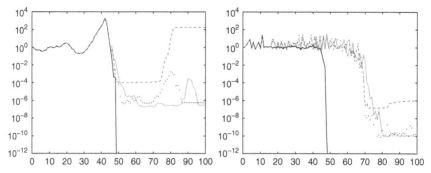

Figure 5.20 Relative residual norm of the exact CG (bold solid line) compared with the finite precision computations using the implementations of Hestenes and Stiefel (solid line), Rutishauser (dotted line), and Hageman and Young (dashed line); see [289]. We can observe a growth of the relative residual norm for many steps (left) as well as its oscillations and a significant delay of convergence (right).

Convergence in the left part of the figure is slightly delayed with insignificant variations between the different implementations. On the other hand, the behaviour of the three implementations for $n \geq 50$ and the maximal attainable accuracy differ substantially. The right part of Figure 5.20 shows the same quantities for CG applied to another system used in [289]. The residual norm behaves rather erratically for almost 70 iterations, and the delay of convergence is much more pronounced than in the previous case. It is worth commenting that there is no simple relationship between the observed phenomena and the condition number of the system matrix. In the first case $\kappa(A) = 1.86 \times 10^{10}$, in the second case $\kappa(A) = 1.97 \times 10^{6}$.

We further illustrate in Figure 5.21 different convergence characteristics of the CG method when applied to the problems used in Figure 5.20. The results were obtained with the standard Hestenes and Stiefel implementation. We plot the relative residual norm $\|b - Ax_n\|/\|r_0\|$ (solid line; coincides with the solid lines in Figure 5.20), the relative A-norm of the error $\|x - x_n\|_A/\|x - x_0\|_A$ (dots),

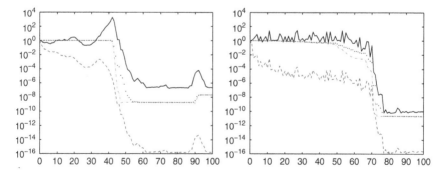

Figure 5.21 Different measures of convergence can give different information. Relative residual norm (solid line), relative A-norm of the error (dots), relative Euclidean norm of the error (dashed-dotted line), and normwise relative backward error (dashed line) for the Hestenes and Stiefel CG implementation (Algorithm 2.5.1).

the relative Euclidean norm of the error $\|x - x_n\|/\|x - x_0\|$ (dashed-dotted line) and the normwise relative backward error $\beta(x_n) = \|b - Ax_n\|/(\|b\| + \|A\|\|x_n\|)$ (dashed line). We can observe a significantly different behaviour of the plotted characteristics.

5.9.4 Back to the Poisson Model Problem

As mentioned several times above, stopping the iteration when the computed approximation is within a user-specified tolerance of the exact solution is crucial for an efficient practical use of iterative methods. The question of how to set an appropriate measure of closeness is, however, intriguing.

Starting from Lanczos, Hestenes and Stiefel in [406, 313], various stopping criteria have been discussed in the literature. In [498], the backward error idea was used (in a more general setting than computations using the CG method) to derive a family of stopping criteria which quantify levels of confidence in A and b. These stopping criteria are implemented in generally available software for solving linear algebraic systems and least squares problems [497]. Other developments and discussion can be found, e.g. in [9], [37, 200], [315, Section 17.5], [12, 13]. As for CG, the simplest (but least recommended; cf. Section 5.9.3 and also the historical note in Section 5.2.1) measure of convergence still is the relative residual norm $\|r_n\|/\|r_0\|$. In addition, although the norm of the forward error $\|x - x_n\|$ can be bounded using perturbation theory (see (5.8.20) and [315, Chapter 7]), the size of the corresponding worst-case bounds (although an important indicator of possible inaccuracies) do not necessarily give (for ill-conditioned problems) reliable information about the actual size of the algebraic error. An advantage of the residual norm is that it can always be computed with a negligible additional inaccuracy. Contrary to this, although the energy norm of the error can only be estimated, it can lead to a much more realistic evaluation of the size of the error; see, e.g. the comparison given in the context of the finite volume discretisation of the second-order elliptic model problem in [368]. The question of whether or not the computed estimate of the energy norm of the error is significantly affected by rounding errors needs, however, careful analysis; see [249, 457, 598, 599, 600]. It is worth noticing that in most publications on estimating the energy norm of the CG error (or on the closely related estimating of various quadratic forms) the crucial point of rounding error analysis is not even mentioned although it is indispensable for the justification of the proposed approaches. The fact that some practically used estimates have been proven numerically unreliable (see [598, 600]) shows that the point is indeed relevant.

When the condition number $\kappa(A)$ is small, the difference between the relative residual norm, the relative Euclidean norm of the error and the relative energy norm of the error is insignificant. This is important whenever preconditioning leads to a small condition number of the preconditioned matrix. The preconditioned relative residual can then give good information on the energy norm of the error for both the preconditioned and the original system; see [599]. A related discussion can also be found, e.g. in [27, 28] and [23].

We have claimed repeatedly throughout the book, and illustrated on a simple two-dimensional Poisson model problem in Section 5.1, that stopping criteria in algebraic computations should be related to the whole context of solving the underlying real-world problem. This should take into account that the local distribution of the algebraic error in the functional space can be very different from the distribution of the discretisation error. In particular, the algebraic error observed in Section 5.1 exhibits significant oscillations which determine the shape of the total error, despite the fact that the size of the algebraic error *measured in norm* is substantially (even orders of magnitude) smaller than the *norm* of the discretisation error. Based on the results presented in Sections 5.6.1 and 5.6.4, we can now give a detailed explanation of this behaviour.

We have seen that the behaviour of CG computations is linked with convergence of Ritz values to the eigenvalues of the system matrix. Assuming exact arithmetic, this link is quantified in Corollary 5.6.2, Theorem 5.6.3, Theorem 5.6.9, and Theorem 5.6.10. The relation (5.6.31) in the last theorem describes the evolution of the components of the error $x - x_n$ in the individual invariant subspaces with increasing iteration step n. Indeed, considering, as above, the spectral decomposition $A = Y \mathrm{diag}(\lambda_j) Y^*$ and using the notation of Theorem 5.6.9,

$$\left| y_j^* (x - x_n) \right|^2 = \left| y_j^* \varphi_n^{CG}(A)(x - x_0) \right|^2 = \left| \varphi_n^{CG}(\lambda_j) y_j^* (x - x_0) \right|^2$$

$$= \left| \varphi_n^{CG}(\lambda_j) \zeta_j \right|^2 = \left(\varphi_n^{CG}(\lambda_j) \right)^2 \frac{\|r_0\|^2}{\lambda_j^2} \omega_j, \qquad (5.9.4)$$

where we have used $\omega_j = |\zeta_j|^2 \lambda_j^2 / \|r_0\|^2$; see (3.5.2). This gives another expression for the right-hand side of (5.6.31),

$$\|x - x_n\|_A^2 = \frac{\|r_0\|^2}{\xi_n^{2n+1}} \sum_{j=1}^{N} \omega_j \prod_{\ell=1}^{n} \left(\lambda_j - \theta_\ell^{(n)} \right)^2$$

$$= \sum_{j=1}^{N} \lambda_j \left| y_j^* (x - x_n) \right|^2 = \|r_0\|^2 \sum_{j=1}^{N} \frac{\omega_j}{\lambda_j} \left(\varphi_n^{CG}(\lambda_j) \right)^2. \qquad (5.9.5)$$

If the algebraic error is now expressed in the *functional space* by summing up the components in the individual invariant subspaces,

$$x - x_n = \sum_{j=1}^{N} \left(\varphi_n^{CG}(\lambda_j) y_j^* (x - x_0) \right) y_j, \qquad (5.9.6)$$

then, using the functional representation of the individual eigenvectors y_j, $j = 1, 2, \ldots, N$, the squared size of the components (see (5.9.4)) determine, together with the shape of the individual eigenvectors, the local distribution of the algebraic error in the given domain. This is documented in the figures, in which we have used the same data as in the experiment in Section 5.1. In particular, $m = 50$, so that the matrix A is of order 2500.

Figure 5.22 shows the squared size of the individual components of the initial error $x - x_0 = x$ (recall that we use $x_0 = 0$) in the direction of the eigenvectors of

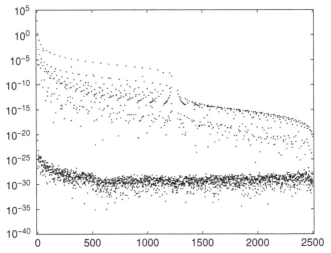

Figure 5.22 Squared size of the components of the initial error $x - x_0$ in the invariant subspaces of A in the increasing ordering of the eigenvalues. We can observe few dominating low frequency components and the negligible role of the high frequency components corresponding to the upper part of the spectrum.

A with increasing ordering of the eigenvalues. Because of the very smooth solution, several low frequency components dominate, and the high frequency components do not play a role. Also note that due to the shape of the solution many eigenvectors are not 'active', i.e. the solution does not have a component in their direction (here the corresponding computed components are nonzero but close to machine precision).

The four most dominating eigenvectors are visualized in Figure 5.23. We see the increasing oscillation of the eigenvectors as the eigenvalues increase; see also [66]. As the CG iterations proceed, the pattern of dominating oscillations changes. The contribution of low frequency eigenvectors is gradually eliminated and their role is taken over by more oscillating eigenvectors corresponding to larger eigenvalues. This is illustrated for the iteration step $n = 35$, i.e. for stopping the CG iteration with $\gamma = 1.0$ (see Section 5.1), in Figures 5.24 and 5.25. The linear combination of the four dominating eigenvectors (see (5.9.6)) closely approximates the algebraic error $u_h - u_h^{(35)}$. This is illustrated in Figure 5.26; see also Figure 5.1. Figure 5.27 illustrates the development of the squared size of the dominating components using the iterations 27, 35, 42, 50, and 56.

Finally, the Ritz values $\theta_\ell^{(n)}$ determined throughout the CG computations and the CG convergence behaviour are presented in Figures 5.28 and 5.29, respectively. Since in this experiment the Ritz values do not approximate (in the observed range of iterations) the eigenvalues of A with the accuracy approaching machine precision, the computation is not significantly affected by rounding errors; see the right part of Figure 5.29. Within the given iteration range there is no observable delay of convergence, and the maximal attainable accuracy is reached even before

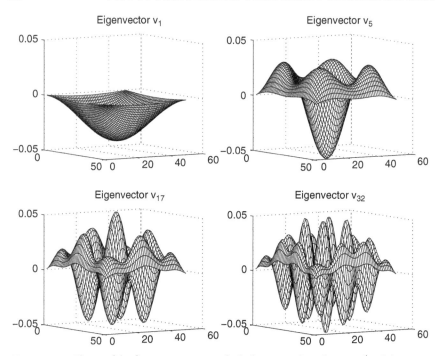

Figure 5.23 Shape of the four eigenvectors which dominate the solution x (and the initial error $x - x_0$, $x_0 = 0$). The eigenvectors are ordered according to the size of the corresponding components.

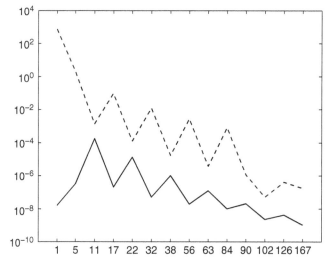

Figure 5.24 Squared size of the several dominating components of the initial error $x - x_0$ (dashed line) and of the error $x - x_{35}$ (solid line) in the invariant subspaces of A. We can observe the gradual elimination of the lowest frequency components; see also Figure 5.22.

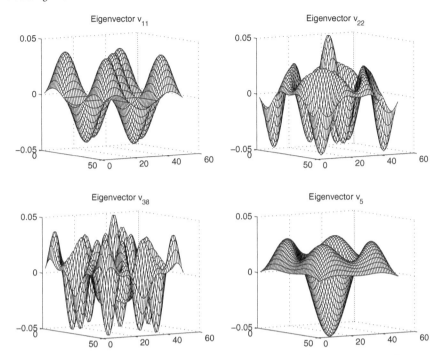

Figure 5.25 Shape of the four eigenvectors which dominate the error $x - x_{35}$. The eigenvectors are ordered according to the size of the corresponding components. We can observe the increasingly oscillating pattern of the dominating eigenvectors in comparison with the initial error; see Figure 5.23.

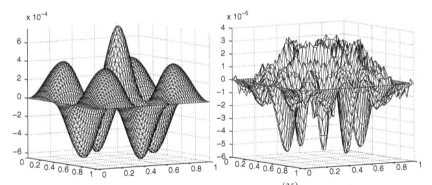

Figure 5.26 Approximation of the algebraic error $u_h - u_h^{(35)}$ using the four most dominating eigenvectors (left; the vertical axis is scaled by 10^{-4}) and the error of this approximation (right; the vertical axis is scaled by 10^{-5}).

the loss of linear independence among the computed Lanczos vectors (CG residuals) occurs. The loss of orthogonality at the iteration step 110 measured by $\|I - V_n^* V_n\|_F$, where V_n is the $N \times n$ matrix containing the computed Lanczos vectors, is about 10^{-3}. This is due to the distribution of eigenvalues of A and due to a relatively small number of iterations performed.

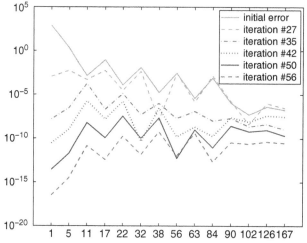

Figure 5.27 Squared size of the several dominating components of the initial error $x - x_0$ and of the errors $x - x_{27}, x - x_{35}, x - x_{42}, x - x_{50}$ and $x - x_{56}$ in the invariant subspaces of A. We can observe the gradual elimination of the low frequency components and the relative increase of the more oscillating ones. This explains the increasingly oscillating pattern of the error $u_h - u_h^{(n)}$ as the iteration step n increases; see Section 5.1.

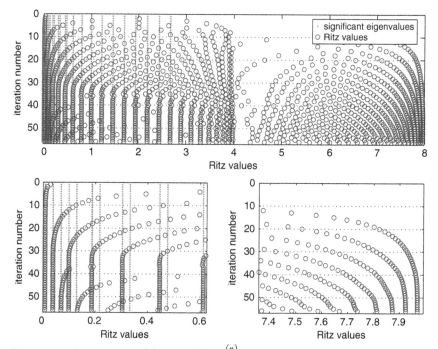

Figure 5.28 Convergence of the Ritz values $\theta_\ell^{(n)}$ to the eigenvalues λ_j of the matrix A as the iteration proceeds, with a magnification of the lower part and the upper part of the spectrum. The dominating eigenvalues are depicted by vertical dotted lines.

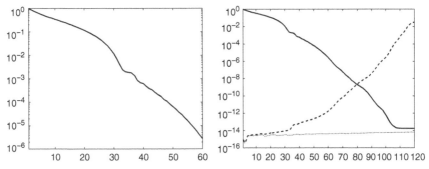

Figure 5.29 Left: The relative energy norm of the CG error $\|x - x_n\|_A / \|x - x_0\|_A$; observe an interesting irregularity in the rate of convergence most visible between iterations 30 and 40. Right: The loss of orthogonality among the computed Lanczos vectors (dashed line) and the analogous loss of orthogonality in the CG computations with double reorthogonalised residual vectors (dots). The relative energy norm of the error corresponding to the reorthogonalised CG computation (dash-dotted line) coincides with the relative energy norm of the error for the standard CG computation (solid line).

Figure 5.29 shows another interesting point. In the context of Krylov subspace methods it has been observed that convergence occurs in three phases: *sublinear, linear, and superlinear*; see [477, Section 1.8]. As mentioned in [477, p. 9]: 'In practice all phases need not be identifiable, nor need they appear only once and in this order.' In Section 5.7.4 we saw, for example, that in the GMRES method any nonincreasing sequence of N real numbers can be a convergence curve for a matrix having any N eigenvalues. Therefore the basic question is *whether and when* any of the three phases actually appear in solving practical (discretised and therefore finite-dimensional) problems. In the GMRES behaviour described in Section 5.7.4, the departure of the matrix A from normality can play a role. Figure 5.29 gives an example with a symmetric positive definite matrix A. We can observe 'delays' of convergence, which are most visible between the iteration steps 30 and 40. This can be explained by the clustered character of the spectrum of the standard 2D discretised Laplacian (see, e.g. [290, Chapter 4]), analogously to the explanation given in Section 5.9.1 for the (true) delays due to rounding errors. Here, however, we deal with convergence behaviour that is not affected in an observable way by rounding errors.

In conclusion, this example illustrates our general point that in order to evaluate cost of computations using the CG method, it is necessary to consider the whole context of the problem to be solved, from the mathematical formulation of the model through its discretisation to the properties of the computational methods and their algorithmic implementations. Only then is it possible to look for a relevant measure of convergence, suggest a meaningful stopping criterion and evaluate the cost of computation with the particular data. The presented example points out, in particular, a need for investigating the local distribution of the total error (including its algebraic part) throughout the domain on which the problem is going to be solved.

5.10 ROUNDING ERRORS IN THE GMRES METHOD

In Section 2.5.5, GMRES was presented as Algorithm 2.5.2 and it was attributed to Saad and Schultz, whose paper from 1986 [544] is highly recommended to all readers of this book. Algorithm 2.5.2 is based on computing (in exact arithmetic) an orthonormal basis of the Krylov subspace $\mathcal{K}_n(A, r_0)$ using the Arnoldi algorithm, and then solving the least squares problem (2.5.55). Many mathematically equivalent variants of this algorithm exist. Some of them were known even before the work of Saad and Schultz on GMRES, for example including the iterative least-square method of Khabaza [381], ORTHOMIN and ORTHODIR due to Vinsome, Young, and Jea [642, 685], GCR and its generalisations due to Axelsson [20, 21], and Eisenstat, Elman, and Schultz [156]. As an invaluable source describing the state-of-the-art before the appearance of the GMRES paper [544] we recommend the PhD thesis of Elman from 1982 [157]; see also the comments in Section 5.7.2 and the review in [134].

Here we are not interested in a systematic comparison of different implementations. We will consider GMRES and some of its modifications which are based on the full Arnoldi recurrence computing a *single* sequence of basis vectors. Other implementations, although they can have their merits and advantages (as, in particular, GCR), suffer in comparison with GMRES from additional instabilities, which can be benign or serious, depending on the problem to be solved. We refer interested readers to, e.g. [367, 366] for details. Our goal in this section is to investigate to what extent methods based on the full Arnoldi recurrence, where the newly computed Arnoldi vector v_{n+1} is obtained by the matrix–vector multiplication $A v_n$ with the subsequent orthogonalisation against *all* previously computed basis vectors $v_1, \ldots, v_n$, behave *qualitatively* differently (with respect to rounding errors) from methods based on short recurrences. In brief, we will abstract from numerical differences due to different ways of dealing with the projected problem (see, e.g. [574, 609]), and concentrate on the benefits and costs related to the use of full recurrences.

5.10.1 The Choice of the Basis Affects Numerical Stability

Assume, for the moment, exact arithmetic. In the GMRES method of Saad and Schultz (Algorithm 2.5.2) the Arnoldi recurrence for the orthogonal basis vectors $v_1, \ldots, v_{n+1}$ can be written as the matrix form of the QR factorisation

$$[r_0, AV_n] = V_{n+1} [\|r_0\| e_1, \underline{H}_{n,n}], \quad V_{n+1}^* V_{n+1} = I; \qquad (5.10.1)$$

see (2.5.53) and [422, Section 4.1]. Then the approximate solution x_n is determined by

$$x_n = x_0 + V_n t_n, \quad t_n = \underset{t}{\operatorname{argmin}} \left\| \|r_0\| e_1 - \underline{H}_{n,n} t \right\|; \qquad (5.10.2)$$

see (2.5.63). This means that x_n can not be easily updated using the previous approximate solution x_{n-1}, which complicates the implementation. This matter is not resolved, but the situation is somewhat simplified using the orthonormal basis vectors $w_1, \ldots, w_n$ of the Krylov subspace $A\mathcal{K}_n(A, r_0)$, which can be determined by the Arnoldi recurrence written again as the matrix QR factorisation

$$A[r_0, W_{n-1}] = W_n S_n, \quad W_n^* W_n = I; \qquad (5.10.3)$$

see [422, Section 4.3]. This gives the approximate solution

$$x_n = x_0 + [r_0, W_{n-1}]\tilde{t}_n, \quad \tilde{t}_n = \underset{t}{\operatorname{argmin}} \|r_0 - W_n S_n t\| \qquad (5.10.4)$$

which is mathematically equivalent to the approximate solution obtained via (5.10.2). The least squares problem in (5.10.4) can be solved using the linear algebraic system

$$S_n \tilde{t}_n = W_n^* r_0. \qquad (5.10.5)$$

Since S_n is upper triangular, this solution is a bit simpler than the one for t_n in (5.10.2), which led Walker and Zhou to call their resulting algorithm 'Simpler GMRES' [656]. Note that (5.10.5) represents nothing but solving the least squares problem (5.10.4) via the QR factorisation. Forming the right-hand side by the matrix–vector multiplication $W_n^* r_0$ would cause unnecessary numerical errors (see [536]). Instead, it should be computed in the process of the QR factorisation of the matrix $[Ar_0, AW_{n-1}]$; see (5.10.3). Assuming that the vectors $w_1, \ldots, w_n$ are explicitly available during the computation (as in the modified Gram–Schmidt implementation of Simpler GMRES; see [422, Section 7]), one can write

$$r_1 = r_0 - (w_1^* r_0) w_1, \ldots, r_n = r_{n-1} - (w_{n-1}^* r_{n-1}) w_{n-1},$$

i.e.

$$r_0 = r_j + \sum_{\ell=1}^{j} (w_\ell^* r_{\ell-1}) w_\ell, \quad j = 1, \ldots, n;$$

see also (5.7.18). Then $w_{j+1}^* r_0 = w_{j+1}^* r_j$, and (still assuming exact arithmetic)

$$W_n^* r_0 = \left[w_1^* r_0, w_2^* r_1, \ldots, w_n^* r_{n-1}\right]^T, \qquad (5.10.6)$$

which should be used for computation of the right-hand side in (5.10.5).[15]

The example of Simpler GMRES illustrates that the development of new algorithms or implementations should always be accompanied by rounding error analysis: Although an immediate comparison of the standard and Simpler GMRES

---

15. Here we do not consider other algorithms based, e.g. on A^*A-orthogonal bases of $\mathcal{K}_n(A, r_0)$. Such algorithms allow simple step-by-step updates of the approximate solution and residual at the cost of an additional recurrence for a sequence of auxiliary basis vectors. This may cause numerical instabilities which can be, however, kept under control in many practical cases. For details we refer to [157, 367, 422].

does not reveal any danger, a closer look does. Indeed, while the approximate solution $x_n = x_0 + V_n t_n$ in standard GMRES uses (assuming exact arithmetic) orthonormal basis vectors V_n, the Simpler GMRES approximate solution $x_n = x_0 + [r_0, W_{n-1}] \widetilde{t}_n$ uses the basis $[r_0, W_{n-1}]$. This basis becomes *highly ill-conditioned* as r_0 is approximated by the columns of W_{n-1} (which by construction form an orthonormal basis of $AK_{n-1}(A, r_0)$). In particular (see [422, relation (5.18)]),

$$\kappa\left([r_0/\|r_0\|, W_{n-1}]\right) = \frac{\|r_0\| + \left(\|r_0\|^2 - \|r_{n-1}\|^2\right)^{1/2}}{\|r_{n-1}\|}. \qquad (5.10.7)$$

Consequently, the matrix S_n obtained from the QR factorisation of the matrix $A[r_0, W_{n-1}]$ becomes ill-conditioned as $\|r_n\|$ decreases, and the computation of the Simpler GMRES approximate solution x_n suffers from serious numerical difficulties. A comprehensive summary and further details on potential instability of Simpler GMRES can be found in [597, Section 4.4.1].

A partial remedy was analysed in [367]. Using instead of $[r_0, W_{n-1}]$ the matrix containing the normalised residual vectors,

$$\left[\frac{r_0}{\|r_0\|}, \frac{r_1}{\|r_1\|}, \ldots, \frac{r_{n-1}}{\|r_{n-1}\|}\right],$$

leads to the 'RB-Simpler GMRES method' that is closely related to the GCR method of Eisenstat, Elman, and Schultz [156], and that gives good numerical results provided that there is a sufficiently large decrease of the residual norm *at each iteration*. In order to overcome this restriction, Jiránek and Rozložník suggested an adaptive choice of the basis vectors which combines the approaches of Simpler GMRES and RB-Simpler GMRES. A detailed analysis and numerical experiments can be found in [366].

5.10.2 Does the Orthogonalisation Algorithm Matter?

With the crucial observation that the first n steps of the Arnoldi algorithm can be written as the QR factorisation (5.10.1) of the matrix $[v_1, AV_n]$ (see [654, 655]), one can apply the results described in Section 5.8.1. Unlike the decomposed matrix B in that section, the matrix $[v_1, AV_n]$ is not known *a priori*. If the QR factorisation is performed column by column, then the fact that the $(n+1)$th column Av_n is not available before the $(n+1)$th step does not matter. Before that step this column vector is simply not needed. Depending on the orthogonalisation used within the Arnoldi algorithm, one can obtain different GMRES implementations where the *same* (assuming exact arithmetic) basis can have very different numerical properties due to rounding errors.

An implementation based on the Householder Arnoldi algorithm was proposed by Walker [654]. If the matrix A is not close to singular (the details of this technical assumption can be found in [146, Theorem 4.1 and Corollary 4.2]), then a rather tedious but more or less straightforward analysis using (5.8.5)–(5.8.7) shows that the Householder Arnoldi GMRES is backward stable in the normwise sense, i.e.

the normwise relative backward error in the Nth GMRES step is proportional to machine precision,

$$\frac{\|b - Ax_N\|}{\|b\| + \|A\| \, \|x_N\|} \approx \varepsilon.$$

The proportionality factor is bounded by a (positive) low-degree polynomial in N. The proof was published in 1995, with references to earlier work by Karlson [376], Chatelin, Frayssè, Braconnier, and others [101, 102]; see [146]. Since the orthogonality of the Arnoldi basis vectors orthogonalised via the Householder reflections remains close to machine precision level, i.e.

$$\|I - V_N^* V_N\| \approx \varepsilon,$$

the proof consists of a sequence of technical steps showing that the quantities computed in finite precision arithmetic do not substantially depart from their exact counterparts. As a consequence, the normwise relative backward error does not significantly depart from machine precision.

We will now focus on the GMRES implementation based on the modified Gram–Schmidt implementation of the Arnoldi algorithm for generating (in exact arithmetic) an orthonormal basis of $\mathcal{K}_n(A, r_0)$; see Algorithm 2.4.2 in this book. We will refer to this implementation as MGS GMRES. For details on GMRES using the classical Gram–Schmidt orthogonalisation (where the orthogonality is lost rather quickly) we refer to [146, 233, 234].

When the column-by-column MGS (see, e.g. [279]) is applied to the QR factorisation in (5.10.1), the orthogonality of the MGS Arnoldi vectors V_{n+1} computed in finite precision arithmetic is bounded by (see (5.8.11))

$$\|I - V_{n+1}^* V_{n+1}\| \leq \frac{\varepsilon \, \nu_7(N, n) \, \kappa([v_1, AV_n])}{1 - \varepsilon \, \nu_7(N, n) \, \kappa([v_1, AV_n])}.$$

(Here we rescale the first column, i.e. we use $v_1 = r_0 / \|r_0\|$.) Let us comment on this bound. In exact arithmetic, the matrix V_{n+1} is invariant with respect to the column scaling of the matrix $[v_1, AV_n]$, i.e. with respect to the choice of $\gamma > 0$ and a positive diagonal matrix D_n such that the matrix $[v_1 \gamma, AV_n D_n]$ is used in the QR factorisation instead of $[v_1, AV_n]$. Using an argument attributed to Bauer (see [43] and the discussion in [193, Chapter 11]), it can be argued that, ignoring a small error proportional to machine precision ε, the same is true also in finite precision computations; see [500, Section 5]. Assuming that $\varepsilon \nu_7(N, n) \kappa([v_1 \gamma, AV_n D_n]) \ll 1$, one can therefore write

$$\|I - V_{n+1}^* V_{n+1}\| \approx \kappa([v_1 \gamma, AV_n D_n]) \, \varepsilon, \qquad (5.10.8)$$

where the factor of proportionality is bounded by a (positive) low degree polynomial in N. We can observe that the bound for the step $n + 1$ contains the condition number of the matrix depending on V_n computed throughout the preceding n steps. This tremendously complicates the further analysis.

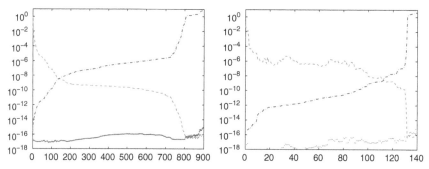

Figure 5.30 Results of MGS GMRES computations with the matrices Sherman2 (left) and West132 (right) from Matrix Market. Throughout the computation the product (dots) of the normwise relative backward error (dashed line) and the loss of orthogonality among the MGS Arnoldi vectors (dashed-dotted line) are close to (or below) machine precision.

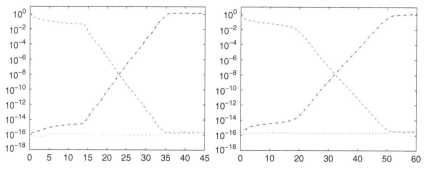

Figure 5.31 The same as in Figure 5.30 for matrices from the convection–diffusion model problem with dominating convection (see Section 5.7.5), the discontinuous inflow boundary conditions, the vertical wind (left) and the curl wind (right).

As suggested by (5.10.8), orthogonality among the computed Arnoldi vectors is gradually lost as n increases. Most remarkably, it has been observed in numerical experiments that the gradual *loss of orthogonality is almost inversely proportional to the normwise backward error* of the computed approximate solution. This observation is illustrated in Figures 5.30 and 5.31 (also see, e.g. [279], where the relative residual norm was used instead of the normwise relative backward error). The first figure shows numerical results for MGS GMRES applied to a linear algebraic systems with the matrices Sherman2 (left) and West132 (right) taken from MatrixMarket;[16] the second one for matrices that arise from the SUPG discretisation of the convection–diffusion model problem with dominating convection; see Section 5.7.5 and, e.g. [425]. Here we use $m = 15$ gridpoints in each direction (resulting in $N = 225$), the diffusion parameter $\nu = 0.01$ and the discontinuous inflow boundary conditions introduced by Raithby [516] (see also [425, relations (2.22)–(2.23)]), the vertical wind equal to $[0; 1]$ (left) and the curl wind

---

16. http://math.nist.gov/MatrixMarket/

$[\sin(\pi/4); \cos(\pi/4)]$ (right). The figures show the normwise relative backward error $\|b - Ax_n\|/(\|b\| + \|A\|\|x_n\|)$ (dashed line), the loss of orthogonality among the computed MGS Arnoldi basis vectors (stored in the matrix V_n) measured in the Frobenius norm $\|I - V_n^* V_n\|_F$ (dashed-dotted line), and the product of the two (dots). Throughout the computation this product is close to, or even below the machine precision. When the orthogonality is lost completely, the normwise relative backward error is therefore proportional to machine precision. We will investigate this relationship in the next section.

5.10.3 MGS GMRES is Backward Stable

An important step in the rounding error analysis of MGS GMRES was made in [279]. Using the secular equation in order to describe the decrease of the smallest singular value of a matrix when appended by an additional column (see [279, Theorem 2.1]; an analogous technique was also used later in [501] mentioned below), it was proven that the loss of orthogonality among the computed MGS Arnoldi vectors (measured in the Euclidean norm) is bounded by

$$\left\| I - V_{n+1}^* V_{n+1} \right\| \leq \mathcal{O}\left(nN^{3/2}\right) \varepsilon \kappa(A) \frac{\|r_0\|}{\|r_n\|}. \tag{5.10.9}$$

Consequently, the full loss of orthogonality (accompanied by the subsequent loss of numerical linear independence among the columns of V_{n+1}) can take place *only after* the relative residual norm $\|r_n\|/\|r_0\|$ becomes proportional to $\varepsilon \kappa(A)$. The paper [279] did not incorporate scaling of the matrix $[r_0, AV_n]$ and, as mentioned above, it related the loss of orthogonality among the MGS Arnoldi vectors to the relative GMRES residual norms instead of the normwise relative backward error. Its ideas (including the use of the secular equation technique), however, triggered extensive further investigation which resulted in a (still growing) number of papers [501, 499, 500, 502, 495, 326, 324]. They link the investigation of *finite precision behaviour* of GMRES with the seemingly unrelated mathematical (i.e. exact precision) concept of solutions in total least squares problems; see [637]. In this way, rounding error analysis of GMRES has led to revisiting the theory of the total least squares problem [501] through introduction of the so-called *core problem* [502, 326, 324]; see also, e.g. [246, Chapter 14]. This link, which had not been anticipated when the investigation was initiated, perhaps illustrates a bit of the unpredictable beauty and efficiency of mathematics. In this context we wish also to recall the paper by Kasenally and Simoncini, which investigated a somewhat related problem of minimising the normwise relative backward error (instead of residual norm) over Krylov subspaces [377]. Although not directly related, the idea as well as personal communications with Valeria Simoncini were inspirational.

The results outlined above can easily form the material for another book. Here we will only make a few points, and comment on the main result on backward stability of MGS GMRES.

In [500], results of the more general works [501, 499] are applied to GMRES, and in particular this paper focuses on the following observation illustrated in Figures 5.30 and 5.31:

> In finite precision MGS GMRES computations the product of the normwise relative backward error $\|b - Ax_n\|/(\|b\| + \|A\|\,\|x_n\|)$ and the loss of orthogonality $\|I - V_{n+1}^* V_{n+1}\|_F$ is almost constant and close to machine precision ε.

As we have seen in (5.10.8) (the difference between the Euclidean and Frobenius norm of $I - V_{n+1}^* V_{n+1}$ is in our context insignificant), the loss of orthogonality is essentially controlled by the condition number $\kappa([v_1\gamma, AV_n D_n])$. Let $\delta_n = \sigma_{n+1}([v_1\gamma, AV_n D_n])/\sigma_n([AV_n D_n])$. Assuming exact arithmetic and using the scaling $\gamma = \|r_0\|/\|b\|$, $D_n = \|A\|^{-1}I$, the main result in [500, Theorem 5.1, Propositions 6.1 and 6.2] states (using $x_0 = 0, r_0 = b$) that

$$\frac{1}{\sqrt{2}} \leq \kappa\left([v_1\gamma, AV_n D_n]\right) \cdot \frac{\|b - Ax_n\|}{\|b\| + \|A\|\,\|x_n\|}$$

$$\leq \sqrt{2} \left\{ \frac{\|b\|^2 + \|A\|^2 \|x_n\|^2 \left(1 - \delta_n^2\right)^{-1}}{\|b\|^2 + \|A\|^2 \|x_n\|^2} \right\}^{\frac{1}{2}}.$$

For $\delta_n < 1/\sqrt{2}$ this can be simplified to

$$\frac{1}{\sqrt{2}} \leq \kappa\left([v_1\gamma, AV_n D_n]\right) \cdot \frac{\|b - Ax_n\|}{\|b\| + \|A\|\,\|x_n\|} \leq 2. \qquad (5.10.10)$$

The paper [500] anticipated that the finite precision analogue of (5.10.10) would be proven in a subsequent paper that would theoretically justify the observed phenomenon

$$\|I - V_{n+1}^* V_{n+1}\|_F \cdot \frac{\|b - Ax_n\|}{\|b\| + \|A\|\,\|x_n\|} \approx \mathcal{O}(1)\,\varepsilon$$

(see Figures 5.30 and 5.31) and, consequently, prove that the full loss of orthogonality among the MGS Arnoldi vectors implies the normwise backward stability of the computed MGS GMRES approximate solution. That would complement the result (5.10.9), but it would not complete the proof of backward stability of MGS GMRES. A technically complicated question on the situation when $\|I - V_{n+1}^* V_{n+1}\|_F$ does not become $\mathcal{O}(1)$ would remain unresolved.

The final step was made with the substantial contribution of Paige in [495]. Using the augmented backward stability result of Björck and Paige mentioned in Section 5.8.1 (this result was first observed by Sheffield; see [62]) and the results of Giraud and Langou [232], it was shown that under the standard assumption on nonsingularity of the matrix A (see [495, relations (1.1) and (1.2)]) MGS GMRES computes a backward stable approximate solution x_m (with the normwise relative backward error close to machine precision) for some $m \leq N$. Until the step m is reached, the condition number of the matrix containing the MGS Arnoldi vectors is bounded as $\kappa(V_m) \leq 4/3$.

In conclusion, unless the matrix A is close to singular, MGS GMRES provides, despite the (gradual) loss of orthogonality among the computed MGS Arnoldi vectors, an approximate solution with normwise relative backward error comparable to the Householder Arnoldi GMRES.

5.11 OMITTED ISSUES AND AN OUTLOOK

In this final section of the book we will comment on omitted issues and point out some open questions that we consider important for further investigations.

As mentioned in the Preface, we do not study preconditioning in our book. However, we would like to make a few comments on this subject, without attempting a systematic coverage or claiming completeness. For comprehensive surveys on preconditioning techniques we refer to [272, Part II], [452, Chapters 8–10], [543, Chapters 9–14], [545, Section 8], and, in particular, to the excellent survey [54]. An operator-based approach to preconditioning, which underlines the link between the discretised linear algebraic systems and the properties of the underlying PDEs is described, e.g. in [26, 323, 443]; see also [160]. Even from the titles of these few references it seems clear that construction of efficient preconditioners requires an insightful combination of mathematical areas and methods. It must also take into account highly nontrivial issues of practical computer implementations like sparsity of data and intermediate results, computational cost, and memory requirements.

The name 'preconditioning' suggests that these techniques aim at improving the *conditioning* of a given problem, and in particular at reducing the matrix condition number. This appears to be mainly motivated by the worst-case convergence bounds for the CG method based on the matrix condition number; see in particular the bound (5.6.23). Here we consider the difficult cases when the condition number of the preconditioned matrix is *not* small (when it is small, no further discussion is needed). If the eigenvalues of the original and the preconditioned matrix are *fairly uniformly distributed*, then in the worst possible case the CG iteration polynomials are in some sense 'close' to the scaled and shifted Chebyshev polynomials on the spectral interval, and the worst-case convergence bounds for CG can offer, *to some extent*, reasonably descriptive information (compare [545, p. 13]). Reducing the condition number by preconditioning in such cases improves the worst-case convergence behaviour of CG. This can be analysed and successfully applied for discretisations of model problems such as the Poisson equation on a rectangular domain (this problem has appeared in Sections 2.5.2, 5.1, and 5.9.4 in this book).

In general, as we have seen in Section 5.6.2, the matrix condition number does not give enough information for describing the convergence of the CG method (even in the worst case), let alone of methods for non-Hermitian matrices like GMRES; see Section 5.7. The *technique* of preconditioning must be (and actually can be) extended beyond simple model problems and the CG method. Although extensions to other cases keep the same name, it should not be forgotten that this name conforms to some specific assumptions, namely, that the conditioning of the system matrix describes the practical behaviour of an iterative method. Computational techniques can be extended more easily than a mathematical theory, which would cover behaviour of preconditioned Krylov subspace methods in a general non-Hermitian case. This fact is well known and well articulated by many authors; see, e.g. [452, Section 8.1]. In practice, however, it is not always reflected. The condition number and the eigenvalue distribution (in particular the 'clustering' of the eigenvalues) of the preconditioned system matrix are considered, often without any justification, indicators for the rate of convergence of preconditioned Krylov subspace methods. Examples presented throughout this chapter and the whole book suggest, on the other hand, that linking spectral information and convergence needs *substantial further work*.

It seems reasonable to argue that understanding of preconditioning requires an understanding of the methods which are (in theory) applied to the preconditioned systems, i.e. which are (in practice) preconditioned. A contribution towards better understanding of Krylov subspace methods therefore helps in development of preconditioning techniques used with them. In our opinion, application of well understood and relatively simple Krylov subspace methods combined with efficient preconditioning, which efficiently exploit knowledge about the particular method and the structure of the particular problem to be solved (e.g. the mathematical model and its discretisation), is a promising approach in practical computations.

The problem we have portrayed in the last chapter of the book seems intractable:

> Evaluate the computational cost of a relatively small number of nonlinear iterations which in principle (assuming exact arithmetic) form, when continued, a finite process that can not be treated as a sequence of linearisations, and which are (possibly strongly) affected by rounding errors.

However, not all hope is lost. In theory and practice it can be useful to ignore a part of the difficulties, for example by focusing, where appropriate, only on convergence in exact arithmetic, or by considering quantitative descriptions only of a significant part of the rounding error effects. One must always, however, avoid simplifications that substantially change (even qualitatively) the investigated phenomena. The point is expressed by a quote attributed to Einstein and presented as a motto of Chapter 0 in the *Oxford User's Guide to Mathematics*, edited by Eberhard Zeidler with the help of Wolfgang Hackbusch and Hans Rudolf Schwarz [688]:

Everything should be made as simple as possible, but not simpler.

In order to make a simplification right, it is essential to consider the whole picture with links between its parts. We have tried to present and emphasise this view in this book.

A possibility for simplification is to use, as we saw above, model problems. Two basic reasons for considering a model problem instead of a 'practical' problem have been pointed out by Kratzer, Parter, and Steuerwalt [395, pp. 256–257]. The theory can be developed in great generality but is easier to explain for a simplified model ('pedagogical aspect'), and, on the other hand, the theory can be developed only for a model but the results nevertheless give insight into more general problems ('generalisability aspect').

The issues in the analysis of Krylov subspace methods are so involved that rarely is one able to use model problems for pedagogical reasons only. In the context of Krylov subspace methods the use of model problems is mostly justified with the 'generalisability aspect'. Moreover, good model problems represent 'common structural denominators' [46, p. 30] of real-world applications. Verification and validation in scientific computing [30, 31] gives an example why considering model problems is important. In our opinion, evaluation of *computational cost* should focus on *particular computations* instead of the most general or worst-case settings. This can be done by studying particular model problems.

The key issues for the computational cost of Krylov subspace methods identified above, namely data-dependency and the inherent link between convergence and numerical stability, are brought to the point in the following quote from an essay by Baxter and Iserles [46, p. 27]:

> Clearly, good computation requires the algorithm to respond to the data it is producing and change the allocation of computing resources (step size, size of the grid, number of iterations, even the discretisation method itself) accordingly.

This remark can be addressed directly to the use of Krylov subspace methods or iterative methods in general. Yet, it is made in a discussion of the concept of *adaptivity* and the meaning of the word 'computational' in 'computational mathematics'. This shows that the issues in the context of computational cost of Krylov subspace methods are *typical* for modern computational mathematics, to which the essay by Baxter and Iserles is devoted. Challenges of computational mathematics deal more and more with the uncovering and exploitation of the inner structure of the particular problem (or a class of problems) and data. Here we can point back to the motto of this book: the same idea was expressed in the letters of Lanczos and Einstein when they discussed the 'inner nature of the problem'.

In several aspects this book can be considered a basis for further investigations. It seems therefore appropriate to formulate and comment on some open questions that we consider important. Some of them are related to material in the book, others to topics that were omitted.

1. What does spectral information tell us about the behaviour of Krylov subspace methods? By spectral information we do not mean solely the distribution of eigenvalues, but the whole structure of invariant subspaces, and the distribution of ill-conditioning among the individual eigenvalues, eigenvectors, and the bases of invariant subspaces. We know that for GMRES any nonincreasing convergence curve is possible for a matrix having any eigenvalues. If a 'good' distribution of eigenvalues is linked in a practical problem with a 'poor' convergence (or vice versa), what does this tell us about the mathematical model and its discretisation?

Convection–diffusion model problems with dominating convection seem to be a good starting point for investigations.

2. We have seen that rounding errors in the CG method typically cause delay of convergence due to loss of orthogonality (and, consequently, the repeated loss of rank) among the computed residuals (Lanczos vectors). In MGS GMRES, on the other hand, the orthogonality among the computed Arnoldi vectors is not lost until the maximal attainable accuracy of the approximate solution is reached, and there is no delay of convergence due to rounding errors. Results on *inexact Krylov subspace methods* (see, e.g. [73, 632, 569]) show to what extent one can relax the accuracy of some local operations (like matrix–vector multiplications) without affecting the *maximal attainable accuracy*. How much can we relax the accuracy of local operations (or, stated differently, how much can we perturb the Krylov subspace method recurrences; see, e.g. [268, 269, 690]) without affecting the *rate of convergence*, i.e. without causing an unwanted delay?

3. Because of rounding errors it is, in methods based on short recurrences, practically impossible to keep (in general) the computed basis vectors sufficiently well-conditioned. Hence, as mentioned in Section 2.5.6, look-ahead techniques in non-Hermitian Lanczos methods can not guarantee well-conditioning of the computed basis vectors, even if there is no (near) occurrence of the serious breakdown. What is the interplay of look-ahead techniques and rounding errors in practical computations?

4. How can we build-up an analytic justification of deflation and recycling techniques (see the end of Section 5.6.4)? How can the effect of rounding errors be included into such analysis?

5. Following the points made in Sections 5.1 and 5.9.4, how can the algebraic error be included into locally efficient and fully computable *a-posteriori* error bounds in adaptive PDE solvers? How can the algebraic error be translated into a functional backward error that can be interpreted as a physically meaningful perturbation of the original mathematical model?

Krylov subspace methods can be viewed as methods for *model reduction*. We have emphasised this point and the corresponding *matching moment properties*. Since moments represent very general objects that occur in many areas of mathematics and beyond, we believe that Krylov subspace methods represent tools that are significantly more general than their immediate context of solving linear algebraic systems (or eigenvalue problems). Since virtually every solution process of difficult mathematical problems requires some form of model reduction, we believe that Krylov subspace methods will continue to find applications in the foreseeable future.

The list of omitted issues and important work that is not covered or even mentioned in this book is certainly very long. This is not due to negligence by its authors, but due to the lack of physical and mental abilities to do better. We do hope that despite its insufficiencies and failures, this text will be received by the reader with kindness and understanding.

References

1. *Biographical Dictionary of Mathematicians. Vol. 3*, Charles Scribner's Sons, New York, 1991. Thomas Kirkman–Isaac Newton, Reference biographies from the *Dictionary of Scientific Biography*.
2. L. V. AHLFORS, *Complex Analysis*, McGraw-Hill Book Co., New York, third ed., 1978.
3. L. V. AHLFORS AND OTHERS, *On the mathematics curriculum of the high school*, Amer. Math. Monthly, 69 (1962), pp. 189–193.
4. M. A. AIZERMAN AND OTHERS, *Felix Ruvimovich Gantmakher (obituary)*, Russ. Math. Surv., 20 (1965), pp. 143–151.
5. N. I. AKHIEZER, *The Classical Moment Problem and Some Related Questions in Analysis*, Hafner Publishing Co., New York, 1965. Translated by N. Kemmer.
6. N. I. AKHIEZER AND I. M. GLAZMAN, *Theory of Linear Operators in Hilbert Space*, Dover Publications Inc., New York, 1993. Translated from the Russian and with a preface by Merlynd Nestell, reprint of the 1961 and 1963 translations, two volumes bound as one.
7. A. C. ANTOULAS, *Approximation of Large-Scale Dynamical Systems*, vol. 6 of Advances in Design and Control, Society for Industrial and Applied Mathematics (SIAM), Philadelphia, PA, 2005. With a foreword by Jan C. Willems.
8. M. ARIOLI, *A stopping criterion for the conjugate gradient algorithms in a finite element method framework*, Numer. Math., 97 (2004), pp. 1–24.
9. M. ARIOLI, I. DUFF, AND D. RUIZ, *Stopping criteria for iterative solvers*, SIAM J. Matrix Anal. Appl., 13 (1992), pp. 138–144.
10. M. ARIOLI, E. H. GEORGOULIS, AND D. LOGHIN, *Convergence of inexact adaptive finite element solvers for elliptic problems*, RAL-TR 2009-021, Rutherford Appleton Laboratory (RAL), Didcot, UK, 2009.
11. M. ARIOLI AND D. LOGHIN, *Stopping criteria for mixed finite element problems*, Electron. Trans. Numer. Anal., 29 (2007/08), pp. 178–192.
12. M. ARIOLI, D. LOGHIN, AND A. J. WATHEN, *Stopping criteria for iterations in finite element methods*, Numer. Math., 99 (2005), pp. 381–410.
13. M. ARIOLI, E. NOULARD, AND A. RUSSO, *Stopping criteria for iterative methods: applications to PDE's*, Calcolo, 38 (2001), pp. 97–112.
14. M. ARIOLI, V. PTÁK, AND Z. STRAKOŠ, *Krylov sequences of maximal length and convergence of GMRES*, BIT, 38 (1998), pp. 636–643.
15. G. ARNOLD, N. CUNDY, J. VAN DEN ESHOF, A. FROMMER, S. KRIEG, T. LIPPERT, AND K. SCHÄFER, *Numerical methods for the QCD overlap*

operator. II. Optimal Krylov subspace methods, in QCD and Numerical Analysis III, vol. 47 of Lect. Notes Comput. Sci. Eng., Springer, Berlin, 2005, pp. 153–167.

16. W. E. ARNOLDI, *The principle of minimized iteration in the solution of the matrix eigenvalue problem*, Quart. Appl. Math., 9 (1951), pp. 17–29.

17. S. F. ASHBY, T. A. MANTEUFFEL, AND P. E. SAYLOR, *A taxonomy for conjugate gradient methods*, SIAM J. Numer. Anal., 27 (1990), pp. 1542–1568.

18. W. ASPRAY AND M. GUNDERLOY, *Early computing and numerical analysis at the National Bureau of Standards*, Annals Hist. Comput., 11 (1989), pp. 3–12.

19. O. AXELSSON, *A class of iterative methods for finite element equations*, Comput. Methods Appl. Mech. Engrg., 9 (1976), pp. 123–127.

20. ———, *Conjugate gradient type methods for unsymmetric and inconsistent systems of linear equations*, Linear Algebra Appl., 29 (1980), pp. 1–16.

21. ———, *A generalized conjugate gradient, least square method*, Numer. Math., 51 (1987), pp. 209–227.

22. ———, *Iterative Solution Methods*, Cambridge University Press, Cambridge, 1994.

23. ———, *Optimal preconditioners based on rate of convergence estimates for the conjugate gradient method*, Numer. Funct. Anal. Optim., 22 (2001), pp. 277–302.

24. O. AXELSSON AND V. A. BARKER, *Finite Element Solution of Boundary Value Problems: Theory and Computation*, vol. 35 of Classics in Applied Mathematics, SIAM, Philadelphia, PA, 2001. Reprint of the 1984 original.

25. O. AXELSSON AND I. KAPORIN, *Error norm estimation and stopping criteria in preconditioned conjugate gradient iterations*, Numer. Linear Algebra Appl., 8 (2001), pp. 265–286.

26. O. AXELSSON AND J. KARÁTSON, *Equivalent operator preconditioning for elliptic problems*, Numer. Algorithms, 50 (2009), pp. 297–380.

27. O. AXELSSON AND G. LINDSKOG, *On the eigenvalue distribution of a class of preconditioning methods*, Numer. Math., 48 (1986), pp. 479–498.

28. ———, *On the rate of convergence of the preconditioned conjugate gradient method*, Numer. Math., 48 (1986), pp. 499–523.

29. I. BABUŠKA, *Numerical stability in problems of linear algebra*, SIAM J. Numer. Anal., 9 (1972), pp. 53–77.

30. ———, *Mathematics of the verification and validation in computational engineering*, in Proceedings of the Conference on Mathematical and Computer Modelling in Science and Engineering, M. Kočandrlová and V. Kelar, eds., Union of Czech Mathematicians and Physicists, Prague, 2003, pp. 5–12.

31. I. BABUŠKA AND J. T. ODEN, *Verification and validation in computational engineering and science: basic concepts*, Comput. Methods Appl. Mech. Engrg., 193 (2004), pp. 4057–4066.

32. I. BABUŠKA AND T. STROUBOULIS, *The Finite Element Method and its Reliability*, Numerical Mathematics and Scientific Computation, The Clarendon Press Oxford University Press, New York, 2001.

33. J. BAGLAMA AND L. REICHEL, *Augmented GMRES-type methods*, Numer. Linear Algebra Appl., 14 (2007), pp. 337–350.

34. Z. BAI, *Krylov subspace techniques for reduced-order modeling of large-scale dynamical systems*, Appl. Numer. Math., 43 (2002), pp. 9–44.

35. M. B. BALK, *Polyanalytic functions*, vol. 63 of Mathematical Research, Akademie-Verlag, Berlin, 1991.

36. W. BANGERTH AND R. RANNACHER, *Adaptive Finite Element Methods for Differential Equations*, Lectures in Mathematics ETH Zürich, Birkhäuser Verlag, Basel, 2003.
37. R. BARRETT, M. BERRY, T. F. CHAN, ET AL., *Templates for the Solution of Linear Systems: Building Blocks for Iterative Methods*, SIAM, Philadelphia, PA, 1994.
38. T. BARTH, *Implementation of the conjugate gradient method using short multiple recursions*, PhD thesis, University of Colorado at Denver, Denver, 1996.
39. T. BARTH AND T. MANTEUFFEL, *Variable metric conjugate gradient methods*, in Advances in Numerical Methods for Large Sparse Sets of Linear Equations, Number 10, Matrix Analysis and Parallel Computing, PCG 94, M. Natori and T. Nodera, eds., Keio University, Yokohama, 1994, pp. 165–188.
40. ———, *Conjugate gradient algorithms using multiple recursions*, in Proceedings of the AMS-IMS-SIAM Summer Research Conference held at the University of Washington, Seattle, WA, July 9–13, 1995, L. Adams and J. L. Nazareth, eds., SIAM, Philadelphia, PA, 1996, pp. 107–123.
41. ———, *Multiple recursion conjugate gradient algorithms. I. Sufficient conditions*, SIAM J. Matrix Anal. Appl., 21 (2000), pp. 768–796.
42. ———, *Multiple recursion conjugate gradient algorithms. II. Necessary conditions*, unpublished manuscript, (2000).
43. F. L. BAUER, *Optimally scaled matrices*, Numer. Math., 5 (1963), pp. 73–87.
44. F. L. BAUER AND A. S. HOUSEHOLDER, *Moments and characteristic roots*, Numer. Math., 2 (1960), pp. 42–53.
45. ———, *In memoriam: J. H. Wilkinson (1919–1986)*, Numer. Math., 51 (1987), pp. 1–2.
46. B. J. C. BAXTER AND A. ISERLES, *On the foundations of computational mathematics*, in Handbook of Numerical Analysis, Vol. XI, North-Holland, Amsterdam, 2003, pp. 3–34.
47. R. BECKER AND S. MAO, *Convergence and quasi-optimal complexity of a simple adaptive finite element method*, M2AN Math. Model. Numer. Anal., 43 (2009), pp. 1203–1219.
48. B. BECKERMANN, *The condition number of real Vandermonde, Krylov and positive definite Hankel matrices*, Numer. Math., 85 (2000), pp. 553–577.
49. B. BECKERMANN AND E. BOURREAU, *How to choose modified moments?*, J. Comput. Appl. Math., 98 (1998), pp. 81–98.
50. B. BECKERMANN AND A. B. J. KUIJLAARS, *On the sharpness of an asymptotic error estimate for conjugate gradients*, BIT, 41 (2001), pp. 856–867.
51. ———, *Superlinear convergence of conjugate gradients*, SIAM J. Numer. Anal., 39 (2001), pp. 300–329.
52. ———, *Superlinear CG convergence for special right-hand sides*, Electron. Trans. Numer. Anal., 14 (2002), pp. 1–19.
53. B. BECKERMANN AND L. REICHEL, *The Arnoldi process and GMRES for nearly symmetric matrices*, SIAM J. Matrix Anal. Appl., 30 (2008), pp. 102–120.
54. M. BENZI, *Preconditioning techniques for large linear systems: a survey*, J. Comput. Phys., 182 (2002), pp. 418–477.
55. M. BENZI, G. H. GOLUB, AND J. LIESEN, *Numerical solution of saddle point problems*, Acta Numer., 14 (2005), pp. 1–137.
56. M. BENZI AND V. SIMONCINI, *On the eigenvalues of a class of saddle point matrices*, Numer. Math., 103 (2006), pp. 173–196.

57. R. BHATIA, *Matrix Analysis*, vol. 169 of Graduate Texts in Mathematics, Springer-Verlag, New York, 1997.
58. J. BIENAYMÉ, *Considérations à l'appui de la découverte de Laplace sur la loi de probabilité dans la méthode des moindres carrés*, J. math. pures et appl., Sér. 2, 12 (1867), pp. 158–176.
59. Å. BJÖRCK, *Solving linear least squares problems by Gram–Schmidt orthogonalization*, Nordisk Tidskr. Informations-Behandling, 7 (1967), pp. 1–21.
60. ———, *Numerical Methods for Least Squares Problems*, SIAM, Philadelphia, PA, 1996.
61. Å. BJÖRCK, T. ELFVING, AND Z. STRAKOŠ, *Stability of conjugate gradient and Lanczos methods for linear least squares problems*, SIAM J. Matrix Anal. Appl., 19 (1998), pp. 720–736.
62. Å. BJÖRCK AND C. C. PAIGE, *Loss and recapture of orthogonality in the modified Gram–Schmidt algorithm*, SIAM J. Matrix Anal. Appl., 13 (1992), pp. 176–190.
63. A. BLAIR, N. METROPOLIS, J. VON NEUMANN, A. H. TAUB, AND M. TSINGOU, *A study of a numerical solution to a two-dimensional hydrodynamical problem*, Report LA-2165, Los Alamos Scientific Laboratory of the University of California, Los Alamos, NM, 1958. Report written September 28, 1957, and distributed October 8, 1958.
64. ———, *A study of a numerical solution to a two-dimensional hydrodynamical problem*, Math. Tables Aids Comput., 13 (1959), pp. 145–184.
65. L. BLUM, *Computing over the reals: where Turing meets Newton*, Notices Amer. Math. Soc., 51 (2004), pp. 1024–1034.
66. D. BOFFI, *Finite element approximation of eigenvalue problems*, Acta Numer., 19 (2010), pp. 1–120.
67. B. BOHNHORST, *Beiträge zur numerischen Behandlung des unitären Eigenwertproblems*, PhD thesis, Fakultät für Mathematik, Universität Bielefeld, 1993.
68. D. BOLEY AND G. H. GOLUB, *A survey of matrix inverse eigenvalue problems*, Inverse Problems, 3 (1987), pp. 595–622.
69. D. L. BOLEY, D. G. TRUHLAR, Y. SAAD, R. E. WYATT, AND L. E. COLLINS, eds., *Practical Iterative Methods for Large Scale Computations*, North Holland, Amsterdam, 1989.
70. J. A. M. BOLLEN, *Round-off error analysis of descent methods for solving linear equations*, PhD thesis, Technische Hogeschool Eindhoven, Eindhoven, 1980.
71. ———, *Numerical stability of descent methods for solving linear equations*, Numer. Math., 43 (1984), pp. 361–377.
72. ———, *Round-off error analysis of the gradient method*, in Computational Mathematics (Warsaw, 1980), vol. 13 of Banach Center Publ., PWN, Warsaw, 1984, pp. 589–606.
73. A. BOURAS AND V. FRAYSSÉ, *Inexact matrix–vector products in Krylov methods for solving linear systems: a relaxation strategy*, SIAM J. Matrix Anal. Appl., 26 (2005), pp. 660–678.
74. F. BRECHENMACHER, *Algebraic generality vs arithmetic generality in the controversy between C. Jordan and L. Kronecker (1874)*, Technical Report, Laboratoire de Mathématiques Lens, arXiv:0712.2566, 2007.
75. S. C. BRENNER AND L. R. SCOTT, *The Mathematical Theory of Finite Element Methods*, vol. 15 of Texts in Applied Mathematics, Springer-Verlag, New York, third ed., 2008.

76. C. BREZINSKI, *A Bibliography on Continued Fractions, Padé Approximation, Sequence Transformation and Related Subjects*, vol. 3 of Ciencias, Prensas Universitarias de Zaragoza, Zaragoza, 1991.
77. ———, *History of Continued Fractions and Padé Approximants*, vol. 12 of Springer Series in Computational Mathematics, Springer-Verlag, Berlin, 1991.
78. ———, *The methods of Vorobyev and Lanczos*, Linear Algebra Appl., 234 (1996), pp. 21–41.
79. ———, *Projection Methods for Systems of Equations*, vol. 7 of Studies in Computational Mathematics, North-Holland Publishing Co., Amsterdam, 1997.
80. C. BREZINSKI AND M. REDIVO-ZAGLIA, *The PageRank vector: properties, computation, approximation, and acceleration*, SIAM J. Matrix Anal. Appl., 28 (2006), pp. 551–575.
81. C. BREZINSKI, M. REDIVO ZAGLIA, AND H. SADOK, *Avoiding breakdown and near-breakdown in Lanczos type algorithms*, Numer. Algorithms, 1 (1991), pp. 261–284.
82. C. BREZINSKI AND H. SADOK, *Avoiding breakdown in the CGS algorithm*, Numer. Algorithms, 1 (1991), pp. 199–206.
83. M. S. BRODSKIĬ, *Triangular and Jordan Representations of Linear Operators*, vol. 32 of Translations of Mathematical Monographs, American Mathematical Society, Providence, RI, 1971. Translated from the Russian by J. M. Danskin.
84. A. N. BROOKS AND T. J. R. HUGHES, *Streamline upwind/Petrov–Galerkin formulations for convection dominated flows with particular emphasis on the incompressible Navier–Stokes equations*, Comput. Methods Appl. Mech. Engrg., 32 (1982), pp. 199–259.
85. P. N. BROWN, *A theoretical comparison of the Arnoldi and GMRES algorithms*, SIAM J. Sci. Statist. Comput., 12 (1991), pp. 58–78.
86. C. G. BROYDEN, *A new taxonomy of conjugate gradient methods*, Comput. Math. Appl., 31 (1996), pp. 7–17.
87. A. BULTHEEL AND M. VAN BAREL, *Linear Algebra, Rational Approximation and Orthogonal Polynomials*, vol. 6 of Studies in Computational Mathematics, North-Holland Publishing Co., Amsterdam, 1997.
88. J. R. BUNCH, *The weak and strong stability of algorithms in numerical linear algebra*, Linear Algebra Appl., 88/89 (1987), pp. 49–66.
89. A. BUNSE-GERSTNER AND H. FASSBENDER, *Error bounds in the isometric Arnoldi process*, J. Comput. Appl. Math., 86 (1997), pp. 53–72.
90. P. BÜRGISSER, M. CLAUSEN, AND M. A. SHOKROLLAHI, *Algebraic Complexity Theory*, vol. 315 of Grundlehren der Mathematischen Wissenschaften, Springer-Verlag, Berlin, 1997. With the collaboration of Thomas Lickteig.
91. J. V. BURKE AND A. GREENBAUM, *Characterizations of the polynomial numerical hull of degree k*, Linear Algebra Appl., 419 (2006), pp. 37–47.
92. C. BURSTEDDE AND A. KUNOTH, *Fast iterative solution of elliptic control problems in wavelet discretization*, J. Comput. Appl. Math., 196 (2006), pp. 299–319.
93. ———, *A wavelet-based nested iteration-inexact conjugate gradient algorithm for adaptively solving elliptic PDEs*, Numer. Algorithms, 48 (2008), pp. 161–188.
94. P. BUTZER AND F. JONGMANS, *P. L. Chebyshev (1821–1894). A guide to his life and work*, J. Approx. Theory, 96 (1999), pp. 111–138.
95. E. CAHILL, A. IRVING, C. JOHNSON, AND J. SEXTON, *Numerical stability of Lanczos methods*, Nucl. Phys. Proc. Suppl., 83 (2000), pp. 825–827.

96. D. CALVETTI, S.-M. KIM, AND L. REICHEL, *Quadrature rules based on the Arnoldi process*, SIAM J. Matrix Anal. Appl., 26 (2005), pp. 765–781.
97. A. L. CAUCHY, *Sur l'équation à l'aide de laquelle on détermine les inégalités séculaires des mouvements des planètes*, Exercices de mathématiques, 4 (1829). Reprinted in Oeuvres Complètes, Sér. II, Tome IX (Gauthier-Villars, Paris, 1891), pp. 174–195.
98. A. CAYLEY, *A memoir on the theory of matrices*, Philos. Trans. Roy. Soc. London, 148 (1858), pp. 17–37.
99. L. CESARI, *Sulla risoluzione dei sistemi di equazioni lineari per approssimazioni successive*, Memoria estratta della Rassegna delle poste, dei telegrafi e dei telefoni, 4 (1937), p. 37.
100. ———, *Sulla risoluzione dei sistemi di equazioni lineari per approssimazioni successive*, Rendic. Reale Accademia Nazionale dei Lincei, Classe scienze fis., mat., natur., Ser. 6a, 25 (1937), pp. 422–428.
101. F. CHAITIN-CHATELIN AND V. FRAYSSÉ, *Lectures on finite precision computations*, Software, Environments, and Tools, Society for Industrial and Applied Mathematics (SIAM), Philadelphia, PA, 1996. With a foreword by Iain S. Duff.
102. F. CHAITIN-CHATELIN, V. FRAYSSÉ, AND T. BRACONNIER, *Computations in the neighbourhood of algebraic singularities*, Numer. Funct. Anal. Optim., 16 (1995), pp. 287–302.
103. R. CHANDRA, *Conjugate gradient methods for partial differential equations*, PhD thesis, Yale University, New Haven, 1978.
104. P. L. CHEBYSHEV, *Sur les fractions continues*, (1855). Reprinted in Oeuvres I, 11 (Chelsea, New York, 1962), pp. 203–230.
105. ———, *Le développement des fonctions à une seule variable*, (1859). Reprinted in Oeuvres I, 19 (Chelsea, New York, 1962), pp. 501–508.
106. ———, *Sur l'interpolation par la méthode des moindres carrés*, (1859). Reprinted in Oeuvres I, 18 (Chelsea, New York, 1962), pp. 473–498.
107. ———, *Sur les valeurs limites des intégrales*, J. math. pures et appl., Sér. 2, 19 (1874), pp. 157–160. Reprinted in Oeuvres II, 10 (Chelsea, New York, 1962), pp. 183–185.
108. ———, *Sur l'interpolation des valeurs équidistantes*, (1875). Reprinted in Oeuvres II, 12 (Chelsea, New York, 1962), pp. 219–242.
109. ———, *Sur la représentation des valeurs limites des intégrales par des résidus intégraux*, Acta Math., (1886), pp. 35–56. French translation of the 1885 Russian original by Sophie Kowalevski. Reprinted in Oeuvres II, 22 (Chelsea, New York, 1962), pp. 421–440.
110. ———, *The Theory of Probability*, Hänsel-Hohenhausen, Egelsbach, 1999. Lectures delivered in 1879–1880 as taken down by A. M. Liapunov, with a foreword by A. Krylov, translated from the 1936 Russian edition and with an introduction by Oscar Sheynin.
111. T. S. CHIHARA, *An Introduction to Orthogonal Polynomials*, vol. 13 of Mathematics and its Applications, Gordon and Breach Science Publishers, New York, 1978.
112. A. W. CHOU, *On the optimality of Krylov information*, J. Complexity, 3 (1987), pp. 26–40.
113. E. B. CHRISTOFFEL, *Über die Gaußische Quadratur und eine Verallgemeinerung derselben*, J. Reine Angew. Math., 55 (1858), pp. 61–82. Reprinted in Gesammelte mathematische Abhandlungen I (B. G. Teubner, Leipzig, 1910), pp. 65–87.

114. ———, *Sur une classe particulière de fonctions entières et de fractions continues*, Annali di Matematica Pura ed Applicata, 8 (1877), pp. 1–10. Reprinted in Gesammelte mathematische Abhandlungen II (B. G. Teubner, Leipzig, 1910), pp. 42–50.
115. M. T. CHU AND G. H. GOLUB, *Inverse Eigenvalue Problems: Theory, Algorithms, and Applications*, Numerical Mathematics and Scientific Computation, Oxford University Press, New York, 2005.
116. B. A. CIPRA, *The best of the 20th century: Editors name top 10 algorithms*, SIAM News, 33 (2000).
117. H. COHN, R. KLEINBERG, B. SZEGEDY, AND C. UMANS, *Group-theoretic algorithms for matrix multiplication*, in Proceedings of the 46th Annual IEEE Symposium on Foundations of Computer Science, IEEE Computer Society, Pittsburgh, PA, 2005, pp. 379–388.
118. H. COHN AND C. UMANS, *A group-theoretic approach to fast matrix multiplication*, in Proceedings of the 44th Annual IEEE Symposium on Foundations of Computer Science, IEEE Computer Society, Cambridge, MA, 2003, pp. 438–449.
119. D. COPPERSMITH AND S. WINOGRAD, *Matrix multiplication via arithmetic progressions*, J. Symbolic Comput., 9 (1990), pp. 251–280.
120. R. COTES, *Harmonia Mensurarum*, Cantabrigiæ (Cambridge), 1722. Edited by Robert Smith.
121. J. CULLUM AND A. GREENBAUM, *Relations between Galerkin and norm-minimizing iterative methods for solving linear systems*, SIAM J. Matrix Anal. Appl., 17 (1996), pp. 223–247.
122. J. H. CURTISS, *The National Applied Mathematics Laboratories of the National Bureau of Standards: a progress report covering the first five years of its existence*, tech. report, Dept. of Commerce, National Bureau of Standards, 1953. Reprinted in Annals Hist. Comput., 11 (1989), pp. 69–98.
123. ———, *The National Applied Mathematics Laboratories – A prospectus*, Annals Hist. Comput., (1989), pp. 69–98. Reprint of the internal report of the National Bureau of Standards, Washington, D.C., 1947.
124. G. DAHLQUIST AND Å. BJÖRCK, *Numerical Methods in Scientific Computing. Vol. I*, SIAM, Philadelphia, PA, 2008.
125. G. DAHLQUIST, S. C. EISENSTAT, AND G. H. GOLUB, *Bounds for the error of linear systems of equations using the theory of moments*, J. Math. Anal. Appl., 37 (1972), pp. 151–166.
126. G. DAHLQUIST, G. H. GOLUB, AND S. G. NASH, *Bounds for the error in linear systems*, in Semi-Infinite Programming (Proc. Workshop, Bad Honnef, 1978), vol. 15 of Lecture Notes in Control and Information Sci., Springer, Berlin, 1979, pp. 154–172.
127. J. W. DANIEL, *The conjugate gradient method for linear and nonlinear operator equations*, SIAM J. Numer. Anal., 4 (1967), pp. 10–26.
128. ———, *Convergence of the conjugate gradient method with computationally convenient modifications*, Numer. Math., 10 (1967), pp. 125–131.
129. P. J. DAVIS AND P. RABINOWITZ, *Methods of Numerical Integration*, Computer Science and Applied Mathematics, Academic Press Inc., Orlando, FL, second ed., 1984.
130. C. DE BOOR AND G. H. GOLUB, *The numerically stable reconstruction of a Jacobi matrix from spectral data*, Linear Algebra Appl., 21 (1978), pp. 245–260.
131. J. DEMMEL, I. DUMITRIU, O. HOLTZ, AND R. KLEINBERG, *Fast matrix multiplication is stable*, Numer. Math., 106 (2007), pp. 199–224.

132. J. W. DEMMEL, *Trading off parallelism and numerical stability*, in Linear Algebra for Large Scale and Real-time Applications (Leuven, 1992), vol. 232 of NATO Adv. Sci. Inst. Ser. E Appl. Sci., Kluwer Acad. Publ., Dordrecht, 1993, pp. 49–68.
133. ———, *Applied Numerical Linear Algebra*, SIAM, Philadelphia, PA, 1997.
134. J. E. DENNIS, JR. AND K. TURNER, *Generalized conjugate directions*, Linear Algebra Appl., 88/89 (1987), pp. 187–209.
135. P. DEUFLHARD, *Cascadic conjugate gradient methods for elliptic partial differential equations: algorithm and numerical results*, in Domain Decomposition Methods in Scientific and Engineering Computing (University Park, PA, 1993), vol. 180 of Contemp. Math., American Mathematical Society, Providence, RI, 1994, pp. 29–42.
136. R. A. DEVORE AND G. G. LORENTZ, *Constructive approximation*, vol. 303 of Grundlehren der Mathematischen Wissenschaften, Springer-Verlag, Berlin, 1993.
137. J. DIEUDONNÉ, *History of Functional Analysis*, vol. 49 of North-Holland Mathematics Studies, North-Holland Publishing Co., Amsterdam, 1981.
138. ———, *O. Toeplitz's formative years*, in Toeplitz centennial (Tel Aviv, 1981), vol. 4 of Operator Theory: Adv. Appl., Birkhäuser, Basel, 1982, pp. 565–574.
139. G. L. DIRICHLET, *Gedächtnisrede auf Carl Gustav Jacob Jacobi*, J. Reine Angew. Math., 52 (1856), pp. 193–217.
140. J. DONGARRA AND F. SULLIVAN, *The Top 10 Algorithms (Guest editors' introduction)*, Comput. Sci. Eng., 2 (2000), pp. 22–23.
141. J. J. DONGARRA, I. S. DUFF, D. C. SORENSEN, AND H. A. VAN DER VORST, *Numerical Linear Algebra for High-Performance Computers*, Software, Environments, and Tools, SIAM, Philadelphia, PA, 1998.
142. J. J. DONGARRA, H. W. MEUER, H. D. SIMON, AND E. STROHMAIER, *Recent trends in high performance computing*, in The Birth of Numerical Analysis, A. Bultheel and R. Cools, eds., World Scientific, Singapore, 2010, pp. 93–107.
143. Z. DOSTÁL, *Conjugate gradient method with preconditioning by projector*, International Journal of Computer Mathematics, 23 (1988), pp. 315–323.
144. A. DRAUX, *Polynômes Orthogonaux Formels - Applications*, vol. 974 of Lecture Notes in Mathematics, Springer-Verlag, Berlin, 1983.
145. T. A. DRISCOLL, K.-C. TOH, AND L. N. TREFETHEN, *From potential theory to matrix iterations in six steps*, SIAM Rev., 40 (1998), pp. 547–578.
146. J. DRKOŠOVÁ, A. GREENBAUM, M. ROZLOŽNÍK, AND Z. STRAKOŠ, *Numerical stability of GMRES*, BIT, 35 (1995), pp. 309–330.
147. Z. DRMAČ AND K. VESELIĆ, *New fast and accurate Jacobi SVD algorithm. I*, SIAM J. Matrix Anal. Appl., 29 (2007), pp. 1322–1342.
148. ———, *New fast and accurate Jacobi SVD algorithm. II*, SIAM J. Matrix Anal. Appl., 29 (2007), pp. 1343–1362.
149. J. DUINTJER-TEBBENS, J. LIESEN, AND Z. STRAKOŠ, *Analysis of the second phase of the GMRES convergence for a convection–diffusion model problem*, Preprint 861, DFG Research Center MATHEON, 2012.
150. J. DUINTJER-TEBBENS AND G. MEURANT, *Any Ritz value behavior is possible for Arnoldi and GMRES*, SIAM J. Matrix Anal. Appl., (2012), to appear.
151. C. DUNECZKY, R. E. WYATT, D. CHATFIELD, K. HAUG, D. W. SCHWENKE, D. G. TRUHLAR, Y. SUN, AND D. J. KOURI, *Iterative methods for solving the non-sparse equations of quantum mechanical reactive scattering*, Comp. Phys. Comm., 53 (1989), pp. 357–379.

152. H.-D. EBBINGHAUS, H. HERMES, F. HIRZEBRUCH, M. KOECHER, K. MAINZER, J. NEUKIRCH, A. PRESTEL, AND R. REMMERT, *Numbers*, vol. 123 of Graduate Texts in Mathematics, Springer-Verlag, New York, 1991. With an introduction by K. Lamotke, translated from the second 1988 German edition by H. L. S. Orde, translation edited and with a preface by J. H. Ewing.
153. M. EIERMANN, *Fields of values and iterative methods*, Linear Algebra Appl., 180 (1993), pp. 167–197.
154. M. EIERMANN AND O. G. ERNST, *Geometric aspects of the theory of Krylov subspace methods*, Acta Numer., 10 (2001), pp. 251–312.
155. M. EIERMANN, O. G. ERNST, AND O. SCHNEIDER, *Analysis of acceleration strategies for restarted minimal residual methods*, J. Comput. Appl. Math., 123 (2000), pp. 261–292.
156. S. C. EISENSTAT, H. C. ELMAN, AND M. H. SCHULTZ, *Variational iterative methods for nonsymmetric systems of linear equations*, SIAM J. Numer. Anal., 20 (1983), pp. 345–357.
157. H. C. ELMAN, *Iterative methods for large sparse nonsymmetric systems of linear equations*, PhD thesis, Yale University, New Haven, 1982.
158. H. C. ELMAN AND A. RAMAGE, *An analysis of smoothing effects of upwinding strategies for the convection–diffusion equation*, SIAM J. Numer. Anal., 40 (2002), pp. 254–281.
159. ———, *A characterisation of oscillations in the discrete two-dimensional convection-diffusion equation*, Math. Comp., 72 (2003), pp. 263–288.
160. H. C. ELMAN, D. J. SILVESTER, AND A. J. WATHEN, *Finite Elements and Fast Iterative Solvers: With Applications in Incompressible Fluid Dynamics*, Numerical Mathematics and Scientific Computation, Oxford University Press, New York, 2005.
161. L. ELSNER AND K. D. IKRAMOV, *On a condensed form for normal matrices under finite sequences of elementary unitary similarities*, Linear Algebra Appl., 254 (1997), pp. 79–98.
162. J. F. ENCKE, *Gesammelte mathematische und astronomische Abhandlungen. Erster Band*, Ferd. Dümmlers Verlagsbuchhandlung, Berlin, 1888.
163. M. ENGELI, T. GINSBURG, H. RUTISHAUSER, AND E. STIEFEL, *Refined iterative methods for computation of the solution and the eigenvalues of self-adjoint boundary value problems*, Mitt. Inst. Angew. Math. Zürich, 8 (1959).
164. H. W. ENGL, M. HANKE, AND A. NEUBAUER, *Regularization of Inverse Problems*, vol. 375 of Mathematics and its Applications, Kluwer Academic Publishers Group, Dordrecht, 1996.
165. B. ENGQUIST AND G. GOLUB, *From numerical analysis to computational science*, in Mathematics Unlimited—2001 and Beyond, Springer, Berlin, 2001, pp. 433–448.
166. J. ERHEL AND F. GUYOMARC'H, *An augmented conjugate gradient method for solving consecutive symmetric positive definite linear systems*, SIAM J. Matrix Anal. Appl., 21 (2000), pp. 1279–1299.
167. K. ERIKSSON, D. ESTEP, P. HANSBO, AND C. JOHNSON, *Computational Differential Equations*, Cambridge University Press, Cambridge, 1996.
168. O. G. ERNST, *Residual-minimizing Krylov subspace methods for stabilized discretizations of convection-diffusion equations*, SIAM J. Matrix Anal. Appl., 21 (2000), pp. 1079–1101.
169. B. C. EU, *Method of moments in the collision theory*, The Journal of Chemical Physics, 48 (1968), pp. 5611–5622.

170. L. EULER, *An essay on continued fractions*, Math. Systems Theory, 18 (1985), pp. 295–328. Translated from the Latin by B. F. Wyman and M. F. Wyman.
171. ――――, *Introduction to Analysis of the Infinite. Book I*, Springer-Verlag, New York, 1988. Translated from the Latin and with an introduction by John D. Blanton.
172. V. FABER, A. GREENBAUM, AND D. E. MARSHALL, *The polynomial numerical hulls of Jordan blocks and related matrices*, Linear Algebra Appl., 374 (2003), pp. 231–246.
173. V. FABER, W. JOUBERT, E. KNILL, AND T. MANTEUFFEL, *Minimal residual method stronger than polynomial preconditioning*, SIAM J. Matrix Anal. Appl., 17 (1996), pp. 707–729.
174. V. FABER, J. LIESEN, AND P. TICHÝ, *The Faber–Manteuffel theorem for linear operators*, SIAM J. Numer. Anal., 46 (2008), pp. 1323–1337.
175. ――――, *On orthogonal reduction to Hessenberg form with small bandwidth*, Numer. Algorithms, 51 (2009), pp. 133–142.
176. V. FABER AND T. MANTEUFFEL, *Necessary and sufficient conditions for the existence of a conjugate gradient method*, SIAM J. Numer. Anal., 21 (1984), pp. 352–362.
177. V. FABER AND T. A. MANTEUFFEL, *Orthogonal error methods*, SIAM J. Numer. Anal., 24 (1987), pp. 170–187.
178. D. K. FADDEEV AND V. N. FADDEEVA, *Computational Methods of Linear Algebra*, W. H. Freeman and Co., San Francisco, 1963.
179. D. FASINO, *Spectral properties of Hankel matrices and numerical solutions of finite moment problems*, J. Comput. Appl. Math., 65 (1995), pp. 145–155.
180. H. FASSBENDER AND K. D. IKRAMOV, *SYMMLQ-like procedure for Ax = b where A is a special normal matrix*, Calcolo, 43 (2006), pp. 17–37.
181. E. FEIREISL, *Dynamics of Viscous Compressible Fluids*, vol. 26 of Oxford Lecture Series in Mathematics and its Applications, Oxford University Press, Oxford, 2004.
182. E. FEIREISL AND A. NOVOTNÝ, *Singular Limits in Thermodynamics of Viscous Fluids*, Advances in Mathematical Fluid Mechanics, Birkhäuser Verlag, Basel, 2009.
183. B. FISCHER, *Polynomial Based Iteration Methods for Symmetric Linear Systems*, Wiley-Teubner Series Advances in Numerical Mathematics, John Wiley & Sons Ltd., Chichester, 1996.
184. B. FISCHER AND R. FREUND, *On the constrained Chebyshev approximation problem on ellipses*, J. Approx. Theory, 62 (1990), pp. 297–315.
185. B. FISCHER AND R. W. FREUND, *On adaptive weighted polynomial preconditioning for Hermitian positive definite matrices*, SIAM J. Sci. Comput., 15 (1994), pp. 408–426.
186. B. FISCHER, A. RAMAGE, D. J. SILVESTER, AND A. J. WATHEN, *Minimum residual methods for augmented systems*, BIT, 38 (1998), pp. 527–543.
187. ――――, *On parameter choice and iterative convergence for stabilised discretisations of advection–diffusion problems*, Comput. Methods Appl. Mech. Engrg., 179 (1999), pp. 179–195.
188. H.-J. FISCHER, *On generating orthogonal polynomials for discrete measures*, Z. Anal. Anwendungen, 17 (1998), pp. 183–205.
189. D. A. FLANDERS AND G. SHORTLEY, *Numerical determination of fundamental modes*, J. Appl. Phys., 21 (1950), pp. 1326–1332.

190. R. FLETCHER, *Conjugate gradient methods for indefinite systems*, in Numerical Analysis (Proc. 6th Biennial Dundee Conf., Univ. Dundee, Dundee, 1975), vol. 506 of Lecture Notes in Mathematics, Springer, Berlin, 1976, pp. 73–89.

191. G. E. FORSYTHE, *Gauss to Gerling on relaxation*, Math. Tables Aids Comput., 8 (1951), pp. 255–258. Translation, with notes, of a letter by Gauss.

192. ———, *Solving linear algebraic equations can be interesting*, Bull. Amer. Math. Soc., 59 (1953), pp. 299–329.

193. G. E. FORSYTHE AND C. B. MOLER, *Computer Solution of Linear Algebraic Systems*, Prentice-Hall Inc., Englewood Cliffs, N.J., 1967.

194. G. E. FORSYTHE AND W. R. WASOW, *Finite-Difference Methods for Partial Differential Equations*, Applied Mathematics Series, John Wiley & Sons Inc., New York, 1960.

195. L. FOX, *A short account of relaxation methods*, Quart. J. Mech. Appl. Math., 1 (1948), pp. 253–280.

196. L. FOX, H. D. HUSKEY, AND J. H. WILKINSON, *Notes on the solution of algebraic linear simultaneous equations*, Quart. J. Mech. Appl. Math., 1 (1948), pp. 149–173.

197. J. G. F. FRANCIS, *The QR transformation: a unitary analogue to the LR transformation. I*, Comput. J., 4 (1961/1962), pp. 265–271.

198. ———, *The QR transformation. II*, Comput. J., 4 (1961/1962), pp. 332–345.

199. S. P. FRANKEL, *Convergence rates of iterative treatments of partial differential equations*, Math. Tables and Other Aids to Computation, 4 (1950), pp. 65–75.

200. V. FRAYSÉ, L. GIRAUD, S. GRATTON, AND J. LANGOU, *Algorithm 842: A set of GMRES routines for real and complex arithmetics on high performance computers*, ACM Trans. Math. Software, 31 (2005), pp. 228–238.

201. R. W. FREUND, *Quasi-kernel polynomials and their use in non-Hermitian matrix iterations*, J. Comput. Appl. Math., 43 (1992), pp. 135–158.

202. ———, *Model reduction methods based on Krylov subspaces*, Acta Numer., 12 (2003), pp. 267–319.

203. R. W. FREUND, G. H. GOLUB, AND N. M. NACHTIGAL, *Iterative solution of linear systems*, Acta Numer., 1 (1992), pp. 57–100.

204. R. W. FREUND, M. H. GUTKNECHT, AND N. M. NACHTIGAL, *An implementation of the look-ahead Lanczos algorithm for non-Hermitian matrices*, SIAM J. Sci. Comput., 14 (1993), pp. 137–158.

205. R. W. FREUND AND M. HOCHBRUCK, *Gauss quadratures associated with the Arnoldi process and the Lanczos algorithm*, in Linear Algebra for Large Scale and Real-Time Applications, M. S. Moonen, G. H. Golub, and B. L. R. de Moor, eds., Kluwer Academic Publishers, 1993, pp. 377–380.

206. R. W. FREUND AND N. M. NACHTIGAL, *QMR: a quasi-minimal residual method for non-Hermitian linear systems*, Numer. Math., 60 (1991), pp. 315–339.

207. V. M. FRIDMAN, *The method of minimum iterations with minimum errors for a system of linear algebraic equations with a symmetrical matrix*, USSR Comput. Math. Math. Phys., 2 (1963), pp. 362–363.

208. K. FRIEDRICHS AND G. HORVAY, *The finite Stieltjes momentum problem*, Proc. Nat. Acad. Sci. U.S.A., 25 (1939), pp. 528–534.

209. G. F. FROBENIUS, *Ueber lineare Substitutionen und bilineare Formen*, J. Reine Angew. Math., 84 (1878), pp. 1–63. Reprinted in Gesammelte Abhandlungen 1 (Springer-Verlag, Berlin, 1968), pp. 343–405.

210. A. FROMMER AND V. SIMONCINI, *Stopping criteria for rational matrix functions of Hermitian and symmetric matrices*, SIAM J. Sci. Comput., 30 (2008), pp. 1387–1412.
211. K. GALLIVAN, E. GRIMME, AND P. VAN DOOREN, *Asymptotic waveform evaluation via a Lanczos method*, Appl. Math. Lett., 7 (1994), pp. 75–80.
212. M. J. GANDER AND G. WANNER, *From Euler, Ritz and Galerkin to modern computing*, SIAM Rev., (2012), to appear.
213. W. N. GANSTERER, Y. BAI, R. M. DAY, AND R. C. WARD, *A framework for approximating eigenpairs in electronic structure computations*, Comput. Sci. Eng., 6 (2004), pp. 50–59.
214. W. N. GANSTERER, R. C. WARD, R. P. MULLER, AND W. A. GODDARD, III, *Computing approximate eigenpairs of symmetric block tridiagonal matrices*, SIAM J. Sci. Comput., 25 (2003), pp. 65–85.
215. F. R. GANTMACHER, *On the algebraic analysis of Krylov's method of transforming the secular equation*, Trans. Second Math. Congress, II (1934), pp. 45–48. In Russian. Title translation as in [216].
216. ———, *The Theory of Matrices. Vols. 1, 2*, Chelsea Publishing Co., New York, 1959.
217. F. R. GANTMACHER AND M. G. KREIN, *Oscillation Matrices and Kernels and Small Vibrations of Mechanical Systems*, AMS Chelsea Publishing, Providence, RI, revised ed., 2002. Translation based on the 1941 Russian original, edited and with a preface by Alex Eremenko.
218. A. GAUL, M. H. GUTKNECHT, J. LIESEN, AND R. NABBEN, *A framework for deflated and augmented Krylov subspace methods*. Technical Report, arXiv:1206.1506, 2012.
219. C. F. GAUSS, *Methodus nova integralium valores per approximationem inveniendi*, Commentationes Societatis Regiae Scientiarum Gottingensis, (1814), pp. 39–76. Reprinted in Werke, Band III (Göttingen, 1876), pp. 163–196.
220. ———, *Werke. Band IX*, Georg Olms Verlag, Hildesheim, 1973. Reprint of the 1903 original.
221. W. GAUTSCHI, *Construction of Gauss–Christoffel quadrature formulas*, Math. Comp., 22 (1968), pp. 251–270.
222. ———, *A survey of Gauss–Christoffel quadrature formulae*, in E. B. Christoffel (Aachen/Monschau, 1979), Birkhäuser, Basel, 1981, pp. 72–147.
223. ———, *On generating orthogonal polynomials*, SIAM J. Sci. Statist. Comput., 3 (1982), pp. 289–317.
224. ———, *Is the recurrence relation for orthogonal polynomials always stable?*, BIT, 33 (1993), pp. 277–284.
225. ———, *Numerical Analysis: An Introduction*, Birkhäuser Boston Inc., Boston, MA, 1997.
226. ———, *The interplay between classical analysis and (numerical) linear algebra— a tribute to Gene H. Golub*, Electron. Trans. Numer. Anal., 13 (2002), pp. 119–147.
227. ———, *Orthogonal Polynomials: Computation and Approximation*, Numerical Mathematics and Scientific Computation, Oxford University Press, New York, 2004. Oxford Science Publications.
228. W. GAUTSCHI AND R. S. VARGA, *Error bounds for Gaussian quadrature of analytic functions*, SIAM J. Numer. Anal., 20 (1983), pp. 1170–1186.
229. M. K. GAVURIN, *The application of polynomials of best approximation to the improvement of the convergence of iterative processes*, in Four Articles on Numerical

Matrix Methods, NBS Rep. 2007, U.S. Department of Commerce National Bureau of Standards, Washington, D.C., 1952, pp. 44–50. Translated by C. D. Benster.

230. B. GELLAI, *The Intrinsic Nature of Things. The Life and Science of Cornelius Lanczos*, American Mathematical Society, Providence, RI, 2010.

231. L. GEYER, *Sharp bounds for the valence of certain harmonic polynomials*, Proc. Amer. Math. Soc., 136 (2008), pp. 549–555.

232. L. GIRAUD AND J. LANGOU, *When modified Gram–Schmidt generates a well-conditioned set of vectors*, IMA J. Numer. Anal., 22 (2002), pp. 521–528.

233. L. GIRAUD, J. LANGOU, AND M. ROZLOŽNÍK, *The loss of orthogonality in the Gram–Schmidt orthogonalization process*, Comput. Math. Appl., 50 (2005), pp. 1069–1075.

234. L. GIRAUD, J. LANGOU, M. ROZLOŽNÍK, AND J. VAN DEN ESHOF, *Rounding error analysis of the classical Gram–Schmidt orthogonalization process*, Numer. Math., 101 (2005), pp. 87–100.

235. W. GIVENS, *A method of computing eigenvalues and eigenvectors suggested by classical results on symmetric matrices*, in Simultaneous Linear Equations and the Determination of Eigenvalues, National Bureau of Standards Applied Mathematics Series, No.29, U.S. Government Printing Office, Washington, D.C., 1953, pp. 117–122.

236. M. S. GOCKENBACH, *Understanding and Implementing the Finite Element Method*, SIAM, Philadelphia, PA, 2006.

237. S. K. GODUNOV, *Modern Aspects of Linear Algebra*, vol. 175 of Translations of Mathematical Monographs, American Mathematical Society, Providence, RI, 1998. Translated from the 1997 Russian original by Tamara Rozhkovskaya.

238. I. GOHBERG AND S. GOLDBERG, *A simple proof of the Jordan decomposition theorem for matrices*, Amer. Math. Monthly, 103 (1996), pp. 157–159.

239. I. GOHBERG, P. LANCASTER, AND L. RODMAN, *Invariant Subspaces of Matrices with Applications*, vol. 51 of Classics in Applied Mathematics, SIAM, Philadelphia, PA, 2006. Reprint of the 1986 original.

240. H. H. GOLDSTINE, *A History of Numerical Analysis from the 16th through the 19th Century*, vol. 2 of Studies in the History of Mathematics and Physical Sciences, Springer-Verlag, New York, 1977.

241. G. H. GOLUB, *The use of Chebyshev matrix polynomials in the iterative solution of linear equations compared to the method of successive over-relaxation*, PhD thesis, University of Illinois, Urbana, 1959.

242. ———, *Matrix computation and the theory of moments*, in Proceedings of the International Congress of Mathematicians, Vol. 1, 2 (Zürich, 1994), Birkhäuser, Basel, 1995, pp. 1440–1448.

243. ———, *Milestones in Matrix Computation: Selected Works of Gene H. Golub, with Commentaries*, Oxford Science Publications, Oxford University Press, Oxford, 2007. Edited by Raymond H. Chan, Chen Greif, and Dianne P. O'Leary.

244. G. H. GOLUB AND G. MEURANT, *Matrices, moments and quadrature*, in Numerical analysis 1993 (Dundee, 1993), vol. 303 of Pitman Res. Notes Math. Ser., Longman Sci. Tech., Harlow, 1994, pp. 105–156.

245. G. H. GOLUB AND G. MEURANT, *Matrices, moments and quadrature. II. How to compute the norm of the error in iterative methods*, BIT, 37 (1997), pp. 687–705.

246. ———, *Matrices, Moments and Quadrature with Applications*, Princeton Series in Applied Mathematics, Princeton University Press, Princeton, NJ, 2010.

247. G. H. GOLUB AND D. P. O'LEARY, *Some history of the conjugate gradient and Lanczos algorithms: 1948–1976*, SIAM Rev., 31 (1989), pp. 50–102.
248. G. H. GOLUB, M. STOLL, AND A. WATHEN, *Approximation of the scattering amplitude and linear systems*, Electron. Trans. Numer. Anal., 31 (2008), pp. 178–203.
249. G. H. GOLUB AND Z. STRAKOŠ, *Estimates in quadratic formulas*, Numer. Algorithms, 8 (1994), pp. 241–268.
250. G. H. GOLUB AND C. F. VAN LOAN, *Matrix Computations*, Johns Hopkins Studies in the Mathematical Sciences, Johns Hopkins University Press, Baltimore, MD, third ed., 1996.
251. G. H. GOLUB AND R. S. VARGA, *Chebyshev semi-iterative methods, successive over-relaxation iterative methods, and second order Richardson iterative methods. I*, Numer. Math., 3 (1961), pp. 147–156.
252. ———, *Chebyshev semi-iterative methods, successive over-relaxation iterative methods, and second order Richardson iterative methods. II*, Numer. Math., 3 (1961), pp. 157–168.
253. G. H. GOLUB AND J. H. WELSCH, *Calculation of Gauss quadrature rules*, Math. Comp. 23 (1969), 221-230; addendum, ibid., 23 (1969), pp. A1–A10.
254. S. GOOSSENS AND D. ROOSE, *Ritz and harmonic Ritz values and the convergence of FOM and GMRES*, Numer. Linear Algebra Appl., 6 (1999), pp. 281–293.
255. R. G. GORDON, *Error bounds in equilibrium statistical mechanics*, J. Math. Phys., 9 (1968), pp. 655–663.
256. W. B. GRAGG, *Matrix interpretations and applications of the continued fraction algorithm*, Rocky Mountain J. Math., 4 (1974), pp. 213–225.
257. ———, *The QR algorithm for unitary Hessenberg matrices*, J. Comput. Appl. Math., 16 (1986), pp. 1–8.
258. ———, *Positive definite Toeplitz matrices, the Arnoldi process for isometric operators, and Gaussian quadrature on the unit circle*, J. Comput. Appl. Math., 46 (1993), pp. 183–198.
259. W. B. GRAGG AND W. J. HARROD, *The numerically stable reconstruction of Jacobi matrices from spectral data*, Numer. Math., 44 (1984), pp. 317–335.
260. W. B. GRAGG AND A. LINDQUIST, *On the partial realization problem*, Linear Algebra Appl., 50 (1983), pp. 277–319.
261. W. B. GRAGG AND L. REICHEL, *On the application of orthogonal polynomials to the iterative solution of linear systems of equations with indefinite or non-Hermitian matrices*, Linear Algebra Appl., 88/89 (1987), pp. 349–371.
262. J. P. GRAM, *Om Rækkeudviklinger, bestemte ved Hjælp af de mindste Kvadraters Methode*, PhD thesis, Kjöbenhavn, 1879.
263. ———, *Ueber die Enwickelung reeller Functionen in Reihen mittelst der Methode der kleinsten Quadrate*, J. Reine Angew. Math., 94 (1883), pp. 41–73.
264. A. GRAMA, A. GUPTA, G. KARYPIS, AND V. KUMAR, *Introduction to Parallel Computing*, Addison-Wesley, Essex, second ed., 2003.
265. J. F. GRCAR, *How ordinary elimination became Gaussian elimination*, Historia Math., 38 (2011), pp. 163–218.
266. ———, *John von Neumann's analysis of Gaussian elimination and the origins of modern numerical analysis*, SIAM Rev., 53 (2011), pp. 607–682.
267. ———, *Mathematicians of Gaussian elimination*, Notices Amer. Math. Soc., 58 (2011), pp. 782–792.
268. A. GREENBAUM, *Comparison of splittings used with the conjugate gradient algorithm*, Numer. Math., 33 (1979), pp. 181–193.

269. ———, *Behavior of slightly perturbed Lanczos and conjugate-gradient recurrences*, Linear Algebra Appl., 113 (1989), pp. 7–63.
270. ———, *The Lanczos and conjugate gradient algorithms in finite precision arithmetic*, in Proceedings of the Cornelius Lanczos International Centenary Conference (Raleigh, NC, 1993), SIAM, Philadelphia, PA, 1994, pp. 49–60.
271. ———, *Estimating the attainable accuracy of recursively computed residual methods*, SIAM J. Matrix Anal. Appl., 18 (1997), pp. 535–551.
272. ———, *Iterative Methods for Solving Linear Systems*, vol. 17 of Frontiers in Applied Mathematics, SIAM, Philadelphia, PA, 1997.
273. ———, *Generalizations of the field of values useful in the study of polynomial functions of a matrix*, Linear Algebra Appl., 347 (2002), pp. 233–249.
274. ———, *Card shuffling and the polynomial numerical hull of degree k*, SIAM J. Sci. Comput., 25 (2003), pp. 408–416.
275. ———, *Some theoretical results derived from polynomial numerical hulls of Jordan blocks*, Electron. Trans. Numer. Anal., 18 (2004), pp. 81–90.
276. ———, *Upper and lower bounds on norms of functions of matrices*, Linear Algebra Appl., 430 (2009), pp. 52–65.
277. A. GREENBAUM AND L. GURVITS, *Max-min properties of matrix factor norms*, SIAM J. Sci. Comput., 15 (1994), pp. 348–358.
278. A. GREENBAUM, V. PTÁK, AND Z. STRAKOŠ, *Any nonincreasing convergence curve is possible for GMRES*, SIAM J. Matrix Anal. Appl., 17 (1996), pp. 465–469.
279. A. GREENBAUM, M. ROZLOŽNÍK, AND Z. STRAKOŠ, *Numerical behaviour of the modified Gram-Schmidt GMRES implementation*, BIT, 37 (1997), pp. 706–719.
280. A. GREENBAUM AND Z. STRAKOŠ, *Predicting the behavior of finite precision Lanczos and conjugate gradient computations*, SIAM J. Matrix Anal. Appl., 13 (1992), pp. 121–137.
281. ———, *Matrices that generate the same Krylov residual spaces*, in Recent Advances in Iterative Methods, vol. 60 of IMA Vol. Math. Appl., Springer, New York, 1994, pp. 95–118.
282. A. GREENBAUM AND L. N. TREFETHEN, *GMRES/CR and Arnoldi/Lanczos as matrix approximation problems*, SIAM J. Sci. Comput., 15 (1994), pp. 359–368.
283. E. J. GRIMME, *Krylov projection methods for model reduction*, PhD thesis, University of Illinois at Urbana-Champaign, Urbana, 1997.
284. S. GUGERCIN AND K. WILLCOX, *Krylov projection framework for Fourier model reduction*, Automatica J. IFAC, 44 (2008), pp. 209–215.
285. M. GUTKNECHT, *The Lanczos process and Padé approximation*, in Proceedings of the Cornelius Lanczos International Centenary Conference (Raleigh, NC, 1993), SIAM, Philadelphia, PA, 1994, pp. 61–75.
286. M. H. GUTKNECHT, *A completed theory of the unsymmetric Lanczos process and related algorithms. I*, SIAM J. Matrix Anal. Appl., 13 (1992), pp. 594–639.
287. ———, *A completed theory of the unsymmetric Lanczos process and related algorithms. II*, SIAM J. Matrix Anal. Appl., 15 (1994), pp. 15–58.
288. ———, *Spectral deflation in Krylov solvers: A theory of coordinate space based methods*, Electron. Trans. Numer. Anal., 39 (2011), pp. 156–185.
289. M. H. GUTKNECHT AND Z. STRAKOŠ, *Accuracy of two three-term and three two-term recurrences for Krylov space solvers*, SIAM J. Matrix Anal. Appl., 22 (2000), pp. 213–229.
290. W. HACKBUSCH, *Elliptic Differential Equations. Theory and Numerical Treatment*, vol. 18 of Springer Series in Computational Mathematics, Springer-Verlag, Berlin,

1992. Translated from the author's revision of the 1986 German original by Regine Fadiman and Patrick D. F. Ion.

291. ——, *Iterative Solution of Large Sparse Systems of Equations*, vol. 95 of Applied Mathematical Sciences, Springer-Verlag, New York, 1994. Translated and revised from the 1991 German original.

292. L. A. HAGEMAN AND D. M. YOUNG, *Applied Iterative Methods*, Academic Press Inc., New York, 1981. Computer Science and Applied Mathematics.

293. S. W. R. HAMILTON, *Lectures on Quaternions*, Hodges and Smith, Dublin, 1853.

294. P. C. HANSEN, *Rank-Deficient and Ill-Posed Problems. Numerical Aspects of Linear Inversion*, SIAM Monographs on Mathematical Modeling and Computation, SIAM, Philadelphia, PA, 1998.

295. ——, *Discrete Inverse Problems. Insight and Algorithms*, vol. 7 of Fundamentals of Algorithms, SIAM, Philadelphia, PA, 2010.

296. P. C. HANSEN, M. E. KILMER, AND R. H. KJELDSEN, *Exploiting residual information in the parameter choice for discrete ill-posed problems*, BIT, 46 (2006), pp. 41–59.

297. H. HARBRECHT AND R. SCHNEIDER, *On error estimation in finite element methods without having Galerkin orthogonality*, Berichtsreihe des SFB 611, No. 457, Universität Bonn, 2009.

298. E. HARMS, *Application of the method of moments to three-body scattering*, Physics Letters B, 41 (1972), pp. 26–28.

299. T. HAWKINS, *Cauchy and the spectral theory of matrices*, Historia Math., 2 (1975), pp. 1–29.

300. ——, *Another look at Cayley and the theory of matrices*, Arch. Internat. Histoire Sci., 27 (1977), pp. 82–112.

301. ——, *Weierstrass and the theory of matrices*, Arch. Hist. Exact Sci., 17 (1977), pp. 119–163.

302. ——, *Frobenius and the symbolical algebra of matrices*, Arch. Hist. Exact Sci., 62 (2008), pp. 23–57.

303. E. HEINE, *Handbuch der Kugelfunctionen. Theorie und Anwendungen. Erster Band*, Zweite umgearbeitete und vermehrte Auflage, Reimer, Berlin, 1878.

304. ——, *Handbuch der Kugelfunctionen. Theorie und Anwendungen. Zweiter Band*, Zweite umgearbeitete und vermehrte Auflage, Reimer, Berlin, 1881.

305. D. HELLER, *A survey of parallel algorithms in numerical linear algebra*, SIAM Rev., 20 (1978), pp. 740–777.

306. E. HELLINGER AND O. TOEPLITZ, *Zur Einordnung der Kettenbruchtheorie in die Theorie der quadratischen Formen von unendlichvielen Veränderlichen*, J. Reine Angew. Math., 144 (1914), pp. 212–238, 318.

307. ——, *Integralgleichungen und Gleichungen mit unendlichvielen Unbekannten*, in Encyklopädie der mathematischen Wissenschaften mit Einschluss ihrer Anwendungen, Band II, 3. Teil, 2. Hälfte, Teubner, Leipzig, 1927, pp. 1395–1597.

308. S. HELSEN, A. B. J. KUIJLAARS, AND M. VAN BAREL, *Convergence of the isometric Arnoldi process*, SIAM J. Matrix Anal. Appl., 26 (2005), pp. 782–809.

309. J. J. HENCH AND Z. STRAKOŠ, *The RCWA method—a case study with open questions and perspectives of algebraic computations*, Electron. Trans. Numer. Anal., 31 (2008), pp. 331–357.

310. P. HENRICI, *The quotient-difference algorithm*, Nat. Bur. Standards Appl. Math. Ser., 49 (1958), pp. 23–46.

311. K. HESSENBERG, *Behandlung linearer Eigenwertaufgaben mit Hilfe der Hamilton-Cayleyschen Gleichung*, Numerische Verfahren, 1. Bericht, Technische Hochschule Darmstadt, Institut für praktische Mathematik, 1940.
312. M. R. HESTENES, *Conjugacy and gradients*, in A History of Scientific Computing (Princeton, NJ, 1987), ACM Press Hist. Ser., ACM, New York, 1990, pp. 167–179.
313. M. R. HESTENES AND E. STIEFEL, *Methods of conjugate gradients for solving linear systems*, J. Research Nat. Bur. Standards, 49 (1952), pp. 409–436.
314. M. R. HESTENES AND J. TODD, *Mathematicians Learning to Use Computers*, National Institute of Standards and Technology Special Publication, 730, U.S. Department of Commerce National Institute of Standards and Technology, Washington, D.C., 1991. With a foreword by J. Barkley Rosser.
315. N. J. HIGHAM, *Accuracy and Stability of Numerical Algorithms*, SIAM, Philadelphia, PA, second ed., 2002.
316. ———, *Functions of Matrices. Theory and Computation*, SIAM, Philadelphia, PA, 2008.
317. N. J. HIGHAM AND P. A. KNIGHT, *Componentwise error analysis for stationary iterative methods*, in Linear Algebra, Markov Chains, and Queueing Models (Minneapolis, MN, 1992), vol. 48 of IMA Vol. Math. Appl., Springer, New York, 1993, pp. 29–46.
318. ———, *Finite precision behavior of stationary iteration for solving singular systems*, Linear Algebra Appl., 192 (1993), pp. 165–186.
319. D. HILBERT, *Grundzüge einer allgemeinen Theorie der linearen Integralgleichungen (Vierte Mitteilung)*, Nachr. Ges. Wiss. Göttingen, Math.-phys. Kl., (1906), pp. 157–227.
320. ———, *Grundzüge einer allgemeinen Theorie der linearen Integralgleichungen*, Teubner, Leipzig und Berlin, 1912.
321. D. HILBERT, J. VON NEUMANN, AND L. NORDHEIM, *Über die Grundlagen der Quantenmechanik*, Math. Ann., 98 (1928), pp. 1–30.
322. R. O. HILL, JR. AND B. N. PARLETT, *Refined interlacing properties*, SIAM J. Matrix Anal. Appl., 13 (1992), pp. 239–247.
323. R. HIPTMAIR, *Operator preconditioning*, Comput. Math. Appl., 52 (2006), pp. 699–706.
324. I. HNĚTYNKOVÁ, D. M. PLEŠINGER, MARTIN SIMA, Z. STRAKOŠ, AND S. V. HUFFEL, *The total least squares problem in $AX \approx B$: A new classification with the relationship to the classical works*, SIAM J. Matrix Annal. Appl., 32 (2011), pp. 748–770.
325. I. HNĚTYNKOVÁ, M. PLEŠINGER, AND Z. STRAKOŠ, *The regularizing effect of the Golub–Kahan iterative bidiagonalization and revealing the noise level in the data*, BIT, 49 (2009), pp. 669–696.
326. I. HNĚTYNKOVÁ AND Z. STRAKOŠ, *Lanczos tridiagonalization and core problems*, Linear Algebra Appl., 421 (2007), pp. 243–251.
327. M. HOCHBRUCK AND C. LUBICH, *Error analysis of Krylov methods in a nutshell*, SIAM J. Sci. Comput., 19 (1998), pp. 695–701.
328. U. HOCHSTRASSER, *Die Anwendung der Methode der konjugierten Gradienten und ihrer Modifikationen auf die Lösung linearer Randwertprobleme*, PhD thesis, Eidgenössische Technische Hochschule, Zürich, 1954.
329. R. W. HOCKNEY, *Characterization of parallel computers and algorithms*, Computer Phys. Comm., 26 (1982), pp. 285–291.

330. R. W. HOCKNEY, *Synchronization and communication overheads on the LCAP multiple FPS-164 computer-system*, Parallel Comp., 9 (1989), pp. 279–290.
331. ———, *Performance parameters and benchmarking of supercomputers*, Parallel Comp., 17 (1991), pp. 1111–1130.
332. R. W. HOCKNEY AND I. J. CURRINGTON, *F1/2 – a parameter to characterize memory and communication bottlenecks*, Parallel Comp., 10 (1989), pp. 277–286.
333. R. W. HOCKNEY AND C. R. JESSHOPE, *Parallel Computers 2*, Adam Hilger, Bristol and Philadelphia, 1988.
334. M. HOEMMEN, *Communication-avoiding Krylov subspace methods*, PhD thesis, University of California, Berkeley, 2010.
335. K. HOFFMAN AND R. KUNZE, *Linear Algebra*, Prentice-Hall Inc., Englewood Cliffs, NJ, second ed., 1971.
336. O. HOLTZ AND N. SHOROM, *Computational complexity and numerical stability of linear problems*, in European Congress of Mathematics, Amsterdam, 14–18 July, 2008, European Mathematical Society, 2010, pp. 381–400.
337. J. HORÁČEK, *A note on iterative solution of scattering equations*, Acta Phys. Pol., A71 (1987), pp. 1–8.
338. J. HORÁČEK AND T. SASAKAWA, *Method of continued fractions with application to atomic physics*, Phys. Rev. A (3), 28 (1983), pp. 2151–2156.
339. ———, *Method of continued fractions with application to atomic physics. II*, Phys. Rev. A (3), 30 (1984), pp. 2274–2277.
340. ———, *Method of continued fractions for on- and off-shell t matrix of local and nonlocal potentials*, Phys. Rev. C, 32 (1985), pp. 70–75.
341. R. A. HORN AND C. R. JOHNSON, *Matrix Analysis*, Cambridge University Press, Cambridge, 1985.
342. ———, *Topics in Matrix Analysis*, Cambridge University Press, Cambridge, 1991.
343. G. HORVAY, *Solution of large equation systems and eigenvalue problems by Lanczos' matrix iteration method*, Report KAPL-1004, United States Atomic Energy Commission, Technical Information Service, Oak Ridge, Tennessee, 1953.
344. H. HOTELLING, *Some new methods in matrix calculation*, Ann. Math. Statistics, 14 (1943), pp. 1–34.
345. A. S. HOUSEHOLDER, *Principles of Numerical Analysis*, Dover Publications Inc., New York, 1974. Unabridged, corrected version of the 1953 edition.
346. ———, *The Theory of Matrices in Numerical Analysis*, Dover Publications Inc., New York, 1975. Reprint of the 1964 edition.
347. A. S. HOUSEHOLDER AND F. L. BAUER, *On certain methods for expanding the characteristic polynomial*, Numer. Math., 1 (1959), pp. 29–37.
348. T. J. R. HUGHES AND A. BROOKS, *A multidimensional upwind scheme with no crosswind diffusion*, in Finite Element Methods for Convection Dominated Flows (Papers, Winter Ann. Meeting Amer. Soc. Mech. Engrs., New York, 1979), vol. 34 of AMD, Amer. Soc. Mech. Engrs. (ASME), New York, 1979, pp. 19–35.
349. M. HUHTANEN, *Combining normality with the FFT techniques*, Research Report A451, Helsinki University of Technology, Institute of Mathematics, 2002.
350. ———, *Orthogonal polyanalytic polynomials and normal matrices*, Math. Comp., 72 (2003), pp. 355–373.
351. M. HUHTANEN AND R. M. LARSEN, *Exclusion and inclusion regions for the eigenvalues of a normal matrix*, SIAM J. Matrix Anal. Appl., 23 (2002), pp. 1070–1091.

352. K. HWANG AND F. A. BRIGGS, *Computer Architecture and Parallel Processing*, McGraw-Hill series in Computer Organization and Architecture, McGraw-Hill, Inc., New York, 1984.
353. K. D. IKRAMOV, *The matrix adjoint to a normal matrix A as a polynomial in A*, Dokl. Akad. Nauk, 338 (1994), pp. 304–305.
354. I. C. F. IPSEN AND C. D. MEYER, *The idea behind Krylov methods*, Amer. Math. Monthly, 105 (1998), pp. 889–899.
355. E. ISAACSON AND H. B. KELLER, *Analysis of Numerical Methods*, John Wiley & Sons Inc., New York, 1966.
356. C. G. J. JACOBI, *Ueber Gauss neue Methode, die Werthe der Integrale näherungsweise zu finden*, J. Reine Angew. Math., 1 (1826), pp. 301–308. Reprinted in Gesammelte Werke, 6. Band (Reimer, Berlin, 1891), pp. 3–11.
357. ———, *Über ein leichtes Verfahren die in der Theorie der Säcularstörungen vorkommenden Gleichungen numerisch aufzulösen*, J. Reine Angew. Math., 30 (1846), pp. 51–94. Reprinted in Gesammelte Werke, 7. Band (Reimer, Berlin, 1891), pp. 97–144.
358. ———, *Über die Reduction quadratischer Formen auf die kleinste Anzahl Glieder*, Bericht über die zur Bekanntmachung geeigneten Verhandlungen der Königl. Preuss. Akad. Wiss. Berlin, (1848), pp. 414–417.
359. ———, *Über die Reduction der quadratischen Formen auf die kleinste Anzahl Glieder*, J. Reine Angew. Math., 39 (1850), pp. 290–292. Reprinted in Gesammelte Werke, 6. Band (Reimer, Berlin, 1891), pp. 318–320.
360. ———, *Über eine elementare Transformation eines in Bezug auf jedes von zwei Variablen-Systemen linearen und homogenen Ausdrucks*, J. Reine Angew. Math., 53 (1857), pp. 265–270. Aus den hinterlassenen Papieren von C. G. J. Jacobi mitgetheilt durch C. W. Borchardt. Reprinted in Gesammelte Werke, 3. Band (Reimer, Berlin, 1884), pp. 583–590.
361. ———, *Gesammelte Werke*, Herausgegeben auf Veranlassung der Königlich Preussischen Akademie der Wissenschaften, Reimer, Berlin, 1881–1891. Edited by C. W. Borchardt (1. Band, 1881), K. Weierstraß (2. Band, 1882; 3. Band, 1884; 4. Band, 1886; 5. Band, 1890; 6. Band, 1891; 7. Band, 1891).
362. C. JAGELS AND L. REICHEL, *The isometric Arnoldi process and an application to iterative solution of large linear systems*, in Iterative Methods in Linear Algebra (Brussels, 1991), North-Holland, Amsterdam, 1992, pp. 361–369.
363. C. F. JAGELS AND L. REICHEL, *A fast minimal residual algorithm for shifted unitary matrices*, Numer. Linear Algebra Appl., 1 (1994), pp. 555–570.
364. A. JENNINGS, *Influence of the eigenvalue spectrum on the convergence rate of the conjugate gradient method*, J. Inst. Math. Appl., 20 (1977), pp. 61–72.
365. A. JENNINGS AND G. M. MALIK, *The solution of sparse linear equations by the conjugate gradient method*, Internat. J. Numer. Methods Engrg., 12 (1978), pp. 141–158.
366. P. JIRÁNEK AND M. ROZLOŽNÍK, *Adaptive version of simpler GMRES*, Numer. Algorithms, 53 (2010), pp. 93–112.
367. P. JIRÁNEK, M. ROZLOŽNÍK, AND M. H. GUTKNECHT, *How to make simpler GMRES and GCR more stable*, SIAM J. Matrix Anal. Appl., 30 (2008), pp. 1483–1499.
368. P. JIRÁNEK, Z. STRAKOŠ, AND M. VOHRALÍK, *A posteriori error estimates including algebraic error and stopping criteria for iterative solvers*, SIAM J. Sci. Comput., 32 (2010), pp. 1567–1590.

369. C. JORDAN, *Traité des substitutions et des équations algébriques*, Paris, 1870.
370. ———, *Sur les polynômes bilinéaires*, Comptes rendus, 77 (1873), pp. 1478–1491. Reprinted in Oeuvres III (Paris, 1962), pp. 7–11.
371. T. F. JORDAN, *Linear Operators for Quantum Mechanics*, John Wiley & Sons Inc., New York, 1969.
372. W. JOUBERT, *A robust GMRES-based adaptive polynomial preconditioning algorithm for nonsymmetric linear systems*, SIAM J. Sci. Comput., 15 (1994), pp. 427–439.
373. R. E. KALMAN, *On partial realizations, transfer functions, and canonical forms*, Acta Polytech. Scand. Math. Comput. Sci. Ser., 31 (1979), pp. 9–32.
374. S. KANIEL, *Estimates for some computational techniques in linear algebra*, Math. Comp., 20 (1966), pp. 369–378.
375. S. KARLIN AND L. S. SHAPLEY, *Geometry of moment spaces*, Mem. Amer. Math. Soc., 1953 (1953), p. 93.
376. R. KARLSON, *A study of some roundoff effects of the GMRES method*, Technical Report Li-TH-MAT-R-1990-11, University of Linköping, 1991.
377. E. M. KASENALLY AND V. SIMONCINI, *Analysis of a minimum perturbation algorithm for nonsymmetric linear systems*, SIAM J. Numer. Anal., 34 (1997), pp. 48–66.
378. J. KAUTSKÝ AND G. H. GOLUB, *On the calculation of Jacobi matrices*, Linear Algebra Appl., 52/53 (1983), pp. 439–455.
379. C. T. KELLEY, *Iterative Methods for Linear and Nonlinear Equations*, vol. 16 of Frontiers in Applied Mathematics, SIAM, Philadelphia, PA, 1995.
380. M. D. KENT, *Chebyshev, Krylov, Lanczos: Matrix relationship and computations*, PhD thesis, Stanford University, Stanford, 1989.
381. I. M. KHABAZA, *An iterative least-square method suitable for solving large sparse matrices*, Comput. J., 6 (1963/1964), pp. 202–206.
382. D. KHAVINSON AND G. NEUMANN, *On the number of zeros of certain rational harmonic functions*, Proc. Amer. Math. Soc., 134 (2006), pp. 1077–1085.
383. D. KHAVINSON AND G. ŚWIĄTEK, *On the number of zeros of certain harmonic polynomials*, Proc. Amer. Math. Soc., 131 (2003), pp. 409–414.
384. S. KHRUSHCHEV, *Orthogonal Polynomials and Continued Fractions. From Euler's Point of View*, vol. 122 of Encyclopedia of Mathematics and its Applications, Cambridge University Press, Cambridge, 2008.
385. R. A. KIRSCH, *Computer development at the National Bureau of Standards*, in A Century of Excellence in Measurements, Standards, and Technology, U.S. Department of Commerce, National Bureau of Standards and Technology, 2001, pp. 86–89. NIST Special Publication 958.
386. T. H. KJELDSEN, *The early history of the moment problem*, Historia Math., 20 (1993), pp. 19–44.
387. A. KLAWONN AND G. STARKE, *Block triangular preconditioners for nonsymmetric saddle point problems: field-of-values analysis*, Numer. Math., 81 (1999), pp. 577–594.
388. A. KLEPPNER, *The cyclic decomposition theorem*, Integr. Equ. Oper. Theory, 25 (1996), pp. 490–495.
389. P. KNUPP AND K. SALARI, *Verification of Computer Codes in Computational Science and Engineering*, Discrete Mathematics and its Applications, Chapman & Hall/CRC, Boca Raton, FL, 2003.

390. L. KOENIGSBERGER, *Carl Gustav Jacob Jacobi*, B. G. Teubner, Leipzig, 1904.
391. A. N. KOLMOGOROV AND S. V. FOMĪN, *Introductory Real Analysis*, Dover Publications Inc., New York, 1975. Translated from the second Russian edition and edited by Richard A. Silverman, corrected reprinting.
392. A. N. KOLMOGOROV AND A. P. YUSHKEVICH, eds., *Mathematics of the 19th Century*, Birkhäuser Verlag, Basel, 1998. Translated from the 1987 Russian original by Roger Cooke.
393. Z. KOPAL, *Numerical Analysis: With Emphasis on the Application of Numerical Techniques to Problems of Infinitesimal Calculus in Single Variable*, John Wiley & Sons Inc., New York, second ed., 1961.
394. A. KOWALEWSKI, *Newton, Cotes, Gauss, Jacobi*, von Veit & Comp., Leipzig, 1917.
395. D. KRATZER, S. V. PARTER, AND M. STEUERWALT, *Block splittings for the conjugate gradient method*, Comput. & Fluids, 11 (1983), pp. 255–279.
396. M. G. KREĬN, *The ideas of P. L. Čebyšev and A. A. Markov in the theory of limiting values of integrals and their further development*, Uspehi Matem. Nauk (N.S.), 6 (1951), pp. 3–120. English translation in Amer. Math. Soc. Transl., Series 2, 12 (1959), pp. 1–121.
397. L. KRONECKER, *Sur les faisceaux de formes quadratiques et bilinéaires*, Comptes rendus, 78 (1874), pp. 1181–1182. Reprinted in Werke I (B. G. Teubner, Leipzig, 1895), pp. 415–419.
398. A. N. KRYLOV, *On the numerical solution of the equation by which the frequency of small oscillations is determined in technical problems*, Izv. Akad. Nauk SSSR, Ser. Fiz.-Mat., 4 (1931), pp. 491–539. In Russian. Title translation as in [216].
399. ———, *Moi vospominaniia: The world of the Russian Naval Reformer A. N. Krylov: Reminiscences on Russian Life at the Turn of the Century*, Edwin Mellen Press, Lampeter, Wales, 1999. In Russian.
400. V. N. KUBLANOVSKAYA, *Certain algorithms for the solution of the complete problem of eigenvalues*, Dokl. Akad. Nauk SSSR 136 (1961), pp. 26–28. In Russian. Translated as Soviet Math. Dokl., 2 (1961), pp. 17–19.
401. ———, *Some algorithms for the solution of the complete problem of eigenvalues*, Ž. Vyčisl. Mat. i Mat. Fiz., 1 (1961), pp. 555–570. In Russian. Translated as *On some algorithms for the solution of the complete eigenvalue problem*, USSR Comput. Math. and Math. Phys., 1 (1962), pp. 637–657.
402. A. B. J. KUIJLAARS, *Convergence analysis of Krylov subspace iterations with methods from potential theory*, SIAM Rev., 48 (2006), pp. 3–40.
403. A. B. J. KUIJLAARS AND E. A. RAKHMANOV, *Zero distributions for discrete orthogonal polynomials*, J. Comput. Appl. Math., 99 (1998), pp. 255–274.
404. ———, *Corrigendum to: "Zero distributions for discrete orthogonal polynomials"*, J. Comput. Appl. Math., 104 (1999), p. 213.
405. C. LANCZOS, *An iteration method for the solution of the eigenvalue problem of linear differential and integral operators*, J. Research Nat. Bur. Standards, 45 (1950), pp. 255–282.
406. ———, *Solution of systems of linear equations by minimized iterations*, J. Research Nat. Bur. Standards, 49 (1952), pp. 33–53.
407. ———, *Chebyshev polynomials in the solution of large-scale linear systems*, in Proceedings of the Association for Computing Machinery, Toronto, 1952, Sauls Lithograph Co. (for the Association for Computing Machinery), Washington, D.C., 1953, pp. 124–133.

408. ——, *Applied Analysis*, Prentice-Hall Inc., Englewood Cliffs, NJ, 1956.
409. ——, *The inspired guess in the history of physics*, Studies: An Irish Quarterly Review, 53 (1964), pp. 398–412.
410. ——, *Collected Published Papers with Commentaries*, North Carolina State University, Raleigh, NC, 1998.
411. J. LANGOU, *Translation and modern interpretation of Laplace's Théorie Analytique des Probabilités, pages 505-512, 516-520*, Technical Report, arXiv:0907.4695v1, 2009.
412. P.-S. de LAPLACE, *Théorie Analytique des Probabilités*, Paris, 1820.
413. A. J. LAUB AND A. LINNEMANN, *Hessenberg and Hessenberg/triangular forms in linear system theory*, Internat. J. Control, 44 (1986), pp. 1523–1547.
414. D. P. LAURIE, *Accurate recovery of recursion coefficients from Gaussian quadrature formulas*, J. Comput. Appl. Math., 112 (1999), pp. 165–180.
415. ——, *Computation of Gauss-type quadrature formulas*, J. Comput. Appl. Math., 127 (2001), pp. 201–217.
416. A. W. LEISSA, *The historical bases of the Rayleigh and Ritz methods*, Journal of Sound and Vibration, 287 (2005), pp. 961–978.
417. R.-C. LI, *On Meinardus' examples for the conjugate gradient method*, Math. Comp., 77 (2008), pp. 335–352.
418. R.-C. LI AND W. ZHANG, *The rate of convergence of GMRES on a tridiagonal Toeplitz linear system*, Numer. Math., 112 (2009), pp. 267–293.
419. ——, *The rate of convergence of GMRES on a tridiagonal toeplitz linear system. II*, Linear Algebra Appl., 431 (2009), pp. 2425–2436.
420. J. LIESEN, *When is the adjoint of a matrix a low degree rational function in the matrix?*, SIAM J. Matrix Anal. Appl., 29 (2007), pp. 1171–1180.
421. J. LIESEN AND B. N. PARLETT, *On nonsymmetric saddle point matrices that allow conjugate gradient iterations*, Numer. Math., 108 (2008), pp. 605–624.
422. J. LIESEN, M. ROZLOŽNÍK, AND Z. STRAKOŠ, *Least squares residuals and minimal residual methods*, SIAM J. Sci. Comput., 23 (2002), pp. 1503–1525.
423. J. LIESEN AND P. E. SAYLOR, *Orthogonal Hessenberg reduction and orthogonal Krylov subspace bases*, SIAM J. Numer. Anal., 42 (2005), pp. 2148–2158.
424. J. LIESEN AND Z. STRAKOŠ, *Convergence of GMRES for tridiagonal Toeplitz matrices*, SIAM J. Matrix Anal. Appl., 26 (2004), pp. 233–251.
425. ——, *GMRES convergence analysis for a convection–diffusion model problem*, SIAM J. Sci. Comput., 26 (2005), pp. 1989–2009.
426. ——, *On optimal short recurrences for generating orthogonal Krylov subspace bases*, SIAM Rev., 50 (2008), pp. 485–503.
427. J. LIESEN AND P. TICHÝ, *Convergence analysis of Krylov subspace methods*, GAMM Mitt. Ges. Angew. Math. Mech., 27 (2004), pp. 153–173.
428. ——, *The worst-case GMRES for normal matrices*, BIT, 44 (2004), pp. 79–98.
429. ——, *On the worst-case convergence of MR and CG for symmetric positive definite tridiagonal Toeplitz matrices*, Electron. Trans. Numer. Anal., 20 (2005), pp. 180–197.
430. R. K. LIVESLEY, *The analysis of large structural systems*, Comput. J., 3 (1960/1961), pp. 34–39.
431. L. A. LJUSTERNIK, *Solution of problems in linear algebra by the method of continued fractions*, Trudy Voronezskovo Gosudarstvennovo Instituta, Voronezh, 2 (1956), pp. 85–90. In Russian.

432. L. LORENTZEN AND H. WAADELAND, *Continued Fractions with Applications*, vol. 3 of Studies in Computational Mathematics, North-Holland Publishing Co., Amsterdam, 1992.

433. J. LU AND D. L. DARMOFAL, *A quasi-minimal residual method for simultaneous primal–dual solutions and superconvergent functional estimates*, SIAM J. Sci. Comput., 24 (2003), pp. 1693–1709.

434. D. G. LUENBERGER, *Hyperbolic pairs in the method of conjugate gradients*, SIAM J. Appl. Math., 17 (1969), pp. 1263–1267.

435. ———, *Optimization by Vector Space Methods*, John Wiley & Sons Inc., New York, 1969.

436. ———, *The conjugate residual method for constrained minimization problems*, SIAM J. Numer. Anal., 7 (1970), pp. 390–398.

437. ———, *Introduction to Linear and Nonlinear Programming*, Addison-Wesley, New York, 1973.

438. N. N. LUZIN, *On Krylov's method for transforming the secular equation*, Izv. Akad. Nauk SSSR, Ser. Fiz.-Mat., 7 (1931), pp. 903–958. In Russian. Title translation as in [216].

439. J. MÁLEK AND K. R. RAJAGOPAL, *On the modeling of inhomogeneous incompressible fluid-like bodies*, Mechanics of Materials, 38 (2006), pp. 233–242.

440. L. MANSFIELD, *On the use of deflation to improve the convergence of conjugate gradient iteration*, Communications in Applied Numerical Methods, 4 (1988), pp. 151–156.

441. T. A. MANTEUFFEL AND J. S. OTTO, *On the roots of the orthogonal polynomials and residual polynomials associated with a conjugate gradient method*, Numer. Linear Algebra Appl., 1 (1994), pp. 449–475.

442. R. F. MARCIA, *On solving sparse symmetric linear systems whose definiteness is unknown*, Appl. Numer. Math., 58 (2008), pp. 449–458.

443. K.-A. MARDAL AND R. WINTHER, *Preconditioning discretizations of systems of partial differential equations*, Numer. Linear Algebra Appl., 18 (2011), pp. 1–40.

444. A. MARKOFF, *Démonstration de certaines inégalités de M. Tchébychef*, Math. Ann., 24 (1884), pp. 172–180.

445. ———, *Sur la méthode de Gauss pour le calcul approché des intégrales*, Math. Ann., 25 (1885), pp. 427–432.

446. W. MARKOFF, *Über Polynome, die in einem gegebenen Intervalle möglichst wenig von Null abweichen*, Math. Ann., 77 (1916), pp. 213–258.

447. V. MEHRMANN AND H. XU, *Numerical methods in control*, J. Comput. Appl. Math., 123 (2000), pp. 371–394.

448. D. MEIDNER, R. RANNACHER, AND J. VIHHAREV, *Goal-oriented error control of the iterative solution of finite element equations*, J. Numer. Math., 17 (2009), pp. 143–172.

449. G. MEINARDUS, *Über eine Verallgemeinerung einer Ungleichung von L. V. Kantorowitsch*, Numer. Math., 5 (1963), pp. 14–23.

450. L. A. M. MELLO, E. DE STURLER, G. H. PAULINO, AND E. C. N. SILVA, *Recycling Krylov subspaces for efficient large-scale electrical impedance tomography*, Comput. Methods Appl. Mech. Engrg., 199 (2010), pp. 3101–3110.

451. N. METROPOLIS, J. HOWLETT, AND G.-C. ROTA, eds., *A History of Computing in the Twentieth Century. A Collection of Essays*, Academic Press Inc., New York, 1980.

452. G. MEURANT, *Computer Solution of Large Linear Systems*, vol. 28 of Studies in Mathematics and its Applications, North-Holland Publishing Co., Amsterdam, 1999.

453. ———, *The Lanczos and Conjugate Gradient Algorithms. From Theory to Finite Precision Computations*, vol. 19 of Software, Environments, and Tools, SIAM, Philadelphia, PA, 2006.

454. ———, *The complete stagnation of GMRES for $n \leq 4$*, Electron. Trans. Numer. Anal., 39 (2012), pp. 75–101.

455. ———, *Necessary and sufficient conditions for GMRES complete and partial stagnation*, submitted, (2011).

456. ———, *On the residual norm in FOM and GMRES*, SIAM J. Matrix Anal. Appl., 32 (2011), pp. 394–411.

457. G. MEURANT AND Z. STRAKOŠ, *The Lanczos and conjugate gradient algorithms in finite precision arithmetic*, Acta Numer., 15 (2006), pp. 471–542.

458. H.-D. MEYER AND S. PAL, *A band-Lanczos method for computing matrix elements of a resolvent*, J. Chem. Phys., 91 (1989), pp. 6195–6204.

459. A. MIĘDLAR, *Inexact adaptive finite element methods for elliptic PDE eigenvalue problems*, PhD thesis, Institut für Mathematik, Technische Universität Berlin, 2010.

460. P. MICHAUD, Y. SAZEIDES, AND A. SEZNEC, *Proposition for a sequential accelerator in future general-purpose manycore processors and the problem of migration-induced cache misses*, in Proceedings of the 7th ACM International Conference on Computing Frontiers (Bertinoro, Italy, 2010), ACM, Philadelphia, PA, 2010, pp. 237–246.

461. L. M. MILNE-THOMSON, *The Calculus of Finite Differences*, MacMillan & Co., London, 1933.

462. M. MOHIYUDDIN, M. HOEMMEN, J. DEMMEL, AND K. YELICK, *Minimizing communication in sparse matrix solvers*, in Proceedings of the Conference on High Performance Computing Networking, Storage and Analysis (Portland, OR), 2009.

463. B. C. MOORE, *Principal component analysis in linear systems: controllability, observability, and model reduction*, IEEE Trans. Automat. Control, 26 (1981), pp. 17–32.

464. G. MORO AND J. H. FREED, *Calculation of ESR spectra and related Fokker-Planck forms by the use of the Lanczos algorithm*, J. Chem. Phys., 74 (1981), pp. 3757–3773.

465. K. W. MORTON, *Numerical Solution of Convection–Diffusion Problems*, vol. 12 of Applied Mathematics and Mathematical Computation, Chapman & Hall, London, 1996.

466. R. MURPHY, *Second memoir on the inverse method of definite integrals*, Trans. Cambridge Phil. Soc., 5 (1835), pp. 113–148.

467. I. P. MYSOVSKIH, *On the construction of cubature formulas with the smallest number of nodes*, Dokl. Akad. Nauk SSSR, 178 (1968), pp. 1252–1254.

468. N. NACHTIGAL, *A look-ahead variant of the Lanczos algorithm and its application to the quasi-minimal residual method for non-Hermitian linear systems*, PhD thesis, Massachusetts Institute of Technology, Cambridge, 1991.

469. N. M. NACHTIGAL, S. C. REDDY, AND L. N. TREFETHEN, *How fast are nonsymmetric matrix iterations?*, SIAM J. Matrix Anal. Appl., 13 (1992), pp. 778–795.

470. A. E. NAIMAN, I. M. BABUŠKA, AND H. C. ELMAN, *A note on conjugate gradient convergence*, Numer. Math., 76 (1997), pp. 209–230.
471. A. E. NAIMAN AND S. ENGELBERG, *A note on conjugate gradient convergence. II, III*, Numer. Math., 85 (2000), pp. 665–683, 685–696.
472. S. G. NASH, ed., *A History of Scientific Computing*, ACM Press History Series, ACM Press, New York, 1990. Papers from the Conference on the History of Scientific and Numeric Computation held at Princeton University, Princeton, New Jersey, 1987.
473. NATIONAL BUREAU OF STANDARDS, *Projects and Publications of the National Applied Mathematics Laboratory. A Quarterly Report*, no. January through March, 1949.
474. ———, *Projects and Publications of the National Applied Mathematics Laboratory. A Quarterly Report*, no. July through September, Washington, D.C., 1949.
475. T. NEEDHAM, *Visual Complex Analysis*, The Clarendon Press Oxford University Press, New York, 1997.
476. P. G. NEVAI, *Orthogonal Polynomials*, no. 213 in Memoirs of the American Mathematical Society, American Mathematical Society, Providence, RI, 1979.
477. O. NEVANLINNA, *Convergence of Iterations for Linear Equations*, Lectures in Mathematics ETH Zürich, Birkhäuser Verlag, Basel, 1993.
478. M. NEWMAN AND J. TODD, *The evaluation of matrix inversion programs*, J. Soc. Indust. Appl. Math., 6 (1958), pp. 466–476.
479. I. NEWTON, *Philosophiæ Naturalis Principia Mathematica*, Sumptibus Societatis, Amstælodami (Amsterdam), 1723.
480. C. M. M. NEX, *Estimation of integrals with respect to a density of states*, J. Phys. A, 11 (1978), pp. 653–663.
481. ———, *The block Lanczos algorithm and the calculation of matrix resolvents*, Comp. Phys. Comm., 53 (1989), pp. 141–146.
482. R. A. NICOLAIDES, *Deflation of conjugate gradients with applications to boundary value problems*, SIAM J. Numer. Anal., 24 (1987), pp. 355–365.
483. Y. NOTAY, *On the convergence rate of the conjugate gradients in presence of rounding errors*, Numer. Math., 65 (1993), pp. 301–317.
484. J. T. ODEN, I. BABUŠKA, F. NOBILE, Y. FENG, AND R. TEMPONE, *Theory and methodology for estimation and control of errors due to modeling, approximation, and uncertainty*, Comput. Methods Appl. Mech. Engrg., 194 (2005), pp. 195–204.
485. J. T. ODEN, J. C. BROWNE, I. BABUŠKA, K. M. LIECHTI, AND L. F. DEMKOWICZ, *A computational infrastructure for reliable computer simulations*, in Lecture Notes in Computer Science, vol. 2660, Springer-Verlag, Heidelberg, 2003, pp. 385–392.
486. W. OETTLI AND W. PRAGER, *Compatibility of approximate solution of linear equations with given error bounds for coefficients and right-hand sides*, Numer. Math., 6 (1964), pp. 405–409.
487. D. P. O'LEARY, Z. STRAKOŠ, AND P. TICHÝ, *On sensitivity of Gauss–Christoffel quadrature*, Numer. Math., 107 (2007), pp. 147–174.
488. E. P. OZHIGOVA, *The part played by the Petersburg Academy of Sciences (the Academy of Sciences of the USSR) in the publication of Euler's collected works*, in Euler and Modern Science, N. N. Bogolyubov, G. K. Mikhaĭlov, and A. P. Yushkevich, eds., Mathematical Association of America, Washington, D.C., 2007, pp. 53–74. Translated from the 1988 Russian original by Robert Burns.

489. C. C. PAIGE, *The computation of eigenvalues and eigenvectors of very large and sparse matrices*, PhD thesis, London University, London, England, 1971.
490. ———, *Computational variants of the Lanczos method for the eigenproblem*, J. Inst. Math. Appl., 10 (1972), pp. 373–381.
491. ———, *Error analysis of the Lanczos algorithm for tridiagonalizing a symmetric matrix*, J. Inst. Math. Appl., 18 (1976), pp. 341–349.
492. ———, *Accuracy and effectiveness of the Lanczos algorithm for the symmetric eigenproblem*, Linear Algebra Appl., 34 (1980), pp. 235–258.
493. ———, *An augmented stability result for the Lanczos Hermitian matrix tridiagonalization process*, SIAM J. Matrix Anal. Appl., 31 (2010), pp. 2347–2359.
494. C. C. PAIGE, B. N. PARLETT, AND H. A. VAN DER VORST, *Approximate solutions and eigenvalue bounds from Krylov subspaces*, Numer. Linear Algebra Appl., 2 (1995), pp. 115–133.
495. C. C. PAIGE, M. ROZLOŽNÍK, AND Z. STRAKOŠ, *Modified Gram–Schmidt (MGS), least squares, and backward stability of MGS-GMRES*, SIAM J. Matrix Anal. Appl., 28 (2006), pp. 264–284.
496. C. C. PAIGE AND M. A. SAUNDERS, *Solution of sparse indefinite systems of linear equations*, SIAM J. Numer. Anal., 12 (1975), pp. 617–629.
497. ———, *Algorithm 583 LSQR: Sparse linear equations and sparse least squares problem*, ACM Trans. Math. Software, 8 (1982), pp. 195–209.
498. ———, *LSQR: An algorithm for sparse linear equations and sparse least squares*, ACM Trans. Math. Software, 8 (1982), pp. 43–71.
499. C. C. PAIGE AND Z. STRAKOŠ, *Bounds for the least squares distance using scaled total least squares*, Numer. Math., 91 (2002), pp. 93–115.
500. ———, *Residual and backward error bounds in minimum residual Krylov subspace methods*, SIAM J. Sci. Comput., 23 (2002), pp. 1898–1923.
501. ———, *Scaled total least squares fundamentals*, Numer. Math., 91 (2002), pp. 117–146.
502. ———, *Core problems in linear algebraic systems*, SIAM J. Matrix Anal. Appl., 27 (2005), pp. 861–875.
503. L. J. PAIGE AND O. TAUSSKY, eds., *Simultaneous Linear Equations and the Determination of Eigenvalues*, no. 29 in Applied Mathematics Series, National Bureau of Standards, Washington, D.C., 1953.
504. L. PAINVIN, *Sur un certain système d'équations linéaires*, J. Math. Pures Appl., 3 (1858), pp. 41–46.
505. B. N. PARLETT, *The contribution of J. H. Wilkinson to numerical analysis*, in A History of Scientific Computing (Princeton, NJ, 1987), ACM Press Hist. Ser., ACM, New York, 1990, pp. 17–30.
506. ———, *Reduction to tridiagonal form and minimal realizations*, SIAM J. Matrix Anal. Appl., 13 (1992), pp. 567–593.
507. ———, *Do we fully understand the symmetric Lanczos algorithm yet?*, in Proceedings of the Cornelius Lanczos International Centenary Conference (Raleigh, NC, 1993), SIAM, Philadelphia, PA, 1994, pp. 93–107.
508. ———, *The Symmetric Eigenvalue Problem*, vol. 20 of Classics in Applied Mathematics, SIAM, Philadelphia, PA, 1998. Corrected reprint of the 1980 original.
509. B. N. PARLETT, D. R. TAYLOR, AND Z. A. LIU, *A look-ahead Lanczos algorithm for unsymmetric matrices*, Math. Comp., 44 (1985), pp. 105–124.

510. R. V. PATEL, A. J. LAUB, AND P. M. VAN DOOREN, eds., *Numerical Linear Algebra Techniques for Systems and Control*, IEEE Press, New York, 1994.
511. D. PELLETIER AND P. J. ROACHE, *Verification and validation of computational heat transfer*, in Handbook of Numerical Heat Transfer, W. J. Minkowycz, E. M. Sparrow, and J. Y. Murthy, eds., John Wiley & Sons, Inc., New York, second ed., 2009, pp. 417–442.
512. W. P. PETERSEN AND P. ARBENZ, *Introduction to Parallel Computing*, Oxford University Press, Oxford, 2004.
513. J. L. PHILLIPS, *The triangular decomposition of Hankel matrices*, Math. Comp., 25 (1971), pp. 559–602.
514. J. D. PRYCE, *Numerical Solution of Sturm–Liouville Problems*, Monographs on Numerical Analysis, The Clarendon Press Oxford University Press, New York, 1993.
515. A. QUARTERONI AND A. VALLI, *Numerical Approximation of Partial Differential Equations*, vol. 23 of Springer Series in Computational Mathematics, Springer-Verlag, Berlin, 1994.
516. G. D. RAITHBY, *Skew upstream differencing schemes for problems involving fluid flow*, Comput. Methods Appl. Mech. Engrg., 9 (1976), pp. 153–164.
517. R. RANNACHER, A. WESTENBERGER, AND W. WOLLNER, *Adaptive finite element solution of eigenvalue problems: balancing of discretization and iteration error*, J. Numer. Math., 18 (2010), pp. 303–327.
518. L. RAYLEIGH, *On the calculation of Chladni's figures for a square plate*, Phil. Mag., 6th series, 22 (1911), pp. 225–229.
519. B. D. REDDY, *Introductory Functional Analysis. With Applications to Boundary Value Problems and Finite Elements*, vol. 27 of Texts in Applied Mathematics, Springer-Verlag, New York, 1998.
520. L. REICHEL, *Fast QR decomposition of Vandermonde-like matrices and polynomial least squares approximation*, SIAM J. Matrix Anal. Appl., 12 (1991), pp. 552–564.
521. J. K. REID, *On the method of conjugate gradients for the solution of large sparse systems of linear equations*, in Large Sparse Sets of Linear Equations (Proc. Conf., St. Catherine's Coll., Oxford, 1970), Academic Press, London, 1971, pp. 231–254.
522. W. P. REINHARDT, L^2 *discretization of atomic and molecular electronic continua: Moment, quadrature and J-matrix techniques*, Comp. Phys. Comm., 17 (1979), pp. 1–21.
523. S. H. RHIE, *n-point gravitational lenses with $5(n-1)$ images*, Technical Report, arXiv:astro-ph/0305166, 2003.
524. L. F. RICHARDSON, *The approximate arithmetical solution by finite differences of physical problems involving differential equations, with an application to the stresses in a masonry dam*, Phil. Trans. Roy. Soc. London. A, 210 (1911), pp. 307–357.
525. F. RIESZ, *Sur les opérations fonctionelles linéaires*, C. R. Acad. Sci. Paris, 149 (1909), pp. 974–977.
526. J.-L. RIGAL AND J. GACHES, *On the compatibility of a given solution with the data of a linear system*, J. Assoc. Comput. Mach., 14 (1967), pp. 543–548.
527. W. RITZ, *Über eine neue Methode zur Lösung gewisser Variationsprobleme der mathematischen Physik*, J. Reine Angew. Math., 135 (1908), pp. 1–61.
528. T. J. RIVLIN, *Chebyshev Polynomials*, Pure and Applied Mathematics, John Wiley & Sons Inc., New York, second ed., 1990.

529. P. J. ROACHE, *Quantification of uncertainty in computational fluid dynamics*, in Annual Review of Fluid Mechanics, vol. 29, Annual Reviews, Palo Alto, CA, 1997, pp. 123–160.
530. ———, *Verification and Validation in Computational Science and Engineering*, Hermosa Publishers, Albuquerque, New Mexico, 1998.
531. ———, *Building PDE codes to be verifiable and validatable*, Comput. Sci. Eng., 6 (2004), pp. 30–38.
532. S. ROBINSON, *Toward an optimal algorithm for matrix multiplication*, SIAM News, 38 (2005).
533. S. ROMAN, *Advanced Linear Algebra*, vol. 135 of Graduate Texts in Mathematics, Springer-Verlag, New York, third ed., 2008.
534. H.-G. ROOS, M. STYNES, AND L. TOBISKA, *Numerical Methods for Singularly Perturbed Differential Equations*, vol. 24 of Springer Series in Computational Mathematics, Springer-Verlag, Berlin, 1996.
535. J. B. ROSSER, *Rapidly converging iterative methods for solving linear equations*, in Simultaneous linear equations and the determination of eigenvalues, U.S. Government Printing Office, Washington, D.C., 1953, pp. 59–64. Proceedings of a symposium held August 23–25, 1951, in Los Angeles, California, USA, under the sponsorship of the National Bureau of Standards, in cooperation with the Office of Naval Research.
536. M. ROZLOŽNÍK AND Z. STRAKOŠ, *Variants of the residual minimizing Krylov space methods*, in Proceedings of the 11th Summer School on Software and Algorithms of Numerical Mathematics, West Bohemia University, Plzeň, 1995, pp. 208–225.
537. T. RUSTEN AND R. WINTHER, *A preconditioned iterative method for saddlepoint problems*, SIAM J. Matrix Anal. Appl., 13 (1992), pp. 887–904.
538. H. RUTISHAUSER, *Beiträge zur Kenntnis des Biorthogonalisierungs-Algorithmus von Lanczos*, Z. Angew. Math. Physik, 4 (1953), pp. 35–56.
539. ———, *Solution of eigenvalue problems with the LR-transformation*, Nat. Bur. Standards Appl. Math. Ser., 49 (1958), pp. 47–81.
540. ———, *Bestimmung der Eigenwerte orthogonaler Matrizen*, Numer. Math., 9 (1966), pp. 104–108.
541. Y. SAAD, *Krylov subspace methods for solving large unsymmetric linear systems*, Math. Comp., 37 (1981), pp. 105–126.
542. ———, *The Lanczos biorthogonalization algorithm and other oblique projection methods for solving large unsymmetric systems*, SIAM J. Numer. Anal., 19 (1982), pp. 485–506.
543. ———, *Iterative Methods for Sparse Linear Systems*, SIAM, Philadelphia, PA, second ed., 2003.
544. Y. SAAD AND M. H. SCHULTZ, *GMRES: a generalized minimal residual algorithm for solving nonsymmetric linear systems*, SIAM J. Sci. Statist. Comput., 7 (1986), pp. 856–869.
545. Y. SAAD AND H. A. VAN DER VORST, *Iterative solution of linear systems in the 20th century*, J. Comput. Appl. Math., 123 (2000), pp. 1–33.
546. Y. SAAD, M. YEUNG, J. ERHEL, AND F. GUYOMARC'H, *A deflated version of the conjugate gradient algorithm*, SIAM J. Sci. Comput., 21 (2000), pp. 1909–1926.
547. R. A. SACK AND A. F. DONOVAN, *An algorithm for Gaussian quadrature given modified moments*, Numer. Math., 18 (1971/72), pp. 465–478.

548. T. SAUER AND U. MAJER, eds., *David Hilbert's Lectures on the Foundations of Physics 1915-1927*, vol. 5 of David Hilbert's Foundational Lectures, Springer-Verlag, Berlin, 2009.

549. V. K. SAUL'YEV, *Integration of Equations of Parabolic Type by the Method of Nets*, vol. 54 of International Series of Monographs in Pure and Applied Mathematics, Pergamon Press, London, 1960. Translated from the Russian by G. J. Tee, translation edited and editorial introduction by K. L. Stewart.

550. P. E. SAYLOR AND D. C. SMOLARSKI, *Addendum to: "Why Gaussian quadrature in the complex plane?"*, Numer. Algorithms, 27 (2001), pp. 215–217.

551. ———, *Why Gaussian quadrature in the complex plane?*, Numer. Algorithms, 26 (2001), pp. 251–280.

552. L. SCHLESSINGER AND C. SCHWARTZ, *Analyticity as a useful computational tool*, Phys. Rev. Lett., 16 (1966), pp. 1173–1174.

553. E. SCHMIDT, *Entwickelung willkürlicher Funktionen nach Systemen vorgeschriebener*, PhD thesis, Georg-August-Universität, Göttingen, 1905.

554. ———, *Zur Theorie der linearen und nichtlinearen Integralgleichungen. I. Teil: Entwicklung willkürlicher Funktionen nach Systemen vorgeschriebener*, Math. Ann., 63 (1907), pp. 433–476.

555. ———, *Antrittsrede*, Sitzungsber. Preuß. Akad. Wiss. Berlin, (1919), pp. 564–566.

556. M. SCHULTZ, ed., *Numerical Algorithms for Modern Parallel Computer Architectures*, vol. 13 of The IMA Volumes in Mathematics and Its Applications, Springer-Verlag, New York, 1988.

557. D. S. SCOTT, *Analysis of the symmetric Lanczos algorithm*, PhD thesis, University of California, Berkeley, 1978.

558. ———, *How to make the Lanczos algorithm converge slowly*, Math. Comp., 33 (1979), pp. 239–247.

559. ———, *The Lanczos algorithm*, in Sparse Matrices and Their Use, I. S. Duff, ed., The Institute of Mathematics and Its Applications Conference Series, Academic Press, New York, 1981, pp. 139–159.

560. T. SHEIL-SMALL, *Complex Polynomials*, vol. 75 of Cambridge Studies in Advanced Mathematics, Cambridge University Press, Cambridge, 2002.

561. J. A. SHOHAT AND J. D. TAMARKIN, *The Problem of Moments*, vol. I of American Mathematical Society Mathematical Surveys, American Mathematical Society, New York, 1943.

562. H. J. SILVERSTONE, M.-L. YIN, AND R. L. SOMORJAI, *Energy calculation by the method of local moments*, The Journal of Chemical Physics, 47 (1967), pp. 4824–4827.

563. D. J. SILVESTER AND V. SIMONCINI, *An optimal iterative solver for symmetric indefinite systems stemming from mixed approximation*, ACM Trans. Math. Software, 37 (2011), pp. 42:1–42:22.

564. B. SIMON, *CMV matrices: five years after*, J. Comput. Appl. Math., 208 (2007), pp. 120–154.

565. H. D. SIMON, *The Lanczos algorithm for solving symmetric linear systems*, PhD thesis, University of California, Berkeley, 1982.

566. ———, *Analysis of the symmetric Lanczos algorithm with reorthogonalization methods*, Linear Algebra Appl., 61 (1984), pp. 101–131.

567. ———, *The Lanczos algorithm with partial reorthogonalization*, Math. Comp., 42 (1984), pp. 115–142.

568. V. SIMONCINI, *On a non-stagnation condition for GMRES and application to saddle point matrices*, Electron. Trans. Numer. Anal., 37 (2010), pp. 202–213.

569. V. SIMONCINI AND D. B. SZYLD, *Theory of inexact Krylov subspace methods and applications to scientific computing*, SIAM J. Sci. Comput., 25 (2003), pp. 454–477.

570. ———, *Recent computational developments in Krylov subspace methods for linear systems*, Numer. Linear Algebra Appl., 14 (2007), pp. 1–59.

571. ———, *New conditions for non-stagnation of minimal residual methods*, Numer. Math., 109 (2008), pp. 477–487.

572. G. L. G. SLEIJPEN AND H. A. VAN DER VORST, *A Jacobi–Davidson iteration method for linear eigenvalue problems*, SIAM Rev., 42 (2000), pp. 267–293.

573. G. L. G. SLEIJPEN, H. A. VAN DER VORST, AND D. R. FOKKEMA, *BiCGstab(l) and other hybrid Bi-CG methods*, Numer. Algorithms, 7 (1994), pp. 75–109.

574. G. L. G. SLEIJPEN, H. A. VAN DER VORST, AND J. MODERSITZKI, *Differences in the effects of rounding errors in Krylov solvers for symmetric indefinite linear systems*, SIAM J. Matrix Anal. Appl., 22 (2000), pp. 726–751.

575. S. SMALE, *Complexity theory and numerical analysis*, Acta Numer., 6 (1997), pp. 523–551.

576. P. SONNEVELD, *CGS, a fast Lanczos-type solver for nonsymmetric linear systems*, SIAM J. Sci. Statist. Comput., 10 (1989), pp. 36–52.

577. D. A. SPIELMAN AND J. WOO, *A note on preconditioning by low-stretch spanning trees*, Technical Report, arXiv:0903.2816v1, 2009.

578. G. STARKE, *Field-of-values analysis of preconditioned iterative methods for nonsymmetric elliptic problems*, Numer. Math., 78 (1997), pp. 103–117.

579. L. A. STEEN, *Highlights in the history of spectral theory*, Amer. Math. Monthly, 80 (1973), pp. 359–381.

580. E. STEIN, ed., *Error-Controlled Adaptive Finite Elements in Solid Mechanics*, John Wiley & Sons Ltd., Chichester, 2003.

581. E. L. STIEFEL, *Kernel polynomials in linear algebra and their numerical applications*, Nat. Bur. Standards Appl. Math. Ser., 49 (1958), pp. 1–22.

582. T. J. STIELTJES, *Sur certaines inégalités dues à M. P. Tchebychef*. Published posthumously in Oeuvres II (P. Noordhoff, Groningen, 1918), pp. 586–593.

583. ———, *Sur l'évaluation approchée des intégrales*, C. R. Acad. Sci. Paris, 97 (1883), pp. 740–742, 798–799. Reprinted in Oeuvres I (P. Noordhoff, Groningen, 1914), pp. 314–316, 317–318.

584. ———, *Note sur quelques formules pour l'évaluation de certaines intégrales*, Bul. Astr. Paris, 1 (1884), p. 568. Reprinted in Oeuvres I (P. Noordhoff, Groningen, 1914), pp. 426–427.

585. ———, *Quelques recherches sur la théorie des quadratures dites mécaniques*, Ann. Sci. École Norm. Sup. (3), 1 (1884), pp. 409–426. Reprinted in Oeuvres I (P. Noordhoff, Groningen, 1914), pp. 377–394.

586. ———, *Sur une généralisation de la théorie des quadratures mécaniques*, C. R. Acad. Sci. Paris, 99 (1884), pp. 850–851. Reprinted in Oeuvres I (P. Noordhoff, Groningen, 1914), pp. 428–429.

587. ———, *Note à l'occasion de la réclamation de M. Markoff*, Ann. Sci. École Norm. Sup., Serie 3, 2 (1885), pp. 183–184. Reprinted in Oeuvres I (P. Noordhoff, Groningen, 1914), pp. 430–431.

588. ———, *Recherches sur les fractions continues*, Ann. Fac. Sci. Toulouse Sci. Math. Sci. Phys., 8 (1894), pp. J. 1–122. Reprinted in Oeuvres II (P. Noordhoff, Groningen, 1918), pp. 402–566. English translation *Investigations on continued fractions* in

Thomas Jan Stieltjes, Collected Papers, Vol. II (Springer-Verlag, Berlin, 1993), pp. 609–745.

589. J. STOER, *Solution of large linear systems of equations by conjugate gradient type methods*, in Mathematical programming: the state of the art (Bonn, 1982), Springer, Berlin, 1983, pp. 540–565.

590. J. STOER AND R. FREUND, *On the solution of large indefinite systems of linear equations by conjugate gradient algorithms*, in Computing Methods in Applied Sciences and Engineering, V (Versailles, 1981), North-Holland, Amsterdam, 1982, pp. 35–53.

591. M. STOLL, *Solving linear systems using the adjoint*, PhD thesis, Oxford University, Oxford, 2009.

592. M. H. STONE, *Linear Transformations in Hilbert Space*, vol. 15 of American Mathematical Society Colloquium Publications, American Mathematical Society, Providence, RI, 1990. Reprint of the 1932 original.

593. Z. STRAKOŠ, *Effectivity and optimizing of algorithms and programs on the host-computer/array-processor system*, Parallel Comp., 4 (1987), pp. 189–207.

594. ———, *On the real convergence rate of the conjugate gradient method*, Linear Algebra Appl., 154–156 (1991), pp. 535–549.

595. ———, *Convergence and numerical behaviour of the Krylov space methods*, in Algorithms for large scale linear algebraic systems (Gran Canaria, 1996), vol. 508 of NATO Adv. Sci. Inst. Ser. C Math. Phys. Sci., Kluwer Acad. Publ., Dordrecht, 1998, pp. 175–196.

596. Z. STRAKOŠ AND A. GREENBAUM, *Open questions in the convergence analysis of the Lanczos process for the real symmetric eigenvalue problem*, IMA Preprint Series 934, Institute of Mathematics and Its Application (IMA), University of Minnesota, 1992.

597. Z. STRAKOŠ AND J. LIESEN, *On numerical stability in large scale linear algebraic computations*, ZAMM Z. Angew. Math. Mech., 85 (2005), pp. 307–325.

598. Z. STRAKOŠ AND P. TICHÝ, *On error estimation in the conjugate gradient method and why it works in finite precision computations*, Electron. Trans. Numer. Anal., 13 (2002), pp. 56–80.

599. ———, *Error estimation in preconditioned conjugate gradients*, BIT, 45 (2005), pp. 789–817.

600. ———, *On efficient numerical approximation of the bilinear form $c^*A^{-1}b$*, SIAM J. Sci. Comput., 33 (2011), pp. 565–587.

601. E. STUDY, *Einleitung in die Theorie der Invarianten linearer Transformationen auf Grund der Vektorenrechnung*, Friedr. Vieweg & Sohn Akt.-Ges., Braunschweig, 1923.

602. M. STYNES, *Steady-state convection-diffusion problems*, Acta Numer., 14 (2005), pp. 445–508.

603. O. SZÁSZ, *Az Hadamard-féle determinánstétel egy elemi bebizonyítása*, Mathematikai és Physikai Lapok, 19 (1910), pp. 221–227.

604. ———, *A végtelen determinánsok elméletéhez*, Mathematikai és Physikai Lapok, 21 (1912), pp. 223–295.

605. ———, *Ein elementarer Beweis des Hadamardschen Determinantensatzes*, in Mathematische und Naturwissenschaftliche Berichte aus Ungarn, R. Baron Eötvös and J. König, eds., vol. 27 (1909), B. G. Teubner, Leipzig, 1913, pp. 172–180.

606. O. SZÁSZ, *Collected Mathematical Papers*, Department of Mathematics, University of Cincinnati, Cincinnati, Ohio, 1955. Edited by H. D. Lipsich.
607. G. SZEGÖ, *Orthogonal Polynomials*, vol. XXIII of American Mathematical Society Colloquium Publications, American Mathematical Society, New York, 1939.
608. D. B. SZYLD AND O. B. WIDLUND, *Variational analysis of some conjugate gradient methods*, East-West J. Numer. Math., 1 (1993), pp. 51–74.
609. S.-C. T. CHOI, C. C. PAIGE, AND M. A. SAUNDERS, *MINRES-QLP: a Krylov subspace method for indefinite or singular symmetric systems*, SIAM J. Sci. Comput., 33 (2011), pp. 1810–1836.
610. D. R. TAYLOR, *Analysis of the look ahead Lanczos algorithm*, PhD thesis, University of California, Berkeley, 1982.
611. H. J. J. TE RIELE, T. J. DEKKER, AND H. A. VAN DER VORST, eds., *Algorithms and Applications on Vector and Parallel Computers*, Special Topics in Supercomputing, North Holland, Amsterdam, 1987.
612. P. TICHÝ, J. LIESEN, AND V. FABER, *On worst-case GMRES, ideal GMRES, and the polynomial numerical hull of a Jordan block*, Electron. Trans. Numer. Anal., 26 (2007), pp. 453–473.
613. J. TODD, *Experiments on the inversion of a 16×16 matrix*, in Simultaneous Linear Equations and the Determination of Eigenvalues, U.S. Government Printing Office, Washington, D.C., 1953, pp. 59–64. Proceedings of a symposium held August 23–25, 1951, in Los Angeles, California, USA, under the sponsorship of the National Bureau of Standards, in cooperation with the Office of Naval Research.
614. ———, *Numerical analysis at the National Bureau of Standards*, SIAM Rev., 17 (1975), pp. 361–370. Presented on October 17, 1972, at the SIGNUM-SIAM Panel on the 25th Aniversary of Modern Numerical Mathematics.
615. O. TOEPLITZ, *Die Jacobische Transformation der quadratischen Formen von unendlichvielen Veränderlichen*, Nachr. Ges. Wiss. Göttingen, Math.-phys. Kl., (1907), pp. 101–109.
616. ———, *Zur Theorie der quadratischen Formen von unendlichvielen Veränderlichen*, Nachr. Ges. Wiss. Göttingen, Math.-phys. Kl., (1910), pp. 489–506.
617. ———, *Das algebraische Analogon zu einem Satze von Fejér*, Math. Z., 2 (1918), pp. 187–197.
618. K.-C. TOH, *GMRES vs. ideal GMRES*, SIAM J. Matrix Anal. Appl., 18 (1997), pp. 30–36.
619. K. C. TOH, M. J. TODD, AND R. H. TÜTÜNCÜ, *SDPT3—a MATLAB software package for semidefinite programming, version 1.3*, Optim. Methods Softw., 11/12 (1999), pp. 545–581.
620. S. TOMOV, J. DONGARRA, AND M. BABOULIN, *Towards dense linear algebra for hybrid GPU accelerated manycore systems*, Parallel Comput., 36 (2010), pp. 232–240.
621. L. N. TREFETHEN, *Approximation theory and numerical linear algebra*, in Algorithms for Approximation, II (Shrivenham, 1988), Chapman and Hall, London, 1990, pp. 336–360.
622. ———, *The definition of numerical analysis*, SIAM News, 25 (1992).
623. ———, *Pseudospectra of linear operators*, SIAM Rev., 39 (1997), pp. 383–406.
624. ———, *Numerical analysis*, in Princeton Companion to Mathematics, Princeton University Press, Princeton, NJ, 2008.

625. L. N. TREFETHEN AND D. BAU, III, *Numerical Linear Algebra*, SIAM, Philadelphia, PA, 1997.
626. L. N. TREFETHEN AND M. EMBREE, *Spectra and Pseudospectra. The Behavior of Nonnormal Matrices and Operators*, Princeton University Press, Princeton, NJ, 2005.
627. A. M. TURING, *Rounding-off errors in matrix processes*, Quart. J. Mech. Appl. Math., 1 (1948), pp. 287–308.
628. E. E. TYRTYSHNIKOV, *How bad are Hankel matrices?*, Numer. Math., 67 (1994), pp. 261–269.
629. ———, *A Brief Introduction to Numerical Analysis*, Birkhäuser Boston Inc., Boston, MA, 1997.
630. G. VALENT AND W. VAN ASSCHE, *The impact of Stieltjes' work on continued fractions and orthogonal polynomials: additional material*, J. Comput. Appl. Math., 65 (1995), pp. 419–447.
631. W. VAN ASSCHE, *The impact of Stieltjes' work on continued fractions and orthogonal polynomials*, in Thomas Jan Stieltjes, Collected Papers, Vol. I, G. van Dijk, ed., Springer-Verlag, Berlin, 1993, pp. 5–37.
632. J. VAN DEN ESHOF AND G. L. G. SLEIJPEN, *Inexact Krylov subspace methods for linear systems*, SIAM J. Matrix Anal. Appl., 26 (2004), pp. 125–153.
633. A. VAN DER SLUIS AND H. A. VAN DER VORST, *The rate of convergence of conjugate gradients*, Numer. Math., 48 (1986), pp. 543–560.
634. ———, *The convergence behavior of Ritz values in the presence of close eigenvalues*, Linear Algebra Appl., 88/89 (1987), pp. 651–694.
635. H. A. VAN DER VORST, *Krylov subspace iteration*, Comput. Sci. Eng., 2 (2000), pp. 32–37.
636. ———, *Iterative Krylov Methods for Large Linear Systems*, vol. 13 of Cambridge Monographs on Applied and Computational Mathematics, Cambridge University Press, Cambridge, 2003.
637. S. VAN HUFFEL AND J. VANDEWALLE, *The Total Least Squares Problem. Computational Aspects and Analysis*, vol. 9 of Frontiers in Applied Mathematics, Society for Industrial and Applied Mathematics (SIAM), Philadelphia, PA, 1991. With a foreword by Gene H. Golub.
638. R. S. VARGA, *A comparison of the successive overrelaxation method and semi-iterative methods using Chebyshev polynomials*, J. Soc. Indust. Appl. Math., 5 (1957), pp. 39–46.
639. ———, *Matrix Iterative Analysis*, Prentice-Hall Inc., Englewood Cliffs, NJ, 1962.
640. ———, *Matrix Iterative Analysis*, vol. 27 of Springer Series in Computational Mathematics, Springer-Verlag, Berlin, expanded ed., 2000.
641. C. D. VILLEMAGNE AND R. E. SKELTON, *Model reduction using a projection formulation*, Int. J. Control, 46 (1987), pp. 2141–2169.
642. P. K. W. VINSOME, *Orthomin, an iterative method for solving sparse sets of simultaneous linear equations*, in Proceedings of the Fourth Symposium on Reservoir Simulation, Society of Petroleum Engineers of AIME, Los Angeles, 1976, pp. 149–159.
643. V. V. VOEVODIN, *On methods of conjugate directions*, U.S.S.R. Comput. Maths. Phys., 19 (1979), pp. 228–233.
644. ———, *The question of non-self-adjoint extension of the conjugate gradients method is closed*, U.S.S.R. Comput. Maths. Phys., 23 (1983), pp. 143–144.

645. V. V. VOEVODIN AND E. E. TYRTYSHNIKOV, *On generalization of conjugate direction methods*, in Numerical Methods of Algebra (Chislennye Metody Algebry), Moscow State University Press, Moscow, 1981, pp. 3-9. (English translation provided by E. E. Tyrtyshnikov).
646. M. VOHRALÍK, *A Posteriori Error Estimates, Stopping Criteria, and Inexpensive Implementations for Error Control and Efficiency in Numerical Simulations*, Habilitation thesis, Université Pierre et Marie Curie, 2010.
647. J. VON NEUMANN, *Mathematische Begründung der Quantenmechanik*, Nachr. Ges. Wiss. Göttingen, Math.-phys. Kl., (1927), pp. 1-57.
648. ———, *Mathematical Foundations of Quantum Mechanics*, Princeton Landmarks in Mathematics, Princeton University Press, Princeton, NJ, 1996. Translated from the 1932 German original and with a preface by Robert T. Beyer.
649. J. VON NEUMANN AND H. H. GOLDSTINE, *Numerical inverting of matrices of high order*, Bull. Amer. Math. Soc., 53 (1947), pp. 1021-1099.
650. Y. V. VOROBYEV, *Operator orthogonal polynomials and approximate methods of determination of the spectrum of linear bounded operators*, Uspehi Matem. Nauk (N.S.), 9 (1954), pp. 83-90. In Russian.
651. ———, *Methods of Moments in Applied Mathematics*, Gordon and Breach Science Publishers, New York, 1965. Translated from the Russian by Bernard Seckler.
652. A. VUCINICH, *Science in Russian Culture 1861-1917*, Stanford University Press, Stanford, CA, 1970.
653. C. VUIK, A. SEGAL, AND J. A. MEIJERINK, *An efficient preconditioned CG method for the solution of a class of layered problems with extreme contrasts in the coefficients*, J. Comput. Phys, 152 (1999), pp. 385-403.
654. H. F. WALKER, *Implementation of the GMRES method using Householder transformations*, SIAM J. Sci. Statist. Comput., 9 (1988), pp. 152-163.
655. ———, *Implementations of the GMRES method*, Comput. Phys. Comm., 53 (1989), pp. 311-320.
656. H. F. WALKER AND L. ZHOU, *A simpler GMRES*, Numer. Linear Algebra Appl., 1 (1994), pp. 571-581.
657. H. S. WALL, *Analytic Theory of Continued Fractions*, AMS Chelsea Publishing, Providence, RI, 2000. Reprint of the 1948 original published by D. Van Nostrand Co.
658. K. F. WARNICK AND W. C. CHEW, *Numerical simulation methods for rough surface scattering*, Waves Random Media, 11 (2001), pp. R1-R30.
659. A. WATHEN AND D. SILVESTER, *Fast iterative solution of stabilised Stokes systems. I. Using simple diagonal preconditioners*, SIAM J. Numer. Anal., 30 (1993), pp. 630-649.
660. D. S. WATKINS, *Some perspectives on the eigenvalue problem*, SIAM Rev., 35 (1993), pp. 430-471.
661. ———, *Unitary orthogonalization processes*, J. Comput. Appl. Math., 86 (1997), pp. 335-345.
662. ———, *Fundamentals of Matrix Computations*, Pure and Applied Mathematics, John Wiley & Sons, Inc., New York, third ed., 2010.
663. K. WEIERSTRASS, *Zur Theorie der bilinearen und quadratischen Formen*, Monatsberichte der Königlichen Preußischen Akademie der Wissenschaften zu Berlin, (1868), pp. 311-338. Reprinted in Mathematische Werke 2 (Mayer & Müller, Berlin, 1895), pp. 19-44.

664. ———, *Über die analytische Darstellbarkeit sogenannter willkürlicher Funktionen einer reellen Veränderlichen*, Sitzungsberichte der Königlichen Preußischen Akademie der Wissenschaften zu Berlin, (1885), pp. 633–639; 789–805. Reprinted (with some changes) in Mathematische Werke 3 (Mayer & Müller, Berlin, 1903), pp. 1–37.
665. B. WENDROFF, *On orthogonal polynomials*, Proc. Amer. Math. Soc., 12 (1961), pp. 554–555.
666. H. WEYL, *David Hilbert and his mathematical work*, Bull. Amer. Math. Soc., 50 (1944), pp. 612–654.
667. H. S. WILF, *Mathematics for the Physical Sciences*, Dover Publications Inc., New York, 1978. Reprint of the 1962 original.
668. J. H. WILKINSON, *Error analysis of direct methods of matrix inversion*, J. Assoc. Comput. Mach., 8 (1961), pp. 281–330.
669. ———, *Modern error analysis*, SIAM Rev., 13 (1971), pp. 548–568.
670. ———, *The Algebraic Eigenvalue Problem*, Monographs on Numerical Analysis, The Clarendon Press Oxford University Press, New York, 1988. 1st ed. published 1963.
671. ———, *Rounding Errors in Algebraic Processes*, Dover Publications Inc., New York, 1994. Reprint of the 1963 original published by Prentice-Hall, Englewood Cliffs, NJ.
672. J. H. WILKINSON AND C. REINSCH, *Handbook for Automatic Computation. Vol. II. Linear Algebra*, vol. 186 of Grundlehren der Mathematischen Wissenschaften, Springer-Verlag, New York, 1971.
673. K. WILLCOX AND A. MEGRETSKI, *Fourier series for accurate, stable, reduced-order models in large-scale linear applications*, SIAM J. Sci. Comput., 26 (2005), pp. 944–962.
674. A. WINTNER, *Spektraltheorie der unendlichen Matrizen. Einführung in den analytischen Apparat der Quantenmechanik*, S. Hirzel, Leipzig, 1929.
675. B. I. WOHLMUTH AND R. H. W. HOPPE, *A comparison of a posteriori error estimators for mixed finite element discretizations by Raviart-Thomas elements*, Math. Comp., 68 (1999), pp. 1347–1378.
676. H. WOŹNIAKOWSKI, *Round-off error analysis of iterations for large linear systems*, Numer. Math., 30 (1978), pp. 301–314.
677. ———, *Roundoff-error analysis of a new class of conjugate-gradient algorithms*, Linear Algebra Appl., 29 (1980), pp. 507–529.
678. W. WÜLLING, *On stabilization and convergence of clustered Ritz values in the Lanczos method*, SIAM J. Matrix Anal. Appl., 27 (2005), pp. 891–908.
679. ———, *The stabilization of weights in the Lanczos and conjugate gradient method*, BIT, 45 (2005), pp. 395–414.
680. D. YOUNG, *Iterative methods for solving partial difference equations of elliptic type*, Trans. Amer. Math. Soc., 76 (1954), pp. 92–111.
681. ———, *On Richardson's method for solving linear systems with positive definite matrices*, J. Math. Physics, 32 (1954), pp. 243–255.
682. ———, *A historical overview of iterative methods*, Comput. Phys. Commun., 53 (1989), pp. 1–17.
683. D. M. YOUNG, *Iterative methods for solving partial difference equations of elliptic type*, PhD thesis, Harvard University, Cambridge, MA, 1950.
684. ———, *Iterative Solution of Large Linear Systems*, Academic Press, New York, 1971.

685. D. M. YOUNG AND K. C. JEA, *Generalized conjugate-gradient acceleration of non-symmetrizable iterative methods*, Linear Algebra Appl., 34 (1980), pp. 159–194.
686. I. ZAVORIN, *Analysis of GMRES convergence by spectral factorization of the Krylov matrix*, PhD thesis, University of Maryland, College Park, 2001.
687. I. ZAVORIN, D. P. O'LEARY, AND H. ELMAN, *Complete stagnation of GMRES*, Linear Algebra Appl., 367 (2003), pp. 165–183.
688. E. ZEIDLER, ed., *Oxford Users' Guide to Mathematics*, Oxford University Press, Oxford, 2004. Translated from the 1996 German original by Bruce Hunt.
689. J.-P. M. ZEMKE, *Krylov subspace methods in finite precision: A unified approach*, PhD thesis, Arbeitsbereich Technische Informatik, Technische Universität Hamburg-Harburg, 2003.
690. ———, *Abstract perturbed Krylov methods*, Linear Algebra Appl., 424 (2007), pp. 405–434.
691. H. ZHA AND Z. ZHANG, *The Arnoldi process, short recursions, and displacement ranks*, Linear Algebra Appl., 249 (1996), pp. 169–188.

Index of Historical Personalities

Bauer, Friedrich Ludwig (born 10.06.1924), German computer scientist, 67, 146, 341
Borchardt, Karl Wilhelm (22.02.1817–27.06.1880), German mathematician, 132
Brouncker, William (1620–05.04.1684), Irish mathematician, 107
Cauchy, Augustin Louis (21.08.1789–23.05.1857), French mathematician, viii, 93, 132, 188
Cayley, Arthur (16.08.1821–26.01.1895), English mathematician, 189
Chebyshev, Pafnuty Lvovich (16.05.1821–08.12.1894), Russian mathematician, viii, 34, 65, 96, 100, 101, 104, 108, 133, 145, 146, 160, 163
Christoffel, Elwin Bruno (10.11.1829–15.03.1900), German mathematician and physicist, viii, 8, 88, 101, 105, 160, 163
Cotes, Roger (10.07.1682–05.06.1716), English mathematician, 87
Darboux, Jean Gaston (14.08.1842–23.02.1917), French mathematician, 105
Dirichlet, Johann Peter Gustav Lejeune (13.02.1805–05.05.1859), German mathematician, 131
Einstein, Albert (14.03.1879–18.04.1955), German physicist, vi, 228, 240, 250, 346
Encke, Johann Franz (23.09.1791–26.08.1865), German astronomer, 87
Euler, Leonhard (15.04.1707–18.09.1783), Swiss mathematician and physicist, 65, 96, 104, 163
Fejér, Leopold (09.02.1880–15.10.1959), Hungarian mathematician, 35
Frobenius, Ferdinand Georg (26.10.1849–03.08.1917), German mathematician, 189
Galle, Johann Gottfried (09.06.1812–10.07.1910), German astronomer, 88
Gantmacher, Felix Rumvimovich (23.02.1908–16.05.1964), Russian mathematician, viii, 19, 34, 66
Gauss, Johann Carl Friedrich (30.04.1777–23.02.1855), German mathematician, astronomer, and physicist, viii, 8, 79, 87, 108, 160, 308
Goldstine, Herman Heine (13.09.1913–16.06.2004), American mathematician and computer scientist, 315
Gram, Jørgen Pedersen (27.06.1850–29.04.1916), Danish actuary and mathematician, 34
Hamilton, Sir William Rowan (04.08.1805–02.09.1865), Irish physicist, astronomer, and mathematician, 189
Heine, Heinrich Eduard (15.03.1821–21.10.1881), German mathematician, 34, 88, 105, 133, 160
Heisenberg, Werner Karl (05.12.1901–01.02.1976), German physicist, 136
Hellinger, Ernst David (30.09.1883–28.03.1950), German mathematician, 134

Hestenes, Magnus Rudolph (13.02.1906–31.05.1991), American mathematician, vii, 7, 36, 66, 69, 96, 138, 146, 160, 161, 169, 241, 309, 325, 331
Hilbert, David (23.01.1862–14.02.1943), German mathematician, viii, 35, 134, 140
Householder, Alston Scott (05.05.1904–04.07.1993), American mathematician, 67, 146, 238
Jacobi, Carl Gustav Jacob (10.12.1804–18.02.1851), German mathematician, viii, 8, 87, 107, 108, 130, 160
Jordan, Marie Ennemond Camille (05.01.1838–22.01.1922), French mathematician, 188
Korkin, Aleksandr Nikolaevich (15.03.1837–01.09.1908), Russian mathematician, 65
Kowalewski, Arnold Christian Felix (27.11.1873–1945), German philosopher, 87
Krein, Mark Grigorievich (03.04.1907–17.10.1989), Russian mathematician, 20
Kronecker, Leopold (07.12.1823–29.12.1891), German mathematician, 189
Krylov, Alexei Nikolaevich (15.08.1863–26.10.1945), Russian naval engineer and applied mathematician, viii, 19, 64, 161, 184
Lanczos, Cornelius (02.02.1893–25.06.1974), Hungarian mathematician and physicist, vi, vii, 7, 29, 31, 35, 66, 69, 146, 160, 161, 185, 228, 240, 241, 250, 255, 309, 325, 331, 347
Laplace, Pierre-Simon (23.03.1749–05.03.1827), French mathematician, physicist, and astronomer, 33
Liouville, Joseph (24.03.1809–08.09.1882), French mathematician, 34
Markov, Andrey Andreyevich (14.06.1856–20.07.1922), Russian mathematician, viii, 76, 84, 104, 146, 160, 163
Newton, Sir Isaac (04.01.1643–31.03.1727), English physicist, mathematician, astronomer, and philosopher, 65, 87
Painvin, Louis Félix (1826–1875), French mathematician, 133
Riesz, Frigyes (22.01.1880–28.02.1956), Hungarian mathematician, 135
Ritz, Walther (22.02.1878–07.07.1909), Swiss physicist, 42, 49
Rutishauser, Heinz (30.01.1918–10.11.1970), Swiss mathematician and computer scientist, 41, 117, 138, 243, 251, 254, 255, 279
Schmidt, Erhard (13.01.1876–06.12.1959), German mathematician, 33
Schrödinger, Erwin Rudolf Josef Alexander (12.08.1887–04.01.1961), Austrian physicist, 136
Stiefel, Eduard Ludwig (21.04.1909–25.11.1978), Swiss mathematician, vii, 7, 36, 51, 66, 69, 96, 138, 146, 160, 161, 169, 241, 309, 325, 331
Stieltjes, Thomas Joannes (29.12.1856–31.12.1894), Dutch mathematician, viii, 9, 73, 86, 96, 101, 104, 135, 143, 146, 160, 163, 307
Sturm, Jacques Charles François (29.09.1803–15.12.1855), French mathematician, 34
Szegö, Gábor (20.01.1885–07.08.1985), Hungarian mathematician, 108, 214
Szász, Otto (11.12.1884–19.12.1952), Hungarian mathematician, 35
Toeplitz, Otto (01.08.1881–15.02.1940), German mathematician, 35, 134, 296
Turing, Alan Mathison (23.06.1912–07.06.1954), English mathematician and computer scientist, 315
von Neumann, John (28.12.1903–08.02.1957), Hungarian mathematician and physicist, viii, 136, 140, 255, 315
Vorobyev, Yuri Vasilevich (20th century), Russian mathematician, 36, 68, 72, 145–147, 149, 152, 154, 156, 159, 160, 185, 245, 267
Wallis, John (23.11.1616–28.10.1703), English mathematician, 107

Weierstrass, Karl Theodor Wilhelm (31.10.1815–19.02.1897), German mathematician, 86, 188

Weyl, Hermann Klaus Hugo (09.11.1885–08.12.1955), German mathematician, 136

Wilkinson, James Hardy (27.09.1919–05.10.1986), English mathematician and computer scientist, 33, 68, 315

Index of Technical Terms

$(s+2, t)$-term recurrence, 216
B-adjoint of a matrix, 204
B-inner product and B-norm, 10
B-normal(s) matrix, 205
B-orthogonality, 11
ϵ-pseudospectrum, 295

algebraic error, 46
Arnoldi algorithm, 26
   classical Gram–Schmidt, 27
   isometric, 214
   linear operator version, 190
   modified Gram–Schmidt, 28
asymptotic convergence factor, 307

backward error analysis, 230, 319, 341, 343, 345
   componentwise backward error, 233
   history, 315
   normwise backward error, 231
band Hessenberg matrix, 191

Cayley–Hamilton Theorem, 176, 189
CG method, 24
   algorithm, 41
   bounds for the A-norm of the error, 266, 267, 269, 276, 326
   clustered eigenvalues, 282
   condition number convergence bound, 267
   delay of convergence, 320, 330
   expressions for the A-norm of the error, 141, 261–264, 279, 326, 332
   finite precision behaviour, 323
   history, 66, 241, 254

influence of the initial error, 272, 325
iteration polynomial, 260
maximal attainable accuracy, 330
minimisation of quadratic functional, 49
moment matching properties, 137
orthogonality properties, 40, 263
outlying eigenvalues, 275
relation to Galerkin finite element method, 47
relation to Gauss–Christoffel quadrature, 139, 141, 283
Stiefel's tactic, 51
strictly monotonically decreasing error norm, 54, 261, 264
superlinear convergence, 270, 274, 278
three-term variant, 41
worst-case behaviour, 266, 271
Chebyshev polynomial, 252
Chebyshev semi-iterative method, 253, 255
Cholesky factorisation, 33, 144
Christoffel numbers, 85
communication-avoiding algorithms, 246
complexity theory, 248
composite polynomial, 275, 327
computational cost, 245, 248
conjugate basis, 38
conjugate gradient like descent method, 170
constraints space, 13
continued fraction, 97
convection–diffusion model problem, 303, 342
cyclic decomposition theorem, 183

Index of Technical Terms

cyclic subspace, 174
   Jordan basis, 186

degree of a polynomial, 174
delay of convergence, 317, 320
discretisation error, 45
distribution function, 73
divided difference, 77, 84
division theorem for polynomials, 82, 111, 174

element growth factor, 311
energy norm, 44
error vector, 13
Euclidean inner product, 13
exponent of matrix multiplication, 249

Faber–Manteuffel Theorem, 172, 205
field of values, 296
forward error bound, 319

Galerkin finite element method, 43, 229
   algebraic error, 46, 229
   discretisation error, 45, 229
   energy norm, 44
   finite-dimensional weak formulation, 43
   orthogonality property, 45, 234
   weak formulation, 42
Gauss–Christoffel quadrature, 82
   Arnoldi generalisation, 154
   convergence, 85
   error term, 84
   history, 87
   most compact version, 112
   non-Hermitian Lanczos generalisation, 154
   sensitivity, 283
generalised Lanczos algorithm, 220
GMRES method, 24
   algorithm, 60
   and non-normality, 292, 294, 303
   asymptotic convergence analysis, 306
   backward stability, 340, 345
   choice of the basis, 338
   complete stagnation, 298
   convergence bounds, 290, 291, 294–296
   convergence for convection–diffusion model problem, 305
   eigenvalues and convergence, 300, 303
   exact stagnation, 287
   finite precision behaviour, 338
   Householder Arnoldi implementation, 340
   ideal GMRES, 294, 303, 306
   iteration polynomial, 286
   modified Gram–Schmidt (MGS) implementation, 341
   nonzero initial approximation, 318
   roots of iteration polynomials, 287, 289
   Simpler GMRES, 339
   worst-case GMRES, 294
grade of a vector, 19, 173
Gram–Schmidt method, 26
   classical (CGS), 313
   for normalised orthogonal polynomials, 89
   history, 33
   modified (MGS), 313

Hankel matrix, 142
harmonic polynomial, 209
harmonic rational function, 218
Hegedüs trick, 318
Hermitian Lanczos algorithm, 29
Hessenberg matrix, 28

interlacing property, 92
interpolatory quadrature, 77
invariant subspace of an operator, 173
irreducible polynomial, 177
isometric Arnoldi algorithm, 214

Jacobi matrix, 30, 90, 108
   definition, 30
   history, 130
   LDL^T factorisation, 37
   LQ factorisation, 55
   persistence theorem, 123
Jordan canonical form, 187

kernel of an operator, 173
Krylov sequence, 19, 67
   maximal length, 188
Krylov subspace, 20, 69

Krylov subspace method, 22
  BiCG, 61, 288
  CG, *see* CG method
  FOM, 61, 288
  GCR, 57, 338
  GMRES, *see* GMRES method
  MINRES, 24, 57
  ORTHODIR, 57, 338
  ORTHORES, 61
  QMR, 62, 288
  SYMMBK, 54
  SYMMLQ, 24, 54
Krylov's method, 66, 184

Lanczos algorithm, 26
  generalised, 220
  Hermitian, 29, 135
  look-ahead, 33
  lucky breakdown, 32
  non-Hermitian, 31
  serious breakdown, 33
LDL^T factorisation, 37
LQ factorisation, 55
LU factorisation, 62, 249, 311

Markov parameters, 163
matrix function, 225
matrix of moments, 142
maximal attainable accuracy, 317, 328
McMillan degree, 218
minimal partial realisation, 162
minimal polynomial
  of a vector, 19, 173
  of an operator, 175
MINRES method, 24, 57
moment, 73
moment problem, 73
  relation to quadrature, 76
  simplified Stieltjes, 74
  solution via Gauss–Christoffel
    quadrature, 85
  Stieltjes' formulation, 73
  Vorobyev's formulation and Arnoldi, 152
  Vorobyev's formulation and
    non-Hermitian Lanczos, 149
  Vorobyev's original formulation, 147

Newton–Lagrange interpolation
  polynomial, 77, 196
non-Hermitian Lanczos algorithm, 31
nonderogatory operator, 176
normal(s) operator, 194–197
normwise backward error, 231

optimal $(s+2)$-term recurrence, 192
orthogonal polynomials, 79
  interlacing property, 92
  location of roots, 81
  norm minimisation property, 80
  normalised, 89
  Stieltjes recurrence, 89
  three-term recurrence, 89

parallel computations, 245
Petrov–Galerkin method, 14
point of increase, 74
Poisson model problem, 42, 232, 331
polyanalytic polynomial, 220
polynomial numerical hull, 296
preconditioning, 269, 345
projection process, 13
  and model reduction, 71
  constraints space, 13
  error vector, 13
  finite termination property, 19
  projected matrix, 15
  projected system, 15
  residual vector, 13
  search space, 13
  well-defined, 15

QR factorisation, 33, 35, 58, 63, 212, 249, 312, 338
quadrature, 75
  n-node interpolatory, 78
  n-node mechanical, 75
  algebraic degree of exactness, 75
  Gauss–Christoffel, *see*
    Gauss–Christoffel quadrature

range of an operator, 173
Rayleigh–Ritz method, 49
reducibility to band Hessenberg form, 199
residual vector, 13
resolution of unity, 140

Richardson iteration, 252
Riemann–Stieltjes integral, 73
Ritz method, 49
Ritz value, 259, 288
   and convergence of CG, 263, 279, 322, 325, 332
   as root of CG polynomial, 259
   harmonic, 289
   orthogonality condition, 288
   with respect to a subspace, 288

saddle point problem, 208
search space, 13
Simpler GMRES, 339
stationary iterative method, 250

   asymptotic rate of convergence, 251
   Richardson iteration, 252
   SOR, 251, 255
Stieltjes algorithm, 29
Stieltjes recurrence, 89
Sturm sequence, 93
SYMMLQ method, 24, 54

three-term recurrence, 29, 89, 172
transfer function, 162

verification and validation, 3, 230

Weierstrass canonical form, 189